ISBN 978-1-331-99311-7
PIBN 10264433

IZ111115J
1146

Despatch Note

Order Number 206-0631311-9863542

Supplied by: UKPaperbackshop

Catalogue Number   Title and Artist   Qty

= 9781331993117 A Complete Treatise on the Electro-Deposition of M ...   1

|||||||||||| *206-0631311-9863542* ||||||||||||

SOME ITEMS MAY BE SHIPPED SEPARATELY

Order Processed: 11/11/2015 08:49:

Payment has been received from: Dr Robert M Perkin

1 MONTH OF
FREE
READING

at
www.ForgottenBooks.com

By purchasing this book you are eligible for one month membership to ForgottenBooks.com, giving you unlimited access to our entire collection of over 700,000 titles via our web site and mobile apps.

To claim your free month visit:
www.forgottenbooks.com/free264433

Similar Books Are Available from
www.forgottenbooks.com

PREFACE TO THE SEVENTH AMERICAN EDITION.

THE number of American editions through which Dr. George Langbein's work, *Handbuch der elecktrolytischen Metall-Niederschläge*, has passed in rapid succession, and the continued demand for it, may be accepted as evidence that the book, written from a scientific, as well as practical, standpoint, has been found to fulfill the purpose for which it is primarily intended, namely to serve as a ready book of reference and practical guide to the electroplater, who, if he would be a master of his art, must be conversant with the scientific principles upon which it rests.

In this the seventh American edition, now presented to the public, the general scheme and scope of the sixth edition have been retained, but a thorough revision has been made, and a good deal of new matter has been added.

Due attention has been paid to all important innovations, and it has been endeavored to include all practical methods of plating which have become known since the publication of the sixth edition, as well as the most recent machinery and apparatus.

The editor is under obligations to The Hanson & Van Winkle Co., of Newark, N. J., the well-known manufacturers of, and dealers in, electroplaters' supplies and to The Egyptian Lacquer Manufacturing Co., of New York, for valuable information and engravings. He has also diligently consulted the leading trade journals and freely quoted from them, due credit having been given in the text; but he would acknowledge his special indebtedness to " The Metal Industry "

The publishers have spared no expense in the proper il-

(iii)

355432

lustration and the mechanical production of the work, and, like the previous editions, it has been provided with such a copious table of contents and very full index as to render reference to any subject prompt and easy.

W. T. B.

PHILADELPHIA, OCTOBER 15, 1913.

PREFACE TO THE FIRST AMERICAN EDITION.

THE art of the electro-deposition of metals has during recent years attained such a high degree of development that it was felt that a comprehensive and complete treatise was needed to represent the present advanced state of this important industry. In furtherance of this object, a translation of Dr. George Langbein's work, *Vollständiges Handbuch der Galvanischen Mettall-Niederschläge*, is presented to the English-reading public with the full confidence that it will not only fill a useful place in technical literature, but will also prove a ready book of reference and a practical guide for the workshop. In fact, it is especially intended for the practical workman, wherein he can find advice and information regarding the treatment of the objects while in the bath, as well as before and after electroplating. The author, Dr. George Langbein, is himself a master of the art, being the proprietor of an extensive electroplating establishment combined with a manufactory of chemical products, machinery and apparatus used in the industry.

The results yielded by the modern dynamo-electric ma chines, to which the great advance in the electro-plating art it largely due, are in every respect satisfactory, and the more so since the need of accurate, and at the same time handy, measuring instruments has also been supplied. With the assistance of such measuring instruments, the establishment of fixed rules regarding the current-conditions for an electroplating bath has become possible, so that good results are guaranteed from the start. While formerly the electro-plater had to determine the proper current-strength for the depositions in an empirical manner, by time-consuming experiments, to-day, by duly observing the determined conditions, and pro-

(v)

vided with well-working measuring instruments, he can at once produce beautiful and suitable deposits of the various metals.

The data referring to these current-conditions, according to measurements by Dr. Langbein, are given as completely as possible, while for the various baths, only formulæ yielding entirely reliable results have been selected. To most of the baths a brief review of their mode of action and of their advantages for certain uses is added, thus enabling the operator to select the bath most suitable for his special purpose. To the few formulæ which have not been tested, a note to that effect is in each case appended, and they are only given with due reserve.

To render the work as useful as possible, the most suitable formulæ for plating by contact and immersion, as well as the best methods for coloring the metals, and the characteristic properties of the chemicals used in the industry, are given. However, the preparation of the chemicals has been omitted, since they can be procured at much less expense from chemical works than it would be possible for the electro-plater to make them in small quantities, even if he possessed the necessary apparatus and the required knowledge of chemistry and skill in experimenting.

It is hoped that the additions made here and there by the translator, as well as the chapter on "Apparatus and Instruments," and that on "Useful Tables," added by him, may contribute to the usefulness of the treatise.

Finally, it remains only to be stated that the publishers have spared no expense in the proper illustration and the mechanical production of the book ; and, as is their universal practice, have caused it to be provided with a copious table of contents, and a very full index, which will add additional value by rendering any subject in it easy and prompt of reference.

W. T. B.

Philadelphia, July 1, 1891.

CONTENTS.

I.

HISTORICAL PART

CHAPTER I.

HISTORICAL REVIEW OF ELECTRO-METALLURGY.

(vii)

II.

THEORETICAL PART

CHAPTER II.

MAGNETISM AND ELECTRICITY.

Magnetism.

Electro-Magnetism.

Induction.

Fundamental Principles of Electro-Chemistry.

III.

SOURCES OF CURRENT

CHAPTER III.

VOLTAIC CELLS, THERMO-PILES, DYNAMO-ELECTRIC MACHINES, ACCUMULATORS.

IV.

PRACTICAL PART.

CHAPTER IV.

ARRANGEMENT OF ELECTRO-PLATING ESTABLISHMENTS IN GENERAL.

CHAPTER V.

PREPARATION OF THE METALLIC OBJECTS.

A. MECHANICAL TREATMENT PREVIOUS TO ELECTRO-PLATING.

CHAPTER VI.

DEPOSITION OF NICKEL AND COBALT.

I. DEPOSITION OF NICKEL.

2. DEPOSITION OF COBALT.

CHAPTER VII.

DEPOSITION OF COPPER, BRASS AND BRONZE.

I. DEPOSITION OF COPPER.

CHAPTER VII.

DEPOSITION OF SILVER.

CHAPTER IX.

DEPOSITION OF GOLD.

CHAPTER XII.

DEPOSITION OF ANTIMONY, ARSENIC, ALUMINIUM.

1. DEPOSITION OF ANTIMONY.

2. DEPOSITION OF ARSENIC.

3. DEPOSITION OF ALUMINIUM.

4. DEPOSITION UPON ALUMINIUM.

CHAPTER XIII.

DEPOSITION BY CONTACT, BY BOILING, AND BY FRICTION.

CHAPTER XIV.

COLORING OF METALS.

CHAPTER XVII.

GALVANOPLASTY (REPRODUCTION).

CHAPTER XVIII.

CHEMICALS USED IN ELECTRO-PLATING AND GALVANOPLASTY.

ELECTRO-DEPOSITION OF METALS.

I.

HISTORICAL PART.

CHAPTER I.

HISTORICAL REVIEW OF ELECTRO-METALLURGY.

In reviewing the history of the development of electrolysis, *i. e.*, the reduction of a metal or a metallic alloy from the solution of its salts by the electric current, the simple reduction which takes place by the immersion of one metal in the solution of another, may be omitted. This mode of reduction was well known to the alchemist Zozimus, who described the reduction of copper from its solutions by means of iron, while Paracelsus speaks of coating copper and iron with silver by simple immersion in a silver solution.

Before the discovery, in 1789, of contact-electricity by Luigi Galvani, there was nothing like a scientific reduction of metals by electricity ; and only in 1799 did Alexander Volta, of Pavia, succeed in finding the true causes of Galvani's discovery. Galvani observed, while dissecting a frog on a table, whereon stood an electric machine, that the limbs suddenly became convulsed by one of his pupils touching the crural nerve with the dissecting-knife at the instant of taking a spark from the conductor of the machine. The experiment was several times repeated, and it was found to answer in all cases when a metallic conductor was connected with the nerve, but not otherwise. He observed that muscular contractions were produced by forming a connection between two different metals, one of which was applied to the nerve, and the other to the muscles of the leg. Similar phenomena having been found to arise when the leg of the frog was connected with the electric machine, it could scarcely be doubted that in both cases the muscular contractions were produced by the same agent. From a course of experiments, Galvani drew the erroneous inference that these muscular contractions were caused by a fluid having its seat in the nerves, which through the metallic connections flowed over upon the muscles. Everywhere, in Germany, England and France, eminent scientists hastened to repeat Galvani's experiments, in the hope of discovering in the organism a fluid which they considered the vital principle ; but it was reserved to Volta to throw light upon the prevailing darkness. In his repeated experiments this eminent philosopher observed that one circumstance had been entirely overlooked, namely, that in order to produce strong muscular contractions in the frog-leg experiment, it was absolutely necessary for the metallic connection to consist of two different metals coming in contact with each other. From this he drew the inference that the agent producing the muscular contractions was not a nerve-fluid, but was developed by the contact of dissimilar metals, and identical with the electricity of the electric machine.

This discovery led to the construction of what is known as the *pile of Volta,* or the *voltaic pile.* The same philosopher found that the development of electricity could be produced by building up in regular order a pile of pairs of plates of dissimilar metals, each pair being separated on either side from the adjacent pairs by pieces of moistened card-board or felt. On account of various defects of the voltaic pile, Cruikshank soon afterwards devised his well-known *trough battery,* which consisted of square plates of copper and zinc soldered together, and so arranged and fastened in parallel order in a wooden box that between each pair of plates a sort of trough was formed, which was filled with acidulated water.

Nicholson and Carlisle, in 1800, were the first to decompose water electrolytically into hydrogen and oxygen, using a Volta pile. The method has only acquired practical importance during the last few years. Wollaston, in 1801, found that if a piece of silver in contact with a more positive metal, for instance, zinc, be immersed in copper solution, the silver will be coated with copper, and this coating will stand burnishing.

Cruikshank, in 1803, investigated the behavior of solutions of nitrate of silver, sulphate of copper, acetate of lead, and of several other metallic salts, towards the galvanic current, and found that the metals were so completely reduced from their solutions by the current as to suggest to him the analysis of minerals by means of the electric current.

To Brugnatelli we owe the first practical results in electro-gilding. In 1805, he gilded two silver medals by connecting them by means of copper wire with the negative pole of the pile, and allowing them to dip in a solution of fulminating gold in potassium cyanide, while a piece of metal was suspended in the solution from the positive pole. He also observed that the positive plate, if it consisted of an oxidizable metal, was dissolved.

One of the greatest discoveries connected with the subject, however is that of Sir Humphry Davy, in 1807, when by decomposing potassium hydroxide and sodium hydroxide by means of a powerful electric current he obtained the metals potassium and sodium.

Prof. Oersted, of Copenhagen, in 1820, found that the magnetic needle is deflected from its direction by the electric current. It was known long before this that powerful electric discharges affect the magnetic needle. It had, for instance, been observed that the needle of a ship's compass struck by lightning had lost its property of indicating the North Pole, and several physicists, among them Franklin, had succeeded in producing the same phenomena by heavy discharges of the electrical machine, but they were satisfied with the supposition that the electric current acted mechanically, like the blow of a hammer. Oersted first perceived that electricity must be in a state of motion in order to act upon magnetism. This led to the construction of the galvanoscope or galvanometer, an instrument which indicates whether the cells or other source of current furnish a current or not, and by which the intensity of the source of current may also to a certain degree be recognized.

Ohm, in 1827, discovered the law named after him, that *the strength of a continuous current is directly proportional to the difference of potential or electro-motive force in the circuit*, and *inversely proportional to the resistance of the circuit*. This law will be more fully discussed in the theoretical part.

Ohm's discovery was succeeded, in 1831, by the important discovery of *electric induction* by Faraday. By induction is understood the production of an electric current in a closed circuit which is in the immediate proximity of a current-carrying wire. Faraday further found that the current induced in the contiguous wire is not constant, because after a few oscillations the magnetic needle returned to the position occupied by it before a current was passed through the current-carrying wire; whilst, when the current was broken, the needle deflected in the opposite direction.

In the year following the discovery of Faraday, Pixii, of Paris, constructed the *first electro-magnetic induction machine.*

Faraday's electrolytic law of the proportionality of the current-strength and its chemical action, and that the quantities of the various substances which are reduced from their combinations by the same current are proportional to their chemical equivalents, was laid down and proved in 1833, and upon this Faraday based the measurements of the current-strength by chemical deposition, as, for instance, that of water, in the voltmeter.

Of the practical electro-chemical discoveries there remains to be mentioned the production of iridescent colors, in 1826, by Nobili, and the production of the amalgams of potassium and sodium, in 1853, by Bird.

The actual galvanoplastic process, however, dates from 1838. In the spring of that year Prof. Jacobi made known to the Academy of Sciences of St. Petersburg his discovery of the utility of galvanic electricity as a means of reproducing objects of metal. He produced an exact mould of metals and artistic objects by means of wax or plaster, and then coated every detail of the surface of this mould with very fine graphite, thus rendering it electrically conductive. He then suspended the mould from the negative pole (cathode) of an electrolytic bath containing a suitable metallic salt, and formed the positive pole of the same metal; on passing an electric current through this bath the mould became lined with very fine particles of metal, forming a continuous and compact surface. The metal forming the anode was gradually dissolved in the bath as fast as it was deposited on the cathode. Hence, Jacobi must be considered the father of galvanoplasty in so far as he was the first to utilize and give practical form to the discoveries made up to that time.

Though Jacobi's process was published in the English periodical, " The Athenæum," of May 4, 1839, Mr. T. Spencer, who read a paper on the same subject, September 13, 1839, before the Liverpool Polytechnic Society, claimed priority of invention, as was also done by Mr. C. J. Jordan, who, on May 22, 1839, sent a letter to the " London Mechanical Magazine," which was published on June 8, 1839.

From this time forward the galvanoplastic art made rapid progress, and by the skill and enterprise of such men as the Elkingtons, of Birmingham, and De Ruolz, of Paris, it was speedily added to the industrial arts.

Though copies of metallic objects by means of galvanoplasty could now be made, the employment of the process was restricted to metallic objects of a form suitable for the purpose, until, in 1840, Murray succeeded in making non-metallic surfaces conductive by the application of graphite (black lead, plumbago), which rendered the production of galvanoplastic copies of wood-cuts, plaster-of-Paris casts, etc., possible.

Dr. Montgomery, in 1843, sent to England samples of gutta-percha, which was soon found to be a suitable material for the production of negatives of the original models to be reproduced by galvanoplasty.

Though it was now understood how to produce heavy deposits of copper, those of gold and silver could only be obtained in very thin layers. Scheele's observations on the solubility of the cyanide combinations of gold and silver in potassium cyanide, led Wright, a co-worker of the Elkingtons, to employ, in 1840, such solutions for the deposition of gold and silver, and it was found that deposits produced from these solutions could be developed to any desired thickness. The use of solutions of metallic cyanides in potassium cyanide prevails at the present time, and the results obtained thereby have not been surpassed by any other practice.

From the same year also dates the patent for the deposition of nickel from solution of nitrate of nickel, which, however, did not attract any special attention. This may have been chiefly due to the fact that the deposition of nickel from its nitrate solution is the most imperfect and the least suitable for the practice.

To Mr. Alfred Smee we owe many discoveries in the deposition of antimony, platinum, gold, silver, iron, lead, copper, and zinc. In publishing his experiments, in 1841, he originated the very appropriate term "electro-metallurgy" for the process of working in metals by means of electrolysis.

Prof. Bœttger, in 1842, pointed out that dense and lustrous depositions of nickel could be obtained from its double salt, sulphate of nickel with sulphate of ammonium, as well as from ammoniacal solution of sulphate of nickel; and that such deposits, on account of their slight oxidability, great hardness, and elegant appearance, were capable of many applications. However, Bœttger's statements fell into oblivion, and only in later years, when the execution of nickeling was practically taken up in the United States, his labors in this department were remembered in Germany. To Bœttger we are also indebted for directions for coating metals with iron, cobalt, platinum, and various patinas.

In the same year, De Ruolz first succeeded in depositing metallic alloys—for instance, brass—from the solutions of the mixed metallic salts.   In 1843, the first use of thermo-electricity appears to have been made by Moses Poole, who took out a patent for the use of a thermo-electric pile instead of a voltaic battery for depositing purposes.

From this time forward innumerable improvements in existing processes were made ; and also the first endeavors to apply Faraday's discoveries to practical purposes.

The invention of depositing metals by means of a permanent current of electricity obtained from steel magnets was perfected and first successfully worked by Messrs. Prime & Son, at their large silverware works, Birmingham, England, and the original machine constructed by Woolrych in 1844—the first magnetic machine that ever deposited silver on a practical scale—is still preserved.   It is now owned by the Corporation of Birmingham, England.   The Woolrych machine stands 5 feet high, 5 feet long, and $2\frac{1}{2}$ feet wide.

As early as 1854, Christofle & Co. endeavored to replace their batteries by magnetic-electrical machines, and used the Holmes type, better known as the Alliance machine, which, however, did not prove satisfactory; and besides, the prices of these machines were, in comparison with their efficiency, exorbitant.   The machine constructed by Wilde proved objectionable on account of its heating while working, and the consequent frequent interruptions in the operations.

In 1860 Dr. Antonie Pacinotti, of Pisa, suggested the use of an iron ring wound around with insulated wire, in place of the cylinder.   This ring, named after its inventor, has, with more or less modifications, become typical of many machines of modern construction.   In the construction of all older machines, steel magnets had been used, and their magnetism not being constant, the effect of the machine was consequently also not constant.   Furthermore, they generated alternately negative and positive currents, which, by means of commutators, had to be converted into currents of the same direction; and this, in consequence of the vigorous formation of sparks, caused the rapid wearing-out of the commutators.

These defects led to the employment of continuous magnetism in the iron cores of the electro-magnets, the first machine based upon this principle being introduced in 1866, by Siemens, which, in 1867, was succeeded by Wheatstone's

However, the first useful machine was introduced in 1871, by Zenobe Gramme, who in its construction made use of Paci notti's ring. This machine was, in 1872, succeeded by Hefner-Alteneck's, of Berlin. In both machines the poles of the electro-magnet exert an inducing action only upon the outer wire wrappings of the revolving ring, the other portions being scarcely utilized, which increases the resistance and causes a useless production of heat. This defect led to the construction of flat-ring machines, in which the cylindrical ring is replaced by one of a flat shape and of a larger diameter, thus permitting the induction of both flat sides. Such a machine was, in 1874, built by Siemens & Halske, of Berlin; and in the same year by S. Schuckert, of Nüremberg. In Schuckert's machines nearly three-quarters of all the wire wrappings were under the inducing influence of both of the large pole shoes of the electro-magnets. The flat-ring armature was later on replaced by the drum armature, and the more modern machines are almost without exception of the drum-armature type.

By the construction of suitable dynamo-machines a mighty impetus was given to the electro-plating industry. They supplanted the ordinary cell apparatus formerly used and rendered possible the production of electrolytically nickeled, coppered and brassed sheet-steel and tin-plate, as well as that of electrolytically zincked sheets, wire, building materials, etc. All these processes will be fully discussed in the practical part of this work.

THEORETICAL PART.

CHAPTER II.

MAGNETISM AND ELECTRICITY.

Magnetism.

FOR the better understanding of the electrolytic laws it will be necessary to commence with the phenomena presented by magnetism, and to consider them somewhat more closely

A particular species of iron ore is remarkable for its property of attracting small pieces of iron and causing them to adhere to its surface. This iron ore is a combination of ferric oxide with ferrous oxide (Fe_3O_4), and is called loadstone or magnetic iron ore. Its properties were known to the ancients, who called it magnesian stone, after Magnesia, a city in Thessaly, in the neighborhood of which it was found. In the tenth or twelfth century it was discovered that this stone has the property of pointing north and south when suspended by a thread. This property was turned to advantage in navigation and the term load stone ("leading stone") was applied to the magnesian stone. If a natural loadstone be rubbed over a bar of steel, its characteristic properties will be communicated to the bar, which will then be found to attract iron filings like the loadstone itself. The bar of steel thus treated is said to be magnetized, or to constitute an artificial magnet. The artificial magnets thus produced may be straight, in the shape of a horse-shoe, or annular; but no matter what their form may be, there will always be two regions where the attractive force reaches its maximum, while between these two points there is a region which has no attractive effect whatever upon iron filings. The two ends of the magnet, especially, show the greatest attractive force, and they are called the *magnetic poles*, whilst the line running around the magnet, which possesses no attractive force, is termed the *neutral line* or *neutral zone*. In a closed magnet the poles are situated on the ends of one and the same diameter, while the neutral zones are located on the ends of a diameter standing perpendicular to the first.

When a magnetized bar or natural magnet is suspended at its center in any convenient manner, so as to be free to move in a horizontal plane, it is always found to assume a particular direction with regard to the earth, one end pointing nearly north and the other nearly south. If the bar be removed from this position it will tend to reassume it, and after a few oscillations, settle at rest as before. The direction of the magnetic bar, *i. e.*, that of its longitudinal axis, is called the *magnetic meridian*, while the pole pointing toward the north is usually distinguished as the *north pole* of the bar, and that which points southward as the *south pole.*

A magnet, either natural or artificial, of symmetrical form, suspended in the presence of a second magnet, serves to exhibit certain phenomena of attraction and repulsion, which deserve particular attention. When a north pole is presented to a south pole, or a south pole to a north pole, attraction ensues between them, the ends of the bar approaching each other, and, if permitted, adhering with considerable force. When, on the other hand, a north pole is brought near a second north pole, or a south pole near another south pole, mutual repulsion is observed, and the ends of the bar recede from each other as far as possible. *Poles of an opposite name attract, and poles of a similar name repel each other.*

According to Ampère's theory, each molecule of iron or steel has a current of electricity circulating round it; previous to magnetization these molecules—and hence the currents—

are arranged irregularly; during magnetization they are made to move parallel to one another, and as the magnetization becomes more perfect they gradually assume greater parallelism.

If an iron or steel needle be suspended free in proximity to a magnet it assumes a fixed direction according to its greater or smaller distance from the poles or from the neutral zone. However, before the needle assumes this direction, it swings rapidly with a shorter stroke, or slowly with a longer stroke, according to the greater or smaller attractive force exerted upon it. The space within which the magnetic action of a magnet is exercised is called the *magnetic field*, and the magnetic, as well as the electric, attractions and repulsions are, according to Coulomb, as the densities of the fluids acting upon each other, and inversely as the square of their distance.

As electro-magnets act in exactly the same manner as magnets, their further properties will be discussed in the next section.

Electro-Magnetism.

When a wire through which a current is passing is brought near, and parallel, to a magnetic needle, the latter is deflected from its ordinary position, no matter whether the current-carrying wire be placed alongside, above, or beneath it. The deflection of the needle is always in the same direction, *i. e.*, its north pole is always deflected in one and the same direction.

The direction of the deflection is determined by what is known as Ampère's rule, which is as follows: Suppose an observer swimming in the direction of the current, so that it enters by his feet and emerges by his head: if the observer has his face turned towards the needle, the *north pole is always deflected to his left*.

When the current-carrying wire is coiled in many windings around the needle, the action of the current is increased, because every separate winding deflects the north pole in the same direction. Such instruments are known as *multipliers, or galvanoscopes, or galvanometers,* and are used for recognizing feeble currents. These instruments have been improved by Nobili through the use of a very long coil of wire, and by the addition of a second needle. This instrument is known as the *astatic galvanometer.* The two needles are of equal size and magnetized as nearly as possible to the same extent. They are then immovably fixed together parallel and with their poles opposed, and hung by a long fiber of twisted silk, with the lower needle in the coil and the upper one above it. The advantage thus gained is twofold: The system is *astatic,* unaffected, or nearly so, by the magnetism of the earth ; and the needles being both acted upon in the same manner by the current, are urged with much greater force than one alone would be, all the actions of every part of the coil being strictly concurrent. A divided circle is placed below the upper needle, by which the angular motion can be measured, and the whole is inclosed in glass, to shield the needles from the agitation of the air.

The deflection of the magnetic needle by the electric current has led to the construction of instruments which allow of the intensity of the current being measured by the magnitude of the deflection. Such instruments are, for instance, the *tangent galvanometer,* the *sine galvanometer,* etc., but they are almost exclusively used for scientific measurements, while for the determination of the intensity of current for electro-plating purposes other instruments are employed, which will be described later on. However, the electric current exerts not only a reflecting action on magnetic needles, but is also capable of producing a magnetizing effect on iron and steel. If a bar of iron be surrounded by a coil of wire covered with silk or cotton for the purpose of insulation, it becomes magnetic so long as the current is conducted through the coil. Such iron bars converted into temporary magnets by the action of the current are called *electro-magnets,* and they will be the more highly magnetic, the greater the number of turns of the coil, and the more intense the current passing through the turns.

The magnitude of the magnetizing force of the current is expressed by the product from the number of turns and current-strength passing through the turns, and is called *ampère-turn number*.

By interrupting the current passing through the wire-turns, the magnetism of the iron bar disappears to within a very small quantity, its magnitude depending on the quality of the iron. This remaining magnetism is called *remanent* or *residual* magnetism.

An electro-magnet possesses the same properties as an ordinary magnet, and, like it, has a north pole and a south pole,

FIG. 1.

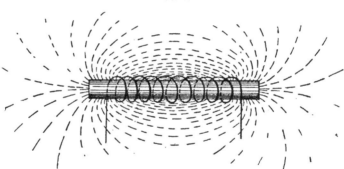

as well as a magnetic field, through which its influence extends. Place a piece of paper above an electro-magnet and sift uniformly iron filings over it. On giving the paper slight taps, the filings arrange themselves in regular groups and lines. Most of the filings collect on the two poles, while, in fixed decreasing proportions, lines of filings are formed from the north pole to the south pole. This experiment demonstrates that the action is strongest on the poles, and decreases towards the center. The entire space in which the magnetic action—the *flow of the magnetic lines of force*—exerts its influence is called the *magnetic field*. The lines of force flow from the north pole to the south pole, where they combine, and flow

back through the iron bar to the north pole, as shown in the
accompanying illustration, Fig. 1.

The dotted lines also take actually their course from one
pole to the other, but by a more circuitous way.   The direction, as well as the magnitude, of the field force (see later on)
varies on all points of the magnet or electro-magnet, with the
sole exception of the symmetrical plane between the two poles,
the latter being on all points struck at right angle by the lines
of force.

By placing a bar of soft iron, $a\,b$, in the proximity of a mag-

FIG. 2.

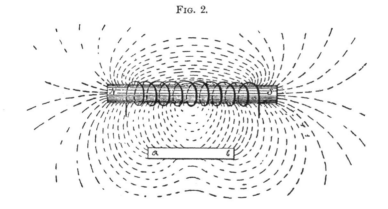

net or electro-magnet $N\,S$, covering both with a sheet of paper
and sifting iron filings upon the latter, delineations, as shown
in Fig. 2, are obtained.

The lines of force gravitate in large numbers towards the
side where the iron bar is, traverse the iron quite compactly,
and while, without the bar, the center of the magnet showed a
feeble magnetic field, the field-force in that place has now
become greater.   Upon the opposite side the density of the
lines of force which pass through the air is less.   The property of a material to be traversed by the lines of force is called
its *permeability.*

The number of lines of force which traverses through 1

square centimeter of cross-section of a material, is called the magnitude of the magnetic induction of the material in question.

Every material opposes a certain fixed resistance to the electrical current, as well as to the magnetic lines of force. Soft iron opposing the least resistance to the lines of force, it is most compactly traversed by them. Air, on the other hand, opposes far greater resistance, and, hence, the density of the lines of force, in Fig. 2, where they pass through the air is much less.

A conducting wire through which passes a powerful current also becomes itself magnetic. If a circular conducting wire, through which a current passes, be suspended so as to move free around its vertical axis, its direction is influenced by the terrestrial magnetism, and it assumes such a position that its plane stands at a right angle upon the plane of the magnetic meridian. By now conducting the current through a spiral wire suspended free—a so-called *solenoid*—the plane of each separate turn will also place itself at a right angle upon the plane of the magnetic meridian, or in other words, the axis of the solenoid will be brought to lie in the magnetic meridian.

In a manner similar to the action upon a magnet by a conducting wire through which a current passes, two conducting wires, through which currents pass, exert attracting and repelling influences one upon the other. Two currents running parallel alongside each other in the same direction attract, but repel, each other, when running in opposite directions.

Induction.

By induction is understood the production of an electric current in a closed conductor which is in the immediate proximity of a current-carrying wire.

Suppose we have two insulated copper-wire coils, a and b, Fig. 3, b being of a smaller diameter and inserted in a. When the two ends of b are connected with the poles of a battery, a current is formed in a the moment the current of b is closed. This current is recorded by the deflection of the magnetic needle of a multiplier, *M*, which is connected with the ends of *a*, the deflection of the needle showing that the current produced in *a* by the current in *b* moves in an opposite direction. The current in *a*, however, is not lasting, because, after a few oscillations, the magnetic needle of the multiplier returns to its previous position and remains there, no matter how long the current may pass through *b*. If, however, the current in *b* be interrupted, the magnetic needle swings to the opposite direction, thus indicating the formation

Fig. 3.

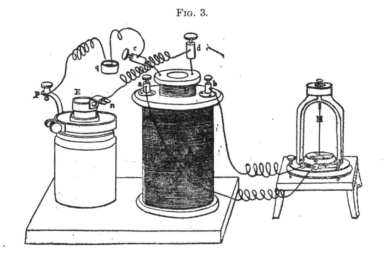

of a current in *a*, which passes through it in the same direction as the interrupted current in *b*.

The current causing this phenomenon is called the *primary*, *inducing* or *main* current, and that produced by it in the closed circuit, the *secondary*, *induced* or *induction*-current. From what has been above said, it is clear that *an electric current at the moment of its formation induces in a contiguous closed circuit a current of opposite direction, but when interrupted, a current of the same direction.*

In the same manner as closing and opening the main current, its sudden augmentation also effects the induction of a current of opposite direction in a contiguous wire, while its sudden weakening induces a current of the same direction. The same effect is also produced by bringing the main current-carrying wire closer to, or removing it further from, the contiguous wire.

It is supposed that by closing the current a magnetic field is formed in the coil b, which sends forth its lines of force radially in an undulating motion. The lines of force cut the turns of the coil, a, which is without current, and thereby induces a current. This current disappears again when the primary current flows in equal force, and re-appears when by the strengthening of the inducing current a change in the number of lines of force takes place by reason of the strengthening of the magnetic field. In the same manner induced currents are also produced by a decrease in the number of lines of force, and hence it follows that the production of induction-currents is always conditional on the change of proportion between the conductor and the magnetic field.

When a magnet or electro-magnet is pushed into a wire coil, an electric current is produced in the turns of the coil so long as the motion of the magnet is continued; when the motion is interrupted, the production of current ceases. If the magnet be now withdrawn from the coil, a current is again formed, which, however, flows in an opposite direction to that formed by pushing the magnet into the coil. The currents produced in the above-mentioned manner are also induction-currents, and their formation is again explained by the fact that the lines of force cut the turns of the conducting wire, and excite thereby a current, the electro-motive force of which increases or decreases with the magnitude of the number of lines of force.

The induced currents follow the law of Ohm (see later on) in precisely the same manner as the inducing currents. A long induction-wire with a small cross-section offers greater resistance than a short wire with a larger cross-section, and

2

consequently, in the first case, the current will be of slighter intensity and higher electro-motive force, and, in the other, of greater intensity and less electro-motive force.

Electro-magnetic alternating actions are the relations which exist between the magnetic field, the conductor, and the motion. The direction of the induced current can readily be followed by Fleming's hand rule, which is as follows: Hold the thumb and the first and the middle fingers of the right hand as nearly as possible at right angles to each other, as shown in Fig. 4, so as to represent three rectangular axes in space. If the thumb points in the direction of motion, and

FIG. 4.

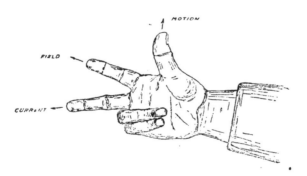

the forefinger points along the direction of the magnetic lines, then the middle finger will point in the direction of the induced electro-motive force.

The mechanism of the formation of the electric current will be fully discussed later on, but it will be necessary to here give the values in which the performances of the current are expressed in order to shape the succeeding chapters more uniformly

Fundamental Principles of Electro-Technics.

Electric Units. For the better comprehension of the properties, effects, and value of the electric current, it has become

customary to compare it with a current of water, and this cus-
tom will here be followed.

Fig. 5 shows a funnel A secured in the stand D, and connected by a tube with the horizontal discharge pipe B.
Underneath B stands the vessel C, which serves for catching
the water. If the funnel be placed in a higher position and
filled with water, the latter runs off more rapidly from the
pipe B, than when the funnel occupies a lower position. If

Fig. 5.

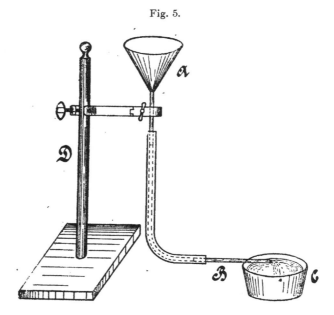

the force of the current of water is expressed according to the
quantity of water which runs out in the time-unit, it follows
that in a certain pipe conduit, the quantity of water which
runs out in the time-unit, increases if there be an increase in
the height of fall.

Suppose it has been determined how many seconds are required for the water in the funnel to run through the pipe B.
If the pipe be now lengthened by joining to it several pipes of

the same cross-section, it will be found that a greater number of seconds are required for emptying the funnel than with the use of only one pipe. From this we learn that with a determined height of fall, the quantity of water which flows in the time-unit through a pipe of determined cross-section decreases when the pipe is lengthened.

If now the discharge pipes used in the last experiment be replaced by pipes of the same length but of smaller cross-sections, it will be found that a greater number of seconds are also required for emptying the funnel than with the use of pipes of larger cross-sections. Hence, at a determined height of fall, the quantity of water which flows through a pipe of fixed length in the time-unit, decreases if the cross-section of the pipe be increased.

The height of fall has to be considered as the motive power, which effects the flow of water. The pipe opposes a resistance to the flowing water, this resistance increasing with the length of the pipe and the reduction of the cross-section, and decreasing as the cross-section becomes larger.

If now these principles be applied to the electric current, by *current-strength* has to be understood the *quantity of electricity* which passes in the time-unit through a conductor.

The unit of the quantity of electricity is called the *coulomb*. Its magnitude results from the fact that for the 1 gramme hydrogen 96,540 coulombs must migrate through the electrolyte.

The unit of current-strength is called the *ampère, i. e.*, a current which every second carries one coulomb through the conductor. The magnitude of an ampère is the current-strength which is capable of separating in one minute 0.01973 gramme of copper, or in one hour 1.184 grammes, from a cupric sulphate solution. In order to separate from an electrolyte 1 gramme of hydrogen, a current of 1 ampère must accordingly pass 96,540 seconds, or 26 hours 49 minutes, through the electrolyte.

The electro-motive force or tension of the electric current corresponds to the height of fall of water. The work an electric current is capable of performing does not only depend on the current-strength, *i. e.*, the quantity of current, which passes in the time-unit through a cross-section of the conductor, but also on the electro-motive force. *The unit of electro-motive force is called the* volt. The material value of a volt is about the electro-motive force of a Daniell's cell (zinc-copper).

In a water conduit the difference in pressure between two points in the pipe is measured according to the difference in the height of the column of water. To this difference in pressure corresponds the *difference of electro-motive force*, also called *difference of potential*, which is expressed by the number of volts.

The product of current-strength in ampères and electro-motive force in volts, which, in so far as an ampère is an electric unit in one second, represents work performed in one second, is called the *volt-ampère* or *watt*, and hence is the *unit of electrical work*.

The electric resistance is similar to the resistance offered by a water-pipe to the flowing water. As previously stated, the quantity of water running out in the time-unit decreases when the pipe is lengthened, as well as when the cross-section is smaller, and, in both cases, the resistance opposed to the water by friction increases. On the other hand, the quantity of water flowing out in the time-unit increases, when the length of pipe is shortened and the cross-section increased, because there is less resistance. The same takes place with the electric current. The quantity of current which can pass through a conductor becomes smaller when the length of the conductor is increased and its cross-section reduced, because the resistance becomes thereby correspondingly greater. It has further been seen that the quantity of flowing water in a certain conduit increases as the height of fall becomes greater. If now the electro-motive force of the electric current be substituted for the height of fall, the current-strength which passes through a conductor will be increased in keeping with the changing

electro-motive force.   From this results the following proposition

*In a determined circuit the current-strength increases at the
same ratio as the electro-motive force which acts upon the circuit.*

If now the current-strength increases proportionally to the
electro-motive force, the expression,

$$\frac{E(= \text{electro-motive force in the circuit})}{J (= \text{current-strength in the circuit}),}$$

must be a fixed value dependent on the magnitude of the
electro-motive force and the current-strength, and this value is
called the *electric resistance of the circuit.*

The unit of electric resistance is called the *ohm*, it having thus
been named after the physicist Ohm, who laid down the rules
known as the laws of Ohm.   The value of the ohm is equal to
the resistance at 0° C. of a column of mercury of one square
millimeter section and one meter long.   A volt is the electro-
motive force which is capable of sending a current-strength of
one ampère through the resistance of one ohm.

Law of Ohm.   It has above been seen that the fraction

(1) $\frac{E}{J}$ — resistance (W),

whereby under E is understood the electro-motive force which
is at disposal in the entire circuit.   The current-strength J is
throughout in all places of the same magnitude, and W indicates the total resistance of the circuit.

From the preceding equation are deduced the following
further equations :

(2) $W. J = E,$

that is, the electro-motive force is equal to the product of
current-strength and resistance ;

(3) $\frac{E}{W} - J,$

that is, the current-strength is equal to the electro-motive
force divided by the resistance.

Example to equation 1.   If through a circuit closed by a long wire and a current-meter, a current of 4 volts and 2 ampères is conducted, the resistance of the circuit is

$$\frac{4 \text{ volts}}{2 \text{ ampères}} = 2 \text{ ohms.}$$

Example to equation 2. 5 ampères are to be conducted through a circuit of 1 ohm resistance, what electro-motive force is required for the purpose?

$$1 \text{ ohm} \times 5 \text{ ampères} = 5 \text{ volt.}$$

Example to equation 3. A current of 10 volts electro-motive force is to be conducted through a circuit with 2 ohms resistance; what current-strength may be looked for?

$$\frac{10 \text{ volts}}{2 \text{ ohms}} = 5 \text{ ampères.}$$

The total resistance, *W*, is composed of the *internal* resistance of the current-source and the *external* resistance which the current in its progression has to overcome. This external resistance is composed of the resistance of the conducting wire, the electrolyte, etc. If the internal resistance be designated W and the external resistances w1 and w2, equation 3 assumes the following aspect:

$$(4) \quad \frac{E}{W + w1 + w2} = J.$$

Hence, the current-strength is equal to the total electro-motive force divided by the sum of the internal and external resistances.

Example to equation 4. A cell possesses an internal resistance of 0.3 ohm and an electro-motive force of 1.8 volts, and the resistance of the conducting wire, *w1*, is 1 ohm and that of the electrolyte 0.5 ohm. The current-strength then amounts to 1 ampère $\left(\dfrac{1.8}{0.3 + 1 + 0.5} = 1\right)$.

If a determined current-strength flows through a resistance, a decrease of electro-motive force results in the resistance, exactly as in a water-conduit the pressure of the column of water is decreased with the length of the pipe, a decrease in pressure taking place. It might be said that the resistance consumes the pressure, and the greater the resistance of a conductor is, the less the current-strength will be, since, if in the equation 3 the divisor grows, the current-strength, J, must become less. According to the law of Ohm, the following proposition here holds good :

The current-strength is inversely proportional to the sum of the resistance of the circuit, or, in other words, the current-strength decreases in the proportion as, with the same electro-motive force, the resistances increase.

The resistance of a wire or of a body increases in proportion to its increase in length, and decreases in proportion to the increase of its cross-section. If the resistance of a conductor be designated W, its length L, and its cross-section Q, then

(5) $W = \dfrac{L}{Q}$.

The decreasing electro-motive force, according to the law of Ohm, is calculated by the following equation, in which a denotes the decrease in electro-motive force, J the current-strength, Wi the internal resistance.

(6) $a = J \times Wi$.

In the example to equation 4, the current-strength amounted to 1, and the internal resistance of the element to 0.3 ohm; this gives a decrease of electro-motive force of $1 \times 0.3 = 0.3$ volt; hence the actual electro-motive force of the current flowing from the cell will only be: $E - a = 1.8 - 0.3 = 1.5$ volts, and this effective electro-motive force is called the *impressed electro-motive force* of the cell or other source of current.

If now the preceding separate propositions of the law of Ohm be collected, the latter reads as follows:

The current-strength is directly proportional to the sum of the electro-motive forces, and inversely proportional to the resistance of the circuit ; however, the resistance of each part of the circuit is proportional to its length, and inversely proportional to its cross-section.

Specific resistances. The resistance of a wire of the same material is consequently proportional to its length and inversely proportional to its cross-section. If now, one after the other, wires of equal length and equal cross-section, but of different materials, be placed between the binding posts of a source of current, different current-strengths are obtained in the wires. From this it follows that every material possesses a definite capacity of its own to conduct the current. Hence, if the resistance is to be calculated from the length of the wire and its cross-section, the magnitude, called the *specific resistance* of the material, has to be taken into consideration. By the specific resistance is to be understood for conductors of the first class, the resistance of a material 1 meter in length and 1 square millimeter cross-section, and for conductors of the second class, the resistance of a cube of fluid of 10 centimeters = 1 decimeter side length.

If the specific resistance be denoted c, the resistance of a wire of L meters length and a cross-section of Q square millimeters cross-section is found from the equation

$$(7) \ W - \frac{L}{Q}. \ c.$$

The specific resistance c of the metals at 59° F., and the coefficient of temperature a (see later on) amount to for ·

	c.	a.
Aluminium . . .	0.029	0.0039
Antimony	0.475	0.0041
Bismuth	1.250	0.0037
Brass	0.10 to 0.071	0.0016.
Copper	0.017	0.0041
German silver . .	0.30 to 0.18	0 0003
Gold	0.024	0.0040
Iron	0.120 to 0.10	0.0048
Lead	0.207	0.0039
Manganin	0.455	0.00002
Mercury	0.953	0.0009
Nickel	0.15	0.0036
Nickelin	0.435 to 0.340	0.000025.
Platinum	0.15 to 0.094	0.0024
Silver	0.016	0.0038
Steel	0.50 to 0.168	0.0040
Tin	0.10	0.0042
Zinc	0.065	0.0040

From the above table it will be seen that silver is the best conductor, then copper, the specific resistance of which is slightly greater, next gold, aluminium, and so on. The greatest specific resistance in descending series have mercury, manganin, nickelin, German silver, these metals or metallic alloys showing at the same time the slightest change in resistance at a higher temperature.

Coefficient of temperature. One and the same material has the same specific resistance only at the same temperature. In conductors of the first class—the metals—the resistance increases, though even only in a slight degree, as the temperature increases. The formula for this is:

(8) $Wt_2 = Wt_1 [+ a (t_2 — t_1)]$,

in which Wt_2 is the resistance at the higher temperature t_2, and Wt_1, the resistance at the lower temperature t_1, and the magnitude a, the number of ohms the resistance increases by a rise of 1° C. in the temperature.

In the conductors of the second class—the electrolytes—the resistance decreases, as a rule quite considerably with a rise in the temperature, and is calculated from the following equation:

(9) $Wt_2 = Wt_2 [1—a (t_2—1_1)]$.

The magnitude a is called the coefficient of temperature of a material, and these coefficients are given in the second column of the above table.

Law of Kirchhoff. From a water-conduit, the water may by means of branch-pipes be conducted to different points. In the same manner, the electric current may be conducted from the main wire by means of different wires to different places. This is called *branching or distributing the current.* The wire from the source of current up to the point of branching is known as the *main wire* and the wires branching off as *branch wires.*

The heavy lines in Fig. 6 represent the main wires; a is the junction from which three wires, 1, 2, and 3, branch off, and b, the junction at which they meet. If a current-meter (see later on) be placed in the main wire, and one in each of the branch wires, it will be found that the sum of the current-quantities flowing through the separate branch wires is equal to the current-quantity in. the main wire. If, however, the current-quantities which flow through the separate branch wires of the same cross-section, 1, 2, 3, are examined, it will be seen that these current-quantities are not the same, but vary one from the other, the current-quantity flowing in the branch wire 1 being greater than that in 2 or 3, while that in 2 is greater than that in 3. These variations are due to the fact

FIG. 6.

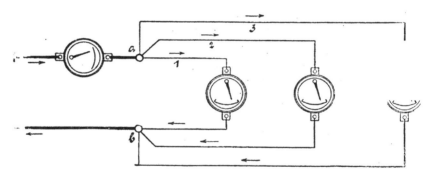

that the branch wire 1 is shorter than 2 or 3, and hence possesses less resistance. Suppose that the longest branch-wire, 3, had a much larger cross-section than the branch-wires 1 and 2. By reason of its slighter resistance more current would flow through it, notwithstanding its length, than through 1 and 2.

Hence the law of Kirchhoff may be summed up as follows:

1. *When the current is branched the sum of the current-strengths in the separate branch wires is exactly as great as the current-strength before and after branching off.*

2. *The current-strengths in the separate branch-wires distribute themselves in inverse proportion to their resistances.*

In the practical part of this work the further conclusions resulting from the law of Kirchhoff will be referred to.

Law of Joule.—If a current flows through a conductor which possesses not too slight a resistance, the latter becomes heated, and, hence, electric energy is converted into heat. It has been shown by experiments that the quantity of heat, which is produced by the passage of a determined current-strength through a determined resistance, increases in the same ratio as the duration of the passage of the current. It has also been shown that by the passage of a determined current-strength through a resistance, the heat produced in the latter in a determined time is proportional to the magnitude of the resistance, and, hence, that the quantity of heat becomes larger as the resistance increases. It has further been established that the quantity of heat produced in a determined resistance during a determined space of time by the current flowing through it, is proportional to the square of the current strength.

From these propositions determined by experiments, the law of Joule may be brought into the formula:

(10) $Q = C . J^2 . W . t$.

If Q is the quantity of heat developed in calories, J is the current-strength in ampères which flows through the resistance, W the resistance through which J flows, and t the space of time in seconds of the passage of the current; C is a constant which by experiments has been ascertained as 0.0002392. In words, Joule's law, therefore, reads: *The quantity of heat produced in t seconds by the passage of a current-strength J through the resistance W is proportional to the expression $J^2 Wt$.*

Frictional Electricity.

In an ordinary state solid bodies exhibit no attractive effect upon such light particles as strips of paper, balls of elderpith, etc., but by being rubbed with a dry cloth or fur, many solid bodies acquire the property of attracting such light bodies as mentioned above. The cause of this phenomenon is called *electricity*, and the bodies which possess this property of becoming electric by friction are termed *idio-electrics*, and those which do not appear to possess it, *non-electrics.* Gray, in 1727, found that all non-electric bodies conduct electricity, and hence are conductors, while those which become electric by friction are non-conductors of electricity. Strictly speaking, there are no non-conductors, because the resins, silk, glass, etc., conduct electricity, though only very badly. It is therefore better to distinguish *good* and *bad* conductors. To test whether a body belongs to the idio-electrics, the so-called electroscope is used, which in its simplest form consists of a glass rod mounted on a stand, and bent at the top into a hook, from which hangs by a silken thread or hair a pith ball. If, on bringing the rubbed body near the pith ball, the latter is attracted, the body is electric ; whilst if the ball is not attracted, the body is either non-electric, or its electricity is too slight to produce an attractive effect.

From the following experiments it was found that there exist *two* kinds of electricity: When a rubbed rod of glass or shellac is brought near the ball of elder-pith suspended to a silk thread, the ball is attracted, touches the rod, adheres for a few moments, and is then repelled. This repulsion is due to the fact that the ball by coming in contact with the rod becomes itself electric, and its electricity must first be withdrawn by touching with the hand before it can again be attracted by the rod. By now taking two such balls, one of which has been made electric by touching with a glass rod, which had been rubbed with silk, and the other by touching with a shellac rod rubbed with cloth, it will be observed that the ball, which is repelled by the glass rod, is attracted by the shellac rod, and *vice versa.* These two kinds of electricity are called *vitreous* or *positive*, and *resinous* or *negative* electricity, and it has been found that *electricities of a similar name attract*, and *electricities of an opposite name repel each other.*

Contact Electricity.

However, a current of electricity is generated not only by friction, but also by the contact of various metals. In the same manner as the copper and iron in Galvani's experiments with the frog-leg, other metals and conductors of electricity also become electric by contact, the electric charges, being, however, stronger or weaker, according to the nature of the metals. If zinc be brought in contact with platinum, it becomes more strongly positively electric than when in contact with copper; whilst, however, copper in contact with zinc is negatively excited, in contact with platinum it becomes positively electric.

The metal which has become positively electric is said to have the higher *potential, i. e.,* it possesses a larger measure of electricity than the metal which has become negatively electric, and as the flow of water from higher to lower points takes place in a larger degree the greater the difference in altitude is, the electric current flows also the more rapidly from a positively charged body—the *positive* pole—to the negatively charged body—the *negative* pole—the greater the difference in their charges is, and this difference in the charges of two bodies is called *difference of potential.*

If now the metals be arranged in a series so that each preceding metal becomes positively electric in contact with the succeeding one, a series of electro-motive force is obtained in which the metals or conductors of electricity stand as follows: *Potassium, sodium, magnesium, aluminium, zinc, cadmium, iron, nickel, lead, tin, copper, silver, mercury, gold, platinum, antimony, graphite.*

While two metals of the series of electro-motive force touching each other. become electrically excited in such a manner that one becomes positively and the other negatively electric, an exchange of the opposite electricities takes place by introducing a conducting fluid between the metals. Thus, if a plate of zinc and a plate of copper connected by a metallic wire are immersed in a conducting fluid, for instance, dilute sulphuric acid, the electricity of the positive zinc passes through the fluid to the negative copper, and returns through the wire—the *closed circuit*—to the zinc. However, in the same degree with which the electricities equalize themselves, new quantities of them are constantly formed on the points of contact of the metals with the conducting fluid ; and, hence, the flow of electricity is continuous. This electric current generated by the contact of metals and fluids is called the *galvanic current;* or, since it is generated by the intervention of fluid conductors, *hydro-electric* current.

A combination of conductors which yield such a galvanic current is called a *galvanic or voltaic cell or battery*, and the production of current from the above-mentioned differences of potential of the metals was formerly explained by the suppositiou that chemical processes take place in the solutions in which the metal plate is immersed. However, as will be seen later on, the production of the current is at present reduced, according to Nernst's theory, to the solution-pressure and the osmotic pressure. It is first of all necessary to explain the fundamental chemical principles, since without a knowledge of them, the subsequent sections could not be understood.

Fundamental Chemical Principles.

The phenomena presented by magnetism and electricity have, so far as required for our purposes, been briefly discussed in the preceding sections. All these phenomena, no matter how much they may vary in their nature, have this in common, that the bodies in which they appear undergo no change in substance and weight, notwithstanding that they acquire the most diverse properties. If, for instance, steel by being rubbed with a magnet has acquired the power of attracting iron articles, and hence has become a magnet itself, no other changes can be noticed in it, even by the most minute examination ; it remains the same steel which had been originally used, it having solely acquired the property of being capable of acting as a magnet.

The phenomena to be treated of in this section devoted to the fundamental chemical principles, are of an entirely different nature, we having constantly to deal with changes in substance, as may be shown by the following examples.

When bright iron or steel is exposed to the action of moist air, it becomes gradually coated with a brown-red powder known as rust, which is formed by the iron combining with the oxygen of the air. On examining this brown-red substance it will be found to possess entirely different properties from iron, and that the latter has undergone a material change. By the absorption of oxygen the iron has been converted into an oxide of iron, and a process known as a chemical process has taken place, whereby from two different substances a third one is formed which possesses other properties, and is of a different composition.

The phenomena which appear in subjecting the well-known red oxide of mercury or red precipitate to the action of heat, furnish another example of a chemical process. If red oxide of mercury be heated in a test-tube, its red color soon disappears, its bulk decreases, and, if heating be for some time continued, it disappears entirely. On the other hand, there will be found deposited upon the upper, cooler portions of the tube, metallic mercury in its characteristic form of globules. If the gaseous products evolved during the process be also caught, a gas, different in its nature from air, is obtained, which will inflame a mere spark on wood. This gas is the well-known *oxygen*, which plays such an important part in the respiratory process of human beings and animals.

While by the formation of a new body in consequence of the combination of different substances, the first example presents a chemical process of a *synthetic, i. e.*, building-up, nature, the second one, shows a process of an *analytical, i. e.*, resolving, nature. We have thus learned the nature of the chemical processes in general, which, no matter how diverse the separate processes may be, consist, in that an alteration in the material nature of the bodies takes place. If the quantities by weight of a substance entering into a chemical change be determined, it will be noticed that in all transpositions, in the decomposition of a compound into its constituents, and in the union of the elements to form compound bodies, loss in weight never occurs. *The weight of the resulting compound is invariably equal to the sum of the weight of the bodies entering into the reaction.* This furnishes proof that the most important law of the indestructibility and non-creation of weighable substance in nature, which is known as the *law of the conservation of matter*, is also valid as regards chemical processes.

Moreover, we find the further conformity to law that the quantities by weight of the substances formed by their mutual action in a chemical process, stand one to the other in a fixed, unchangeable proportion. Thus, for instance, a given quantity by weight of iron can only combine, under the co-operation of water, with an unchangeable quantity of oxygen, to ferric hydroxide (rust); and the quantities by weight of mercury and oxygen formed from red oxide of mercury, must always stand one to the other in an unchangeable proportion.

If now in a similar manner as in the second example, all the bodies offered by nature be decomposed by means of the auxiliary agents at our command, into such constituents as do not allow of a division into further substances, it will be found that there are altogether comparatively few substances which compose the bodies of nature. Such substances are called *chemical elements;* they cannot be converted into each other, but constitute, as it were, the limit of chemical change. At present 79 such elements are known.

The smallest portion of an element, or of a chemical compound, which can exist in a free state, is called a *molecule.* If, for instance, common salt be triturated to such a fine powder that further reduction by mechanical means is impossible, such finest particle represents the molecule. However, common salt consists of two elements, namely, sodium and chlorine. Consequently both these elements must be present in the molecule, and these smallest particles of the elements, which are contained in the molecule, are called *atoms.* Hence the atom of an element is the smallest quantity of it which takes part in chemical combinations. As a rule, the atom is equal to half

3

the molecule. Hence, for the formation of a molecule at least two atoms of an element are required.

The atoms of the elements aggregate according to fixed proportions by weight, and the smallest quantities by weight of the elements which enter into combinations with each other are called their *atomic weights*, the weight of hydrogen, which is the lighest of all the elements, being taken as the unit. It must, however, be stated that a series of elements may unite not only in a single proportion of weight, but also in several different ones, forming thereby combinations of entirely different properties. If, however, these different proportions by weight are more closely compared, they will be found to stand in quite simple relations to each other, the higher being always a simple multiple of the lowest.

In the table below are given the most important chemical elements, together with their atomic weights. In addition the table contains the symbols used for designating the elements. These symbols are formed from the first letters of their names, derived either from the Latin or Greek. Hydrogen is, for instance, represented by the letter H, from the word *Hydrogenium;* Oxygen by O, from *oxygenium ;* Silver by Ag, from *argentum.* If Latin or Greek names of several elements have the same first letters, the latter serves only for the designation of one of these elements, while for the other elements, the first letter is furnished with an additional characteristic letter. Thus, for instance, boron is represented by the letter B ; barium by Ba ; bismuth by Bi ; bromine by Br.

INTERNATIONAL TABLE OF THE ATOMIC WEIGHTS OF THE MOST IMPORTANT ELEMENTS, (1911).

Name of Element.	Symbol.	Atomic Weight.	Name of Element.	Symbol.	Atomic Weight.
Aluminium	Al	27.1	Lead	Pb	207.10
Antimony. . . .	Sb	120.2	Magnesium . .	Mg	24.32
Arsenic	As	74.96	Manganese . .	Mn	54.93
Barium	Ba	137.37	Mercury . . .	Hg	200.0
Bismuth	Bi	208.0	Nickel	Ni	58.68
Boron	B	11.0	Nitrogen . . .	N	14.01
Bromine	Br	79.92	Osmium . . .	Os	190.9
Cadmium	Cd	112.40	Oxygen	O	16.00
Calcium . . .	Ca	40.09	Phosphorus	P	31.04
Carbon	C	12.0	Platinum	Pt	195.2
Chlorine	Cl	35.46	Potassium	K	39.10
Chromium. . . .	Cr	52.0	Selenium . . .	Se	79.2
Cobalt	Co	58.97	Silicon	Si	28.3
Copper	Cu	63.57	Silver	Ag	107.88
Fluorine	F	19.0	Sodium	Na	23.00
Gold	Au	197.2	Sulphur . . .	S	32.07
Hydrogen. . . .	H	1.008	Tin	Sn	119.0
Iodine . . .	I	126.92	Zinc	Zn	65.37
Iron	Fe	55.85			

The symbols not only represent the elementary bodies, but also their fixed quantities by weight, so that, for instance, the symbol Ni means 58.68 parts by weight of nickel.

Compounds produced by the union of the elements are represented by placing their corresponding symbols together and designating them *chemical formulas.* As previously mentioned, common salt consists of one atom sodium (Na) and one atom chlorine (Cl), and hence its formula has to be written NaCl. The latter shows that one molecule of common salt consists of 23.00 parts by weight of sodium and 35.46 parts by weight of chlorine, which together form 58.46 parts by weight of common salt. If several atoms of an element are present in a compound, this is denoted by numbers which are written to the right of the symbol, below, as proposed by Poggendorf, or above, as proposed by Berzelius, and still used at the present by a few people. Water, for instance, contains 2 atoms hydrogen (H) and one atom oxygen (O), and hence its formula is H_2O, which indicates that 2 parts by weight of hydrogen, together with 16 parts by weight of oxygen, form 18.016 parts by weight of water.

The symbols may be said to constitute the chemical alphabet and the formulas may be considered as the words of the chemical language. By means of the symbols and formulas it is made possible, to express in the most simple manner, the chemical processes by equations, which not only denote the manner of the chemical transposition, but also allow of the calculation of the quantities by weight which have entered into reaction in the transposition of the different substances. If, according to this our former examples, by means of which it has been endeavored to explain the nature of a chemical process, be translated into this chemical language, the equations read as follows :

$$1. \quad 2Fe_2 + 3O_2 + 6H_2O = 4Fe_3(OH)_3.$$
Iron.   Oxygen.   Water.   Ferric hydroxide.

$$2. \quad 2HgO = Hg_2 + O_2.$$
Mercuric oxide. Mercury. Oxygen.

Valence of the elements. If the combinations into which the elements enter one with the other are more closely examined, and their formulas compared, it will be seen that entire groups of combinations are composed in an analogous manner. This analogy of composition appears very plainly in the compounds into which a series of elements enters with hydrogen, and we thus come across four different groups of compounds. The elements of the first group, namely, of the halogens, chlorine, bromine, iodine and fluorine, combine with one atom of hydrogen ; those of the second group, to which belong oxygen and sulphur, are capable of saturating two atoms of hydrogen ; those of the third group, which embraces nitrogen, phosphorus, arsenic and antimony, fix three atoms of hydrogen, and finally, the elements of the fourth group, carbon and silicon, may combine with four atoms of hydrogen. Hence, we must ascribe a particular function of affinity to each element in its relation to hydrogen, and this property is called *valence.*

Now, according as the elements are capable of combining with one, two, three or four atoms of hydrogen, they are designated as univalent, bivalent, trivalent, or quadrivalent; and all elements, which possess the same valence, are called chemically equivalent. In chemical compounds, such equivaleut elements may replace each other atom for atom, such substitution being also possible in elements of dissimilar valence, but it must take place in such a manner that a bivalent atom replaces two. hydrogen atoms, a trivalent atom three hydrogen atoms, so that an equal number of valences is always exchanged. Thus, in accordance with this, one atom of chlorine is equivalent to one atom of hydrogen and hence, when a substitution of hydrogen by chlorine results, it can only be by one atom of chlorine taking the place of one atom of hydrogen. Hence it follows that 35.46 parts by weight of chlorine are equivalent to one part by weight of hydrogen. On the other hand, one atom of oxygen is equivalent to two atoms of hydrogen, or 16 parts by weight of the former are equivalent to 2 parts by weight of the latter. A mutual substitution of these two elements must, therefore, always take place in the proportion of 16 to 2. Since the elements, nitrogen, phosphorus, etc., are capable of fixing 3 hydrogen atoms, mutual substitution must also take place in such a manner that 1 nitrogen atom replaces 3 hydrogen atoms or that $\frac{14.01}{3} = 4.67$ parts by weight of nitrogen are substituted for 1 part by weight of hydrogen. Finally, one atom of carbon or of silicon is equivalent to 4 parts by weight of hydrogen, or 1 part by weight of hydrogen is replaced by 3 parts by weight of carbon. These quantities by weight determined for some of the elements, which are equivalent to 1 part by weight of hydrogen, or, in general, to one part by weight of a univalent element, are called *equivalent weights or combining weights,* and are in a similar manner deduced for all the other elements.

While the elements preserve a constant valence towards hydrogen, many of them show a varying valence, which differs also from the hydrogen-valence towards other elements, so that, for instance, the same element may appear opposite to a second one, trivalent in one combination and quinquivalent in another. Combinations of phosphorus with chlorine may serve as an example. Together they form a combination, PCl_3, as well as one PCl_5; in the first case 3 atoms of chlorine or 3×35.46 parts by weight are equivalent to 1 atom of phosphorus or 31.04 parts by weight. This capacity of different elements of being endowed with totally unequal valence, forces us to the assumption that valence is not a characteristic property of the elements, but is dependent on the nature of the elements combining with each other, and is also influenced by the conditions under which the formation of the chemical combination takes place.

By arranging the most important elements according to their valence, we obtain the following groups

Univalent elements: Hydrogen, chlorine, bromine, iodine, fluorine, potassium, sodium, silver.

Bivalent elements: Oxygen, sulphur, barium, strontium, calcium, magnesium, cadmium, zinc, lead, copper, mercury.

Bivalent and trivalent elements: Iron, cobalt, nickel, manganese.

Trivalent elements: Boron, aluminium, gold.

Trivalent and quinquivalent elements: Oxygen, phosphorus, arsenic, antimony, bismuth.

Quadrivalent elements: Carbon, silicon, tin, platinum.

Later on, in the section on the fundamental principles of electro-chemistry, in speaking of the development of the laws of Faraday, these groups will have to be referred to, and their importance will then become evident.

Metals and non-metals. In accordance with the greater or less conformity of their physical properties, the elements have, for the sake of expediency, been sub-divided into two sections, namely *metals* and *non-metals*, the latter being also called *metalloids*. The first section embraces the elements the principle characteristics of which are that they show metallic luster, are opaque or at the utmost translucent in thin laminae, are, as a rule, fairly malleable and ductile, and with the one exception of mercury, are all solid bodies at ordinary temperatures and pressures, and are good conductors of heat and electricity. All the other elements which have not such physical properties in common are classed as metalloids. The two groups of bodies obtained by this mode of division also show in a chemical respect such similarities as to justify this classification, the metalloids forming with hydrogen readily volatile, mostly gaseous, combinations, while the metals unite more rarely with hydrogen, and, at any rate, do not form volatile combinations with it. The combinations which the metalloids form with oxygen also show, in their behavior towards water, very characteristic phenomena, entirely different from those presented by compounds of the metals with oxygen. These differences will later on be referred to in detail. A very remarkable difference of the utmost importance, especially for our purpose, is in the action of the electric current upon the combinations between metals and metalloids, the metals being always deposited on the electro-negative pole, and the metalloids on the electro-positive pole.

However, notwithstanding these properties, differing on the one hand and corresponding on the other, a sharp separation of the elements based upon the above-mentioned considerations cannot be reached, and the classification as regards some elements turns out different according to whether one or the other behavior is first taken into consideration.

On the other hand, a classification free from ambiguity results from adhering, as is now also done in science, to the behavior of the elements towards salts as the distinctive principle. In this manner two sharply-defined groups are obtainable, one comprising the elements—the metals—capable of evolving hydrogen with the acids, while the elements of the other group do not possess this power, and are classed among the metalloids. From this results the following classification :

Metalloids: Chlorine, bromine, iodine, fluorine, oxygen, sulphur, nitrogen, phosphorus, boron, carbon, silicon.

Metals: Potassium, sodium, lithium, magnesium, calcium, barium, strontium, aluminium, zinc, iron, manganese, chromium, nickel, cobalt, copper, cadmium, arsenic, antimony, tin, lead, bismuth, mercury, silver, gold, platinum.

Acids, bases, salts. Attention has previously been drawn to the difference in behavior towards water of combinations of the metalloids, and of the metals with oxygen, and this behavior will have to be somewhat more closely considered, because we are thereby directed to extremely important classes of chemical combinations.

Oxygen is the most widely distributed element, it forming, together with nitrogen, air, and with hydrogen, water. All the elements, with the exception of fluorine and a few more rare ones, show great affinity for it and enter readily into reaction with it. In the processes enacted thereby, the large class of *oxides* is formed, and the chemical process in which an absorption of oxygen takes place is generally called *oxidation*, while the term *reduction* is applied to the opposite process by which a withdrawal of oxygen from a substance is effected.

If these oxides, with the exception of a few so-called indifferent oxides, be brought together with water, they impart to it either an acid taste, as well as the power to redden blue litmus and to evolve hydrogen with metals, or they give to the water a lye-like taste and the power of restoring the blue color to the litmus previously reddened. The oxides of the first kind are chiefly formed with the co-operation of the elements belonging to the metalloids, while those of the second class contain exclusively metals in addition to oxygen.

These two classes of bodies, which possess entirely different, even directly opposite, properties, are the acids and bases, and will have to be separately discussed.

Acids. As characteristic properties of the acids have been mentioned, their acid taste, their power of reddening blue litmus, and to evolve hydrogen with metals, magnesium being especially suitable for the latter purpose. If now the chemical compositions of all the compounds which possess the above-mentioned properties be more closely examined, they will be found to contain, without exception and without regard to their own constituents, hydrogen which can be displaced by metals. This hydrogen may be present in the combinations in one or more atoms, and according to the number of the hydrogen-atoms present, a distinction is made between mono-basic, dibasic, tribasic, etc., acids.

A further distinction is made between acids containing no oxygen, to which belong the *haloid acids* for instance, hydrochloric acid, and acids containing oxygen, which are therefore called *oxy acids.* The latter group comprises the majority of acids, the well-known sulphuric and nitric acids belonging to it. However, the characteristic feature of the acids consists solely in that they contain hydrogen which can be displaced by metals.

Bases. The second group of oxides imparts to water, as previously mentioned, a lye-like taste and the power to restore the blue color of litmus reddened by acid, and these properties are utilized as valuable agents for the characterization of the substances as bases. Nevertheless, by the above-mentioned definitions the meaning of bases is not unequivocally established, and for a thorough investigation of their material nature, their exact composition has to be determined with the assistance of analysis, as was done with the acids. From this it results that, in addition to metals or metal-like groups of atoms, all basic compounds contain oxygen and hydrogen, the latter elements being always present in the same number of atoms, namely, in the form of *hydroxyl* groups, OH. According to their valence the metals combine with one or more hydroxyl groups to bases.

Salts. The groups of chemical combinations above referred

to, show a very remarkable behavior in so far that by their mutual action they are capable of equalizing or saturating their characteristic features, so that by means of a basic combination the specific properties of an acid can be removed, and by means of an acid the specific properties of a base.

An example will explain this process. If to a certain quantity of hydrochloric acid a few drops of blue litmus be added, the fluid in consequence of its acid properties will change the blue coloring matter, the latter acquiring a red color. By now adding drop by drop dilute soda lye, which is a basic combination, it will be noticed that on the spot where the lye falls upon the acid, the red color disappears momentarily, and gives way to a blue one. If the addition of lye be carefully continued and the fluid constantly stirred, a point is suddenly reached when by a single drop of the lye the red color of the entire fluid is removed and converted into pale blue. If no more lye than exactly necessary for the sudden change in color has been brought into the fluid, the latter now possesses neither the properties of an acid nor of a base, but has become what is called *neutral*. A process of the kind above described, by which the acid character of a combination is equalized by the basic character of another, or *vice versa*, is in chemistry called *neutralization*.

This example shows, that it is frequently of importance to know whether a fluid possesses acid, basic or neutral properties, or as it also expressed, whether it shows an acid, basic, or neutral reaction. For the determination of these properties so-called *reagent-papers* are used. They consist of unsized paper dyed with various organic coloring matters, preferably blue litmus tincture, or the latter slightly reddened by acids. When small strips of such papers are dipped in the fluid to be examined, blue litmus paper will be colored red if the fluid has an acid reaction, and red litmus paper, blue, if it shows an alkaline or basic reaction. Finally, fluids which change neither blue nor red litmus paper react neutral, or they show a neutral reaction. If we now return to our example by which the process of neutralization between acid and base has been described, it will above all be of interest to learn whether this equalization of the mutual properties runs its course according to fixed laws, and what the nature of the latter is. It will be further desirable to gain an insight into the chemical transformations which have taken place in the process, and to learn the products which have been newly formed.

For the elucidation of these questions, let us take a determined quantity of acid and, in the same manner as in the above-described example, add to it lye until the acid is just neutral, this being shown by the sudden change in color of the litmus. If we now take another quantity of the same acid and proceed with it in the same manner, it will be found that the consumed quantities of bases stand in the same proportion to each other as the quantities of acid used, so that if, in one case, for 50 ccm. of acid 30 ccm. of lye were used for neutralization, in the other, with the use of the same acid and the same lye, for 75 ccm. of acid 45 ccm. of lye were required to obtain a neutral solution. By repeating these experiments with any other acids and bases, the same conformity to law will always be found, and it will thus be seen that neutralization between acids and bases runs its course in positively fixed quantities, and that for the neutralization of a certain quantity of an acid, a positively fixed quantity of a base is required, and *vice versa.*

Of this conformity to law much use is made in analytical chemistry by volumetric methods for the determination of the content of an acid by means of a base of known content, and *vice versa.*

In order to learn what new products are formed by the neutralization between acids and bases, the neutral solution obtained, according to our example, is concentrated by evaporation, and it will be found that from the fluid separates a white substance in small crystals which, according to analysis, consists of sodium (Na) and chlorine (Cl), and hence constitutes the well-known common salt (NaCl). However, in addition to the common salt, water (H_2O) has also been formed by the chemical process, as shown by analysis.

If now, as another example, we take as an acid, sulphuric acid (H_2SO_4), neutralize it with caustic soda (KOH), and again determine the products formed, we arrive at a substance, the composition of which, according to analysis, is K_3SO_4, hence represents potassium sulphate, water being again formed as an additional product.   The process of neutralization takes its course in an analogous manner with any kinds of acids and bases, and it will be seen that every neutralization of an acid and a base is accompanied by the formation of water, and further, that after the withdrawal of the hydrogen from the acid, the metal of the bases forms with the remainder a new neutral combination, which is called a *salt*.

These processes are more distinctly presented by bringing them into chemical formulas, and for our examples we have to write

$$HCl \quad + \quad NaOH \ = \ H_2O \quad + \quad NaCl.$$
Hydrochloric acid.   Sodium hydrate.   Water.   Sodium chloride (common salt).

$$H_2SO_4 \ + \ 2KOH \ = \ H_2O \quad + \quad K_2SO_4.$$
Sulphuric acid.   Potassium hydrate.   Water.   Neutral potassium sulphate.

These formulas show plainly the connection which exists be tween the acids, bases and salts.   •

The formation of salts from the acids is thus brought about by the replacement of the hydrogen-atoms of the acids by metals.   However, this replacement of the hydrogen can only take place in accordance with the valence of the metal, so that a univalent metal can take the place of only one hydrogen-atom, a bivalent metal of only two hydrogen-atoms, and so on.   With the use of a monobasic acid, *i. e.*, one in which only one hydrogen-atom is contained in the molecule, salts can only be prepared which, besides metal, contain no free hydrogen-atoms, and salts of the above-mentioned kind, namely, *neutral* salts, are exclusively obtained.   By taking, on the other hand, an acid with several bases, its hydrogen-atoms can be either partly or entirely replaced by metals. In the first case, salts result which still possess an acid character, they containing hydrogen besides a metal, and are called *acid salts*, while in the latter case neutral salts are formed, with which we are already acquainted. Sulphuric acid is a dibasic acid, and, hence, contains two hydrogen-atoms in the molecule. Let us take as an example, the salts which sulphuric acid is capable of forming, and first saturate in it only oné hydrogen-atom by a univalent metal, for instance, sodium, by adding just enough soda lye to the soda to half saturate it. This solution still shows a strong acid reaction, and by sufficiently concentrating it, a salt is separated which throughout possesses acid properties and, as shown by analysis, has the chemical formula $NaHSO_4$. It is different from the neutral sodium sulphate, which is obtained by completely saturating the sulphuric acid with caustic soda, i. e., by compounding the sulphuric acid with caustic soda up to the neutral reaction. The two processes just described are explained by the following equations, which also show distinctly the difference between neutral and acid salts :

$$1.\ H_2SO_4 + NaOH = NaHSO_4 + H_2O.$$

| Sulphuric acid. | Sodium hydrate. | Acid sodium sulphate. | Water. |

$$2.\ H_2SO_4 + 2NaOH = Na_2SO_4 + H_2O.$$

| Sulphuric acid | Sodium hydrate. | Neutral sodium sulphate. | Water. |

In an analogous manner, as a dibasic acid is capable of forming two series of salts, three series of salts may be derived from a tribasic acid, for instance, phosphoric acid, so that in general an acid of several bases can form as many series of acids as it contains hydrogen-atoms in the molecule.

Nomenclature of salts. In conformity with the definition of salts given above, according to which they are derived from the acids by the replacement of the hydrogen by metals, they are classified according to the acids they have in common, the salts derived from sulphuric acid being thus designated sulphates. For the sake of distinguishing the various metallic salts of the same acid, the names of the metals are added. Thus, for instance, the scientific term for white vitriol, formed by the action of sulphuric acid upon zinc, is zinc sulphate. The designations for the salts of the other acids are formed in the same manner ; those derived from nitric acid being called nitrates, from phosphoric acid, phosphates, etc. Salts in which all the hydrogen-atoms of the acids from which they are derived, have been replaced by metal-atoms are called neutral, normal, or primary salts in contradistinction to the acid or secondary salts which, besides metal-atoms, also contain hydrogen-atoms in the molecule. Finally, the salts are also designated by indicating with the assistance of the Greek numerals, mono-, di-, etc., the number of metal-atoms contained in one acid-molecule. With the use of the latter mode of designation, the scientific term for the acid sodium sulphate is sodium mono-sulphate, and for the neutral sodium sulphate, sodium disulphate. —

Fundamental Principles of Electro-Chemistry.

Electrolytes. Solutions of chemical compounds which can be decomposed by the current, are called electrolytes.

A distinction is made between *conductors* and *non-conductors* of electricity, and, as previously mentioned, the metals are conductors, while most of the metalloids, for instance, sulphur, do not transmit the electric current.

The conductors are divided into *conductors of the first class*, to which belong the metals, and *conductors of the second class*, the latter being chiefly the aqueous solutions of metallic salts and certain other substances.

The conductors of the first class do not experience a perceptible material change by the passage of the current, they being at the utmost heated thereby. On the other hand, the conductors of the second class undergo, by the passage of the current, a chemical change is so far as that on the places where the current-carrying metallic conductor enters the solution, the constituents of the latter are decomposed and separated.

This phenomenon of the chemical decomposition of substances or compounds by means of an electric current is called *electrolysis*, and the conductors of the second class which undergo such decomposition, are termed *electrolytes*.

The metal plates through which the current passes in and out of the solution are called *electrodes*, the *positive* electrode through which the current enters being termed *anode*, and the *negative* electrode through which it leaves the electrolyte, *cathode*.

Ions. This term is applied to the constituents into which the combinations present in the solution are decomposed by the current, and carried to the cathodes and anodes.

If a sodium chloride solution be subjected to electrolysis, the sodium chloride is decomposed, chlorine being separated on the positive electrode, and sodium on the negative electrode. Thus chlorine and sodium are the ions of sodium chloride. If an acid be decomposed by the electric current, hydrogen is always separated on the negative electrode, and the other constituent of the acid on the positive electrode.

The ions separated on the negative electrodes are called *cations* and, hence, in the above-mentioned examples, sodium and hydrogen are the cations of sodium chloride, or of the acid. The cations migrate from the positive to the negative electrode.

The remaining ions of the combinations migrate from the negative to the positive electrode (anode), and are there separated. These ions separated on the anode are called *anions*. Thus chlorine is the anion of sodium chloride, as well as of hydrochloric acid and of other chlorine compounds.

The ions exhibit, partly, properties entirely different from the elements the names of which they bear. The hydrogen-ion of the acids, for instance, is not known as a gas, but only in solution, while the element hydrogen is gaseous and but very slightly soluble in water. Further, while the hydrogen-ion determines the characteristic properties of the acids, hydrogen gas exhibits none of these properties, and the hydrogen-ion can only be met with in aqueous solutions of acids in which are at the same time present the other constituents of the acids possessing ion-properties.

If in hydrochloric acid, hydrogen exists as ion, chlorine must be the other ion, because this acid contains no other constituents, and this chlorine-ion possesses the same properties exhibited by the chlorine-ions of other combinations in which it is contained, hence, in all soluble metallic chlorides. These properties of the chlorine-ion, however, differ, entirely from those of chlorine in the ordinary elementary state, it possessing neither its odor nor color; it exists only in solution and has not the bleaching effect of chlorine gas.

These totally different properties thus clearly indicate that the ions have to be considered as *modifications of the elements designated by the same name*, or that the ions have to be thought of as *existing in a condition different from the elementary one;* and the reason for these different conditions and properties will be more accurately known after we have to some extent become acquainted with the

Theory of solutions. A solution is not a mere mechanical mixture of an invisible, finely divided solid body with the solvent, but by solution in a solvent a body partially loses its characteristic properties and acquires new ones, and the dissolving process may be viewed as a chemical process in so far as changes of energy (see later on), for instance, fixation or disengagement of heat, are connected with it.

There are not only solutions of solid substances in liquids, but also solutions of liquids in liquids, of gases in liquids, and of gases in gases. However, the last-mentioned solutions are of interest to us only in so far as it has been shown that the laws which they follow are also valid for solutions of solid bodies in liquids. For the proof of this we are indebted to van't Hoff.

If a layer of a dilute, pale blue cupric sulphate solution be carefully brought, so as to avoid mixing, upon a concentrated cupric sulphate solution of a vivid blue color, and the vessel containing the solutions be allowed to stand quietly, it will be noticed that the pale blue solution gradually acquires a more intense blue color, while the concentrated solution becomes paler. The molecules of the cupric sulphate diffuse from the stronger, into the weaker solution until the liquid has acquired a uniform concentration.

This phenomenon is based upon the same law followed by the gases. A gas endeavors, when occasion is offered, to occupy a larger space; the energy of motion (kinetic energy) inherent in the individual gas molecules propels them until their motion is stopped by the walls of the enlarged space. The molecules in the cupric sulphate solution possess a similar energy of motion and by it, as we have seen, are forced from the concentrated, into the weak solution. This force, which corresponds to the gas pressure, is called

Osmotic pressure. Its presence can readily be demonstrated by the following experiment: Fill a glass-cylinder with saturated sugar solution, close the cylinder air-tight with a semi-permeable bladder, and place it upright in a vessel filled with water so that the latter stands a few centimeters above the bladder; the bladder bulges up in a short time. This phenomenon is caused by the effort of the sugar molecules to diffuse into the surrounding water, being, however, prevented from doing so by the bladder, while water molecules penetrate through the bladder into the cylinder. If the cylinder be removed from the water and the bladder be punctured with a pin, the pressure which had existed becomes plainly perceptible by a jet of fluid being forced upward. By exact investigations of the magnitude of osmotic pressure it has been ascertained that it is proportioned to the number of molecules dissolved in the unit volume, and that the temperature has the same effect upon osmotic pressure as upon gases, conformity with the laws valid for gases being thus proved. According

4

to Avogadro's law equal volumes of different gases under the same conditions of temperature and pressure contain equal numbers of molecules, and the weights of these gases are thus in the same ratio as their respective molecular weights.

Solutions, as has been proved by van't Hoff, follow the same law and, according to van't Hoff, the law applied to them is expressed as follows : Solutions which contain an equal number of dissolved molecules in the same volume of solvent (equimolecular solutions) exert, under the same conditions of temperature, the same osmotic pressure which has the same value as the gas-pressure these bodies, if in a gaseous state, would under the same conditions of temperature exert in a volume of gas equal to the volume of solvent.

It should, however, be borne in mind that the osmotic laws are valid only for dilute solutions, just as the gas-laws hold good only for dilute gases.

Electrolytic dissociation. Clausius originated the idea that the molecules of an electrolyte are dissociated to molecular particles corresponding to our ions. He supposed that the molecules are in constant motion whereby they are partially decomposed, and that the molecular particles formed again attract the molecular particles of opposite names of the non-decomposed aggregate molecules, and thus effect the dissociation of the latter. On the other hand, molecular particles of opposite names will again form, under favorable conditions, aggregate molecules. However, as soon as a current passes through the electrolyte, the irregular and changing movements of the molecular particles will cease, and they will take the direction presented by the action of the current, *i. e.*, the positive molecular particles will wander with the direction of the current to the cathode, and the negative ones to the anode.

The method, discovered by Raoult, of determining the molecular weights of dissolved bodies from the elevation of the boiling point and the depression of the freezing point, caused, in connection with van't Hoff's osmotic laws, a further investigation of the dissociation of electrolytes. It was known that salt solutions possess a higher boiling point than the pure solvent. Further investigations proved the elevation of the boiling point to be proportional to the number of the dissolved molecules, and that equimolecular solutions, *i. e.*, solutions which contain an equal number of dissolved molecules in the same volume of solvent, show the same elevation of the boiling point. On the other hand, the freezing point of solutions is lowered in proportion to the dissolved molecules, and equimolecular solutions show the same depression of the freezing point.

However, not all substances in equimolecular solutions furnished at the same temperature, the same osmotic pressure as sugar solutions or solutions of other organic bodies. Thus, solutions of acids, bases, and salts yielded too high an osmotic pressure, and also showed deviations in so far that, as compared with equimolecular solutions of many organic substances, they caused under entirely equal conditions, a higher elevation of the boiling point or depression of the freezing point. Since, as regards gas-pressure, some gases also do not follow Avogadro's law, and these exceptions were explained by assuming that the molecules decompose to molecular particles, the same assumption was made for solutions of acids, bases and salts.

S. Arrhenius, in 1887, found that all the solutions which formed exceptions to the osmotic law and showed deviating results as regards elevation of the boiling point and depression of the freezing point, possessed the common property *of conducting the electric current*, while solutions of organic bodies which, as above mentioned, followed the laws referred to, were *non-conductors of the electric current*. Arrhenius ascertained that very considerable exceptions appear for water as solvent, since the pressure is greater than van't Hoff's law requires, and it would therefore be but natural to suppose that substances which give too large pressures in aqueous solutions are dissociated. He further found that dissociation increases with increasing dilution, and he established the law that for every dilute solution the ratio of dissociation is equal to the ratio of *molecular conductivity* present to the conductivity of infinite dilution, *i. e.*, to the maximum of molecular conductivity. The independent particles of the molecules formed are the *ions*.

It further follows that it is the ions which take charge of the progressive motion of the current because only ion-forming solutions are capable of conducting the current. The ions are supposed to be charged with a certain quantity of electricity —the cations with positive, the anions with negative, electricity —and so long as no current passes through the electrolyte, they move free in the latter. However, when a current is conducted through the electrolyte, the ions are attracted by the electrodes, the positively charged cations by the negatively charged cathode, and the negatively charged anions by the positively charged anode. By reason of the movements of the ions to the electrodes this phenomenon may be called *migration of the ions*.

The ions on reaching the electrodes are freed of their charge, *i. e.*, they yield their electricity to the electrodes, but they lose thereby their ion-nature and are changed into their respective elementary atoms ; they show no longer the properties of ions but those of the ordinary elements. As is well-known the various modifications of carbon (diamond, graphite) are chemically alike, namely in all cases carbon, but they have entirely different properties, the latter being conditional on an entirely different *content of energy*.

Energy. By energy is understood the work and everything which can be the result from work, and be again converted into work. A distinction is made between various kinds of work. The effect of *mechanical work* expresses itself through the product of force and motion, *i. e.*, the force required to convey a body a certain distance. If we push a wagon, the force with which we push against the wagon multiplied by the motion, *i. e.*, the distance the wagon has been pushed, is the value of the work.

Now a distinction has to be made between the force with .

which a man in pushing presses against the wagon, and that with which the wagon presses against the man, who does the pushing. In comparison we speak of both forces as *force* and *counter-force*, and physics teaches us that force and counter-force are, in all cases, of the same magnitude, but exerted in opposite directions. Both these propositions may be combined to the proposition of the *conservation of force and work*, which reads : *No quantity of force and no quantity of work are lost ; the force and work consumed are always again met with in another definite form.*

When a wagon has been pushed to a higher point of an oblique plane, it has taken up a certain quantity of work corresponding to the value from force multiplied by motion in the direction of the force. It possesses a certain *energy* which it can and does give up when it is released ; the wagon runs down the oblique plane, and the velocity with which it runs down is also a form of energy.

If an article be ground upon an emery wheel, a certain frictional work is performed ; the article becomes warm by friction. Hence the heat which is developed is another form of energy of the frictional work, since according to the law of the conservation of work no quantity of work is lost in nature.

If carbon (C) be burnt in the air, carbonic oxide (CO_2) is formed. The law of the conservation of matter teaches us that no substance is lost, and hence the quantity of carbonic acid which has been formed by combustion must be exactly as large as the quantities of carbon and oxygen of the air which existed previous to combustion. The carbon and oxygen of the air prior to their union to carbonic acid possess a quantity of work or energy differing from that after union, the heat generated by the combustion being a manifestation of energy produced in a chemical way.

Every element has to be thought of as possessing a definite, inherent content of energy which, when the element enters into combination with other elements, may, and generally does, undergo a change. Thus in entering into a combination the elements yield a portion of their content of energy, generally in the form of heat, though sometimes also with luminous phenomena, so that the content of energy in the combination is less than the content of energy of the elements before their union. If now a solution of the combination in water be prepared, a change in the content of energy again takes place by dissociation, the content of energy present in the combination being partially converted into electrical energy. The ions appearing thereby receive electrical charges—the metal-ions positive charges and the other ions negative charges —and the nature of ions may be characterized by saying, they differ from the elementary atoms of similar names in having a *different content of energy.*

Processes on the electrodes. As previously mentioned, in the salts all the metal-ions are positive and the other ions of the metal combination—*the acid residue*—negative.

The ions arriving at the electrodes possess the power of entering into *chemical processes* with the constituents of the electrolyte or with the electrodes, as may be shown by the following examples_

When a solution of potassium disulphate (K_2SO_4) is electrolyzed between unassailable platinum electrodes, the following event takes place. The potassium-ions migrate to the cathode and separate metallic potassium,

$$- \left| \; \xleftarrow{} K_2 \; \mid \; SO_4 \xrightarrow{} \; \right| \; +$$
$$\text{Potassium disulphate.}$$

which, however, as is well known, cannot exist in water, but immediately forms with the solvent potassium hydroxide (caustic potash) according to the following equation ·

$$2K + 2H_2O = 2KOH + H_2$$
$$\text{Potassium} \quad \text{Water.} \quad \text{Caustic potash.} \quad \text{Hydrogen.}$$

That this transposition takes place in the manner described, is shown by the abundance of hydrogen * which escapes in the electrolysis of the potassium disulphate. On the other hand, the acid residue SO_4 migrates to the anode, which, as it consists of insoluble platinum, cannot saturate the acid residue, and the latter is also transposed with water according to the following equation :

$$SO_4 \quad + \quad H_2O \quad - \quad H_2SO_4 \quad + \quad O$$
Sulphuric acid residue. Water. Sulphuric acid. Oxygen.

The oxygen escapes, and the sulphuric acid formed combines again to potassium disulphate with the caustic soda formed on the cathode.

A like liberation of oxygen takes place when very dilute hydrochloric acid † is electrolyzed with platinum anodes. Hydrogen escapes on the cathode, but no chlorine appears on the anode, an equivalent quantity of oxygen being, however, liberated. The water is decomposed by the chlorine, hydrogen chloride and oxygen being formed according to the following equation :

$$Cl_2 \quad + \quad 2H_2O \quad = \quad 4HCl \quad + \quad O_2$$
Chlorine. Water. Hydrogen chloride. Oxygen.

The oxygen appearing in both cases, as well as the hydrogen appearing in the first-mentioned example, are called *secondary* products of electrolysis, because the products first separated under the given conditions could not exist and, in being decomposed or transposed, formed together with the solvent the above-mentioned products.

When sodium hydroxide (caustic soda) is electrolyzed, hydrogen appears on the cathode, because sodium, like potas-

---

* This explanation is here retained, though based upon potential measurements, it may, according to Le Blanc, be supposed that the hydrogen separates primarily and originates from the hydrogen-ions of the dissociated water of the solution.

† The process does not pass off as smoothly as represented by the formula, but the example is given as an illustration according to Ostwald's "Grundlinien der Chemie," I. p. 203.

sium in the former example, cannot exist with water, and sodium hydroxide is again formed, hydrogen being at the same time liberated. The hydrogen-ion which also cannot exist by itself, is discharged on the anode, water and oxygen being formed according to the following equation :

$$4OH = 2H_2O + O_2$$
Hydroxide.　　Water.　　Oxygen.

Let us now turn to the other cases in which the ions, separated on the electrodes, enter with the latter into chemical processes.

A solution of cupric sulphate (blue vitriol) $CuSO_4$ is to be electrolyzed. The copper ions migrate to the cathode

$$- \mid \leftarrow Cu \mid SO_4 \rightarrow \mid +$$
Cupric sulphate.

and deposit their copper in the form of a galvanic deposit, the acid residue—the anion—migrating to the anode. If the latter consists of a soluble metal, for instance, copper, the acid residue becomes saturated with copper, dissolving approximately the same quantity of it as has been deposited upon the cathode. Theoretically the quantity dissolved from the anode by the acid residue should exactly correspond to the quantity of metal separated on the cathode. However, in practice, such is not the case, because the acid residue is partly subject to other decompositions, especially the formation of H_2SO_4, oxygen being at the same time separated.

All other processes in which metallic anodes capable of solution by the acid residue are used, run their course in a manner similar to the electrolysis of blue vitriol.

As previously mentioned, secondary products may be liberated by electrolysis. The separation of metal on the cathode may also be effected in a secondary manner, and in galvanic processes this is mostly brought about on purpose. Referring to the previously mentioned example of the electrolysis of blue vitriol, the copper could only be separated in a primary manner ; by adding, however, a small quantity of sulphuric acid to the blue vitriol solution, separation of copper in a secondary manner takes place. The sulphuric acid being in a diluted state is more strongly dissociated than the blue vitriol solution, the ions of sulphuric acid—hydrogen and acid residue SO_4— first of all taking charge of the conduction of the current, and the hydrogen-ions separate the copper on the cathode according to the following equation :

$$CuSO_4 \quad + \quad H_2 \quad = \quad Cu \quad + \quad H_2SO_4$$
Cupric sulphate.　Hydrogen.　Copper.　Sulphuric acid.

In the electrolysis of a silver bath containing potassium-silver cyanide ($KAgCN_2$), potassium-ions and silver cyanide-ions ($AgCN_2$*) appear :

$$(a) - \mid \; \leftarrow K \mid AgCN_2 \rightarrow \; \mid +$$

From the solution of potassium-silver cyanide the potassium-ions separate secondarily metallic silver on the cathode, potassium cyanide being formed according to the following equation :

$$(b) \; K \quad + \quad KAgCN_2 \quad - \quad Ag + 2KCN.$$
Potassium-ion.　Potassium silver cyanide.　Silver potassium cyanide.

The anions $AgCN_2$ migrate to the anode, are there decomposed to silver cyanide ($AgCN$) and cyanogen (CN), the cyanogen-ions dissolve from the anode silver, silver cyanide being formed, and 2 silver cyanide atoms combine with the above (in b) liberated 2 potassium cyanide, to 2 potassium silver cyanide atoms :

$$(c) \; Ag \quad + \quad CN \quad = \quad AgCN$$
Silver.　Cyanogen.　Silver cyanide.

$$(d) \; 2AgCN \quad + \quad 2KCN \quad = \quad 2KAgCN_2$$
Silver cyanide.　Potassium cyanide.　Potassium silver cyanide.

* Hittorf, Ostwald's Klassiker, 23, § 45.

The quantities of substances separated from the electrolytes by the electric current are subject to fixed laws, which, after their discoverer, are named

Laws of Faraday. These laws are followed by both the primary, as well as secondary products of electrolysis, because the latter are produced by primary separations, the secondary being proportional and chemically equivalent to them. The first of these laws is as follows:

The quantity of substances which is liberated on the electrodes is directly proportional to the strength of the electric current which has been conducted through the electrolytes, and the time.

FIG. 7.

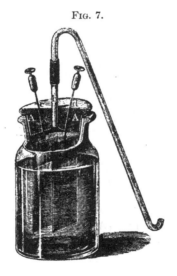

By conducting the current through a closed decomposing cell, Fig. 7, filled with acidulated water and furnished with two platinum electrodes which are connected with the poles of a source of current, oxygen is evolved on the positive electrode, and hydrogen on the negative electrode. If the gas-mixture (oxyhydrogen gas) which is evolved be caught under water in a graduated tube, the quantity of oxyhydrogen gas produced by a current of fixed strength within a determined time can be readily ascertained. If now a current of double the strength be for the same length of time passed through the decomposing cell, the quantity of oxyhydrogen gas produced will be found twice as large as in the first case.

Faraday allowed the same quantity of current to pass through a series of decomposing cells, coupled one after another, which contained electrolytes of different compositions, and determined quantitatively the separations of cations effected in the various cells by an equal quantity of current. Suppose the first cell to be a water-decomposing cell like Fig. 7, let the second cell contain potassium silver cyanide solution with a slight excess of potassium cyanide, the third cell an acidulated solution of cupric sulphate, and the fourth cell a solution of cuprous chloride in hydrochloric acid.

When electrolysis has been carried on for half an hour, the current is interrupted. and the quantity of hydrogen calculated from the measured quantity of oxyhydrogen gas produced. The platinum cathodes, the weight of which has been determined previous to electrolysis, are rinsed in water, next in alcohol, and finally in ether. They are then thoroughly dried and again weighed to determine the quantities of metal separated in the individual cells. The following quantities were found

Electrolyte.	I. Dilute sulphuric acid $1:15$.	II. Potassium silver cyanide $KAgCy_2$.	III. Cupric sulphate $CuSO_4$.	IV. Cuprous chloride $CuCl$.
Quantity of separated cations. .	67 ccm. H = 6.00 mg. H	650 mg. Ag.	190 mg. Cu	380 mg. Cu
For 1 mg. H are separated . . .	1 mg. H	108.33 mg. Ag.	31.66 mg. Cu	63.3 mg. Cu
Atomic weights. .	1	108	63.3	63.3

From this it follows that the separated quantities of cations, referred to one part by weight of hydrogen, represent almost exactly the quantities of metals which correspond to a single valence of their atomic weights. In the electrolytes II and IV, the silver atoms and copper atoms are univalent, and in electrolyte III, bivalent. Hence, in II and IV were separated the quantities of metal, 108.33 mg. silver (error in per cent. 0.33) and 63.3 mg. copper, which corresponds to the univalence of the atoms, and in III only a valence amounting to 31.56 mg. copper which corresponds to the bivalence of the copper atoms in cupric sulphate.

Hence the *second* law of Faraday, as expressed by v. Helmholtz, reads as follows : *The same quantity of current liberates in the different electrolytes an equal number of valences or converts them into other combinations.*

It has previously been mentioned that for the development of 1 g. hydrogen, 96540 coulombs must pass through the electrolyte. According to determinations by F. and W. Kohlrausch, 0.3290 mg. copper is liberated from cupric salts by a quantity of 1 coulomb, or 31.65 g. by 96540 coulombs. This quantity of current is the

Electro-chemical equivalent, i. e., the number of coulombs which split off in one second the portion of atomic weights of the cations (metals) or of the anions referred to a valence and expressed in grammes, *i. e.,* the *gramme-equivalent.* Hence, for the separation of. 1 gramme-equivalent of copper = 31.65, or of 1 gramme-equivalent of silver = 108|96540 coulombs are always required.

From the laws of Faraday results the view previously referred to, that the passage of the current through the electrolyte is confined to the simultaneous movement of the ions, and that no current can pass through the electrolyte if the ions be wanting. Hence the ions of the electrolyte are charged or combined with specified quantities of electricity, and one portion of Faraday's law may, according to Ostwald, be thus expressed : *The quantities of the different ions combined with equal quantities of electricity are proportional to the combining weights of these ions,* and the entire law may be summed up as follows : *In the electrolytes the electricity moves only simultaneously with the constituents of the electrolytes which are the ions. The moved quantities of electricity are proportional to the quantities of ions, and amount to 96540 coulombs, or a multiple of them, for one molecule of any one ion.*

Below is given a table. of the electro-chemical equivalents, and from them will be calculated, in the practical part of the work, the time required for the formation of deposits of a certain specified weight, the current-strength required for the purpose, etc. The specific gravities of metals, which are also required for the above-mentioned calculations, have been added to the table.

	Electro-chemical Equivalent.	Deposit in 1 Ampère-hour	Specific Gravity.
Hydrogen	0.104	0.0375	0.00009
Antimony	0.415	1.4940	6.8
Arsenic	0.258	0.9322	5.7
Cobalt	0.305	1.1001	8.7
Copper from cupric salts . . .	0.329	1.1858	8.8
Copper from cuprous salts . .	0.658	2.3717	8.8
Gold from auric salts	0.681	2,4513	19.2
Gold from aurous salts	2.043	7.3560	19.2
Iron from ferric salts	0.193	0.6950	7.8
Iron from ferrous salts	0.289	1.0423	7.8
Lead	1.071	3.8580	11.3
Nickel	0.304	1.0945	8.6
Platinum	0.504	• 1.8160	21 4
Silver	1.118	4.0248	10.5
Tin from stannic salts . . .	0.308	1.1094	7.3
Tin from stannous salts	0.616	2.2180	7.3
Zinc	0.339	1.2200	7.2

Solution-tension of metals. A fluid evaporates on the surface until the vapor-pressure produced is equal to the evaporation-tension of the fluid. Analogous to this process is the osmotic pressure which a salt exercises when dissolved in water, a pressure which increases with the quantity of the salt until it is in equilibrium with the solution-tension. According to Nernst, every metal when immersed in an electrolyte also possesses the power conditional to its chemical nature to give off metal atoms as ions (cations) to the solution, and this power is called solution-tension.

The solution-tension is the greater the smaller the number of cations which are already present in the electrolyte; if, on the other hand, the electrolyte contains a great number of cations derived from the dissociation of the salt, the osmotic pressure may overbalance the solution-tension, or the osmotic pressure may be equal to the solution-tension.

In the first case, when the solution-tension preponderates, the metal will give up to the solution cations charged with positive electricity, while an equally large quantity of negative electricity remains in the metal. Suppose zinc dipping in water, then the zinc-ions passing into solution will be charged with positive electricity, while the metal is charged with an equally large quantity of negative electricity.

If the water be replaced by a solution of zinc sulphate (white vitriol) which, in consequence of dissociation, already contains a larger number of positive zinc-ions and negative acid-residue-ions, additional positive zinc-ions will be given up to the solution by the zinc so long as the solution-tension of the zinc overbalances the osmotic pressure of the dissolved zinc-ions. When an equilibrium between osmotic pressure and solution-tension is reached, the further formation of zinc-ions ceases.

If the more electro-negative copper be dipped in water it also makes an effort to ionize, i. e., to give up to the solution copper-ions charged with positive electricity. If, however, the water be replaced by cupric sulphate solution, it happens that the osmotic pressure of copper-ions formed by dissociation of the electrolyte is greater than the solution-tension of the copper, and hence not only counteracts the formation of new copper-ions, but carries positive copper-ions from the electrolyte to the copper, the latter receiving thereby a positive charge, while the fluid surrounding the copper becomes negative.

However, no matter whether the solution-tension may considerably overbalance the osmotic pressure, by the mere dipping of the metal in the electrolyte the quantity of ions which are newly formed will always be small, because by reason of the electrostatic attraction of the cations by the negatively-charged metal, there will take place on the contact-surface between the metal and the electrolyte an accumulation of cations, the osmotic pressure of which will consequently be increased, and counteract the solution-tension. The latter can only become again active, when the free electricities are conducted away by a closed circuit, as will be explained in the next section.

Osmotic theory of the production of the current, according to Nernst. The behavior of zinc in a zinc sulphate solution, and that of copper in a cupric sulphate solution, has above been referred to. If a cell be put together of zinc dipping in zinc sulphate solution, and copper in cupric sulphate solution, such as a Daniell cell, in which the two solutions are separated by a porous partition, called a *diaphragm*, the following processes take place:

From the zinc, positive zinc-ions pass into solution so long as the, at first, slighter osmotic pressure of the electrolyte balances the solution-tension; the zinc becomes negatively electric and the electrolyte positively electric on the contact-surface. By the preponderance of the osmotic pressure of the copper sulphate solution over the solution-tension of the copper, positive copper-ions are separated on the copper, and yield their positive charges to the latter. They themselves are transformed from the ion state into the molecular state, thus becoming non-electric, while on the contact-surface the cupric sulphate solution becomes negatively electric. Hence a state of rest supervenes, in which the zinc is charged with negative, and the copper with positive, electricity, while the zinc solution is charged positively and the copper solution negatively. If

now by means of a metallic wire the zinc be outside of the
solutions connected with the copper, thus, establishing a closed
circuit, the following process takes place : The positive electricity in the copper migrates through the wire to the zinc,
and neutralizes the quantity of negative electricity present in
the latter. By the flow of positive electricity from the copper
the state of equilibrium, which existed between copper and
cupric sulphate, is disturbed, and the osmotic power being now
predominant, the solution again gives up copper-ions to the
copper, whereby the latter is again charged with positive
electricity. On the other hand, after the exchange of electricities in the zinc by the solution-tension, fresh zinc-ions can
be brought into solution. Thus a current flows continuously
from copper to zinc until either no more copper-ions are
conveyed from the cupric ·sulphate solution to the copper, or
until all the zinc is ionized, *i. e.*, dissolved.

Nernst's conception of the solution-tension of the metals is
analogous to that of the osmotic pressure, the impelling force
of a Daniell battery having the character of a pressure, and
for that reason Ostwald designates a galvanic battery as a
machine driven by osmotic pressure, eventually by electrolytic
solution-tension.

The electro-motive force of such a cell is mainly determined
by the magnitude of the solution-tension of the metals. In
the closed cell the metal gives up with greater solution-tension
its atoms as ions into the electrolyte in which it is confined,
while the cations of the other electrolyte are discharged
on the metal contiguous to it and pass into the molecule state.
By this, the dissolving metal, to which the anions of the other
electrolyte—the acid residue—migrate, becomes the *anode*, and
the other metal on which the cations of its electrolyte separate
non-electrically, the *cathode*. Since the cations are discharged
on the cathode, the latter is also called the conducting electrode, and the anode which dissolves, the dissolving electrode.

From what has been said, it might appear that the current
in a Daniell cell owes its existence to purely physical forces.

The solution-tension of the metals depends, however, on their chemical affinity, and the current is actually electric energy which has been formed from chemical energy. The solution of the anode-metal is a chemical process, whereby the cations are forced from the electrolyte surrounding the anode; the anode-metal endeavors to expand, and hence the mode of action of chemical affinity in converting chemical into electric energy may be designated as the effect of pressure.

However, additional chemical processes take place in the Daniell cell; the zinc dissolves to zinc sulphate because the anions of the cupric sulphate solution migrate to the zinc, while from this solution a quantity of copper equivalent to the dissolved zinc is deposited on the cathode. By the anion SO_4 of the cupric sulphate solution an oxidation of the zinc takes place, the latter acting therefore as a reducing agent. The cupric sulphate solution, on the other hand, is reduced to copper, and the acid-residue SO_4 being liberated thereby acts as an oxidizing agent, while the copper of the cathode remains chemically unchanged. Since, according to Ostwald, in every chemical process which takes place between an oxidizing and a reducing agent, variations appear in the ion-charges by reason of the varying capacities of the ions to absorb or discharge more quantities of electricity, such cells are also called *oxidizing* and *reducing cells*. *Concentration cells* will later on be referred to.

Polarization. By polarization is understood the appearance of a counter-current passing in a direction opposite to that of the current conducted into an electrolyte; the main current is therefore weakened by this counter-current. Polarization takes place when the current produces substantial changes in the electrolytes or on the electrodes, no matter whether such changes consist in a difference of the nascent concentrations of the electrolyte, or in the formation of gas-cells by the separation of layers of gas on the electrodes, etc.

If a weak current be conducted into a cell filled with standard cupric sulphate solution, both electrodes of which consist

5

of copper, and a galvanometer be placed in the circuit, it will
be noticed that an electrolytic decomposition takes place. The
copper-ions discharged from the copper solution on the electrode connected with the negative pole of the source of current,
pass into the molecular state, and metallic copper separates
upon this electrode, while the anions of the acid-residue SO_4
migrate to the electrode connected with the positive pole,
where they dissolve copper, thus giving up fresh copper-ions
to the solution. Hence the concentration of the electrolyte
remains constant, provided electrolysis lasts not too long, and
the current introduced is not stronger than just necessary for
the decomposition of the cupric sulphate solution ; the nature
of the electrodes themselves remains unchanged. The needle
of the galvanometer makes one deflection and when the current is interrupted returns to the O point, thus indicating the
absence of a counter-current ; the electrodes have proved themselves as non-polarizable.

However, the case is different when an electrolyte is electrolyzed between insoluble electrodes. If a powerful current be
conducted through a platinum anode into standard sulphuric
acid (H_2SO_4), the latter is decomposed into hydrogen-ions
which go to the platinum cathode while the SO_4-ions migrate
to the anode. As previously mentioned, the SO_4-ions cannot
exist in a free state, neither can they dissolve platinum and,
while water is decomposed, sulphuric acid and oxygen-gas are
again formed, the latter being separated on the platinum
anode. The hydrogen separated on the cathode is electro-
positive towards the oxygen separated on the anode, the consequence being that from the hydrogen of the cathode a
counter-current flows to the oxygen of the anode, which is indicated when the primary current is interrupted by the needle
of the galvanometer, instead of merely returning to the O
point, deflecting in a direction opposite to that of the previous
deflection, and returning to the O point only after the equalization of the charges in the electrodes.

The counter-current or polarization-current appears also

when two different metals dip in one electrolyte. In a Volta cup cell, a zinc plate and a copper plate connected by a metallic wire dip in dilute sulphuric acid. A current flows from the copper through the wire to the zinc, and returns from the zinc through the acid to the copper, decomposing thereby the acid into hydrogen and SO_4. The hydrogen separates on the copper, the acid-residue SO_4 on the zinc, and dissolves the latter, zinc sulphate being formed. The separated hydrogen being electro-positive towards the separated acid-residue, a current in the direction from the copper to the zinc is generated, and consequently flows in a direction opposite to that of the main current, which passes from the zinc to the copper. The electro-motive force of the main current is thus decreased by the magnitude corresponding to the electro-motive force of this counter-current.

If a zinc chloride solution be electrolyzed between two platinum electrodes, zinc separates on the cathode while chlorine appears on the anode. If the current be interrupted, a galvanoscope placed on the electrodes indicates a vigorous counter-current which turns from the zinc deposit—hence the cathode —to the anode, therefore opposite to the current at first supplied. This counter-current originates from the tendency of the substances separated on the electrodes to return, in consequence of the solution-tension, to the ion state, and this tendency exists during the entire process of the electrolysis.

The farther the metals in the series of electro-motive force are distant from each other, the greater the electro-motive force which the polarization-current possesses, as will be more particularly shown in the practical part of this work.

Decomposition-pressure. An electric current can only pass through an electrolyte and decompose it, when its electro-motive force possesses a certain minimum magnitude. The characteristic values at which the electrolytes are permanently decomposed are designated, according to Le Blanc, as their *decomposition-values;* the decomposition-pressure being the electro-motive force required for the separation of the electric charge of the ions. The decomposition-values of solutions which separate metals vary. Le Blanc found as decomposition-values of solutions which contained per liter one combining weight of the metallic salts, for

Zinc sulphate, 2.35 volt; Cadmium sulphate, 2.03 volt.

Nickel sulphate, 2.09 volt; Cadmium chloride, 1.88 volt.

Nickel chloride, 1.85 volt; Cobaltous sulphate, 1.92 volt.

Silver nitrate, 0.70 volt; Cobaltous chloride, 1.78 volt.

The difference in the decomposition values of metallic salt solutions explains the feasibility of separating from a solution which contains different metals, the individual metals, one after the other, free from other admixtures.

Velocity of ions. It has previously been shown that no polarization-current is generated when a cupric sulphate solution is for a short time electrolyzed between copper-electrodes. If, however, not too strong a current be for a longer time passed through the solution, a polarization-current appears, the origin of which must be due to another cause than the formation of a gas cell, because no gases are separated with not too strong a current. It has been shown that changes of concentration take place in the solution, concentration becoming greater on the cathode and less on the anode. These changes in concentration have been subjected to a thorough investigation by Hittorf, and it was found that the former view, according to which the number of positive and negative ions which migrate in opposite directions through an electrolyte, must be equal, was an erroneous one. The mobility of the ions varies, and depends on their nature. If, for instance, hydrochloric acid be electrolyzed, the hydrogen-ion migrates about five times as rapidly to the cathode as the chlorine-ion to the anode. The cations and anions of the metallic salts act in a similar manner, and consequently a greater concentration will take place on the cathode and a reduction in the content of metal on the anode, when the anions migrate more slowly than the cations; and *vice versa*, concentration will increase on the anode when the anions migrate more rapidly than the cations.

The middle layer of the electrolyte always remains unchanged and of the same concentration, the changes in concentration being shown in the layers of fluid surrounding the electrodes, and these differences in concentration also effect the formation of a current, which, according to the nature of the electrodes may flow in the sense of the main current or in that of the counter-current.

The quotients obtained by dividing the distances, which the cations and anions perform in the same time, by the total distance of the road traveled by the two ions, Hittorf designates as the transport-values of the respective ions.

We herewith conclude the theoretical considerations, and will later on have occasion to touch upon other fundamental electrolytical principles of less importance.

III.

SOURCES OF CURRENT.

CHAPTER III.

VOLTAIC CELLS, THERMO—PILES, DYNAMO—ELECTRIC MACHINES,
ACCUMULATORS.

THE sources of current which are used for the electro-deposition of metals are : Voltaic cells, thermo-piles, dynamo-electric machines, and accumulators.

A. VOLTAIC CELLS.

It is not within the province of this work to enter into a detailed description of all the forms of voltaic cells, because the number of such constructions is very large, and the number of those which have been successfully and permanently introduced for practical work is comparatively small.

In the theoretical part, we have learned the origin of the current and the explanation of its origin by the solution-tension of the metals or the osmotic pressure of the solutions, and we know further that in a voltaic cell *chemical energy* is converted into *electrical energy*. In speaking of polarization which is formed when two different metals dip in one fluid, we have seen that the hydrogen liberated on the copper in a Volta cup cell generates a counter-current which weakens the principal current. This hydrogen appearing on the positive pole is the cause of a rapid decrease in the efficiency of the cell, and all cells in which the hydrogen on the cathode is not neutralized by suitable means, are called *inconstant* cells, while cells in which the hydrogen is removed in a physical

(70)

way or by chemical agents which oxidize it, are called *constant* cells.

The original form of voltaic cells, the voltaic pile, consisting of zinc and copper plates separated from one another by moist pieces of cloth, has already been mentioned on p. 2, as well as its disadvantages, which led to the construction of the so-called *trough battery*. The separate elements of this battery are square plates of copper and zinc, soldered together, and parallel fixed into water-tight grooves in the sides of a wooden trough so as to constitute water-tight partitions, which are filled with acidulated water. The layer of water serves here as a substitute for the moist pieces of cloth in the voltaic pile.

In other constructions the fluid is in different vessels, each vessel containing a zinc and a copper plate which do not touch one another, the copper plate of the one vessel being connected with the zinc plate of the next, and so on.

In all cells with one exciting fluid, the current is quite strong at first, but decreases rapidly for the reasons given above. On the one hand, during the interruption of the current a change takes place in the fluid by the local effect in the cell, and, on the other, the zinc forms with the impurities contained in it, small voltaic piles with a closed circuit, in consequence of which the cell performs a certain chemical work even when the current is interrupted. The local action can be reduced to a minimum by amalgamating the zinc. Such amalgamation is also a protection against the above-mentioned chemical work of the cell, the hydrogen bubbles adhering so firmly during the interruption of the current to the amalgamated homogeneous surface as to form a layer of gas around the zinc surface by which its contact with the fluid is prevented.

Amalgamation may be effected in various ways. The zinc is either scoured with coarse sand moistened with dilute sulphuric or hydrochloric acid, or pickled in a vessel containing either of the dilute acids. The mercury may be either mixed with moist sand and a few drops of dilute sulphuric acid, and the zinc be amalgamated by applying the mixture by means of a wisp of

straw or a piece of cloth ; or the mercury may be applied by
itself by means of a steel-wire brush, the brush being dipped in
the mercury and what adheres is quickly distributed upon the
zinc by brushing until the entire surface acquires a mirror-like
appearance.   The most convenient mode of amalgamation is
to dip the zinc in a suitable solution of mercury salt and rub
with a woolen rag.   A suitable solution is prepared by dissolving 10 parts by weight of mercurous nitrate in 100 parts of
warm water, to which pure nitric acid is added until the milky
turbidity disappears.   Another solution, which is also highly
recommended, is obtained by dissolving 10 parts by weight of
mercuric chloride (corrosive sublimate) in 12 parts of hydrochloric acid and 100 of water, or by dissolving 10 parts by
weight of potassium mercuric cyanide and 2 parts potassium
cyanide in 100 parts of water.   In order to preserve as much
as possible the coating of mercury upon the zinc, sulphuric
acid saturated with neutral mercuric sulphate is used for the
cells.   For this purpose frequently shake the concentrated
sulphuric acid (before diluting with water) with the mercury
salt.   As much mercuric sulphate or mercuric chloride as will ·
lie upon the point of a knife may also be added in the cells to
the zinc.

Instead of the addition of mercuric sulphate, Bouant recommends to compound the dilute sulphuric acid with 2 per cent.
of a solution obtained as follows : Boil a solution of $3\frac{1}{2}$ ozs. of
nitrate of mercury in 1 quart of water, together with an excess
of a mixture of equal parts of mercuric sulphate and mercuric
chloride, and, after cooling, filter and use the clear solution.

Smee cell.   This cell consists of a zinc plate and a platinized
silver plate dipping into dilute acid.   It may be formed of two
zinc plates mounted with the platinized silver between them
in a wooden frame, which being a very feeble conductor may
carry away a minute fraction of the current, but serves to hold
the metals in position, so that quite a thin sheet of silver may
be employed without fear of its bending out of shape and
making a short circuit.   Platinizing is effected by suspending

the silver plates in a vessel filled with acidulated water, adding some chloride of platinum and placing the vessel in a
porous clay cell filled with acidulated water and containing a
piece of zinc, the latter being connected with the silver plates
by copper wire. The platinum coating obtained in this manner is a black powder which roughens the surfaces, in consequence of which the bubbles of hydrogen become readily detached, and polarization is less than with silver plates not
platinized. The use of electrolytically-prepared copper plates,
which are first strongly silvered and then platinized, is still
more advantageous on account of their greater roughness.
To increase the constancy of the cell, it is advisable to add
some chloride of platinum to the dilute acid of the element.

The Smee cell is still frequently used in England and the
United States with silver and gold plating solutions. Its
electro-motive force is about 0.48 volt

As previously mentioned, polarization can be entirely
avoided only by allowing the electro-negative pole-plate to
dip in a fluid which, by combustion, reduces the hydrogen
evolved to water, or, in other words, which immediately oxidizes the hydrogen to water. From this
conviction originated the so-called *constant
cells* with two fluids, the first of these cells
being, in 1829, constructed by Becquerel,
which, in 1836, was succeeded by the more
effective one of Daniell.

FIG. 8.

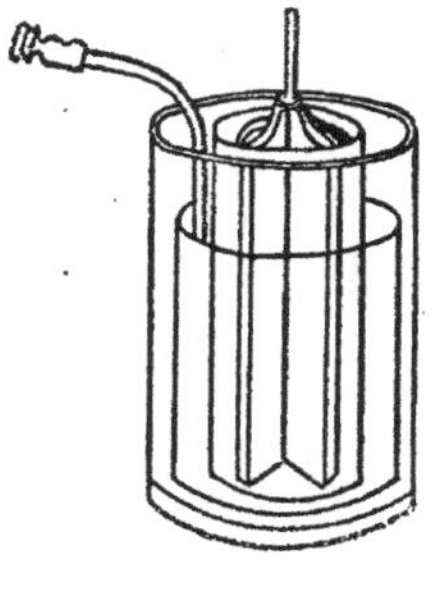

Daniell cell. In its most usual form
Daniell's cell (Fig. 8) consists of a glass
vessel, a copper cylinder, a porous earthenware pot and a zinc rod suspended in the
latter. The glass vessel is filled with
saturated blue vitriol solution, a small piece of blue vitriol
being added, and the porous earthenware pot with dilute
sulphuric acid about 1 part of acid to 12 to 20 parts of water.
The acid residue SO_4 migrates to the positive zinc, and there
forms zinc sulphate, while the hydrogen which is liberated on

the electro-negative copper, reduces from the blue vitriol solution an equivalent quantity of copper, which is deposited upon the electro-negative plate according to the following equation:

$$CuSO_4 \quad + \quad 2H \quad - \quad Cu \quad + \quad H_2SO_4$$

Cupric sulphate.   Hydrogen.   Copper.   Sulphuric acid.

Thus the hydrogen is removed by its combining with the acid-residue SO_4 to sulphuric acid.   A drawback of the Daniell cell is that the blue vitriol solution diffuses into the porous pot, where it is decomposed by the zinc on coming in contact with it, and the copper is separated upon the zinc, the efficiency being thus destroyed, or at least very much weakened.   The electro-motive force of the Daniell cell is quite exactly 1.1 volt.

Meidinger cell.   This may be considered a modified Daniell cell.   Like the Callaud cell, it has no porous partition, the mixture of the two fluids being retarded by their different specific gravities.   The form of the Meidinger cell, as most generally used, is shown in Fig. 9.

FIG. 9.

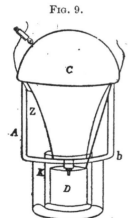

Upon the bottom of a glass vessel, *A*, provided at *b* with a shoulder, stands a small glass cylinder, *K*, which contains the electro-negative copper cylinder *D;* from the latter a conducting wire leads to the exterior.   Upon the shoulder, at *b*, rests the zinc cylinder *Z*, which is also provided with a conducting wire leading to the exterior.   The balloon *C* closes the vessel by being placed upon it.   The balloon is filled with pieces of blue vitriol and Epsom salt solution.   The entire cell is also filled with Epsom salt solution (1 part Epsom salt to 5 water.) In the balloon *C* concentrated solution of blue vitriol is formed which flows into the glass cylinder *K*.   If the battery is not

closed, the concentrated copper solution remains quietly standing in K, its greater specific gravity preventing it from rising
higher and reaching the zinc. If, however, the current be
closed, zinc is dissolved, while metallic copper is separated
from the blue vitriol solution, and concentrated solution flows
from the balloon C to the same extent as the blue vitriol solution in D becomes dilute by the separation of copper. Hence
the action of the cell remains constant for quite a long time,
and of all the modified forms of Daniell's cell consumes the
least blue vitriol for a determined quantity of current. However, in consequence of its great internal resistance (3 to 5
ohms, according to its size) its current-strength is small. The
electro-motive force of the Meidinger cell is 0.95 volt.

Bunsen cell. Bunsen, in 1841, replaced the expensive platinum by prisms cut from gas-carbon, which is still less electro-
negative than platinum, and very hard and solid, so that it
perfectly resists the action of the nitric acid. In place of the
gas-carbon an *artificial carbon* may be prepared by kneading
a mixture of pulverized coal and coke with sugar solution or
syrup, bringing the mass under pressure into suitable iron
moulds and heating it red-hot, the air being excluded. After
cooling the carbon is again saturated with sugar solution
(others use tar, or a mixture of tar and glycerine) and again
heated, the air being excluded. These operations are, if
necessary, repeated once more, especially when great demands
are made on the electro-motive force and solidity of the
artificial carbons.

In the Bunsen cell the zinc electrode dips in dilute sulphuric
acid and the carbon in concentrated nitric acid. Independent
of the fact that by reason of its rough surface, the carbon has
by itself the tendency to repel the hydrogen-bubbles and thus
acts to a certain degree as a depolarizer, *depolarization, i. e.,*
the removal of the hydrogen-bubbles which produce polarization, is most effectively assisted by the nitric acid, the hydrogen being oxidized to water according to the following
equation, while the nitric acid is reduced to nitric oxide:

$$2HNO_3 \quad + \quad 6H \quad = \quad 2NO \quad + \quad 4H_2O$$

Nitric acid.　　　　Hydrogen.　　　Nitric oxide.　　　Water.

The processes which take place in the Bunsen cell are as follows : From the positive carbon a current passes through the wire to the zinc and returns from the latter through the dilute sulphuric acid to the carbon. The sulphuric acid (H_2SO_4), is thereby decomposed to hydrogen and sulphuric acid residue SO_4 the hydrogen migrating to the carbon and is oxidized to water by the nitric acid, while the sulphuric acid residue migrates .to the zinc and combines with it to zinc sulphate ($ZnSO_4$) as illustrated by the following scheme :

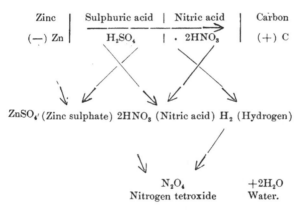

To prevent the two fluids from mixing, the use of a porous partition is required, the same as in Daniell's cell.

Figs. 10, 11 and 12 show three forms of Bunsen's cell generally used.

Fig. 11 is the most convenient and practical form. It consists of an outer vessel of glass or earthenware. In this is placed a cylinder of zinc in which stands a porous clay cup, and in the latter the prism of gas-carbon. This substance is the graphite of the gas retorts. It is not coke. It is easily procurable in lump at a small price, but costs much more when cut into plates, as, when the material is good, it is exceedingly difficult to work. It is generally cut with a thin strip of iron and watered silver-sand. Blocks for Bunsen cells cost less because they are more easily produced. Rods for Bunsen cells should be a few inches longer than the pots to protect the top from contact with the acid. A good carbon is of a clear gray appearance, has a finely granulated surface, and is very hard. A band of copper is soldered or secured by means of a binding-screw to the zinc cylinder, while the prism of gas carbon carries the binding-screw (armature), as seen in Fig. 10 in the upper part of which a copper sheet or wire is fixed for the transmission of the current. The other vessel is filled with

FIG. 10.   FIG. 11.   FIG. 12.

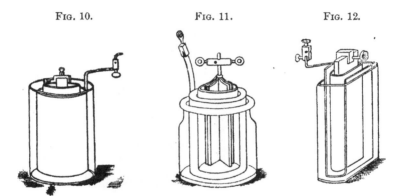

dilute sulphuric acid (1 part by weight of sulphuric acid of 66° Bé.—free from arsenic—and 15 parts by weight of water), and the porous cup with concentrated nitric acid of at least 36° Bé., or still better 40° Bé., care being had that both fluids have the same level.

In Fig. 11 the cylinder of artificial carbon is in the glass vessel, while the zinc, which, in order to increase its surface, has a star-shaped cross-section, is placed in the porous clay cup. In this case the outer vessel is filled with concentrated nitric acid, and the clay cell with dilute sulphuric acid.

The form of the Bunsen cell shown in Fig. 10 is more advantageous, because its effective zinc surface can be kept larger.

Fig. 12 shows a plate cell such as is chiefly used for plunge batteries.

Fig. 13 shows an improved Bunsen cell of great power. It is particularly adapted for use with nickel, copper, brass or bronze solutions. It has an electro-motive force of 1.9 volts. Where the absence of power prevents the use of a dynamo, a battery of these cells is very suitable for nickel plating.

The Bunsen cells are much used for electro-deposition, since they possess a high electro-motive force (1.88 volts), and, on account of slight resistance (0.5 to 0.25 ohm, according to

FIG. 13.

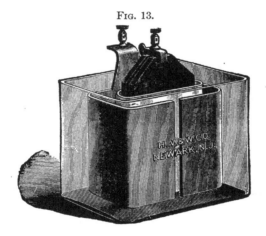

their size), develop considerable current-strength. Like the Grove cells, they have the inconvenience of evolving vapors of nitrogen tetroxide, which are not only injurious to health, but also attack the metallic articles in the workshop. Wherever possible they should be placed in a box at such a height that they may be readily manipulated. The box should have means of ventilation in such a way that the air coming in at the lower part will escape at the top through a flue, and carry away with it the acid fumes disengaged. It is still better to keep the cells in a room separate from that where the baths and metals are located. Furthermore, as the nitric acid be-

comes diluted by the oxidation of the hydrogen, and the sul-.
phuric acid is consumed in the formation of sulphate of zinc,
the acids have to be frequently renewed.

To get rid of the acid vapors, as well as to render the cells
more constant, A. Duprè has proposed the use of a 30 per
cent. solution of bisulphate of potash in water, in place of the
dilute sulphuric acid, and a mixture of water 600 parts, con-
centràted sulphuric acid 400, sodium nitrate 500, and bichromate of potash 60, in place of the nitric acid.

The following method can be recommended: The outer
vessel which contains the zinc cylinder is filled with a moderately concentrated (about 30 per cent.) solution of bisulphate
of potash or soda, and the clay cup with solution of chromic
acid—1 part chromic acid to 5 parts water. As soon as the
electro-motive force of the cell abates, it is strengthened by
the addition of a few spoonfuls of pulverized chromic acid to·
the chromic acid solution. It is preferable to use the chromic·
acid in the form of a powder especially prepared for this pur-·
pose than a chromic acid solution produced by mixing potas-·
sium dichromate solution with sulphuric acid, such a solution.
having a great tendency to form crystals which exerts a dis-·
turbing effect. Solution of sodium dichromate compounded,
with sulphuric acid does not show this drawback.

The efficiency of the chromic acid solution rapidly abates in
a comparatively short time, the electro-motive force of the cell
decreasing in a few hours and chromic acid has frequently. to·
be added, or the cell eventually refilled.

Dr. Langbein has succeeded in preparing a soluble chromium combination which depolarizes rapidly and for a longer
time maintains the efficiency of the cell constant. With a
single filling of this solution, the battery has been kept working for six days, from morning to evening, without refilling
being required. During the night the battery remained
filled, but inactive. The solution is obtained by treating
Langbein's chromic iron powder with concentrated sulphuric·
acid and carefully diluting with water.

The electro-motive force of a cell filled with this solution is 1.8 volts. Considering the lasting quality and great constancy, and consequent cheapness, as well as freedom from odor of this solution, it would appear to be the most suitable.

If nitric acid is used for filling the cells it is advisable in order to decrease the vapors, to cover the acid with a layer of oil $\frac{1}{3}$ to $\frac{3}{4}$ inch deep.

The binding-screws which effect the metallic contacts must of course be frequently inspected and cleaned, the latter being best done by means of a file or emery paper. It is advisable to place a piece of platinum sheet between the binding surface of the carbon armature and the carbon, in order to prevent the acid, rising through the capillarity of the carbon, from acting directly upon the armature (generally brass or copper). To prevent the acid from rising, the upper portions of the carbons may be impregnated with paraffine. For this purpose the carbons are placed $\frac{3}{4}$ to 1 inch deep in melted paraffine and allowed to remain 10 minutes. On the sides where the armature comes in contact with the carbon, an excess of paraffine is removed by scraping with a knife-blade or rasp.

Treatment of Bunsen cells. Before use the zincs should be carefully amalgamated according to one of the methods given on page 71. The nitric acid need not be pure, the crude commercial article answering very well, but it should be as concentrated as possible and show at least 30° Bé. Carbons of hard retort-carbon are to be preferred, although those cut from carbon produced in gas-houses, gasifying coal without an addition of lignite, may also be used. Artificial carbon, if employed, should be examined as to its suitability, the non-success of the plating process being frequently attributed to the composition of the bath, when in fact it is due to the defective carbons of the cells. In order to avoid an unnecessary consumption of zinc and acid, the cells are taken apart when not in use, for instance, over night. Detach the brass armature of the carbon and lay it in water to which some chalk has been added. Lift the carbon from the clay cylinder and place it in a porcelain dish or earthenware pot ; empty the nitric acid of the clay cup into a bottle provided with a glass stopper ; place the clay cup in a vessel of water, and finally take the zinc from the dilute sulphuric acid and place it upon two sticks of wood laid across the glass vessel to drain off. In putting the cells together the reverse order is followed, the zinc being first placed in the glass vessel and then the carbon in the porous clay cup. The latter is then filled about three-quarters full with used nitric acid, and fresh acid is added until the fluid in the clay cup stands at a level with that in the outer vessel. The cleansed brass armature is then screwed upon the carbon. Finally, add to the dilute sulphuric acid in the outer vessel a small quantity of concentrated sulphuric acid saturated with mercury salt.

It is advisable to have at least a duplicate set of porous clay cups, and, in putting the cells together, to use only cups which have been thoroughly soaked in water. The reason for this is as follows: The nitric acid fills the pores of the cup, and, finally reaching the zinc of the outer vessel, causes strong local action and a correspondingly rapid destruction of the zinc. It is, therefore, best to change the clay cups every day, allowing those which have been in use to lie in water the next day with frequent renewal of the water. For the same reason the nitric acid in the clay cup should not be at a higher level than the sulphuric acid in the outer vessel.

When the Bunsen cells are in steady use from morning till night, the acids will have to be entirely renewed every third or fourth day. The solution of sulphate of zinc in the outer vessel, being of no value, is thrown away, while the acid of the clay cells may be mixed with an equal volume of concentrated sulphuric acid, and this mixture can be used as a preliminary pickle for brass and other copper alloys.

The *Leclanché* cell (zinc and carbon in sal-ammoniac solution with manganese peroxide as a depolarizer) need not be further described, it not being adapted for regular use in electroplating. It is in very general use for electric bells, its great

6

recommendation being that, when once charged, it retains its
power without attention for a long time.

Cupric oxide cell.   Lallande and Chaperon have introduced
a cupric oxide cell shown in Fig. 14 which possesses certain
advantages.   It consists of the outer vessel G, of cast-iron or
copper, which forms the negative pole-surface, and to which
the wire leading to the anodes is attached, and a strip of zinc,
Z, coiled in the form of a spiral, which is suspended from an

FIG. 14.

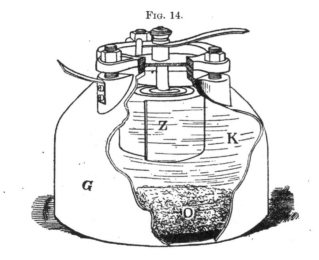

ebonite cover carrying a terminal connected with the zinc.
The hermetical closing of the vessel *G* by the ebonite cover is
effected by means of three screws and an intermediate rubber
plate.   Upon the bottom of the vessel *G* is placed a 3 to 4 inch
deep layer of cupric oxide, *O*, and the vessel is filled with a
solution of 50 parts of caustic potash in 100 of water.   When
the cell is closed, decomposition of water takes place, the oxygen which appears on the zinc forming with the latter zinc
oxide, which readily dissolves in the caustic potash solution,
while the hydrogen is oxidized, and cupric oxide at the same
time reduced to copper.   When the cell is open, *i. e.*, the
circuit not closed, neither the zinc nor the cupric oxide is

attacked, and hence no local action nor any consumption of material takes place. The electro-motive force of this cell is 0.98 volt, and its internal resistance very low. It is remarkably constant, and is well adapted for electro-plating purposes by using two of them for one Bunsen cell. The following rules have to be observed in its use: It is absolutely necessary that the ebonite cover should hermetically close the vessel G, as otherwise the caustic potash solution would absorb carbonic acid from the air, whereby carbonate of potash would be formed, which would weaken the exciting action of the solution. Further, the vessels G which form one of the poles must be insulated one from the other as well as from the ground, as otherwise a loss of current or defective working would be the consequence.

The regeneration of the cuprous oxide or metallic copper formed by reduction from the cupric oxide to cuprous oxide, requires it to be subjected to calcination in a special furnace. The expense connected with this operation is, however, about the same as that of procuring a fresh supply of cupric oxide. Lallande himself, as well as Edison, endeavored to bring the pulverulent cupric oxide into compact plates, but the regeneration of these plates was still more troublesome. By treatment with various chemical agents, Dr. Böttcher, of Leipsic, has succeeded in producing porous plates of cupric oxide which, after subsequent reduction by absorption of oxygen from the air, can be readily re-oxidized to cupric oxide, but as far as we know of, cells with these plates have not been introduced into commerce.

Cupron cell. The cell brought into commerce under this name by Umbreit & Matthes is a modification of the Lallande and Chaperon cell, it being furnished with a cuprous oxide plate. A square glass vessel or vat, furnished with a hard rubber cover, contains two zinc plates and between them the porous cuprous oxide plate. The glass vessel is filled with 20 per cent. caustic soda solution, and the current is delivered by means of two binding screws on the outside of the cover. The zinc dissolves, zinc-oxide-soda being formed according to the following scheme, while the cuprous oxide is reduced to copper:

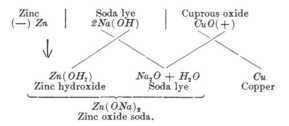

$$\begin{array}{ccccc} \text{Zinc} & | & \text{Soda lye} & | & \text{Cuprous oxide} \\ (-)\ Zn & | & 2Na(OH) & | & CuO(+) \end{array}$$

$$\underbrace{\overset{Zn(OH_2)}{\text{Zinc hydroxide}} \quad \overset{Na_2O + H_2O}{\text{Soda lye}} \quad \overset{Cu}{\text{Copper}}}_{\substack{Zn(ONa)_2 \\ \text{Zinc oxide soda.}}}$$

The reduced positive pole plates are regenerated by rinsing in water and keeping them in a warm place for 20 to 24 hours, it being only necessary to replace the caustic soda solution which has become saturated with zinc oxide. The electromotive force of the cell is 0.8 volt; the standard current-strength, according to the size of the cells, 1, 2, 4, and 8 ampères. Like the Lallande and Chaperon cell, this cell works without giving off any odor and the remarks regarding hermetical closing of the former also apply to the latter. An addition of sodium hyposulphite to the caustic soda solution is recommended as being productive of uniform wear and greater durability of the zinc plates.

According to Jordis' investigations the use of potash lye with 15 per cent. potassium hydrate is more advantageous, as well to heat the plates for the purpose of regeneration to 302° F.

The elements of Mariè, Davy, Náudet, Duchemin, Sturgeon, Trouville, and others, being of little practical value may be passed over.

Plunge batteries. For constructive reasons only one fluid is used into which the zinc plates as well as the carbon plates dip, a solution of chromic acid prepared by dissolving 10 parts of potassium dichromate, or better sodium dichromate, and $\frac{1}{50}$ part of mercuric sulphate in 100 parts of water, and adding 38 parts of pure concentrated sulphuric acid, being employed.

A plunge battery, as constructed by Fein, consists of a wooden box, which contains in two rows six vessels into which dip the zinc and carbon plates. The latter are secured to wooden cross-pieces furnished with handles, and may be maintained at any height desired by the notches in the standards. According to the current-strength required the plates are allowed to dip in more or less deeply.

In using the above-mentioned chromic acid solution origin-

FIG. 15.

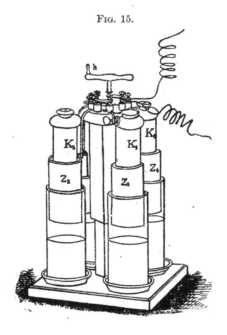

ally recommended by Bunsen, the cells first develop a very strong current, which, however, in a comparatively short time becomes weaker and weaker. The current-strength can be increased by adding at intervals a few spoonfuls of pulverized chromic acid to the chromic acid solution, which, however, finally remains without effect, when the battery has to be freshly filled. Hence these batteries are not suitable for electro-plating operations requiring a constant current for some time, but they may be employed for temporary use.

If plunge batteries are to be used for constant work in electro-plating, it is preferable to use batteries with two acids, namely, dilute sulphuric acid and concentrated nitric acid, or chromic acid.

In Stoehrer's construction (Fig. 15) the porous clay cup is omitted, the massive carbon cylinders K, K, etc., being each provided with a cavity reaching almost to the bottom which is filled with sand and nitric acid. The contact of the carbon and zinc cylinders is prevented by glass beads imbedded in the carbon cylinders.

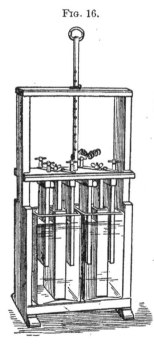

Fig. 16.

Fig. 16 shows a plunge battery manufactured by Dr. G. Langbein & Co., the details of which will be readily understood without further description. The zinc plates dip in the diaphragms, which are filled with a mixture of 26 lbs. of water and 2 lbs. of sulphuric acid free from arsenic, in which $2\frac{3}{4}$ ozs. of amalgamating salt have previously been dissolved. The carbon plates dip into the glass vessels, which contain a solution of commercial crystallized chromic acid in the proportion of 1 part acid to 5 water. In place of this pure chromic acid the following mixture may also be used : Water 10 parts by weight, sodium dichromate 1.5 parts by weight, pure sulphuric acid of 66° Bé. 5 parts by weight.

This solution shows no inclination towards crystallization. In the illustration only two cells are combined to a battery, but in the same manner a plunge battery of four or eight cells may be constructed, the separate cells of which may all be coupled parallel, as well as one after the other, and in mixed groups.

Bichromate cell. For temporary use, for instance by gold-workers and others ; for gilding or silvering small articles, the bottle-form of the bichromate cell (Fig. 17) may be advantageously employed. In the bottle *A* two long strips of carbon united above by a metallic connection are fastened, parallel to one another, to a vulcanite stopper, and are there connected with the binding-screw ; these form the negative element, and pass to the bottom of the bottle. Between them is a short, thick strip of zinc attached to a brass rod passing stiffly through the center of the vulcanite cork, and connected with the binding-screw. The zinc is entirely insulated from the carbon by the vulcanite, and may be drawn out of the solution by means of the brass rod as soon as the services of the cell are no longer required.

Fig 17.

Coupling cells. According to the laws of Ohm, previously discussed, the current-strength J of a cell is equal to its electro motive force E divided by the sum of the internal resistance w and the external resistance wɪ :

$$J = \frac{E}{w + wɪ}$$

By now combining several such cells, say n cells, to a battery, the electro-motive force of the latter will become n.E, but the internal resistance n.w, and with the same closed circuit as the single cell had, the external resistance wɪ will not increase. Hence the current-strength of these n elements has to be written

$$J = \frac{n.E}{n.w. + wɪ}$$

Now it is evident that, if a definite closed circuit with a resistance of wɪ be given, the current-strength cannot be indefinitely raised by increasing the number of n elements. While with an increase in the number of n elements, the electro-motive force to be sure grows as many n times, the internal resistance, w, also grows, so that finally the value wɪ which remains the same disappears for the resistance nw which increases n-times. Thus the current-strength approaches more and more the limit of value

$$\frac{nE}{nw} = \frac{E}{w}$$

On the other hand, the effect can neither be increased at will by enlarging the surface of the pair of plates or decreasing the conducting resistance of the fluid in a given number of cells. Because if wɪ—the external resistance—is large enough so that the internal resistance nw may be disregarded, the current-strength approaches more and more the value $\frac{E}{wɪ}$

Hence, it follows that *the enlargement of the surface of the exciting pair of plates produces an increase in the current-strength only when the external resistance in the closed circuit is small in proportion to the internal resistance of the battery*

If we now apply the results of the above explanations to

Fɪɢ. 18.

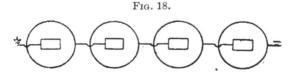

practice, we find that the cells may be coupled in various ways according to requirement.

1. If, for instance, four Bunsen cells (carbon-zinc) are coupled *one after another* in such a manner that the zinc of one cell is connected with the carbon of the next, and so on (Fig. 18), the current passes four times in succession through an

equally large layer of fluid, in consequence of which the internal resistance (4w), is four times greater than that of a single cell, while the resistance of the closed circuit (wi), remains the same. Hence, while the current-strength is thereby not increased, the electro-motive force is, and for this reason this mode of coupling is called the *union* or *coupling of the elements for electro-motive force or tension.*

FIG. 19.

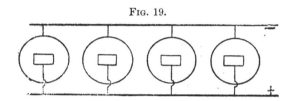

2. By connecting four cells *alongside of each other, i. e.,* all the zinc plates and all the carbon plates one with another (Fig. 19), the current simultaneously passes through the same layer of fluid in four places; the internal resistance of the battery is therefore the same as that of a single cell, and since the surface of the plates is four times as large as that of a single cell, the quantity of current is increased by this mode of coupling. This is called *coupling for quantity of current, or coupling in parallel.*

FIG. 20.

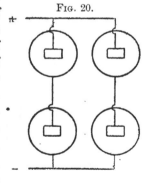

3. Two cells may, however, be connected for electro-motive force or tension, and several such groups coupled alongside of each other, as shown in Fig. 20, whereby, according to what has above been said, the electro-motive force, as well as the current-strength, is increased. This mode of connection is called *mixed coupling,* or *group coupling.*

According to the resistance of the bath as the exterior closed circuit, and according to the surfaces to be plated, the electro-plater may couple his cells in either way. We will here only mention the proposition deduced from Ohm's law, that a *number of voltaic cells yield the maximum of current-quantity when they are so arranged that the internal resistance of the battery is equal to the resistance in the closed circuit.* Hence, when operating with baths of good conductivity and slight resistance, for instance, acid copper baths, silver cyanide baths, etc., with a slight distance between the anodes and the objects, and with a large anode-surface, it will be advantageous to couple the elements *alongside of each other for quantity.* However, for baths with greater resistance and with a greater distance of the anodes from the objects, and with a smaller anode surface, it is best to couple the elements *one after the other for electro-motive force or tension.*

B. Thermo-Electric Piles.

Although thermo-electric piles are only used in isolated

Fig. 21.

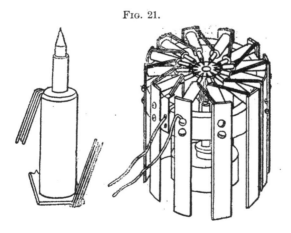

·cases for electro-plating operations, for the sake of completeness their nature and best-known forms will be briefly mentioned.

Professor Seebeck, of Berlin, observed in 1823, that electricity is developed when the soldered joints of two metals are unequally heated; hence, while electricity can be converted into heat, heat *vice versa* can be converted into electricity.

Noë's thermo-electric pile (Fig. 21) consists of a series of small cylinders composed of an alloy of 36½ parts of zinc and 62½ parts of antimony for the positive element and stout German silver as the negative element. The soldering consists of tin. The junctions of the elements are heated by small gas jets, and the alternate junctions are cooled by the heat being conducted away by large blackened sheets of thin copper. A pile of twenty pairs has an electro-motive force of 1.9 volts.

Clamond's thermo-electric pile (Fig. 22) also consists of a zinc-

FIG. 22.

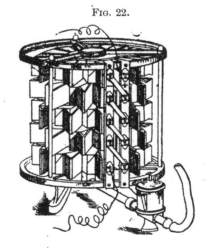

antimony alloy, but in place of German silver, ordinary tinned sheet iron is employed. To insure good contact between the two metals, a strip of tin-plate is bent into a narrow loop at one end. This portion is then placed in a mould and the melted alloy poured around it, so that it is actually imbedded in the casting. The pile shown in the illustration consists of five series, one placed above the other. Each series has ten elements grouped in a circle, and is insulated from the

succeeding series by asbestos disks.   With the consumption of
about 6½ cubic feet of gas per hour, such a pile deposits 0.7 oz.
of copper, which corresponds to an intensity of about 17
ampères.

Gülcher's thermo-electric pile, invented in 1890, is shown in
Fig. 23.   It is arranged for gas-heating, and with a constant
supply of gas requires a pressure-regulator.   The negative
electrodes consist of nickel, and the positive electrodes of an
antimony alloy, the composition of which is kept secret.   The
negative nickel electrodes have the form of thin tubes and are
secured in two rows in a slate plate, which forms the termina-
tion of a gas conduit with a U-shaped cross-section beneath it.
Corresponding openings in the slate plate connect the nickel

FIG 23.

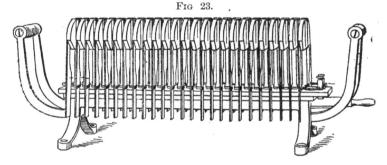

tubes with the gas conduit, the latter being connected by means
of a rubber tube with the pipe supplying the gas.   Thus the
gas first passes into the conduits, next into the nickel tubes, and
leaves the latter through six small holes in a soapstone socket
screwed in the end of each tube.   On leaving these sockets the
gas is ignited and the small blue flames heat the connecting
piece of the two electrodes.   This connecting piece consists of
a circular brass plate placed directly over the soapstone socket.
One end of it is soldered to the nickel tube, while the other
end, towards the top, is in a socket in which are cast the positive electrodes. The latter have the form of cylindrical rods with
lateral angular prolongations.   To the ends of these prolonga-

tions are soldered long copper strips secured in notches in the slate plate. They serve partially for cooling off and partially for connecting the couples. For. the latter purpose each copper strip is connected by a short wire with the lower end of the nickel tube belonging to the next couple. When the pile is to be used, the gas is ignited in one place, the ignition spreading rapidly through the entire series of couples. In about 10 minutes the junctions of the metals have attained their highest temperature and the pile its greatest power, which, with a constant supply of gas, remains unchanged for days or weeks.

In view of the conversion of the heat produced by the combustion of the gas into electricity, the useful effect of the thermo electric pile can be considered only a very slight one. One cubic meter of ordinary coal-gas produces on an average 5200 heat-units, hence 200 litres per hour referred to one second $\frac{1}{5} \cdot \frac{1}{60} \cdot \frac{1}{6}$. $5200 = 0.20$ heat-unit. These correspond to 1208 volt-ampères, 1 volt-ampère being equal to 0.00024 heat-unit. Hence, in Gülcher's thermo-electric pile, which of all known thermo-piles produces the greatest useful effect, not much more than 1 per cent. of heat is utilized in the entire circuit, and about $\frac{1}{2}$ per cent. in the outer circuit.

Although thermo-electric piles may be, and are occasionally, used for electro-plating operations, they cannot compete with dynamo-electric machines driven by steam, which as regards the consumption of heat are at least five times more effective. They can only be used in place of voltaic batteries, having the advantage of being more convenient to put in operation, more simple, cleanly, odorless, and requiring less time for attendance. But, on the other hand, their original cost is comparatively large, it being ten to twenty times that of Bunsen cells.

C. Dynamo-Electric Machines.

While in the voltaic cells, chemical energy is converted into electric energy, and in the thermo-piles, heat into electricity, in the dynamo-electric machine a conversion of mechanical energy into electrical energy takes place.

Fundamental principle of dynamo-electric machines. In the dynamo-electric machines the generation of the current results from induction, and the fundamental principle of such a machine is as follows:

Suppose an iron magnet frame *M*, formed of a powerful horse-shoe magnet, which is provided with two cylindrically-turned planes, and concentrically fixed to these planes, a cylinder *A*, built up of discs of soft iron as shown in Fig. 24.

FIG. 24.

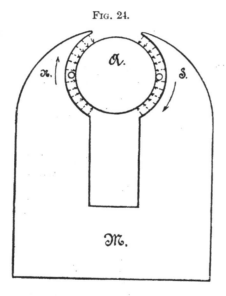

Lines of force running in the direction from the north pole to the south pole permeate the soft-iron cylinder. If in the air-space between the north pole of the magnet and the cylinder, a copper wire, indicated in the illustration by a small circle, be introduced, and moved in such a manner that it cuts the lines of force flowing from the north pole through the air-space to the cylinder, a current is induced, and a certain electro-motive force appears at the ends of the wire. By moving the left-hand wire in the direction indicated by the arrow, the current, according to the hand rule illustrated by Fig. 4 will flow away from the observer into the plane of the illustration, and by moving the right-hand wire in the direction of the arrow, out from the plane of the illustration towards the observer.

Instead of moving the wire in the air-space, it may also be insulated from the soft-iron cylinder and secured to it. If now the cylinder be moved around its axis, the wire cuts the lines of force in exactly the same manner as in its motion in the air-space, the effect remaining the same. If several wires, one alongside the other, be secured upon the cylinder, a corresponding electro-motive force will be produced on the ends of each wire, the positive poles of the wires being then on one side, say the front, of the pole pieces, while the negative poles of all the wires lie upon the other, the rear, side. If now the wires be connected one with the other, so that, when the cylinder is revolved, a positive pole is always attached to a negative pole, the electro-motive force is raised in the same degree as the number of wires coupled one after the other (in series) increases.

These wires fastened upon the iron body are called *windings*, and the term *armature* is applied to an iron body furnished with such windings.

The electro-motive force generated in the windings is the greater, the greater the velocity with which the wires, or conductor forming the windings, are moved through the magnetic field. If the length of the conductors be increased by enlarging the windings, and the velocity with which the armature moves remains the same, the electro-motive force generated in the conductor is proportional to the length of the latter. If, on the other hand, the magnetic field be strengthened, thus increasing the lines of force cut by the conductor during its motion, and the velocity with which the conductor moves, as well as its length, remains the same, the electro motive force is proportional to the number of lines of force, reaching its greatest value when the lines of force are perpendicularly cut by the conductor.

Separate parts of the dynamo-electric machine. The frame.
The production of the magnetic field has for a long time been
effected by electro-magnets. The field magnets of gray cast-
iron or cast-steel are cast in one piece with the gray cast-iron
or cast-steel frame, or screwed to it. These field magnets are
wrapped with wire through which the current, by which they
are magnetically excited, is conducted. This winding is
called *magnet winding* or *field winding.* According to the
number of field magnets, a distinction is made between two-
polar, four-polar, six-polar and multipolar machines.

Fig. 25 shows a two-polar, and Fig. 26 a four-polar type of

FIG. 25.   FIG. 26.

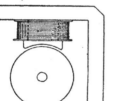
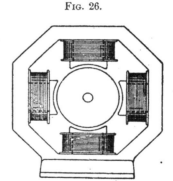

dynamo of the firm of Dr. G. Langbein & Co., Leipsic, Germany. The frame and foundation plate of soft cast-iron are
cast in one single casting; only in larger types is the frame
secured to the foundation plates by screws.

For the production of the magnetic field, the current was
formerly conducted from another source of electricity into the
magnet windings, but since the discovery of the dynamo-
electric principle by W. v. Siemens, the electric current generated in the armature is utilized for the excitation of the magnetic field. The dynamo-electric principle is based upon the
following: Lines of force, few in number, are present from a
previous excitation in every magnet frame, and this is called

remanent magnetism (see p. 13). In revolving the armature the existence of this small number of lines of force suffices for the induction of a weak current which is partly conducted through the magnet winding, the magnetic field being thereby intensified. The effect of this is the generation of currents of considerably greater power in the armature, which again bring about an increase in the current-strength in the magnet winding, until the frame is saturated with ·lines of force. This process is called *self-excitation*, while the term *foreign* or *separate excitation* has been applied to it when the magnetic field is excited by another source of electricity.

Armature or *inductor*. It has already been mentioned that the armature consists of a cylindrical iron body and the windings wrapped around it. The iron body cannot be made of one piece because rotatory currents would be formed in it, which heat the iron very much, and cause a loss of ·current. Hence the body of the armature is built up of thin, soft sheet-iron discs insulated one from the other by discs of paper. The discs are firmly pressed upon the core of the armature and secured by .screws, while the core of the armature itself is wedged upon the shaft by means of a wedge.

According to the manner in which the wire windings are laid around the armature-core, a distinction is made between a *ring armature* and a *drum armature*.

In the ring armature the wire windings are wrapped in a continuous spiral around the armature-core, it being necessary for the latter to have a wide bore in the center through which, in wrapping, the conducting wire may be carried. Fig. 27 represents a scheme of. such ring-winding. *N* and *S* are the two field magnets of the frame. Every two of the continuous wire windings represent a coil, and from the point where the end of one coil is connected with· the commencement of the next coil a conducting wire branches off to the collector. According to what has above been said, induction is greatest when the windings of the wire cut the lines of force at a right angle, this being the case when the windings are directly

under the poles.   In revolving the armature from 0° to 90°,
the generation of current decreases, from 90° to 180° it decreases, from 180° to 270° it increases in a reverse sense, and
from 270° to 300° it again decreases.   Thus, currents flowing
alternately in opposite directions, the so-called *alternating
currents* are generated, and their conversion into **constant**

Fig. 27.

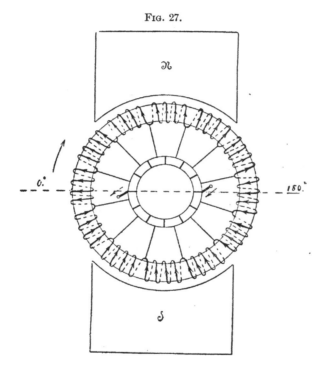

currents of uniform direction is effected by the commutator.
At 0° and at 180°, the generation of current is equal to 0,
and at these points the current changes its direction ; the line
0° to 180° is called the *neutral zone.*

In the *drum armature* the conducting wires are wound upon
the armature-core parallel to its axis, carried on the faces of
the core around the core-shaft, and the ends of every two coils

lying alongside each other on a face are connected, one with the other, and with a segment of the commutator.

Fig. 28 shows the drum winding viewed from the side of the commutator. Each coil is only indicated by a single wire winding, and therefore 8 coils are shown. The full lines indicate the connection of the coils upon the commutator-side

FIG. 28.

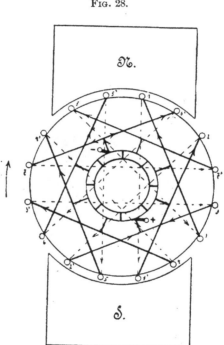

and with the commutator, and the dotted lines, the coil-connections upon the opposite face.

What has been said in regard to the intensity of induction in ring-armatures applies also to drum-armatures.

The chief difference between the modes of winding consists in more wire being required for ring winding, because wires run on the faces as well as in the interior of the bore, which are of no importance as regards the generation of the current by induction, but, on the one hand, materially increase the weight of the armature, and, on the other, enlarge the resistance. As regards these points, drum-winding has much in its favor, and it has the further advantage that the armature-core can be provided, parallel to its axis, with grooves or slots for the reception of the windings, they being thus better protected from injury, and the effect of centrifugal force can in a suitable manner be prevented by bands. In such armatures, even when equipped with thick copper wires or flat copper bands, scarcely any rotatory currents are generated, because the slots are but slightly permeated with lines of force, the latter run-

Fig. 29.

ning rather around the copper wires through the iron. However, the chief advantage of such an armature consists in that the air-space between armature and magnet-pole can be less than in armatures with windings not placed in slots, because the space occupied by the winding of such so-called smooth armatures has to be considered as an air-space and offers the greater magnetic resistance. Hence for armatures furnished with slots, the number of ampère-windings may be less than for smooth amatures. Fig. 29 shows a slotted armature of a dynamo constructed by the firm of Dr. G. Langbein & Co., in which the conductors consist of flat copper rods, connected on the faces by bent copper bands called *evolvents*.

Commutator. This is a cylindrical body built up of segments and fastened to the armature-shaft. It is insulated with mica. The segments consist of copper, tombac, or brass and are insulated from each other as well as from the commutator frame, *ι. e.*, the iron body. The commutator has as many segments as the armature has coils, and every point of junction of two coils is intimately connected by means of copper with a segment. The function of the commutator consists in converting the alternating currents of the windings generated by induction into constant currents of uniform direction. As seen from Fig. 27, currents of opposite directions flow in each half of the windings of the ring-armature. If now sliding contacts be placed on the commutator on the points of the neutral zone, the current of one-half of the windings is carried along as positive current by one of the sliding contacts, and the negative current of the other half by the other sliding contact. The armature winding is divided into two halves by the brushes which are coupled parallel to each other. The induction of each separate coil corresponds to its position for the time being in the magnetic field, the sum of the induction of all the coils in one-half of the armature being equal to that of all the coils in the other half, but as previously shown, the direction of the current in both halves is different.

Brushes. The function of the brushes is to take off the current from the commutator. For such dynamo-electric machines as here come into question, the brushes are of fine copper or brass wire-gauze, or of very thin metal-plate. Carbon brushes are often used for dynamo-electric motors.

The choice of material for the brushes depends on the properties of the material of the commutator. As there should be as little wear as possible of the commutator by the brushes, the material used for the latter should be somewhat softer than that for the former. Copper and brass gauze brushes produce by their wear considerable metallic dust, which settles on all parts of the machine, as well as on the armature and, if not removed by frequent blowing out with a pair of bellows, or a similar instrument, may readily cause short-circuiting.

Brushes of twisted, thin metal-plates (Boudreaux brushes) do not show this disagreeable formation of dust, and cause but little wear of the commutator, rather polishing it. They have, however, the drawback of the portions bearing on the commutator oxidizing readily, in consequence of becoming heated by the large quantities of current. This oxidation is not removed by the friction, and greater resistance is thereby opposed to the passage of the current from the commutator to the brushes. This, on the one hand, results in the commutator and brushes becoming strongly heated and, on the other, causes a decrease in taking off the current.

The bearing surfaces of the brushes should be so large that no heating is caused by the passage of the current, which would increase to a considerable extent the quantity of heat unavoidably formed by the friction, and be a disadvantage as regards the useful effect of the dynamo.

Brush-holders. These serve for securing the brushes and should hold them so as to bear with an even pressure upon the commutator. This is effected by metal springs by means of which the brush-frame, which carries the brush, is elastically connected with the portion of the brush-holder screwed to the bolt of the brush-rocker.

Brush-rocker. This serves for carrying the brush-holder, and for this purpose is furni shed with two thick copperbolts having a cross-section corresponding in size to the quantity of current to be conducted. In multi-polar dynamos, the rockers are equipped with as many bolts as there are poles. These bolts are insulated from the rocker by cases of a good insulating material, and secured to the rocker by insulated nuts.

The rocker is mounted upon the turned end of a bearing, and is concentrically movable to its axis, so that by turning it, the brushes may be shifted into a position at which the dynamo runs with the least sparking. In this position the brush holder is kept by means of an adjusting screw.

The rocker should also be kept free from metal dust, otherwise short-circuiting may readily result.

The other parts of a dynamo, such as bearings, cable, etc., need not be especially referred to, and it only remains to discuss the various types of

Direct current dynamos. If the whole of the current traverses the coil of the field magnet, the dynamo is said to be *series wound;* or if a portion of the current be shunted we have a *shunt-wound* dynamo; or finally there may be a combination of the two in which case the machine is a *compound dynamo.* Whatever be the arrangement, provided the volume of the copper and the density of the current are the same, the same field is always produced.

Nearly all the early types of electric dynamos were what is known as "series wound" machines, where the full current of the armature passed through the field coils. These machines had the very serious disadvantage of possessing poor regulation and being subject to frequent reversal of current direction. The plating dynamos on the market to-day are what is technically known as "shunt-wound" and "compound-wound" machines.

In a shunt-wound dynamo the field magnet coils are placed in a shunt to the armature circuit so that only a portion of the current generated passes through the field magnet coils, but all the difference of potential of the armature acts at the terminals of the field circuit.

In a shunt-wound dynamo, an increase in the resistance of the external circuit increases the electro-motive force, and a decrease in the resistance of the external circuit decreases the electro-motive force. This is just the reverse of the series-wound dynamo.

In a shunt-wound dynamo a continuous balancing of the current occurs. The current dividing at the brushes between the field and the external circuit in the inverse proportion to the resistance of these circuits, if the resistance of the external circuit becomes greater, a proportionately greater current passes through the field magnets, and so causes the electro-motive force to become greater. If, on the contrary, the re-

sistance of the external circuit decreases, less current passes
through the field, and the electro-motive force is proportionately decreased.   Thus, up to a certain degree, a shunt-wound
dynamo regulates itself.

Fig. 30 illustrates a two-pole shunt-wound dynamo, and
Fig. 31 a two-pole shunt-wound dynamo for high current-
strengths

In Fig. 30 the frame is of cast-steel and the bearing plates
are screwed to it.   In Fig. 31 the pillow-blocks and frame are
mounted upon a common cast-iron plate.

The armature is of the slotted drum type described in

FIG.: 30.

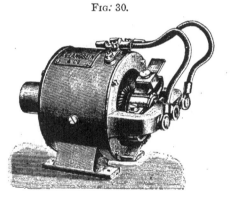

Fig. 29. It is encompassed by two strong field magnets
arranged in vertical position, radially opposite one to the other.
The ends of the field magnets are concentrically turned to the
armature and their oblique tapering shape prevents the jerky
formation or interruption of the current, thus rendering possible a sparkless taking-off of current on the commutator.   The
ends of the armature coils are soldered to the copper segments
of the commutator, loosening of the connecting points being
thus excluded, as is invariably the case with wires secured by
means of screws to the commutator.   An abundance of copper cross-sections being used, the degree of efficiency of the
dynamo is an excellent one.   To decrease friction, the portions

of the steel armature shaft which run in the journal boxes of phosphor-bronze, as well as the latter themselves, are highly polished. The bearings are furnished with automatic ring-lubrication. By reason of the use of large cross-sections of copper upon the armature and magnet winding, the number of revolutions is a moderate one, and consequently the consumption of power and wear of the bearings are slight.

Dynamos which yield high current-strengths are furnished with two commutators to avoid overloading and consequent excessive heating of a single commutator.

FIG. 31.

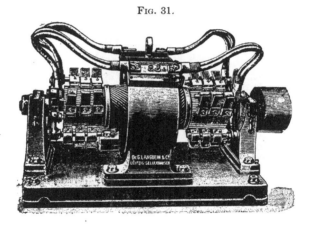

A compound wound dynamo has two distinct windings on its field magnet—one of the very many turns of fine wire, called the shunt winding, and another known as the series winding, which latter consists of a number of turns of heavier gauge wire. The series winding is in series with the vats or external circuit. The current that is used in the vats, passing through this winding, increases the magnetism of the field as the load increases, and thus the drop in voltage, which would otherwise occur by reason of the increased drop in the armature winding and increased magnetic reaction caused by the armature current is provided for.

Fig. 32 shows a multi-polar type of dynamo manufactured by The Hanson & Van Winkle Co., Newark, N. J. The frame is made of a high-grade cast iron, having a high magnetic permeability. The poles are made of soft rolled steel with cast-iron shoes. Field coils are of insulated copper wire wound compactly by machinery, insuring the maximum ampère-turns without great bulk. The whole coil is properly insulated and protected from mechanical injury.

FIG. 32.

The armature, Fig. 33, is of the toothed type. The core is built up of thin soft steel discs, and is insulated on both sides and assembled on a spider constructed to insure the greatest amount of ventilation.

The armature coils are made in a form and perfectly insulated. The slots in which the coils rest are also insulated, so that there is no chance for a ground.

The segments of the commutator are forged from pure copper carefully insulated with the best mica. The radials from the bars are so set that a steady current of air is thrown on the commutator and brushes.

The bearings are self-aligning, boxes are made of special bronze, and are provided with large oil-wells and automatic oiling-rings.

FIG. 33.

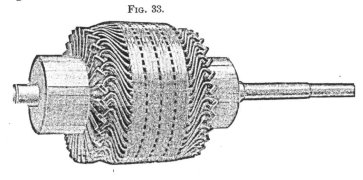

This machine will run continuously under full load with a rise of temperature above the surrounding atmosphere not

FIG. 34.

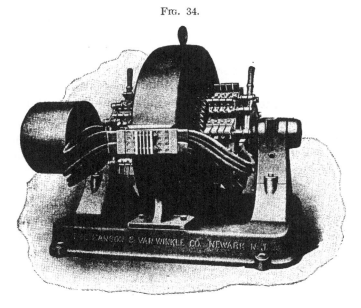

exceeding 55° F. in the accumulator, and something less in windings.

'Fig. 34 shows a separately excited dynamo of the multi-polar type manufactured by The Hanson & Van Winkle Co., Newark, N. J. It is a very popular form of generator, the field being excited from an external circuit, usually 110 or 220 volts D. C. The capacity is 4000 ampères at 6 volts. The commercial efficiency is high, 86 per cent.—the electrical efficiency averages 93 per cent. This form of dynamo is furnished for both two and three wire systems of current distribution. The frame and pole pieces are of steel. For the frame a special, soft grade is used, having a high magnetic permeability. The field coils are made of insulated copper wire wound compactly by machinery, insuring the maximum ampère-turns without great bulk. The whole coil is properly insulated and protected from mechanical injury.

The great advance which has in modern times been made in the art of electro-plating, is largely due to the important improvements that have been made in the construction of dynamo-electric machines, by which mechanical energy generated by the steam-engine or other convenient source of power may be directly converted into electrical energy. Without dynamos it would be impossible to electro-plate large parts of machines, architectural ornaments, etc., which are thus protected from the influence of the weather. They may safely be credited with having called into existence an important branch of the electro-plating art, viz., nickel-plating, and especially the nickel-plating of zinc sheets, as well as sheets of copper, brass, steel, and tin, which would have been impossible if the manufacturer had to rely upon the generation of the electric current by batteries. The latter, at the very best, are troublesome to manage; they only give out their full power when freshly charged, and as the chemical actions upon which they rely for their power progress, they deteriorate in strength and require frequent additions of acids and salts to be freshly charged, and their use demands constant vigilance and attention. Even when working on a small scale, it is cheapest to procure a small gas or other motor for driving a small dynamo, the lathes, and grinding and polishing machines.

Most cities and towns are now supplied with electric light from central stations, and thus the means are furnished to smaller plants to avail themselves of the use of electricity without the necessity of installing their own source of power. From such central stations the conductors are fed with currents of 110 or 220 volts. Hence the wires from the power circuit can be directly connected with a motor-generator, which is constructed for the respective voltage and converts

FIG. 35.

the supply of current into power, driving, for instance a connecting gear, from which the grinding and polishing machines, as well as a dynamo of low voltage, are impelled. The dynamo may be directly connected by means of a flexible or rigid coupling to the motor-generator. The armature of the latter may also be directly placed upon the grinding and polishing shafts, and the magnets arranged around it, so that every working machine becomes a motor-generator.

Fig. 35 shows a 150-ampère motor-generator set, and Fig. 36, a 4000-ampère motor-generator set, manufactured by The Hanson & Van Winkle Co., Newark, N. J. A low voltage dynamo is directly connected to a motor of suitable size, the

whole outfit being mounted on a substantial iron base. There
is no loss of power as in the case when belts are used, so the
full capacity of the generator is available.   In many instances
the plating dynamo is installed some distance from the tank,
and conductors of large cross-sections must be used in order
that there may be no drop in voltage at the tanks.   Th is,of
course, increases the cost of installation.   With the motor-

Fig. 36.

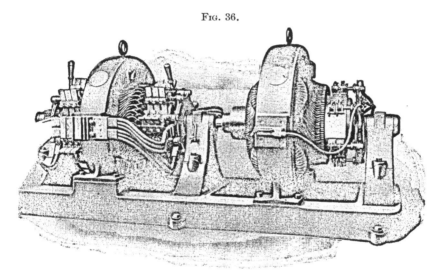

generator set, wires from the power-circuit can be brought to
the plating room and the outfit can be set up near the tanks.
These outfits are made in all sizes, both bipolar generators, as
shown in the illustration, or generators of the multipolar type
being used.

To enable the manufacturer of dynamos to suggest the most
suitable machine the following data should be submitted to
him :

1. Variety, size, and number of the baths which are to be
fed by the machine.

2. The average surface of the articles in the bath, or their
maximum surface, and the metals of which they consist.

3. Whether at one time many, and at another time few, articles are suspended in the bath.

4. The distance at which the machine can be placed from the baths.

5. The power at disposal.

If the establishment is to be electrically-driven by a motor-generator, the machines which, in addition to the dynamo, are to be driven by the motor-generator should be mentioned, as well as the voltage of the power-circuit which is to be used as a supply of electricity.

D. Secondary Cells (Accumulators).

In the theoretical part of this treatise, the polarization-current has been referred to. Although the polarization of metal plates for the production of secondary currents had previously been employed by Ritter, the construction of practically useful accumulators was first accomplished by Planté. He found that lead plates dipping in dilute sulphuric acid were specially well 'adapted for the production of secondary currents, and he arranged the accumulators as follows: In a square glass vessel filled with 10 per cent. sulphuric acid solution, a large number of lead plates were suspended in such a way that all plates with even numbers, 2, 4, 6, and so on, were electrically connected one with the other, while the plates with uneven numbers, hence, 1, 3, 5, and so on, were also in contact with each other. Between the separate plates dipping in the acid was sufficient space to prevent them from touching one another. One series of the plates served as positive, and the other as negative, electrodes. Now by conducting an electric current through the plates, lead peroxide is formed upon the positive electrodes, and by interrupting the current and combining the series of electrodes with each other, the peroxide is reduced to metallic lead, and the negative lead plates are oxidized, whereby an electric discharge takes place, the secondary or accumulator-current passing through the metallic connection of the series of plates from the peroxide to the lead plates.

Hence, in charging, a conversion of electrical energy into chemical energy takes place and, in discharging, a reconversion of the chemical energy into electric energy. A large quantity of the latter can therefore accumulate in the cells, whence the term accumulator is derived.

For the production, in the above-described manner, of currents of high power and longer duration, the plates have to be suspended as closely together as possible without danger of contact, in order to decrease the internal resistance of the element as far as practicable, and also to increase the quantity of lead peroxide.

However, the formation of the layer of lead peroxide upon the lead plates of Planté's accumulator was a slow process, and for this reason Faure used lead grids. The square openings in the negative plates are filled with a paste of litharge and sulphuric acid, and the positive plates with one of minium and sulphuric acid. The current reduces the litharge and peroxidizes the minium.

Planté showed that accumulators *form* by usage—that is to say, that up to a certain point their capacity is greater the more frequently they have been charged and discharged. By repeated oxidation and deoxidation the lead acquires a spongy structure, and gradually a large mass of metal takes part in the reaction. The *formation* is accelerated by immersing the fresh plate for a day or two in nitric acid diluted with its own volume of water.

Chemical processes in the accumulator. Regarding these processes, several theories have been advanced, for instance, by Elbs, Liebenow, and others, but it has not yet been definitely settled which of these views is correct. There can, however, be no doubt that the lead sulphate which is formed by the action of the sulphuric acid upon the lead, plays the principal role, in so far as the charging and discharging of the accumulator are effected only by the decomposition and subsequent reformation of the lead sulphate.

Elb's theory is as follows: As lead is bivalent and quadri-

valent, after the decomposition of the lead sulphate to lead
and sulphuric acid, the latter combines with the lead sulphate,
which remains undecomposed, to lead disulphate. This formation of lead disulphate must chiefly take place on the positive electrodes, since the anion (the sulphuric acid residue)
migrates to the positive pole, and by the action of the water
the lead disulphate is decomposed to lead peroxide and free
sulphuric acid.

If, therefore, the current taken from a dynamo be conducted
into the electrodes of an accumulator, so that the positive
plates are connected with the + pole of the dynamo and the
negative plate with the — pole, decomposition of sulphuric
acid takes place, the hydrogen migrating to the negative electrode, and the sulphuric acid residue to the positive electrode.
On the latter, the sulphuric acid residue forms first of all with
the lead, lead sulphate according to the following equation :

$$SO_4 \ + \ Pb \ = \ PbSO$$

Sulphuric acid residue. Lead.   Lead sulphate.

By the influx of additional SO_4-ions, this lead sulphate is
converted into lead disulphate ·

$$SO_4 \ + \ PbSO_4 \ = \ Pb(SO_4)_2$$

Sulphuric acid residue. Lead sulphate.   Lead disulphate.

However, since the formation of the lead disulphate does
not take place quantitatively, SO_4-ions are simultaneously converted into sulphuric acid, H_2SO_4, oxygen being separated in
the form of gas.

According to Elbs, lead disulphate decomposes with water
to lead peroxide and sulphuric acid according to the following
equation :

$$Pb(SO_4)_2 \ + \ 2H_2O \ - \ PbO_2 \ + \ 2H_2SO_4$$

Lead disulphate.   Water.   · Lead peroxide. Sulphuric acid.

Thus, if the current be interrupted, we have lead peroxide
on the positive electrode, and spongy lead reduced by hydro-

8

gen, on the negative electrode.   If now the positive electrodes be connected with the negative electrodes by a closed wire, a current passes through this wire from the positive lead peroxide electrodes to the negative lead electrodes, and from the latter, through the electrolyte, back to the positive electrodes.

Thus during the discharge, the spongy lead plate becomes the positive electrode and the lead peroxide plate, the negative electrode, in consequence of which, by the decomposition of the sulphuric acid, the anion SO_4 migrates to the positive lead electrode, and forms lead sulphate, while the hydrogen separated on the negative electrode reduces the lead peroxide to lead oxide or to metallic lead.

These processes take place according to the following equations:

On the — electrode $$Pb \quad + \quad SO_4 \quad = \quad PhSO_4$$
Lead.   Sulphuric acid residue.   Lead sulphate.

On the + electrode $$PbO_2 \quad + \quad 2H \quad = \quad PbO \quad + \quad H_2O$$
Lead peroxide.   Hydrogen.   Lead oxide.   Water.

This lead oxide formed on the + electrode also forms lead sulphate with sulphuric acid, and when all the lead peroxide is reduced, the generation of current ceases, the accumulator is exhausted, and has to be recharged, whereby a repetition of the processes above described takes place.   On the spongy lead plate which has now again become the negative electrode, the lead sulphate formed is reduced by the hydrogen to spongy lead and sulphuric acid:

$$PbSO_4 \quad + \quad 2H \quad = \quad Pb \quad + \quad H_2SO_4$$
Lead sulphate.   Hydrogen.   Lead.   Sulphuric acid.

whilst on the positive electrode lead peroxide is formed according to the above-described transpositions.

From these processes it follows that by the discharge of the accumulator, sulphuric acid for the formation of lead sulphate is fixed on the negative, as well as on the positive, electrode. The electrolyte must therefore contain less free sulphuric acid than at the time of charging, during which the lead sulphate at the negative electrode is reduced to lead, and oxidized to lead peroxide on the positive electrode, the sulphuric acid of the sulphate being thus again present in the electrolyte in the form of free sulphuric acid. The specific gravity of the electrolyte will be the higher, the more free sulphuric acid is present, and by determining it by means of a hydrometer it can be seen when charging is finished, the latter being the case when no further increase in the specific gravity is noticed. The completion of charging is further indicated by a copious escape of oxygen on the positive pole plates, which is due to the sulphuric acid residue finding no more material for the formation of lead sulphate, therefore forms sulphuric acid, water being decomposed, while oxygen in the form of gas is liberated.

Liebenow assumes that in charging there are formed by the decomposition of the lead sulphate, sulphuric acid-ions, lead-ions, and, by the co-operation of water, lead peroxide-ions and hydrogen-ions, according to the following equation :

$$2PbSO_4 + 2H_2O = \overset{++}{Pb} + 4\overset{+}{H} + \overset{--}{PbO_2} + 2\overset{--}{SO_4}.$$

The anions sulphuric acid and lead peroxide migrate to the positive pole and the cations lead and hydrogen to the negative pole. However, on both the poles only those ions are separated for the precipitation of which the least work is required, or, in other words, whose decomposition-point is lowest, which in this case are lead peroxide and lead. Since, however, on account of the slight solubility and dissociation of lead salts, the ions in the immediate proximity of the electrodes would soon be exhausted, further charging can only take place when from the lead sulphate formed on the electrodes, fresh molecules are brought into solution, by the dissociation of which the precipitated ions are replaced, and charging is only finished when all the lead sulphate is dissolved and separated as lead peroxide and lead-sponge. With a further passage hydrogen-ions, which possess the next highest decomposition-point, are separated. The above-de-scribed process which in charging takes place by the action of the current, progresses in a reverse sense when, by connecting the positive and negative electrodes, the discharge is rendered possible, whereby the accumulator-current becomes available for exterior work. The lead peroxide is reduced and lead and lead sulphate are formed, while on the negative electrode the lead-sponge is oxidized, sulphate of lead being also formed at the same time.

According to Liebenow's theory the electrolytic process is reversible without loss of energy, while, according to Elbs's, the process is irreversible and connected with a loss of energy. In most recent times, Dolezalek, Nernst, Loeb, and others, have expressed themselves in favor of Liebenow's view, while Le Blanc has discussed the possibility of the formation of lead peroxide-ions alongside of quadrivalent lead-ions. He assumes that at the moment of discharge, the latter are converted into bivalent lead-ions, the dissolving lead peroxide furnishing additional quadrivalent lead-ions, while at the moment of charging the bivalent lead-ions are converted into quadrivalent ones, and form lead peroxide. The view, that instead of one process in the accumulator, several processes are jointly enacted, may prove to be the correct one.

Fig. 37 shows a common form of an accumulator. The separate electrodes are insulated from each other by glass tubes, the entire system being secured by lead springs which press the electrodes against the glass tubes. Small accumulator cells are of glass, hard rubber or celluloid, and larger ones of wood lined with lead.

The sulphuric acid used for filling should be free from chlorine and metallic impurities, and have a specific gravity of 1.18. In a charged state of the accumulator, the specific gravity rises to about 1.21.

Maintenance of accumulators. An accumulator should never be allowed to stand without being charged, since, in such a case, crystals of lead sulphate are formed upon the electrodes, which can only be removed with difficulty, and by this formation of crystals the accumulator acquires a very great resistance. When not in use an accumulator should be freshly charged every two weeks, because it gradually discharges itself.

The acid should be put in the cells to such a height that the electrodes are covered about 5 millimeters deep and, since by the evaporation of water and, especially by the so-called " boiling " of the accumulator, $i.\ e.$, by the escaping gases of oxygen and hydrogen, sulphuric acid is carried along, the fluid has to be brought to its original level by the addition of dilute sulphuric acid of 1.05 specific gravity.

FIG. 37.

By the formation of lead peroxide and its subsequent reduction, the positive electrodes readily undergo changes in volume, they being liable to buckling and the scaling off of active mass ; lead-crystals of considerable length may deposit on the negative electrodes, both these occurrences giving rise to short-circuiting. Hence, the accumulator should be frequently inspected, and the mass collecting on the bottom, as well as the lead-crystals, be removed.

Charging of a cell should always be effected with a higher

voltage than that of the cell, and the dynamo should only be coupled with the accumulator when it furnishes a current of sufficiently good electro-motive force. For a single cell, charging is commenced with an electro-motive force of 2 volts. Towards the completion of charging, the electro-motive force of the charging current should be 2.6 to 2.7 volts. After interrupting the charging current the electro-motive force of each cell falls off to about 2.25 volts.

During the discharge, the electro-motive force of the cells rapidly falls to about 2 volts each, remaining constant at this value for quite a long time, when it falls slowly to 1.8 volts, and rapidly from that point on. The appearance of the last-mentioned occurrence should by all means be avoided and, when the electro-motive force falls to 1.8 volts, discharge of current should be discontinued, as otherwise the electrodes would be subject to rapid destruction.

Coupling accumulators. Like the voltaic cells, the individual accumulators may, according to requirement, be coupled alongside one another (in parallel), or one after the other (in series).

For the production of electrolytic depositions, cells of great capacity have to be taken exclusively into account, that is, cells capable of yielding a great strength of current for a certain number of hours. This value, current-strength × time, is called *ampère-hours capacity.*

If for an electrolytic process a maximum electro-motive force of 1.8 volts is required, one cell may be coupled to the bath, or if its capacity be insufficient, several such cells in parallel. If, on the other hand, the bath requires a greater electro-motive force, two or three cells will have to be coupled one after the other, and an excess of electro-motive force has to be destroyed by a resistance. The cells may be charged and discharged in parallel, or they may be discharged in series by means of a transformer, and *vice versa*, they may be charged in series and discharged in parallel, further details of which will be given in the Practical Part.

IV.

PRACTICAL PART.

CHAPTER IV.

ARRANGEMENT OF ELECTRO–PLATING ESTABLISHMENT IN GENERAL.

ALTHOUGH rules valid for all cases cannot be given, because modifications will be necessary according to the size and extent of the establishment, the nature of the articles to be electro-plated, and the method of the process itself, there are, nevertheless, certain main features which must be taken into consideration in arranging every establishment, be it large or small.

Light in plating rooms. Only rooms with sufficient light should be used, since the eye of the operator is severely taxed in judging whether the articles have been thoroughly freed from fat, in recognizing the different tones of color, etc. A northern exposure is especially suitable, since otherwise the reflection caused by the rays of the sun may exert a disturbing influence. For larger establishments the room containing the baths should, in addition to side-lights, be provided with a sky-light, which, according to the location, is to be protected by curtains from the rays of the sun.

Ventilation. Due consideration must be given to the frequent renewal of the air in the rooms. Often it cannot be avoided that the operations of pickling, etc., must be carried on in the same room in which the baths are located. Especially unfavorable in this respect are smaller establishments working with batteries, in which the vapors evolved from the

(119)

latter are added to the other vapors, and render the atmosphere injurious to health.   Hence, if possible, rooms should be selected having windows on both sides, so that by opening them the air can at any time be renewed, or the baths and batteries should be placed in rooms provided with chimneys. By cutting holes of sufficient size in the chimneys near the ceilings of the rooms, the discharge of injurious vapors will in most cases be satisfactorily effected.

To those working with Bunsen cells, it is recommended to place them in a closet varnished with asphalt or ebonite lacquer, and provided with lock and key.   The upper portion of the closet should communicate by means of a tight wooden flue with a chimney or the open air.

Heating the plating rooms.   Since the baths work with greater difficulty, more slowly and more irregularly below a certain temperature, provision for the sufficient heating of the plating rooms must be made.   Except baths for hot gilding, platinizing, etc., the average temperature of the plating solutions should be from 64.5° to 68° F., at which they work best; it should never be below 59° F., for reasons to be explained later on.   Hence, for large operating rooms such heating arrangements must be made that the temperature of the baths cannot fall below the minimum even during the night, otherwise provision for the ready restoration of the normal temperature at the commencement of the work in the morning has to be made.   Rooms heated during the day with waste steam from the engine, generally so keep the baths during the winter—the only season of the year under consideration—that they show in the evening a temperature of 64.5° to 68° F., and if the room is not too much exposed, the temperature, especially of large baths, will only in rare cases fall below 59° F.   For greater security the heating pipes may be placed in the vicinity of the baths, but if this should not suffice to protect the baths from cooling off too much, it is advisable to locate in the plating room a steam conduit of small cross-section fed from the boiler, and to pass steam for a few minutes through a coil of a metal indifferent to the plating solution suspended in the bath. In this manner baths of 1000 quarts, which on account of several days' interruption in the operation, had cooled to 36° F., were in 10 minutes heated to 68° F.

It has also been tried to heat large baths, for instance, nickel baths, by electrically heated boilers. The consumption of current is, however, very great, and the boilers of nickel sheet thus far do not answer all rational demands, especially as regards durability.

For smaller baths it is advisable to bring a small portion of them in a suitable vessel to the boiling-point over a gas flame, and add it to the cold bath. If, after mixing, the temperature is still too low, repeat the operation.

Renewal of water. Another important factor for the rooms is the convenient renewal of the waters required for rinsing and cleansing. Without water the electro-deposition of metals is impossible; the success of the process depending in the first place on the careful cleansing of the metallic articles to be electro-plated, and for that purpose water, nay, much water, hot and cold, is required, as will be seen in the section treating on the "Preparation of the Articles." Large establishments should, therefore, be provided with pipes for the admission and discharge of water, one conduit terminating as a rose over the table where the articles are freed from grease. In smaller establishments, where the introduction of a system of water-pipes would be too expensive, provision must be made for the frequent renewal of the cleansing water in the various vats.

Floors of the plating rooms. In consequence of rinsing, and transporting the wet articles to the baths much moisture collects upon the floors of the plating rooms. The best material for floors of large rooms is asphalt, it being, when moist, less slippery than cement. A pavement of brick or mosaic laid in cement is also suitable, but has the disadvantage of cooling very much. The pavement of asphalt or cement should have a slight inclination, a collecting basin being located at the lowest point, which also serves for the reception of the rinsing water.   Wood floors cannot be recommended, at least for large establishments, since the constant moisture causes the wood to rot.   However, where their use cannot be avoided, the places where water is most likely to collect should be strewn with sand or sawdust, frequently renewed, or the articles when taken from the rinsing water or bath be conveyed to the next operation in small wooden buckets or other suitable vessels.

Size of plating room.   The plating room should be of such a size as to permit the convenient execution of the necessary manipulations.   Of course, no general rule can be laid down in this respect, as the size of the room required depends on the number of the processes to be executed in it, the size and number of articles to be electroplated daily or within a certain time, etc.   However, there must be sufficient room for the batteries or dynamo, for the various baths, between which there should be a passageway at least twenty inches wide, for the table where the articles are freed from grease, for the lye kettle, hot-water reservoir, sawdust receptacle, tables for tying the articles to hooks, etc.

Grinding and polishing rooms.   The rooms used for grinding, polishing, etc., also require a good light in order to enable the grinder to see whether the article is ground perfectly clean, and all the scratches from the first grinding are removed. Where iron or other hard metals are ground with emery it is advisable to do the polishing in a room separated from the grinding shop by a close board partition ; because in the preparatory grinding with emery, which is done dry, without the use of oil or tallow, the air is impregnated with fine particles of emery, which settle upon the polishing wheels and materials, and in polishing soft metals cause fine scratches and fissures which spoil the appearance of the articles, and can be removed only with difficulty by polishing.   Hence, all operations requiring the use of emery, or coarse grinding powders, should be performed in the actual grinding room, as well as the grinding upon stones and scratch-brushing by means of rapidly revolving steel scratch-brushes. Articles already plated are, of course, scratch-brushed in the plating room itself, either on the table used for freeing the articles from grease, or on a bench especially provided for the purpose. In the polishing room are only placed the actual polishing machines, which by means of rapidly revolving wheels of felt, flannel, etc., and the use of polishing powders, or polishing compositions, impart to the articles the final luster before and after electro-plating. The formation of dust in the polishing rooms is generally overestimated; it is, however, sufficiently serious to render necessary the separation by a close partition of the polishing rooms from the electro-plating room, otherwise the polishing dust might settle upon the baths and give rise to various disturbing phenomena. In rooms in which large surfaces are polished with Vienna lime, as, for instance, nickeled sheets, the dust often seriously affects the health of the polishers, especially in badly ventilated rooms, and in such cases it is advisable to provide a suitable apparatus for keeping the dust out of the room. If this cannot be done, wooden frames covered with packing-cloth, placed opposite the polishing lathes, render good service; the packing-cloth by being frequently moistened with water retains a large portion of the dust.

Many of the states now have laws compelling firms to install some kind of apparatus to keep the dust out of the room. There are many schemes of installing these exhaust fans, the most common of which is, according to T. C. Eichstaedt,* as follows: A fan or blower of sufficient capacity for the number of lathes in use is generally placed at one end of the room, driven by a belt or directly connected with a motor. The latter is the most economical and the better of the two. Then the polishing and buffing lathes are placed in a straight line and a large galvanized iron pipe, having openings with intake pipes and hoods for each wheel, is run to the floor behind each lathe.

* Metal Industry, March, 1913.

Distance between machines. Care should be taken to have sufficient room between the separate machines to prevent the grinders and polishers, when manipulating larger pieces of metal, from inconveniencing each other. Tables for putting down the articles should also be provided.

Transmission. For grinding lathes requiring the belt to be thrown off in order to change the grinding, it is best to place the transmission carrying the belt pulleys at a distance of about three feet from the floor, while for lathes with spindles outside the bearings the transmission may be on the ceiling or wall. The revolving direction of the principal transmission should be such as to render the crossing of the belts to the grinding and polishing machines unnecessary, otherwise the belts on account of the great speed will rapidly wear out.

The more modern electrically-driven grinding and polishing machines are briefly called *grinding* and *polishing motors*, and have decided advantages over machines driven by belts. They will be referred to later on in the section "Mechanical Treatment."

Electro-plating Arrangements in Particular.

The actual electro-plating plant consists of the following parts: 1. *The sources of current* (batteries or dynamo-electric machines) *with auxiliary apparatus.* 2. *The current-conductors.* 3. *The baths,* consisting of the vats, the plating solutions, the anodes, and the conducting rods with their binding-screws. 4. *The apparatus* for cleansing, rinsing, and drying.

Before entering into the discussion of these separate parts of an electro-plating plant, it will first be necessary to speak of the electric conditions in the electrolyte, since what will here be said applies to all electro-plating processes, and will serve for a better comprehension of the succeeding sections.

Current-density. For the result of the electrolytic process, the requisite to be taken first of all into consideration is, that a sufficient quantity of current acts upon the surfaces of the objects to be electro-plated, and next that the current possesses sufficent electro-motive force for the decomposition of the bath. The quantity of current which is necessary for the normal formation of an electro-deposit upon 1 square decimeter = 10 x 10 centimeters (100 square centimeters) is now designated as the *current-density.* In the electro-plating processes to be described later on, the suitable current-density is always given. If, for instance, this normal current-density is for a nickel bath, 0.4 ampère per square decimeter, the electromotive .force 2.5 volts, and the largest object-surface to be nickeled in the bath, 50 cm. x 20 cm. = 1000 square centimeters, a current strength of at least 0.4 x 10 = 4 ampères is required. A Bunsen cell, which furnishes 4 ampères, would therefore suffice if the electro-motive force required for the decomposition of the electrolyte did not amount to 2.5 volts. As previously mentioned, a Bunsen cell furnishes about 1.8 volts, and to attain the greater electro-motive force two cells have to be coupled one after the other. The performance of the battery would then amount ·to 4 ampères and 3.6 volts, and the excess of electro-motive force, which would be an impediment to deposition proceeding in a normal manner, has to be destroyed by a current-regulator to be described later on, in case it is not preferred to increase the object-surface in the bath.

For silvering the current-density amounts to 0.25, and a silver bath with a slight excess of potassium cyanide requires 1 volt. If now, for instance, an object-surface of 55 square decimeters, about equal to 50 large soup spoons, is to be silvered, 55 x 0.25 = 13.75 ampères and 1 volt are required. Hence three cells of 5 ampères each have to be coupled alongside each other to obtain 15-ampères current-quantity.

The abbreviation of ND 100 is used to designate the normal current-density. By multiplying it with the number of square decimeters which the object-surface represents, the current-strength required for the object-surface is found.

When the current-density with which deposition is made is known, the quantity by weight of the deposit effected in a definite time can be readily calculated. The *electro-chemical equivalent* has been referred to on p. 60, and it has been established that it represents the number of coulombs which separates 1 gramme-equivalent of metal per second. When by 1 coulomb, *i. e.*, by 1 ampère, 0.3290 mg. copper per second is separated from cupric oxide salts, 1.184 gr. copper are separated in the ampère-hour (3600 seconds).

For practical purposes the quantities of a metal separated in 1 ampère-hour are designated as the electro-chemical equivalent of the ampère-hour, and the quantities of metal separated with a known current-strength in a definite time are obtained by multiplying the electro-chemical equivalent with the current-strength in ampères and the number of hours.

For calculating the time in which with a known current-strength, a certain quantity by weight is obtained, the latter is simply divided by the weight of the ampère-hours deposit ✕ the current-strength.

Another problem may be to calculate the current-strength which is required for furnishing in a certain time a definite quantity by weight of deposit. For this purpose, divide the quantity by weight by the product of ampère-hours deposit and number of hours.

We will first of all illustrate these calculations by two examples without regard to the current-output. Suppose the time is to be determined during which a square decimeter of surface has to remain in the nickel bath in order to acquire a deposit of $\frac{1}{10}$ millimeter thickness with a current-density of 0.4 ampère. First calculate the weight of the deposit by multiplying the surface in square millimeters with the thickness and specific gravity. One square decimeter is equal to 10,000 square millimeters, which, multiplied by $\frac{1}{10}$ millimeter, gives as a product 1000, which multiplied by the specific gravity of nickel—8.6—gives 8600 milligrammes = 8.6 grammes. Hence a deposit of $\frac{1}{10}$ milligramme thickness upon a surface of 1 square decimeter represents a weight of 8.6 grammes. Since, for the normal deposit per square decimeter, a current-density of 0.4 ampère is required, and 1 ampère deposits, according to the table given on p. 61, 1.1094 grammes in 1 hour, ½ ampère deposits.0.4437 gramme in 1 hour, and, therefore, about 19¾ hours will be required for the deposition of 8.6 grammes.

For calculating the time which one, two or more dozen of knives and forks or spoons, which are to have a deposit of silver of a determined weight, must remain in the bath when the current-density is known, proceed as follows: Suppose 50 grammes of silver are to be deposited upon 1 dozen of spoons, and the most suitable current-density is 0.2 ampère per square decimeter; if the surface of 1 spoon represents 1.10 square decimeters, the surface of 1 dozen spoons of equal size is 13.2 square decimeters.   Hence, they require 13.2 × 0.2 = 2.64 ampères; now, since 1 ampère deposits in 1 hour 4.025 grammes of silver, 2.64 ampères deposit in the same time 10.62 grammes of silver, and with this current, the dozen spoons must remain about 4¾ hours in the bath for the deposition of 50 grammes of silver upon this surface.

However, the figures obtained are correct or approximately correct only when the current-output amounts to 100 per cent., or to approximately this value, as in the case with acid copper baths, silver, gold, zinc and tin baths; with a smaller current-output as yielded by potassium cyanide copper and brass baths, and nickel baths, a suitable correction has to be made.

The current-output of a bath is best determined as follows · Deposit upon an accurately weighed plate (sheet) of metal for several hours with the normal current-density, and note the exact time of deposition and the quantity of current measured by a voltmeter inserted in the circuit.   Rinse the plate first with water and then with alcohol and ether, and dry thoroughly.   Weigh it and by deducting the previous weight, the weight of the deposit is found.   Now calculate from the table of electro-chemical equivalents (p. 61) how much metal should have been precipitated in the time consumed by the current-strength used ; the result will be the theoretical current-output. The practical current-output in per cent. is found by multiplying the weight of the deposit found by 100 and dividing by the calculated weight of the theoretical current-output.

Suppose the plate weighs 12.00 grammes and after having deposited upon it nickel for 3 hours with 1.5 ampère, it weighs 16.45 grammes, which corresponds to a deposit of nickel of 16.45 — 12.00 = 4.45 grammes. Theoretically, 1.5 ampère should separate in 3 hours (1.1094 × 1.5 × 3) 4.923 grammes of nickel. Hence, the practical current-output attained is

$$4.925 : 4.45 = 100 : x$$
$$x = 90.35 \text{ per cent.}$$

In calculating the quantity by weight, the product obtained from electro-chemical equivalent × current-strength × number of hours, would have to be multiplied by the fraction $\frac{\text{Current-output in per cent.}}{100}$; in calculating the time, the result obtained above would have to be multiplied by the fraction $\frac{100}{\text{current-output in per cent.}}$, and for calculating the current-strength the quotient is likewise to be multiplied by the fraction $\frac{100}{\text{current-output in per cent.}}$

Electro-motive force in the bath. It has previously been seen that for the permanent decomposition of an electrolyte, an electromotive force is required which must be large enough to overcome the resistance of the electrolyte, as well as the polarization-current flowing counter to the main current.

The *resistance of the electrolyte* is found by multiplying its specific resistance, *i. e.*, the resistance of a fluid cube of 1 decimeter side-length by the electrode distance in decimeters, and dividing by the object-surface expressed in square decimeters, thus,

$$\text{Resistance of the electrolyte} = \frac{\text{Specific resistance} \times \text{dm. electrode-distance}}{\text{dm. object-surface.}}$$

According to p. 21, the electro-motive force required for sending a certain current-strength through a conductor is equal to the product of current-strength and resistance. To calculate this electro-motive force, the resistance of the electrolyte, i. e., of the bath, as found above, has to be multiplied by the current-density.

For the better understanding, an example may here be given, the problem being to copper in an acid copper-bath an object-surface of 100 square decimeters.

Let the specific resistance of the acid copper-bath of a given composition be 0.92 ohm, the electrode-distance 1.2 decimeters, the normal current-density 1.25 ampères. The required current-strength, J, is found by multiplying the normal density by the object-surface in square decimeters, thus,

$$J = 100 \times 1.25 = 125 \text{ ampères.}$$

From what has above been said, the resistance, W, of the electrolyte is obtained by multiplying the specific resistance by the electrode-distance in decimeters, and dividing the product by the object-surface in square decimeters:

$$W = \frac{0.92 \times 1.2}{100} = 0.01104 \text{ ohm.}$$

From this the electro-motive force, E, required to send the current-strength, J, through the bath is calculated:

$$E = J \times W = 125 \times 0.01104 = 1.38 \text{ volt.}$$

However, this is valid only for the normal temperature of 18° C. (64.40° F.). If the electro-motive force has to be calculated, which is required at a higher temperature, for sending the current-strength of 125 ampères through the bath, we have to fall back upon the temperature-coefficients and the formulas given for them on p. 26, whereby, if the temperature of the bath is 24° C. (75.2° F.), the equation assumes the following form :

9

Specific resistance $= 0.92\ (1 - 0.0113 \times 6) = 0.858$ ohm.

Hence the temperature-coefficient 0.0113 has to be multiplied by the number of degrees C., the bath is warmer than 18° C., the product subtracted from 1, and the remainder multiplied by the specific resistance at 18° C., 0.92 ohm. It will be seen that the specific resistance (Sp. R.), which amounts at 18° C. to 0.92 ohm, amounts at 24° C. only to 0.858 ohm. The resistance, W, of the electrolyte at 24° C. is therefore

$$W = \frac{0.858 \times 1.2}{100} = 0.0103 \text{ ohm,}$$

and the electro-motive force, E, which is capable of forcing 125 ampères through the resistance of 0.0103 ohm :

$$E = J \times W = 125 \times 0.0103 = 1.287 \text{ volt.}$$

If the electrolyte is 6° C. colder than 18° C., the formula is so changed that the temperature-coefficient 0.0113 has to be multiplied by 6, the product added to 1, and the sum multiplied by the specific resistance (Sp. R.) :

$$\text{Sp. R.} = 0.92\,(1 + 0.0113 \times 6) = 0.9824 \text{ ohm ;}$$

the resistance of the bath is then :

$$W = \frac{0.9824 \times 1.2}{100} = 0.01178 \text{ ohm,}$$

the electro-motive force required being therefore :

$$E = J \times W = 125 \times 0.01178 = 1.472 \text{ volt.}$$

Electro-motive counterforce of polarization. In addition to this resistance of the electrolyte, the electro-motive counterforce of the polarization-current has to be taken into consideration. The causes of polarization have been explained on p. 65 ; it being partly due to the formation of gas-cells during electrolysis with insoluble electrodes, especially anodes, partly to changes in concentration in the vicinity of the electrodes, or to oxidizing or reducing processes in the electrolyte. In most cases of electrolysis coming here in question, the dilution formed on the cathodes by the separation of metal will send a polarization-current towards the more concentrated layers of fluid formed by the solution of the anode-metal, to which is added the counter-current formed by the contiguity of fluids with salts of a lower degree of oxidation to fluids with salts of a higher degree of oxidation. The *magnitude of polarization* is materially influenced by the nature of the metals of which the electrodes consist ; the more electro-positive the cathode-metal and the more electro-negative the anode-metal, the greater the electro-motive force of the polarization-current which flows from the more positive cathode to the negative anode, hence in an opposite direction to the main current, which enters at the anode and passes out at the cathode. This explains why in nickeling iron less electro-motive force is required than in nickeling zinc, iron being only to a slight degree more positive than the nickel-metal of the anode, and hence less electro-motive counterforce appears. Zinc, on the other hand, is far more positive than iron, and the electro-motive force of the polarization-current is consequently essentially stronger.

The determination of this electro-motive counterforce is in the most simple manner effected by experiment. If a voltmeter of great resistance be placed at the bath, and the main current which had been passed into the bath be suddenly interrupted by means of a switch, the needle of the voltmeter does not at once return to the O-point, but remains for some time in a position above that point, and then gradually returns to it. The electro-motive force indicated by the needle for the short time after the interruption of the current gives the electro-motive force of the polarization-current.

The electro-motive counterforce is influenced by the magnitude of the current-density, growing and falling with the latter.

When the magnitude of the counterforce has been determined by experiment as above described, the electro-motive force of the main current required for the electrolytic process is made up of the electro-motive force found by multiplying the current-strength by the resistance of the electrolyte plus the electro-motive counterforce of polarization found by experiment.

Proceeding from the opinion that the electric current-lines are subject to scattering similar to the magnetic lines of force, Pfanhauser has taken into account the magnitude of this scattering of the current-lines for the calculation of the resistance of the electrolyte.   When such scattering takes place, the current-lines will not collectively migrate by the shortest road from the anode to the cathode, but describe greater or smaller curves, the cross-section of the fluid which takes part in the conduction of the current, becoming thereby greater than if the current would pass, without deviation whatever, between the elec trodes, and the resistance of the electrolyte consequently becomes smaller.   The least scattering was found with electrodes of the same size, it increasing with the greater distance of the electrodes from each other.   In electro-plating processes running a normal course, the decrease in the resistance of the bath by the scattering of current-lines may practically be disregarded, and it will later on only be referred to in so far as various phenomena which appear in electro-plating have been explained by this scattering.

We will now turn to the discussion of electro-plating installations with the different sources of current, and the arrangement with cells will first be described.   It will be necessary to specify in this section all the laws and rules which are also valid for installations with other sources of current, and the reader is requested thoroughly to study this section, as repetition in subsequent sections is not feasible.

A.   Installations with Cells.

Coupling of cells.   Prior to the time when it became possible to calculate the normal current-strength for a definite object-surface, because the magnitude to which the term current-density has been applied was not known, the transmission of the quantity of current required for the electro-plating processes was effected in a purely empirical manner. The effective zinc surface of the cells was taken as the basis, and it was held that with baths of medium resistance a good deposit is generally effected when the effective zinc-surface of the cells *is of the same size as the object-surface which is to be plated, and as large as the anode-surfaces. The electro-motive force required was obtained by coupling a larger or smaller number of cells one after the other. Suppose we have a nickel bath which requires for its decomposition a current of 2.5 volts of electro-motive force. Now since, according to p. 78, a Bunsen cell develops a current of 1.88 volts, the reduction of the nickel cannot be effected with one such cell alone, but two cells will have to be coupled for electro-motive force one after the other, whereby, leaving the conducting resistance of the wires out of consideration, an electro-motive force of $2 \times 1.88 = 3.76$ volts is obtained, with which the decomposition of the solution can be effected.

If, on the other hand, we have a silver bath which requires only 1 volt for its decomposition, we do not couple two cells one after the other, because the electro-motive force of a single cell suffices for the reduction of the silver. On p. 88 it has been seen that by coupling the elements one after the other (coupling for electro-motive force) the electro-motive force of the battery is increased, but the quantity of current is not increased, and that to attain the latter, the cells must be coupled alongside of one another (coupled for quantity). Hence in a group of, for instance. three cells coupled one after another, only one single zinc surface of the cells can be considered effective in regard to the quantity of current. Now, the larger the area of articles at the same time suspended in the bath is, the greater the number of such effective zinc surfaces of the group of cells to be brought into action must be; and, if for baths with medium resistance, it may be laid down as a rule that the effective zinc surface must be at least as large as the surface of the articles, provided the surface of the anodes is at least equal to the latter, the approximate number of cells and their coupling for a bath can be readily found.

Let us take the nickel bath of medium resistance which, as above mentioned, requires a current of 2.5 volts, and for the decomposition of which two cells must, therefore, be coupled one after the other, and suppose that the zinc surface of the Bunsen cells is 500 square centimeters, then the effective zinc surface of the two cells coupled one after the other will also be 500 square centimeters; hence a brass sheet $20 \times 25 = 500$ centimeters can be conveniently nickeled on one side with

FIG. 38.

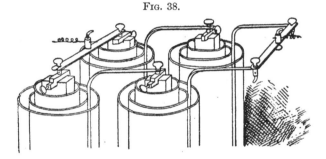

these two cells, or a sheet $10 \times 25 = 250$ centimeters on both sides. Now suppose the surface to be nickeled were twice as large, then the two cells would not suffice, and a second group of two cells, coupled one after the other, would have to be joined to the first group for quantity, as shown in Fig 19, or perspectively in Fig. 38. Three times the object-surface would require three groups of elements, and so on.

In giving these illustrations it is supposed the objects are to have a thick, solid plating. For rapid plating and a thin deposit a different course has to be followed. Only a slight excess of electro-motive force in proportion to the resistance of the bath being in the above-mentioned case present, reduction takes place slowly and uniformly without violent evolution of gas on the objects, and by the process thus conducted, the deposit formed is sure to be homogeneous and dense, since it absorbs but slight quantities of hydrogen, and in most cases it can be obtained of such thickness as to be thoroughly resistant.

For rapid plating, without regard to great solidity and thickness of the deposit, the cells, however, have to be coupled so that the electro-motive force is large as compared with the resistance of the bath, so that the current can readily overcome the resistance. This is accomplished by coupling three, four, or more cells one after the other, as shown in the schéme, Fig. 18. However, special attention has to be drawn to the fact that deposits produced with a large excess of electro-motive force can neither be dense nor homogeneous, because, in accordance with the generally accepted view, the deposits condense and retain relatively large quantities of hydrogen gas, the term *occlusion* being applied to this property.

Current regulation. Only in very rare cases will it be possible to always charge a bath or several baths with the same object-surface; and according to the amount of business, or the preparation of the objects by grinding, polishing and pickling, at one time large, and at another, small surfaces will be suspended in the bath. Now, suppose a battery suitable for a correct deposit upon a surface of, say five square feet, has been grouped together; and, after taking the articles from the bath, a charge of objects only half as large as before is introduced, the current of the battery will, of course, be too strong for this reduced surface, and there will be danger of the deposit not being homogeneous and dense, but forming with a crystalline structure, the consequence of which, in most cases, will be slight adhesiveness, if not absolute uselessness. With sufficient attention the total spoiling of the articles might be prevented by removing the objects more quickly from the bath. But this is groping in the dark, the objects being either taken too soon from the bath, when not sufficiently plated, or too late, when the deposit already shows the consequences of too strong a current.

For the control of the current an instrument called a *current-regulator, resistance board* or *rheostat* has been devised, which allows of the current-strength of a battery being reduced without the necessity of uncoupling cells.   It is obvious that the current of a battery, if too strong, can be weakened by decreasing the number of cells forming the battery, and also by decreasing the surface of the anodes, because the external resistance is thereby increased.   This coupling and uncoupling of cells is, however, not only a time-consuming, but also a disagreeable, labor ; and it is best to use a resistance

Fig. 39.

Fig. 40.

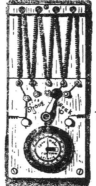

board with which, by the turn of a lever, the desired end is attained.   Figs. 39 and 40 show this instrument.

Its action is based upon the following conditions : As previously explained, the maximum performance of a battery takes place when the external resistance is equal to the internal resistance of the battery.   By increasing the external resistance, the performance is decreased, and a current of less intensity will pass into the bath when resistances are placed in the circuit.   The longer and thinner the conducting wire is, and the less conducting power it possesses, the greater will be the resistance which it opposes to the current.   Hence, the resistance board consists of metallic spirals which lengthen the circuit, contract it by a smaller cross-section, and by the nature of the metallic wire, has a resistance-producing effect. For a slight reduction of the current, copper spirals of various cross-sections are taken, which are succeeded by brass spirals, and finally by German-silver spirals, whose resistance is eleven times greater than that of copper spirals of the same length and cross-section. In Fig. 39 the conducting wire coming from the battery goes to the screw on the left side of the resistance board, which is connected by stout copper wire with the first contact-button on the left; hence by placing the metallic lever upon the button furthest to the left, the current

Fig. 41.

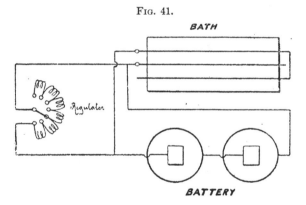

passes the lever without being reduced, and flows off through the conducting wire secured to the setting-screw of the lever. By placing the lever upon the next contact button to the right, two copper spirals are brought into the circuit; by turning the lever to the next button, four spirals are brought into the circuit, and so on. By a proper choice of the cross-sections of the spirals, their length, and the metal of which they are made, the current may be more or less reduced as desired.

In case great current-strengths must flow through the resistance board, it is more advantageous to couple the spirals in parallel, and not one after the other, as in Figs. 41 and 42.

The resistance boards may be placed in the circuit itself in two different ways. If the resistance board is to maintain the electro-motive force of the current at the bath constant at a certain height, it is coupled in series. In this case the same current-strength which is consumed at the bath flows through the resistance. This coupling in series, or one after the other, of the resistance board is shown in Fig. 41.

In the other mode of coupling, Fig. 42, the resistance lies in shunt to the circuit, it being coupled parallel to it. According to Kirchoff's law, if there be a branching-off of the current, the sum of the current-strengths in the separate

Fig. 42.

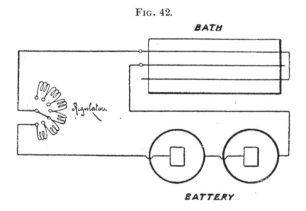

branches is just as great as the current-strength prior to and after branching off, and the current-strengths in the separate branches are inversely proportional to the resistances of the separate branches.

In the case in question the coupling of the resistance-board (Fig. 42) represents such a branching-off of the current; the greater the resistance of the resistance-board, the less the current-strength will be which flows through it; otherwise, a greater resistance in the main circuit, hence in this case in the bath, will cause a portion of the current-strength to flow through the resistance-board, where it is destroyed.

The parallel coupling of the resistance-board with the bath is utilized to remove differences in the operating electromotive force of baths coupled in series, which may appear by electrode-surfaces of uneven size, or by changes in the resistances of the electrolytes.

Current indicator. In order to be able to control the change in the current-conditions which is effected in a circuit by the resistance-board, a *galvanometer* is coupled behind the latter. This instrument consists of a magnetic needle oscillating upon a pin, below which the current is conducted through a strip of copper, or, with weaker currents, through several coils of wire. The electric current deflects the magnetic needle from its north-pole position, and the more so the stronger the current is; hence the current-strength of the battery can be determined by the greater or smaller deflection.

For a weak current, such as, for instance, that yielded by two cells, it is of advantage to use a *horizontal galvanometer* (Fig. 43). It is screwed to a table by means of a few brass screws in such a position that the needle in the north position, which it occupies, points to 0° when no current passes through the instrument. Articles of iron and steel must, of course, be kept away from the instrument.

FIG. 43.

For stronger currents it is better to combine a *vertical galvanometer* with the switch-board and fasten it to the same frame, as shown in Fig. 44. The screw of the lever of the switch-board is connected with one end of the copper strip of the vertical galvanometer, while the other is connected with the screw on the right side of the switch-board, in which is secured the wire leading to the bath. The switch-board and galvanometer are placed in one conducting wire only, either in that of the anodes or of the objects, one of these wires being simply cut, and the end connected to the battery, is secured in the binding-screw on the side of the resistance board marked " strong," while the other end, which is in connection with the bath, is secured in the binding-screw on the

FIG. 44.

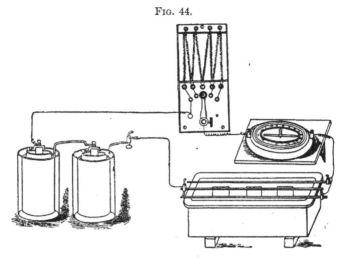

opposite side marked " weak."   The entire arrangement will be perfectly understood from Figs. 44 and 45.

FIG. 45.

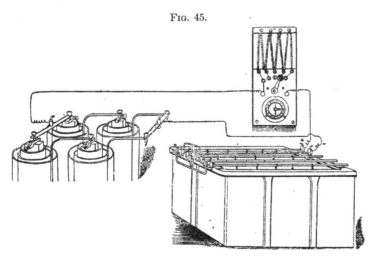

Fig. 46 shows the Hanson & Van Winkle Patent Under

writer's Rheostat. It has twice the carrying capacity of any resistance board, ever made for this purpose, it having sufficient length of wire to allow of turning down the highest electro-motive force used in plating, to the lowest figure called for, without showing heat or any unfavorable symptoms. By the use of this rheostat the output from a plating room using two or more tanks can be doubled, providing the dynamo has the current capacity.

FIG. 46.

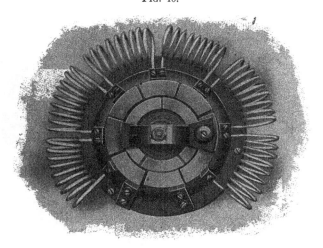

Fig. 47 shows a special rheostat constructed by the Hanson & Van Winkle Co. for use on nickel, copper or brass solutions requiring heavy ampèreage. For the reason that so large an ampère current is used the instrument is especially constructed to withstand any excessive heating to which it may be sub-jected. This rheostat may also be used in the main line to control the voltage of several tanks. It is suitable for solutions containing 175 to 200 square feet of nickel work, or on copper or brass baths of 100 to 125 square feet, or for zinc solution containing 75 feet of work surface.

The advantages derived from the use of a resistance board

having been referred to, it remains to add a few words regarding the indications made by the galvanometer. Since the greater deflection of the needle depends, on the one hand, on the greater current-strength, and on the other, on the slighter resistance of the exterior closed circuit (conducting-wires, baths and anodes), it is evident that a bath with slighter resistance, when worked with the same battery and containing

FIG. 47.

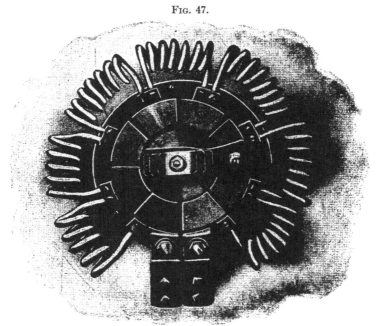

the same surface of anodes and objects, will cause the needle to deflect more than a bath of greater resistance under otherwise equal conditions.

Hence, the deductions drawn from the position of the needle for the electro-plating process are valid only for definite baths and definite equal conditions, but, with due consideration of these conditions, are of great value.

Suppose a nickel bath to work always with the same surfaces

of objects and anodes, and experiments have shown that the suitable current-strength for this surface of objects is that at which the needle stands at 15°; and suppose, further, that the battery has been freshly filled and causes the needle to deflect to 25°, then the lever of the resistance board will have to be turned so far to the right that the needle in consequence of the interposed resistances returns to 15°. Now if, after working for some time, the battery yields a weaker current, the needle, by reason of the resistance remaining the same, will constantly retrograde, and has to be brought back to 15° by turning the lever to the left, when a current of equal strength to the former will again flow into the bath. This manipulation is repeated until finally the lever rests upon the button furthest to the left, at which position the current flows directly into the bath without being influenced by the resistances of the resistance board. If now the needle retrogrades below 15°, it is an indication to the operator that he must renew the filling of the battery if he does not prefer suspending fewer objects in the bath. For this reduced object-surface it is no longer required for the needle to stand at 15° in order to warrant a correct progress of the electric process, since the resistance being in this case greater, a deflection to 10°, or still less, may suffice. This example will make it sufficiently clear that the current-indication by the galvanometer is not and cannot be absolute, but that the deductions must always be drawn with due consideration to the conditions, namely, surfaces of objects and anodes, and distance between them.

It frequently happens that in consequence of defective contacts with the binding-screws of the battery, or by the conductors of the objects and of the anodes touching one another (short circuit with non-insulated conducting wires), no current whatever flows into the bath. Such an occurrence is immediately indicated by the galvanometer, the needle being not at all deflected in the first case, while in the latter the deflection will be much greater than the usual one.

The needle of the galvanometer also furnishes a means of

recognizing the *polarity* of the current. If the galvanometer be placed in the positive (anode) conductor by securing the wire coming from the battery in the binding-screw on the south pole of the galvanometer, and the wire leading to the bath in the binding-screw on the north pole of the needle, the needle, according to Ampère's law, will be deflected in the direction of the hands of a watch, *i. e.*, to the right if the observer stands so in front of the galvanometer as to look from the south pole towards the north pole, because the battery-current flows out from the positive pole through the conducting wire, anodes, and fluid to the objects, and from these back through the object wire to the negative pole of the battery. If now in consequence of the counter-current formed in the bath by the metallic surfaces of dissimilar nature or other causes, and flowing in an opposite direction to that of the battery-current, the latter is weakened, the needle will constantly further retrograde from the zero point, and when the counter- or polarization-current becomes stronger than the battery-current, it will be deflected in an opposite direction as before. Hence, by observing the galvanometer, the operator can avoid the annoying consequences of polarization, which will be further discussed under nickeling.

Measuring instruments. It may here be stated that the use of the galvanometer has been to a great extent abandoned, and measuring instruments are at present generally employed.

For measuring the current-strength, the *ampère-meter* or *ammeter* is employed, and for measuring the electro-motive force of the current, the *volt-meter*, these instruments allowing of the direct reading off of the current-strength in ampères and of the electro-motive force in volts.

Space will not permit us to enter into the different constructions of these measuring instruments, and only the principle of their construction will here in a few words be explained

It has previously been seen that with a given object- and anode-surface, the deposit in the plating bath depends chiefly on the current-strength and electro-motive force of the current. The deposit will turn out most beautiful and most homogeneous only with a definite current-strength, and though the skilled operator may succeed by empirical experiments in obtaining a beautiful deposit without a knowledge of the current-conditions, this mode of working requires far more attention than when by simply reading off the deflection of the needle on the measuring instruments, it can be ascertained that the bath works in the most rational manner, without having first to inspect the objects and the bath itself. Such instruments are a great convenience, especially with a varying size of the object-surface, particularly if each bath is provided with one, because the electro-motive force at the bath changes every time the object-surface is changed. Hence, as previously stated, the current has every time to be regulated before it is allowed to pass into the bath, if the deposit is to be always of the same quality.

While voltmeters allow of a reliable control of the electromotive force in the bath, ammeters serve the purpose of recognizing, on the one hand, whether the current-strength required for a certain object-surface passes into the bath, if the calculation of the total current-strength is based upon the normal current-density. On the other hand, they allow of the determination of the quantities by weight of metal deposited, the weight of the deposit depending solely on the current-strength. Although it is not always necessary to know this, yet it is frequently desirable to ascertain how great the current-strength is, in order to determine what demands are made on the battery or the dynamo.

Notwithstanding their extraordinary simplicity, the instruments constructed according to Hummel's patent, are very sensitive, and do not change in the course of time as is the case with many other constructions. Their mode of action is based upon the phenomenon that soft iron is attracted by a current-conductor. In the scheme, Fig. 48, S is a circular current-conductor, consisting of a greater or smaller number of copper-wire coils. In the interior is a piece of thin sheet-

10

iron, E, connected with an axis of revolution, a. G is a weight which is to be lifted by the attractive force of the cur-

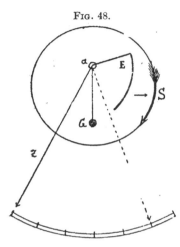

FIG. 48.

rent S upon the iron E. The stronger the current, the greater the attraction of the coils lying next to the sheet-iron, and, hence, the greater the elevation of the weight, G, will be, and the further the indicator, Z, connected with the axis of revolution, and below which a scale is fixed, will deflect.

As regards construction, the voltmeter and ammeter are alike with the exception of the coil S. In the voltmeter it consists of many windings of thin copper wire, and in the ammeter of but a few windings of stout copper wire, or in instruments for great current-strength, of a massive bent piece of copper.

Fig. 49 shows the "Waverly" voltmeter, manufactured by

FIG. 49.

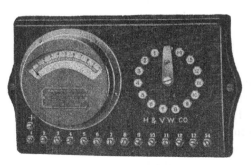

the Hanson and Van Winkle Co., Newark, N. J. It is intended for direct current circuits only; 0 to 10 volts. It is furnished with binding posts for fourteen tanks, thus enabling the operator to use only one instrument in obtaining the read-ing of any number of tanks up to fourteen, by simply moving the switch lever to the tank numbers indicated on the switch of the instrument, and when used in connection with the patent tank rheostat, will enable the operator to reproduce at all times the same electrical conditions which by observation and experience he has found necessary in order to obtain a satisfactory deposit of uniform thickness and color in the shortest possible time.

Fig. 50 shows the Weston ammeter. The ammeter is

FIG. 50.

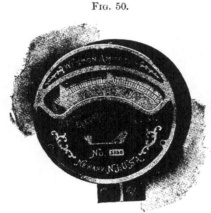

placed in one conductor only, either in that of the objects or of the anode, and thus the whole of the current must pass through it. The voltmeter, however, is connected with both conductors. On the point where the electro-motive force is to be measured, one of the binding posts of the voltmeter is con-nected by means of a copper wire with the object-conductor, and the other, with the anode-conductor.

Fig. 51 illustrates the arrangement of the switch-board and ammeter with a bath operated by means of a battery.

Voltmeter switch. If many baths are in operation in an electro-plating plant, it would be quite an expense to furnish

each bath with a special voltmeter.   However, this is unnecessary, one voltmeter being sufficient for three or four baths.   In
order to be able to read off conveniently on the voltmeter the
electro-motive force passing into one of these baths, a switch is

FIG. 51.

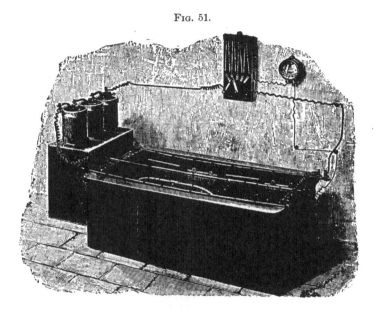

required, the construction of which will be seen from Figs. 52
and 53.

Fig. 52 shows the coupling of the main object-wire (—) and
of the main anode-wire (+), which will be referred to later on,
together with the resistance boards R_1 and R_2, the voltmeter
V, switch U, and the two baths.   In Fig. 53 the coupling is
enlarged, and upon this illustration the following description
is based : Suppose the main object-wire and anode-wire to be
connected with the corresponding poles of a dynamo-machine
or a battery, which for the sake of a clearer view is omitted in
the illustration.   The switch U consists of a brass lever,
mounted with a brass foot, upon a board.   In the foot is a.

screw, with which is connected by a 0.039-inch thick copper wire one of the pole-screws of the voltmeter. The brass handle slides with spring pressure upon contact buttons connected by copper wire with the binding-screws 1, 2, 3, 4, 5 (upon the

FIG. 52.

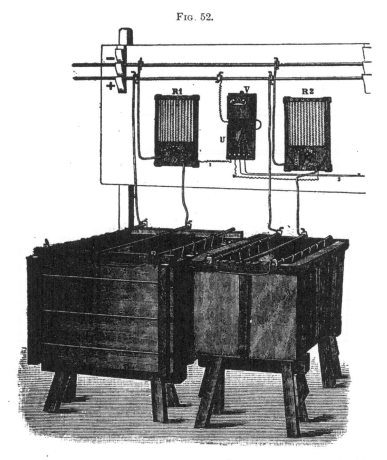

switch), which serve for the reception of the 0.039-inch thick insulated wires 1, 2, 3, 4, for measuring the electro-motive force, which branch off from the various tanks or resistance boards. The other pole-screw of the voltmeter is directly con-

nected with the main anode-wire.   From the main object-wire,
a wire, whose cross-section depends on the strength of the
working current, passes to the screw marked "strong" of the
resistance board R_1; the screw marked "weak" of the resistance board R_1 is connected by a wire of corresponding thickness with the object-wire of bath I, and at the same time with
the binding-screw 1 of the switch.   The resistance board R_2,
of the bath II, is in the same manner connected with the main
object-wire, the bath, and the binding-screw 2 of the switch;

FIG. 53.

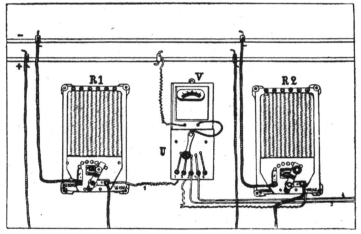

also the resistance boards R_3 and R_4 of the baths III and IV,
which are not shown in the illustration.   With the main
anode-wire each bath is directly connected by conducting the
current to an anode-rod of the bath by means of binding-
screws and a stout copper wire, and establishing a metallic
connection between this anode-rod and the next one.   However, instead of connecting both, the current may also be conducted from the main anode-wire to each anode-rod.

In the illustration, the handle of the switch rests upon the
second contact-button to the left, which is connected with the

binding-screw 2 of the switch. In the latter is secured the wire for measuring the electro-motive force which leads from the resistance board R_2; hence the voltmeter V will indicate the electro-motive force of the current at bath II. Suppose that bath II is full of objects and, with the position of the lever of the resistance board at " weak," as shown in the illustration, the voltmeter indicates 1.5 volts, while the most suitable electro-motive force for the bath is 2.5 volts, the handle of the switch is turned to the left until the needle of the voltmeter indicates the desired 2.5 volts.

If the handle of the switch U be turned to the left so that it rests upon the contact-button 1, the measuring wire of bath II is thrown out, and the voltmeter indicates the electro-motive force, in bath I; if the lever rests upon contact-button 3, the electro-motive force in bath III is indicated, and so on.

Dependence of the current-density on the electro-motive force. If a current of known strength be at the outset conducted through electrodes of a certain size into a bath of determined resistance, and the electrode-surfaces be then doubled, the current-strength must also be doubled in order to maintain the same current-density as before. By increasing the electrode-surfaces to twice their size, the resistance of the bath is, however, reduced one-half the value it amounted to with electrodes of half the size, the increased electrode-surfaces corresponding to a cross-section of the bath-fluid enlarged in the same proportion.

Suppose the resistance of the bath with an electrode-surface of 1 square decimeter amounted to 2.4 ohms, and the current-strength, which in this case also represents the current-density, had been 0.4 ampère, an increase of the electrode-surface to 2 square decimeters will require a current-strength of 0.8 ampère, in order to maintain a current-density of 0.4 ampère per square decimeter. The resistance of the bath then declines from 2.4 ohms to 1.2 ohm. According to the laws of Ohm, the resistance of 2.4 ohms required an electro-motive force of current-strength × resistance, hence of 0.4 × 2.4 = 0.96 volt. After increasing the electrode-surfaces to 2 square decimeters and raising the current-strength to 0.8 ampère, the resistance declined to 1.2 ohm. The electro-motive force then amounts to, $0.8 \times 1.2 = 0.96$ volt, hence to exactly the same as in the first case.

From this it follows, that with an unaltered electrode-distance, the current-density remains unchanged with varying electrode-surfaces, if the electro-motive force at the bath be kept constant at the same height.

It is also obvious that with an increasing electro-motive force at the bath, the current-density must also increase, because, according to the law of Ohm, the current-strength is equal to the electro-motive force divided by the resistance. Since the latter is not changed when the electrode-distance remains the same, the quotient will be adequately larger if the divisor remains the same and the dividend be increased. Hence the current-density becomes greater.

Now, as for the production of a useful deposit, a certain current-density should not be exceeded, the voltmeter furnishes us the means to insure against failure by keeping the electro-motive force at the bath constant with a varying charge of the latter, and such an instrument should not be wanting in an electro-plating establishment.

Conductors. The most suitable material for conducting the current is chemically pure copper, its conducting power being next to that of silver, but the use of the latter noble metal for this purpose is of course excluded by reason of its costliness.

The laws of Ohm have shown us that the current-strength depends on the magnitude of the electro-motive force and the resistance in the circuit; the greater the resistance, the less the current-strength which can flow through the conductor. From this it follows that in order to reduce losses of electro-motive force to a minimum, conductors of adequate cross-sections should be selected.

Conductors which cause a loss of more than 10 per cent. of the electro-motive force have to be considered insufficient as regards dimensions, and it is recommended to entrust the installation of such constructions only to competent hands capable of making the calculations required for the purpose.

In addition to the correct dimensions of the conductors, the mode of mounting them also deserves the greatest attention. All the connections of the conductors, which are called *contacts*, must be made in the most careful manner, since bad contacts cause a transition-resistance, and, in such a case, a large decrease in electro-motive force could not be prevented even by conductors of ample dimensions.

A distinction is made between main conductors and branch conductors, the former effecting the transmission of the current from the source of current to the baths, while the latter branch off from them to the separate baths.

The positive main conductor or anode conductor is connected with the + pole of the source of current, and the negative main conductor or object-conductor with the — pole.

Both bare and insulated conductors are used. For conductors of larger cross-sections, bright electrolytic copper in the form of round bars or flat rails is employed, while for conductors of smaller cross-sections, copper wire covered with an insulating material, such as hemp or jute coated with asphalt or varnish suffices. For connecting certain movable parts with the rigid main conductor, flexible cables of copper wire, either bare or insulated, are very convenient.

Bare conductors must be fixed in such a manner that they do not touch each other, which would cause short-circuiting, and possibly danger of fire, nor come in contact with damp brick-work. This is effected by placing the conductors upon porcelain insulators, to which they are secured by wire.

It is also advisable not to allow even thoroughly insulated conductors to lie directly one upon the other, as the insulation may happen to be damaged, and short-circuiting would result.

As regards the dimensions of the conductors, it should, in view of the slight electro-motive force of the current used for electrolysis, be made a rule to calculate for every ampère current-strength one square millimeter of copper cross-section, if the entire circuit is not over 20 meters long.

Connection of main conductors and branch conductors is effected by inserting the ends of two round conductors in couplings, Fig. 54, securing them by means of screws, and filling any intermediate space with solder.   If the round main conductors are to be run at an angle, the coupling, Fig. 55, is used, and the **T**-coupling, Fig. 56, is employed on the points from which branches are to be run at a right angle from the main conductor.

Flat copper rails are connected in the most simple manner by means of a piece of copper-sheet and screws, the contact surfaces having been first tinned to prevent oxidation.

Fig. 54.   Fig. 55.   Fig. 56.

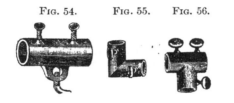

Tanks.   The choice of material for the construction of tanks to hold the plating solutions depends on the nature and properties of the latter.

Solutions containing potassium cyanide require tanks of stoneware, enameled cast-iron or impregnated wood.   Welded steel tanks constructed by the oxy-acetylene welding process are also largely used for cyanide solutions, soap solutions, electric cleaners, etc.   Nickel baths and other baths which do not attack pitch and wood may be kept in wooden tanks lined with pitch.   The best material for wooden tanks without pitch lining is pitch-pine, it containing least tannic acid.   Larch may also be used, but is inferior to pitch-pine.   Wood which contains tannic acid spoils every nickel bath, causing dark nickeling, so that, for instance, an oak tank cannot be used. For smaller baths, up to 300 quarts, the most advantageous tank is one of stoneware or enameled iron.

Wooden tanks must be carefully constructed, and should be securely clamped together with strong iron bars, riveted and bolted, as shown in Fig. 57. The tank is then coated with a mixture of equal parts of pitch and rosin boiled with a small quantity of linseed oil. Another mixture, which has been found to afford a good protective covering to wood, consists of 10 parts of gutta-percha, 3 of pitch, and 1½ each of stearine and linseed oil, melted together and incorporated.

For large acid *copper* and *nickel baths* wooden tanks lined with chemically pure sheet-lead about 0.118-inch thick, and the seams soldered with pure lead, are quite suitable. Care must,.

FIG. 57.

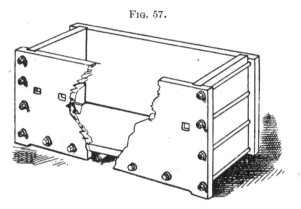

of course, be taken that neither the conducting rods nor the articles suspended in the bath and the anodes come in contact with the lead lining, and therefore the conducting rods should not be laid directly upon the tanks, but placed upon a few thick strips of dry wood. Further, the anodes should be suspended at a sufficient distance from the lead lining, because with too small a distance, metal from the solution is precipitated upon the lead lining. The latter always becomes electric, which, however, does not matter, and if the anodes are at a greater distance from it than the objects no metal is precipitated upon it. If for the better exhaustion of the baths the anodes are suspended at a slight distance from the sides, it is advisable to protect the lead lining with thin wooden boards, or to insulate it by giving it two coats of asphalt-lacquer. However, for this purpose asphalt-lacquer prepared from the residues of the tar industry is not available, and a solution of Syrian asphalt, with a small quantity of Venice turpentine in benzine should be employed.

Based upon careful investigations, such lead-lined tanks have been used for large copper and brass baths containing potassium cyanide without the slightest injury to the baths. If even a film of lead cyanide is formed upon the lead, it is insoluble in excess of potassium cyanide, and hence is entirely indifferent as regards the bath. However, for nickel baths containing large quantities of acetates, citrates and tartrates, these lead-lined tanks cannot be recommended, since these salts possess a certain power of dissolving lead oxide. However, the use of such baths has been almost entirely abandoned, and the small quantities of organic acid which occasionally serve for correct ing the reaction of a nickel bath need not be taken into con sideration. The lead lining might be dispensed with if it were not for the difficulty of keeping wooden tanks tight. Many plating solutions impair the swelling power of the wood, and with even a slight change in the temperature the tanks become pervious, the evil in time increasing. Tanks lined with lead, on the other hand, remain tight, and have the advantage that the baths can be boiled in them by means of steam introduced through a lead coil in the tanks.

For large baths containing potassium cyanide, holders of brick laid in cement may also be used, or holders of boiler-plate lined with a layer of cement. For nickel baths cement-lined tanks cannot be recommended. If a tank of that kind is to be used, direct contact of the nickel solution with the cement lining should be prevented by applying to the latter at least two coats of asphalt-lacquer. Stoneware tanks do not bear heating.

When using lead steam coils or loops in plating tanks or those arranged for electric cleaning, the coil ends entering and returning from the solution should be connected to the heating system with insulating joints, Fig. 58, in order to prevent leakage of the electric current.

FIG. 58.

Conducting fixtures. These include the conducting rods which serve for suspending the objects and anodes, and are laid across the tanks, as well as the binding-posts and screws and copper-connections used for connecting the conducting rods.

The cross-sections of the conducting rods are, on the one hand, dependent on the maximum current-strength which without greater resistance is to pass through them, and, on the other, on the weight of the objects and anodes to be suspended in the bath. The conducting rods may be drawn of hard copper, or for not considerable current-strengths may be made of brass or copper tubing with insertions of iron rods. Bi-metal, *i. e.*, iron rods upon which has been deposited by electrolytic methods a coat of copper adequate to the current-strength, may be highly recommended. By reason of the intimate union of the copper with the iron, the latter takes part in the conduction, which is, as a rule, not the case with copper tubes with insertions of iron rods, in consequence of the formation of oxide and defective contacts.

It is advantageous to provide the narrow sides of the tanks with semicircular notches for the conducting rods to rest in, to prevent their rolling away. When using stoneware tanks the conducting rods are laid directly upon the tanks. Tanks of other material must be provided with an insulated rim of wood, or the rods are insulated by pushing pieces of rubber tubing over their ends. According to the size of the bath, 3, 5, 7, or more conducting rods, best of pure massive copper, or if this is too expensive, of strong brass tubing with iron rods inside, are used.

The rods carrying the anodes, as well as those carrying the objects, must be well connected with each other. This is

·effected by means of binding-posts and screws of the improved
forms shown in Fig. 59, Nos. 1 and 2 being rod connections
for tanks. No. 4, or double connection, is a very convenient
·form, as it can be adapted to so very many changes. The

FIG. 59.

No. 1. No. 3.

No. 4.

No. 2.

three-way connection, No. 3, is so well known that it hardly
needs an explanation.

Arrangement of objects and anodes in the bath. To secure
the uniform coating of the objects with metal they must be
surrounded as much as possible by anodes, *i. e.,* the positive-

pole plates of the metal which is to be deposited. For flat
objects, it suffices to suspend them between two parallel rows
of anodes, the most common arrangement being to place three
rods across the bath, the two outermost of which carry the
anodes, while the objects are secured to the center rod. For
wide baths five conducting rods are frequently used, but they
should always be so arranged that a row of objects is between
two rows of anodes. The arrangement frequently seen with
four rods across the baths, of which the outermost carry anodes,
and the other two, objects, is irrational if the objects are to be
uniformly plated on all sides, because the sides turned towards
the anodes are coated more heavily than those suspended
opposite to the other row of objects.

For large round objects it is better to entirely surround
them with anodes, if. it be not preferred to turn them frequently, so that all sides and portions gradually feel the effect
of the immediate vicinity of the anodes. (See "Nickeling.")

For objects to be plated on one side only the center rod may
be used for the anodes and the two outer ones for the objects;
the surface to be plated being, of course, turned towards the
anodes.

There should be an ample supply of anodes in the bath.
In baths of base metals the anode-surface should at least be
equal in size to the surface to be plated; an exception being
permissible in gold and silver baths.

The anodes should not be too thin, because the thinner
they are, the greater the resistance. Copper, brass and nickel
anodes should not be less than 3 millimeters thick, and the
hooks by which they are suspended should be correspondingly
thick and numerous.

The anodes are suspended from the cross-rods by strong
hooks of the same metal, so that they can be entirely immersed in the bath (Fig. 60). Hooks of another soluble metal
would contaminate the bath by dissolving in it, and this
must be strictly avoided, as it would cause all sorts of
disturbances in the correct working of the bath. In case

hooks of another metal, except platinum, are used, the anodes must be suspended so that they project above the surface of the liquid, and the hooks not being immersed, are therefore, not liable to corrosion; but the anodes are then not completely used up, the portion dipping in the solution being gradually dissolved, whilst the portion projecting above the fluid remains intact. Instead of wire hooks, strips of the same metal as the anodes and fastened to them by a rivet may also be used (Fig. 61).

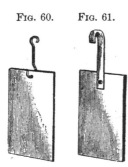

Fig. 60.   Fig. 61.

For suspending the objects, lengths of soft, pure copper wire, technically called *slinging wires*, are used. They are simply suitable lengths of copper wire of a gauge to suit the work in hand, wire No. 20 Birmingham wire gauge being generally employed for such light work as spoons, forks and table utensils. Wire of a larger diameter should be employed for large and heavy goods. The immersed ends of these wires becoming coated with the metal which is being deposited, they should be carefully set aside each time after use, and when the deposit gets thick it should be stripped off in stripping acid, and the wire afterwards annealed and straightened for future use.

To keep the rods clean and to protect them from the fluid draining off from the articles when taken from the bath, it is advisable to cover them with a roof of strips of wood (∧), or a semi-circular strip of zinc coated with ebonite lacquer; by this means the frequent scouring of the rods, which otherwise is necessary in order to secure a good contact with the hooks of the anodes, is done away with.

It need scarcely be mentioned that the anodes and the objects to be plated must not touch each other, as short-circuiting would take place on the point of contact.

The plating solutions, briefly called baths or electrolytes, will be especially discussed in speaking of the various electro-plating processes. Other rules for suspending the objects will be mentioned under " Nickeling," and are valid for all other electro-plating processes.

Apparatus for cleansing and rinsing. It remains to consider the cleansing and rinsing contrivances, without which it would be impossible to carry on electro-plating operations. Every electro-plating establishment, no matter how small, requires at least one tub or vat in which the objects can be rubbed or brushed with a suitable agent in order to free them from grease. This is generally done by placing a small kettle or stoneware pot containing the cleansing material at the right-hand side of the operator alongside the vat or tub. Across the latter, which is half filled with water, is laid a board of soft wood covered with cloth, which serves as a rest for the objects previously tied to wires. The objects are then scrubbed with a brush, or rubbed with a piece of cloth dipped in the cleansing agent. The latter is then removed by rinsing the objects in the water in the tub and drawing them through water in another tub. By this cleansing process a thin film of oxide is formed upon the metals, which would be an impediment to the intimate union of the electro-deposit with the basis-metal. This film of oxide has to be removed by *dipping* or *pickling*, for which purpose another vat or tub containing the *pickle*, the composition of which varies according to the nature of the metal, has to be provided. After dipping, the objects have to be again thoroughly rinsed in water to free them from adhering pickle, so that for the preparatory cleansing processes three vessels with water, which has to be frequently renewed, as well as the necessary pots for pickling solutions, have to be provided.

Larger plants require a special table for freeing the objects from grease. Such a table is shown in Fig. 62. It consists of a box furnished with legs, and is divided by four partitions into two larger and three smaller compartments. Boards covered with cloth are laid over the larger compartments, upon which the objects are brushed with lime-paste for the final thorough

11

freeing from grease. Over each of these compartments is a rose provided with a cock, under which the objects are rinsed with water. The outlets for the waste water from the large compartments are in the bottom of the box and are provided with valves. Of the smaller compartments, the one in the center serves for the reception of the lime-paste (see " Chemical Treatment "), while the others contain each two pots or small stoneware tanks with pickling fluid. In Fig. 66 these

FIG. 62.

tanks are indicated by 11 and 12. The two marked 11 contain dilute sulphuric acid for pickling iron and steel articles, while those marked 12 contain dilute potassium cyanide solution for pickling copper and its alloys, and Britannia, etc. For cleansing smaller articles, four men can at one time work on such a table; but for cleansing larger articles only two. For an establishment which does not require such a large table, one with a larger and two smaller. compartments may be used. The advantages of such a box-table are that everything is handy together; that the pickle, in case a pot should break, cannot run over the floor of the workshop; and that the latter is not ruined by pickle dropping from the objects. The small box on the side of the table serves for the reception of the various scratch-brushes.

After having received the electro-deposit, the objects have to be again rinsed in cold water, which can be done in one of the three tanks or with the rose-jet, and finally have to be immersed in hot water until they have acquired the temperature of the latter. How the water is heated makes no dif-

Fig. 63.

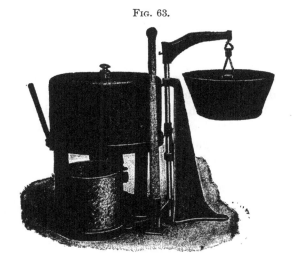

ference, and depends on the size of the establishment. The heated objects are then immediately dried in a box filled with dry, fine sawdust that of boxwood, maple, or other wood free from tannin being suitable for the purpose. A steam sawdust box very suitable for the purpose is made in four removable sections, which consist of a smooth galvanized iron box, hot air chamber with asbestos lining closely built, $\frac{3}{4}$-inch steam radiator, and a rigid stand made of $1\frac{1}{4}$-inch angle iron.

To overcome various troubles and difficulties connected

with drying by means of sawdust mixed with the articles placed in the pan and heated, steam drying barrels have been introduced. One type is practically the same as the oblique tilting tumbling barrel in common use for cleaning metallic surfaces, except that it is jacketed and otherwise constructed to allow a circulation of steam about the inner barrel, automatic ejection of the condensation, still allowing the barrel to be tilted. A barrel load can be thoroughly dried in a few minutes, especially if the work is shaken out of hot water. It will be readily understood that the rolling over and over of the hot barrel thoroughly mixes the work and sawdust, liberates the steam and precludes the possibility of water-marks, etc., and further, brightens the goods at the same time.

A centrifugal dryer for small work, supplied by The Hanson & Van Winkle Co., N. J., is shown in Fig. 63. This machine should be used where mechanical plating apparatus is installed, one to three minutes only being necessary for drying small work. The machine is furnished with or without hoist, and is fitted with ball bearings. It can be supplied with a tapered steel pan or a perforated straight-sided steel or copper basket for holding the work.

B. Installations with Dynamo-Electric Machines.

Setting up and running a dynamo. Most of the troubles with plating-dynamos are caused by neglecting one or more of the conditions necessary for their proper operation, and are not due to any defects in the machines themselves. The troubles most frequently encountered are, in order of their frequency, as follows: First, insufficient or variable speed. Second, improper setting of the brushes, and the use of improper lubricants and cleaning material on the commutator. Third, poor oil, or an insufficient, or too great, an amount of oil in the bearings. Fourth, overloading the machine.

It is important that the dynamo be properly placed, and the following considerations should govern the choice of location : The dynamo should not be exposed to moisture nor to the dirt.

and dust of the polishing room. Cleanliness is a necessity. A cool, well-ventilated room should be chosen. This is important, for a well-ventilated machine will do more work with less wear on parts than one unfavorably placed. The machine should not be boxed in, as this will make it run hotter than it otherwise would. Not only this, the mere fact of having it totally boxed in precludes the probability of receiving the proper amount of attention.

Except on the larger sizes of machines a special foundation is not mechanically necessary, providing the floor is fairly solid. On account, however, of dirt getting into the running parts when the floor is cleaned, it is always well to raise the machine from six to twelve inches above the floor. For a small dynamo a well-made box of two-inch lumber will afford an ample foundation. For the larger sizes two or three strips of 6-inch x 6-inch yellow pine may be used. In either case the box or strips should be solidly nailed or bolted to the floor and the machine secured to its base with four lag screws of the proper size.

The direction of rotation may be ascertained by an inspection of the brushes, the commutator running away from the brushes. One of the troubles mentioned above, namely, variable speed, may be remedied to a large extent by a suitable belt, run in the proper manner. The counter-shaft should never be run directly over the dynamo, but should be placed far enough to one side so that the belt will run *diagonally* and in such a direction that the *under side* of the belt does the work. This is on account of the fact that when the belt is running vertically or diagonally with the upper side doing the work it stretches and sags away from the pulley when a heavy load is thrown on the dynamo, thus giving less pull as the necessity for a greater pull increases. Use good, pliable, single belting with the hair side of the belt to the pulley on smaller and medium-size machines. For the larger sizes a thin, double belt may be used.

After the machine has been properly set and belted, it remains to start it up. Before starting, remove the bearing caps and pour a small quantity of oil on the bearings; loosen the screw holding the rocker-arm in position, and be prepared to shift the rocker-arm backward or forward, so as to get the brushes on the neutral or non-sparking line, as it often happens that the rocker-arm has been shifted from its proper position in transportation.

The proper position for the tips of the brushes on all machines of either the bipolar or multipolar type is about opposite the center of the poles. The tips of the brushes should also be spaced at even intervals, this being, on the two-pole machine, diametrically opposite to each other; on the four-pole machine one-quarter of the circumference from each other; on the six-pole machine, one-sixth of the circumference, and so on. The exact position of the brushes (that is, where they run spark-lessly) can only be ascertained by trial. This adjustment should be made, in case it is necessary, as soon as the machine starts up, for if it is allowed to run any length of time while sparking the commutator will be cut badly, and may necessitate taking out the armature and truing up the commutator. In case this is necessary, a sharp diamond-point tool should be used with a moderate speed, and the commutator should be finished with a fine second-cut file, and then with No. 0 sandpaper and oil. After the proper adjustment of the brushes has been made, take an oil-can, and while the machine is running, pour oil *slowly* into the oil-well until the oil-rings take it up properly and carry it to the top of the bearings, where it enters the distributing slot. If *too little* oil is in the well and the rings do not dip into it sufficiently deep, they will rattle around and spatter oil, whereas if *too much* oil is put in, it will run out at the ends of the bearings and get into the belt, winding and commutator of the machine.

While the commutator should never be allowed to become greasy or dirty, it is equally important that it should not be run perfectly dry, so that the brushes cut. When it becomes dirty, after cleaning with No. 0 sandpaper (emery should never be used) it should be re-oiled by rubbing it with a woolen cloth moistened with kerosene oil, or with the very smallest amount of lubricating oil. The quality and kind of oil used for the bearings is important, and a regular dynamo oil should be used. Under no circumstances should *vegetable or animal oil* (such as castor or sperm oil) be used, but a light grade of mineral dynamo oil.

The brushes should not only be properly set as regards their position around the commutator, but they should have careful individual setting. They should have a fair and even bedding on the commutator, and not touch on the heel or toe or on either edge, as the object is to get full contact surface between the brushes and the commutator. If the commutator is kept in proper shape and the brushes once properly set, it will not be necessary to adjust them often. As it is practically impossible to make a perfectly accurate setting of the brushes, and it takes them some few days to get worn down to a good contact, it will be seen that it does more harm than good to be continually re-setting them. If the ends of the brushes get very ragged, they should, in the case of wire-gauze brushes, be carefully trimmed with a pair of shears, and in the case of strip-copper brushes, filed with a fine second-cut file. The tension spring on the brush holder should be adjusted to make a light but positive contact, for if there is too much pressure, the brushes will cut the commutator, causing it to wear away rapidly.

If there is any doubt about which is the positive and which is the negative pole of the dynamo, the polarity may be readily determined after starting up, by running two small wires from the dynamo and placing the ends in a glass of acidulated water. Around one of these wires more bubbles of gas will be thrown off than around the other, the one evolving the greater amount of gas being the negative pole, to which the work should be attached.

Choice of a dynamo. For electrolytic processes, as previously mentioned, shunt-wound and compound-wound dynamos are at present largely used. Their construction has already been explained, and there remains now only the question what size dynamo, *i. e.*, of what capacity as regards current-strength and electro-motive force, is to be selected for a plant.

We have learned that a certain object-surface requires a certain current-strength. Hence for plants with different baths, it is only necessary to fix the largest object-surface in square decimeters which is to be suspended in the separate baths and to multiply this number of square decimeters by the current-density in ampères, in order to find the supply of current required for each bath. The sum of the current required for the separate bath, with an allowance of 20 to 25 per cent. for an eventual enlargement, gives the current-strength the dynamo must furnish. It must of course be taken into consideration whether all the baths are to be in constant operation at the same time or not. In the latter case a smaller current-strength will of course suffice, and a smaller type of dynamo answer the purpose.

The impressed electro-motive force of the dynamo should be such that, taking into consideration the decline of the electro-motive force in the conductors, it is, at the greatest current-capacity, about $\frac{1}{4}$ to $\frac{1}{2}$ volt greater than the highest electro-motive force of a bath required.

For the purpose of explaining by an example the choice of a suitable dynamo, let us suppose that a nickel bath with an object surface of 50 sq. decimeters; a potassium cyanide copper bath˙ with an object surface of 30 sq. decimeters; a brass bath with an object surface of 40 sq. decimeters; a silver bath with an object surface of 10 sq. decimeters, are to be fed with current.

The standard current-densities and electro-motive forces required for the separate baths are given later on when speaking of them. It will there be found that the current-density for nickeling brass amounts to about 0.4 ampère, the electro-motive force being 2.5 volts; for coppering, 0.35 ampère and 3.0 to 3.5 volts are required; for brassing also 0.35 ampère and 3.0 to 3.25 volts; while for silvering 0.2 ampère and 1 volt are on an average used. This amounts to

For nickel bath 50 sq. decimeters $\times$ 0.4 ampère $=$ 20 ampères.
For copper bath 30 sq. decimeters $\times$ 0.35 ampère $=$ 10.5 ampères.
For brass bath 40 sq. decimeters $\times$ 0.35 ampère $=$ 14 ampères.
For silver bath 10 sq. decimeters $\times$ 0.2 ampère $=$ 2 ampères.

46.5 ampères.

Hence 46.5 ampères are required for the simultaneous operation of these four baths, and a dynamo of 50 ampères. current-strength and 4 volts impressed electro-motive force would have to be selected, since, taking into consideration, a permissible decline of electro-motive force of 10 per cent. $=$ 0.4 volt in the conductors, there are still at disposal 3.6 volts, while the greatest electro-motive force required amounts to 3.5 volts.

Since the various baths of a larger establishment possess different resistances and cannot always be charged with the same object-surfaces, they have to be operated in parallel. This renders it necessary that for each separate bath working with a lower electro-motive force, the excess of electro-motive force as existing in the main conductor has to be destroyed by a resistance, called the main-current regulator or bath-current regulator. Hence as many main-current regulators must be provided as there are baths, and the regulators have to be exactly calculated and constructed for the required effect. Thus in the above-mentioned example, the bath-current regulators, with an electro-motive force of 3.6 volts in the main conductor, must let pass for a nickel bath 20 ampères and destroy 1.1 volts; let pass for a copper bath 10.5 ampères and destroy 0.6 to 0.35 volt; let pass for a brass bath 14 ampères and destroy 0.6 to 0.35 volt; let pass for a silver bath 2 ampères and destroy 2.6 volts.

Since every destruction of electro-motive force means an economic loss, it follows that the impressed electro-motive force of the dynamo should not be greater than absolutely necessary, so that it can be reduced by a regulator to the lowest permissible limit, and this limit should be constantly maintained. Thus, when the electrode surfaces in the bath are changed, and there is consequently also a change in the impressed electro-motive force, the latter can be properly adjusted by the regulator. If this were not done, and the impressed electromotive force would become considerably greater, the bath-current regulators calculated for the destruction of a fixed electro-motive force would no longer be capable of fulfilling their objects From what has been said, it will be seen that voltmeters are indispensable for electro-plating plants in order to be constantly informed as to the electro-motive force prevailing at the baths, and, if necessary, to correct it.

By reason of the economic loss connected with the destruction of an excess of electro-motive force, it may also have to be taken into consideration whether in larger plants it would not be better to use several dynamos with different impressed electro-motive forces than a single dynamo with an impressed electro-motive force required for the greatest electro-motive force for the baths. Suppose, for instance, there are present in a larger plant, in addition to nickel, brass and copper cyanide baths, which require a voltage of up to $3\frac{1}{2}$ volts, a large number of silver and tin baths and acid copper baths for galvanoplasty (with the exception of those for rapid galvanoplasty), for which an impressed electro-motive force of 2 volts is quite sufficient, it would by all means be more judicious to use for the first-named baths a special dynamo with an impressed electro-motive force of 4 volts, and for the last-mentioned baths a dynamo with a voltage of 2 volts.

Another question to be considered in the choice of a dynamo is, whether one or several accumulator cells are to be charged from it. This will be later on referred to.

While, when baths are coupled in parallel, each bath receives its supply of current from the main conductor, and such parallel coupling is always required when baths of different nature, with unequal resistances and unequal electro surfaces, are connected, baths requiring an equal, or approximately equal,. current-strength may be coupled one after the other, i. e., in series. This principle of series-coupling of baths is illustrated by Fig. 64.

The current passes through the anodes of the first bath into. the electrolyte, flows through the latter and passes out through the object-wire. From there it goes through the anodes of the next bath to the objects contained in it, and so on, until it returns through the object-wire of the last bath to the source of current.

Thus for series coupling of the baths, a dynamo with a

FIG. 64.

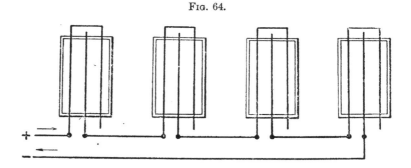

greater impressed electro-motive force than the sum of the electro-motive forces of all the baths coupled one after the other has to be selected. On the other hand, baths coupled one after the other do not require a greater current-strength than a single bath. Suppose. four baths, each charged with 100 square decimeters of cathode- and anode-surfaces are coupled one after the other, and the electro-motive force of one bath amounts to 1.25 volts and the current-density to 2 ampères. Then there will be required for one bath $100 \times 2 = 200$ ampères and 1.25 volts, and for four baths coupled one after another, 200 ampères and $1.25 \times 4 = 5$ volts.

The connection of the baths, resistance boards and measuring instruments to a shunt-wound dynamo is shown in Fig.

65, and requires no further explanation. The resistance board at the right is the field resistance board, the other two belonging to the two baths which are coupled in parallel.

Parallel coupling and series coupling of dynamo-machines. In establishing a larger electro-plating plant, the question may arise whether it would not be advisable to install two smaller dynamos instead of a single larger one capable of filling all demands, even at the busiest season. The installation of two dynamos allows of the business being carried on without serious interruption in case one of the machines requires repairing, and in dull times one dynamo would, as a rule, be sufficient. In case two dynamos are installed, the main conductors must of course have the required cross-sections corresponding to the total current-strength of both machines.

It, however, happens very frequently that as the plant becomes larger by reason of an increase in the number of baths, a larger supply of current will in time be required. The question then arises whether to sell the old dynamo, which may be difficult, especially if it is of an obsolete pattern, or whether to supply the deficit of current by installing an additional dynamo. In such case, if the baths are not to be divided into groups, one of them being furnished with current from one dynamo and the other from the second machine, but both the dynamos are to be connected to a common main conductor, the cross-section of the latter must first of all be increased so as to be capable of carrying the total current-strength of both dynamos without material decrease in electromotive force. Whether for this purpose a new conductor of larger cross-section is to be used, or whether a supplementary conductor is in a suitable manner to be connected with the old one, is best left to the judgment of the person entrusted with the installation.

In coupling several dynamos in parallel to a common conductor, care must in all cases be taken to connect a dynamo to one already in operation only after it had been excited to the same voltage. If this were not done, the current of greater

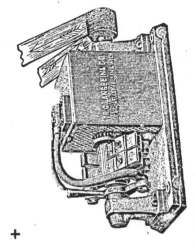

COUPLING SCHEME FOR SHUNT-WOUND DYNO.

electro-motive force of the dynamo in operation would flow from the main conductor to the other dynamo, and the first dynamo would thus be short-circuited by the brushes, commutator and armature of the second one. No current would pass into the baths, but the second dynamo would run as a motor. To prevent this, a switch has to be placed between every dynamo and the main conductor. If one dynamo already furnishes current, the second dynamo has at first to be set in operation with the switch open, until its voltmeter shows the same voltage as possessed by the other dynamo. The switch is then closed, and the desired current-strength generated by means of the shunt-regulator. It is obvious that for coupling in parallel, only dynamos which yield the same voltage are suitable, while a difference in capacity as regards current-strength is no obstacle.

The poles of a similar name of the various machines must of course be connected to one and the same circuit.

Coupling of dynamos in series may become necessary when baths require a greater electro-motive force than can be furnished by a single machine, for instance, in case baths are coupled one after the other. For coupling in series only dynamos which furnish with the same voltage the same current-strength are suitable. Coupling is effected so that the + pole of one dynamo is connected with the — pole of the other one, hence in the same manner as cells and accumulators are coupled.

Coupling in series of dynamos may also be used if there are baths requiring great electro-motive force, for instance, for plating *en masse* in the mechanical apparatus (see later on), while baths requiring a considerably lower electro-motive force are to be fed from the same source of current.

In such case it is advisable to construct the conductors according to the three-wire system. One conductor is branched off from the + pole of one dynamo, the second from the — pole of the other dynamo, and the third, called the neutral or middle conductor, from the junction of the dyna-

mos coupled in series. Between the last-mentioned neutral
conductor and an outside conductor is the lower electro-
motive force as furnished by one dynamo, but between the
two outside conductors, the sum of the electro-motive forces of
both dynamos. Hence the baths requiring a large electro-
motive force are to be coupled between the outside conductors,
and the baths requiring a low electro-motive force between an
outside and the neutral conductor.

Ground plan of an electro-plating plant with dynamo. This
in the most simple form is shown in Fig. 66. In order to
make the sketch more distinct, the measuring instruments
have been omitted. Their arrangement will be understood
from what has been previously said, and from Fig. 66.

NN^1 is a dynamo-electric machine of older construction.
The resistance-board belonging to the machine, which is
placed in the conductor, is indicated by No. 1, and is screwed
to the wall. The main conductors, marked — and +, run
along the wall, from which they are separated by wood, and
consist of rods of pure copper 0.59 inch in diameter. The
rods are connected with each other by brass coupling-boxes
with screws. From the negative pole and the positive pole of
the machine to the object-wire and anode-wire lead two wires,
each 0.27 inch in diameter; one end of each is bent to a flat
loop and secured under the pole-screws of the machine, while
the other ends are screwed into the second bore of the binding-
screws screwed upon each conductor. To the right and left of
the machine the baths are placed, *Zn*, indicating zinc bath;
Ni Ni, nickel baths; *Ku*, copper cyanide bath; *Mg*, brass
bath; *S K*, acid copper bath; *Si*, silver bath; and *Go*, gold
bath. Each of the first-named five baths has its own resistance-board, designated by 2, 3, 4, 5, 6. However, before
reaching the acid copper bath, and the silver and gold baths,
the current is conducted through two resistance-boards, 7 and
8. Since these baths require a current of only slight electro-
motive force, it is necessary to place two, and in many cases
even three or four resistance-boards, one after another, unless

it be preferred to feed these baths with a special **machine of** less voltage.

Fig. 66.

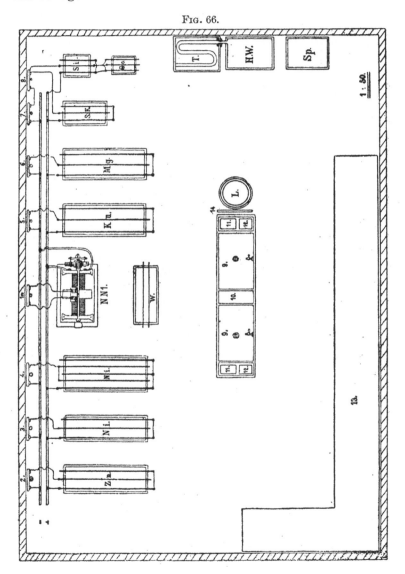

From Fig. 66 it will be seen that the current weakened by the resistance-boards 7 and 8 serves for conjointly feeding the acid-copper, silver, and gold baths.. Hence, practically, only one bath can be allowed to work at one time, as otherwise each bath would have to be provided with as many resistance-boards as would be required for the reduction of the electro-motive force. For want of space the gold bath is placed in the sketch behind the silver bath ; but as their resistances are not the same, they must also be placed parallel.

L is the lye-kettle. It serves for cleansing the objects by means of hot caustic potash or soda lye, from grinding and polishing dirt and oil. For larger plants the use of a jacketed kettle is advisable. By the introduction of steam in the jacket the lye is heated without being diluted. The same object is attained by placing a steam coil upon the bottom of the kettle. Of course, heating may also be effected by a direct fire. Instead of the preparatory cleansing with hot lye, which saponifies the oil, the objects may be brushed off with benzine, oil of turpentine or petroleum, the principal thing being the removal of the greater portion of the grease and dirt, so that the final cleansing, which is effected with lime paste, may not require too much time and labor. It is also advisable to cleanse the objects, in one way or the other, immediately after grinding, as the dirt, which forms a sort of solid crust with the oil, is difficult to soften and to remove when once hard.

The table which serves for the further cleansing of the objects has already been described on p. 161, and illustrated by Fig. 62.

Referring again to Fig. 66, between the lye-kettle L and the box-table, is a frame, 14, for the reception of brass and copper wire hooks of various sizes and shapes suitable for suspending the objects in the bath.

The reservoir W, filled with water, standing in front of the machine, serves for the reception of the cleansed and pickled objects, if for some reason or other they cannot be immediately brought into the bath.

12

$H\ W$ is the hot-water reservoir in which the plated objects are heated to the temperature of the hot water, so that they may quickly dry in the subsequent rubbing in the saw-dust

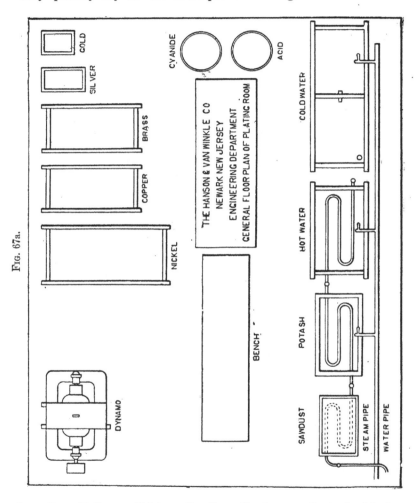

Fig. 67a.

box Sp. Before polishing the deposits, iron and steel objects are thoroughly dried in the drying chamber T (Fig. 66), heated either by steam or direct fire. By finally adding to the appliances a large table, 13, for sorting and tying the objects on the copper wires, and a few shelves not shown in the illustration, everything necessary for operating without disturbance will have been provided.

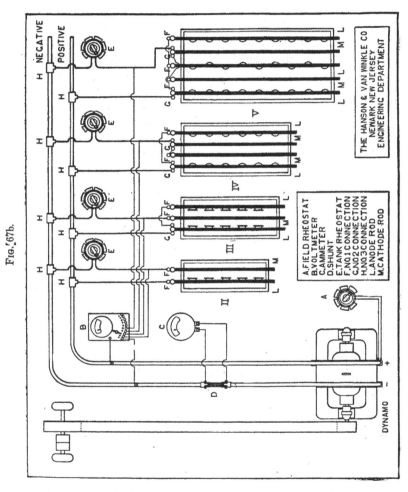

Fig. 67b.

Figs. 67a and 67b show a plating room and method of connecting dynamo, tanks and instruments according to the two-wire system as fitted up by The Hanson & Van Winkle Co., Newark, N. J. The arrangement will be readily understood from the illustrations, so that a detailed description is not necessary.

The three-wire system of current distribution has been generally adopted in the larger plants where a variety of solutions are in use. The necessity of shortening time for deposit without deterioration of the quality of work has been apparent; this condition is effected through the agitation of the solution, and the consequent employment of a higher voltage, with proportionate increase in the ampère current. The majority of plating dynamos in use are capable of delivering 4 to 6 volts only, and their use precludes the adoption of the newer labor-saving method. To meet the demands for a generator that will deliver a higher range of voltage, dynamos operating on the three-wire system are built which will deliver a range of voltage up to 12 volts or higher, if so desired. By the use of these dynamos it is possible to take from the machine voltages of two different strengths at the same time, the higher voltage being double that of the lower, and thus provide a high pressure for mechanical plating apparatus, basket work or agitated solutions, and at the same time operate solutions of a low voltage.

In wiring for this system, three main line conductors are used, the positive and negative, or outside lines, and the neutral or middle line. In this method of wiring there is a saving of over 37 per cent. effected in the cost of copper, as it is not necessary to use conductors of so large a cross-section as would be the case in the ordinary two-wire system.

Figs. 68a and 68b illustrate a three-wire system showing plating room wired for the usual plating tanks, and also mechanical plating apparatus. All necessary voltmeters, ammeters and rheostats are shown.

Switch-boards. In the sketch, Fig. 66, the resistance-board belonging to each bath is secured to the wall in the immediate neighborhood of the bath. This arrangement has the advan-

tage that the operator can, directly after suspending the objects,
conveniently effect regulation from the bath itself.   The

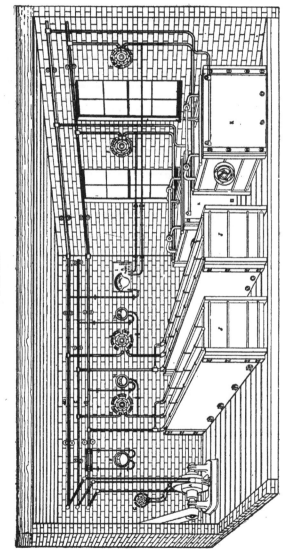

Fig. 68a.

resistances and measuring instruments, as well as the switches, may, however, be also arranged alongside each other on a switch-board. Where a large number of baths are in opera-

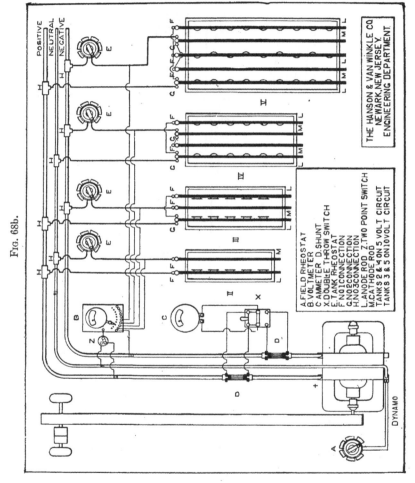

FIG. 68b.

tion, several such switch-boards will, of course, have to be provided to avoid the necessity of the operator, having to walk too great a distance from the bath to the switch-board.

Fig. 69 shows such a switch-board, upon which are mounted the dynamo resistance, the resistance for the accumulator, two resistances for two baths, three ampèremeters, a voltmeter with switch, current indicator, as well as switches for engaging and disengaging the machine as well as the accumulator. For the sake of neatness the conductors and connections are on the back of the switch-board.

A marble slab with a wood rim is the best material for a switch-board. Marble or slate is absolutely required if the arrangements for starting the electro-motors of the aggregates

FIG. 69.

or transformers are mounted upon the switch-board. If, in such a case, the latter were of wood, there would be danger of ignition by reason of the heating of the spirals, etc. Wooden switch-boards may, however, be used for measuring instrument and resistances not especially subject to heating.

The suggestions and directions given in the section " Installation with Cells " as regards regulation of current, measuring instruments, conductors, tanks for solutions, etc., apply also to installations with dynamos, and the reader is referred to that section.

C. Installations and Accumulators.

Only in rare cases will an electro-plating plant be operated by an accumulator alone.

For larger establishments, where deposits requiring many hours for finishing are made, for instance, in copper and nickel galvanoplasty, silvering by weight, etc., the use of an accumulator, in addition to a dynamo, is of great advantage, since the process of depositing need not be interrupted during the noon intermission or, in case it is not finished in the evening, during the night, such interruptions affecting in various ways the quality of the deposit.

However, even for depositing processes requiring a shorter time, for instance nickeling, an accumulator gives the opportunity of turning the working power to better advantage. Suppose, for instance, that bicycle parts which are to be solidly nickeled have to remain in the bath for $1\frac{1}{2}$ hours. However, the steam engine, and consequently the dynamo, are stopped at 12 M. (noon), and hence no objects can be brought into the bath after 10:30 a. m., as otherwise they would not be finished by noon, and an interruption in the nickeling process has to be avoided. If, however, an accumulator can during the noon hour be used for feeding the baths, objects can be suspended up to that time in the baths and taken from them finished when the noon-hour expires and operations are recommenced. The same can be done in the evening, and thus by the use of an accumulator the producing power of an electro-plating plant can be materially increased.

If an accumulator is thus to be made use of, a dynamo of an adequately larger size has to be selected so that in addition to the depositing work, the accumulator can at the same time be charged by direct current from the dynamo.

In establishments where the current is generated by a motor-generator or transformer fed from a central station, an accumulator is a good investment, since in this case the operating current is constantly at disposal and the motor-generators and ransformers can consequently run day and night.

Instead of feeding the baths and the accumulator simultaneously with current from a larger dynamo, two dynamos may, of course, also be used, one of them supplying the baths and the other the accumulator.

The magnitude of the performance of an accumulator depends on the current-strength which it is to yield for a certain time. As previously stated, the value ampère-strength × hours is called the ampère-hour capacity of an accumulator. Hence the question arises for how long the accumulator is to do the work of the dynamo while the latter is not running, and what current-strength is during this time to be transmitted from the accumulator to the bath. If now this ampère-hour capacity is known, as well as the maximum current-strength required for feeding the bath from the dynamo, we also know the current-strength which the dynamo must have.

To explain this by an example, we will suppose that the bath requires at a maximum 200 ampères, and that the dynamo has to directly feed the bath for four hours and at the same time charge a cell, which, when the dynamo stops, is to discharge for two hours with 200 ampères to the bath. The cell must therefore have a capacity of 400 ampère-hours, and taking into consideration the fact that for charging at least 10 per cent. more charging current is necessary than corresponds to the discharging current, 440 ampères will have to be used for one hour in order to charge the cell, or 220 ampères for two hours, 140 ampères for four hours. Since the dynamo, previous to being stopped, has directly to yield for four hours to the bath, 200 ampères, there are four hours at disposal for charging the cell, and the dynamo must therefore have a capacity of 200 + 110 = 310 ampères.

The diagram Fig. 69 shows the connection of a plant as installed by the Electro-Chemical Storage Battery Co., of New York.

By suitable manipulation of the switches and rheostats it is possible to make the following connections: 1. The dynamo alone can be used on the baths. 2. The batteries alone can

Fig. 70.

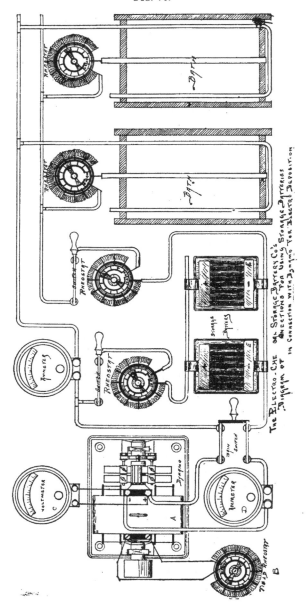

be used on the baths. 3. The dynamo can be used on the baths and the batteries charged with the excess-current, while at the same time steadying the dynamo current. 4. The dynamo and batteries can be used in multiple on the baths, giving a greatly increased capacity.

CHAPTER V.

As previously stated, the metallic objects to be plated have to undergo both a *mechanical* and *chemical* preparation, and each of these processes will be considered separately.

A. MECHANICAL TREATMENT PREVIOUS TO ELECTRO-PLATING.

If the objects are not to be plated while in a crude state, which is but rarely feasible, the mechanical treatment consists in imparting to them a *cleaner surface by scratch-brushing*, or a *smoother* and *more lustrous* one by *grinding* and *polishing*. It may here be explicitly stated that scratch-brushing of plated objects is not to be considered a part of their preparation, since such scratch-brushing is executed in the midst of, or after the plating process, its object being to effect an alteration of the electro-deposits in more than one direction, and not the cleansing of the surface of the metallic base. The following directions, therefore, apply only to scratch-brushing of objects not yet plated. The scratch-brushing of electro-deposits will be considered later on. In regard to grinding, we have to deal with the subject only in so far as it relates to smoothing rough surfaces by the use of grinding powders possessing greater hardness than the metal to be ground. With grinding in the sense of instrument-grinding, the primary object of which is to provide the instrument with a cutting edge, we have nothing to do.

As some platers seem to have wrong ideas regarding the electro-plating process, it may here be mentioned that the deposit is formed exactly in correspondence with the surface of the basis-metal. If the latter has been made perfectly smooth

(188)

by grinding and polishing, the deposit will be of the same nature; but if the basis-surface is rough, the deposit also will be rough. Hence it is wrong to suppose that by electro-plating a rough surface can be converted into a lustrous one, and that pores, holes or scratches in the basis-metal can be filled by plating. In order to obtain a deposit which is to acquire high luster by polishing, it is absolutely necessary to bring the basis into a polished state by mechanical treatment. In doing this it is not necessary to go so far as to produce high

FIG. 71. FIG. 72. FIG. 73. FIG. 74.

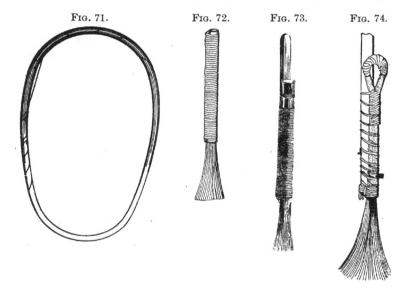

luster, but fine scratches, which would be an impediment to attaining high luster after plating, must be removed.

Scratch-brushing may be effected either by hand or by a scratch-brush lathe. For hand-work, scratch-brushes of more or less hard brass or steel wire, according to the hardness of the metal to be manipulated, are used. Various forms of brushes are employed, the most common ones being shown in the accompanying illustrations (Figs. 71 to 79).

Fig. 78 shows a swing brush for frosting or satin finish, and

Fig. 79 a goblet brush without stem of bristle and wire for use on inside of goblets, pitchers, urns, hollow ware, etc.

In scratch-brushing it is recommended first to remove, or at least to soften, the uppermost hard and dirty crust (the scale) by immersing the objects in a pickle, the nature of which depends on the variety of metal, so that a complete removal of all impurities and non-metallic substances may be effected by means of the scratch-brush in conjunction with sand, pumice-stone, powder, or emery. The composition of pickles will be

FIG. 75.   FIG. 76.   FIG. 77.

FIG 78.

FIG 79.

given later on. Scratch-brushing is complete only when the article shows a clean metallic surface, otherwise the brushing (scouring) must be continued. Scratch-brushes must be carefully handled and looked after, and their wires kept in good order. When they become bent they have to be straightened, which is most readily effected by several times drawing the brush, held in a slanting position, over a sharp grater such as is used in the kitchen. By this means the wires become disentangled and straightened out.

Hand scratch-brushing being slow and tedious work, large establishments use circular scratch-brushes which are attached to the spindle of a lathe. These circular brushes consist of round wooden cases in which, according to requirement, 1 to 6 or more rows of wire bundles (see Fig. 80) are inserted.

Brushes with wooden cases are, however, more suitable for scratch-brushing deposits than for cleansing the metallic base, since for the latter purpose a more energetic pressure is usually applied, in consequence of which the bundles bend and even break off, if the wire is anyways brittle. For cleansing purposes a circular scratch-brush, which the workman can readily refurnish with new bundles of wire, deserves the preference. It is constructed as follows: A round iron disk about 0.11 inch thick, and from 5¾ to 7¾ inches in diameter, is provided in

FIG. 80.

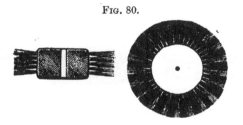

the center with a hole so that it can be conveniently placed upon the spindle of the lathe. At a distance of from 0.19 to 0.31 inch from the periphery of the disk, holes 0.079 to 0.11 inch in diameter are drilled, so that between each two holes is a distance of 0.15 inch. Draw through these holes bundles of wire about 3.93 inches long, so that they project an equal distance on both sides. Then bend the bundles towards the periphery, and on each side of the iron disk place a wooden disk 0.31 to 0.39 inch thick. The periphery of the wooden disk, on the side next to the iron disk, should be turned semi-annular, so that the wooden disks when secured to the spindle press very lightly upon the wire bundles, and the latter remain very mobile. When a circular scratch-brush constructed in this manner and secured to the lathe is allowed to make from 1800 to 2000 revolutions per minute, the bundles of wire, in consequence of the centrifugal force, stand very rigid, but being mobile will give way under too strong a pressure without breaking off, and can thus be utilized to the utmost. When required, the iron disk can be refurnished with wires in less than half an hour. An error frequently committed is that the objects to be cleansed are pressed with too heavy a pressure against the wire brushes. This is useless, since only the sharp points of the wire are effective, the lateral surfaces of the bundles removing next to nothing from the articles.

Brushes. A definition of these instruments is unnecessary, and we shall simply indicate the various kinds suitable to the different operations.

The fire-gilder employs, for equalizing the coating of amal-

FIG. 81.   FIG. 82.   FIG. 83.

gam, a long-handled brush, the bristles of which are long and very stiff. The electro-gilder uses a brush (Fig. 81) with long and flexible bristles.

For scouring with sand and pumice-stone alloys containing nickel, such as German silver, which are difficult to cleanse in acids, the preceding brush, with smaller and stiffer bristles, is used.

The gilder of watch-works has an oval brush (Fig. 82), with stiff and short bristles for graining the silver.

The galvanoplastic operator, for coating moulds with black-lead, besides a number of pencils, uses also three kinds of brushes—the watchmaker's (Fig. 83), a hat brush, and a blacking-brush. The bronzer uses all kinds of brushes.

Brushes are perfectly freed from adherent grease by washing with benzine or carbon disulphide.

In large establishments engaged in electro-plating cast-iron without previous grinding, the use of the sand-blast, in place of the circular wire brush, has been introduced with great advantage. Objects with deep depressions, which cannot be reached with the scratch-brush, as well as small objects, which cannot be conveniently held in the hand and pressed against the revolving scratch-brush, can be brought by the sand-blast into a state of sufficient metallic purity for the electro-plating process.

However, while circular scratch-brushes impart to the ob-

FIG. 84.

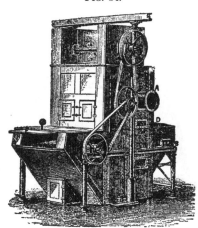

jects a certain, though not very great luster, the metal portions are matted by the sand-blast, and the latter is frequently employed for matting entire lustrous surfaces or for producing contrasts, for instance, mat designs upon lustrous grounds, or *vice versa*.

A large variety of types of sand-blasting machines have been introduced, a number of them having been designed for use in cleansing large iron castings for engineering work. A sand-blast suitable for the electro-plater's purpose is shown in Fig. 84. It is very compact and convenient for use in a

14

limited floor space. The necessary pressure is obtained by
compressed air. The compressed air, the pressure of which
should be at least equal to that of a column of water 18½
inches high, passes through the blast-pipe A into a nozzle
running horizontally through the machine, carries along a
jet of sand, and hurls the latter upon the objects placed underneath the nozzle. The objects are placed upon sheet-iron
plates or in sheet-iron boxes and very slowly passed below the
nozzle, the motion being effected by the shafts BB with the
use of rubber belts. To avoid dust, the machine is provided
with a jacket of sheet-iron or wood ; a few windows enable the
operator to watch the progress of the operation.

The uses to which a sand-blast can be put are very numer-

FIG. 85.

ous. The frosting or satin-finishing of silverware and other
articles, engraving or stenciling of metal or glass, inlaying,
removing scale, etc., the nature of the work being governed
by the fineness of the sand used as well as the pressure.

If a clean metallic surface is at one time to be given to a
large number of small articles, such as buckles, steel beads,
metal buttons, steel watch chains, ferrules, etc., a *tumbling
barrel* or *drum* is frequently used (Fig. 85). It generally consists of a cylindrical or polygonal box having a side door for
the introduction of the work, together with sharp sand or
emery, and is mounted horizontally on an axis furnished with
a winch or pulley, so as to be revolved either by hand or
power, as may be desired. In order to prevent certain objects,

like hooks for ladies' dresses and the like, from catching each other and combining into a mass, a number of nails or wooden pegs are fixed in the interior of the drum.

For ordinary polishing .the articles are brought into the tumbling barrel together with small pieces of leather waste (leather shavings), and taken out in one or two days. However, to produce an actually good polish a somewhat more complicated method has to be pursued. The articles are first freed from adhering scale by washing in water containing 5 per cent. of sulphuric acid, then rinsed and dried in a drying chamber, or in a pan over a fire. They are next brought into the tumbling barrel together with sharp sand, such as is used in glass-making, and revolved for about 12 hours, when they are taken from the barrel and freed from the admixed sand by sifting. They are then returned to the barrel, together with soft, fibrous sawdust, to free them from adhering sand, and at the same time to give them a smoother surface. They are now again taken from the barrel, freed from sawdust, and returned to the barrel, together with leather shavings. They now remain in the barrel until they have acquired the desired polish, which, according to the size and shape of the articles and the degree of polish required, may frequently take two weeks or more. Articles of different shapes and sizes are best treated together, time being thereby saved. The process is also accelerated by adding some fat oil to the leather shavings, which, of course, must be omitted when, after long use, the shavings have become quite greasy. The barrel should be filled about half full, otherwise the articles do not roll freely, and polishing is retarded. On the other hand, when the barrel is less than half full there is danger of the articles bending, or in case they are hardened, for instance buckles, of breaking.

For many purposes polishing in the tumbling barrel is of great advantage, since, independent of its cheapness, the sharp edges of the articles are at the same time rounded off. However, with articles the edges of which have to remain sharp, the process cannot be employed.

The tumbling barrel in which the articles are treated with sand cannot be used for polishing with leather shavings, it being next to impossible to free it entirely from sand. The barrels should make from 50 to 70 revolutions per minute; if allowed to revolve more rapidly, the articles take part in the revolutions without rolling together, which, of course, would prevent polishing.

The brightening of articles of iron and steel may be simplified by using water to which 1 per cent. of sulphuric acid has been added. The barrel used for the purpose must, of course, be water-tight. By the addition of sand the process is accelerated. Nickel and copper blanks for coins are also cleansed in this manner. They are brought into the tumbling barrel, together with a pickling fluid, and, when sufficiently treated, are taken out, rinsed, dried in sawdust, and finally stamped. Fig. 86 shows a form of an adjustable, oblique tumbling

FIG. 86.

barrel, adapted to clean, smooth brighten, and polish nearly every variety of iron and brass goods. The simplicity and durability of the construction and the rapidity with which the work is done are distinct advantages. The machine can be used wet or dry. It is adjustable by screw and wheel to any working elevation up to 50°. The machine shown in the illustration is designed to carry a barrel 24 inches in diameter, but larger or smaller barrels can be used.

Grinding. Wooden wheels covered with leather coated with emery of various degrees of fineness are almost exclusively used for grinding metallic objects preparatory to the plating process. The wooden wheels are made of thoroughly-seasoned poplar, in the manner shown in Fig. 87. The separate pieces are radially glued together, and upon each side in the center a strengthening piece is glued and secured with screws, so that each segment of the wheel is connected with the strengthening piece. The center of the wheel is then provided with a hole corresponding to the diameter of the spindle of the grinding lathe, to which it is secured by means of wedges. The periphery, as well as the sides, is then turned smooth. A good quality of leather, previously soaked in water and cut into strips corresponding to the width of the wheel is then glued to the periph-

Fig. 87.

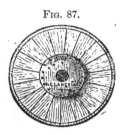

ery, and still further secured by pins of soft wood. When the glue is dry the wheel is again wedged upon the spindle and the leather case fully turned; it is then ready for coating with emery.

With the use of grinding wheels of oak or walnut, covering with leather may be omitted, and the emery can be applied directly to the wheels. However, leather-covered wheels are to be preferred since, by reason of their elasticity, better results in grinding are obtained than with uncovered wheels of the above-mentioned varieties of wood.

For grinding soft metals, hard, impregnated felt wheels "set up" with glue and emery are also employed.

For grinding profiled articles preference should be given to wheels without leather covering, and the grinding surface should be fitted to the profile of the article to be ground by cutting with a turning tool.

Grinding wheels of paste-board and of cork waste have recently been introduced. The former are made by coating on both sides thin, round disks of paste-board with glue mixed with emery, and then gluing a sufficient number of such disks one upon the other to form a wheel of the desired width. The wheel is finally subjected to strong pressure under a hydraulic press, and dried. However, as these wheels have disappeared from commerce, it may be assumed that they have not stood the test in practice. The same may be said of cork wheels. The so-called elastic wheel has also not answered the demands made in practice. The cementing material in the case consisted of a gum or rubber-like mass, which to be sure imparted great elasticity to the wheel, but when the latter became hot during grinding, the mass softened and smeared.

FIG. 88.

The so-called *reform wheel*, Fig. 88, has a better prospect of general introduction.. The leather covering does not consist of a single strap, but pieces of leather, 3 to 5 millimeters thick, are placed alongside each other and secured by means of a sort of dove-tailing to an iron rim, which is screwed upon the wooden disk. According to the length of the pieces of leather a greater or smaller degree of elasticity is attained. One covering lasts at least five to eight times as long as the covering with leather straps, and leather-waste, otherwise of scarcely any value, may be employed for the covering.

For grinding soft metals, hard impregnated felt wheels coated by means of glue with emery are also used.

For gluing with emery three different kinds of emery are used, a coarse quality (Nos. 60 to 80) for preparatory grinding, a finer quality (No. 00) for fine grinding, and the finest quality (No. 0000) for imparting luster. The wheels thus coated are termed respectively "roughing wheel," "medium wheel," and "fine wheel." With the first the surface of the objects are freed from the rough crust. The coarse-grained emery used for this purpose, however, leaves scratches, which have to be removed by grinding upon the medium wheel until the surfaces of the objects show only the marks due to the finer quality of emery, which are in their turn removed by the fine wheel.

In most cases brushing with a circular bristle brush may be substituted for the last grinding, the articles being moistened with a mixture of oil and emery No. 0000. Care must be had not to execute the brushing nor the grinding with the finer quality of emery in the same direction as the preceding grinding, but in a right angle to it.

Treatment of the grinding wheels.—The coating of the roughing wheels with emery is effected by applying to them a good quality of glue and rolling them in dry, coarse emery powder. For the medium and fine wheels, however, the emery is mixed with the glue and the mixture applied to the leather. When the first coat is dry, a second is applied, and finally a third. The whole is then thoroughly dried in a warm place. Before use, a piece of tallow is held to the revolving disc for the purpose of imparting a certain greasiness to it, and in order to remove any roughness due to an unequal application of the emery, it is smoothed by pressing a smooth stone against it. While the preparatory grinding upon the roughing wheel is executed dry, *i. e.*, without the use of oil or fat, in fine grinding, the objects are frequently moistened with a mixture of oil and the corresponding No. of emery. When the layer of emery is used up, the remainder is soaked with warm water and scraped off with a dull knife. The leather of the disks on which oil or tallow has been used is then thoroughly rubbed with caustic lime or Vienna lime * to remove the greasiness, which would prevent the adherence of the layer of glue and emery to be applied later on. When the leather is thoroughly dry a fresh layer of emery may at once be applied.

To prevent the leather from absorbing an excess of water when moistening the old layer of glue and emery for the purpose of softening it, it is advisable to apply moderately wet

---

* Vienna lime is prepared from a variety of dolomite which is first burned, then slaked, and finally ignited for a few hours. It consists of lime and magnesia, and should be kept in well-closed cans, as otherwise it absorbs carbonic acid and moisture from the air, and becomes useless.

clay to the layer and allow it to remain for a few hours when the emery can be readily scraped off.

A very useful machine for removing emery and glue from worn, leather-covered wood polishing wheels is shown in Fig. 89. The compartment is filled with water until it just touches the lower part of the rollers. Then by placing the worn wheels on the rollers and allowing the machine to run for a short time all the glue and emery will be removed without damaging or loosening the leather covering. The rollers carry just enough water to properly feed the face of the wheels, and the friction caused by the weight of the wheels revolving

FIG. 89.

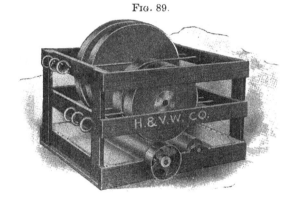

on the rollers quickly forces off the emery and glue. Allow the wheels to dry in the ordinary temperature; do not subject them to heat.

Grinding lathes. For use, the grinding wheels are wedged upon a conical cast-steel spindle provided with a pulley and running in bearings, as plainly shown in Fig. 90. The cast-iron standards are screwed to the floor; the wooden bearings can be shifted forward and backward by wedges and secured in a determined position by a set-screw, thus facilitating the removal of the spindle after throwing off the belt. The wheels being wedged upon a conical spindle, always run centrically.   Changing of the wheels requires but a few

FIG. 90.

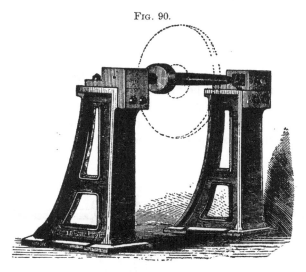

seconds, and on account of the slight friction of the points of

FIG. 91.

the spindle in the wooden bearings, the consumption of power
is very slight.

To avoid the necessity of throwing off the belt while chang
ing the grinding wheels, double machines (Fig. 91) are used,
the principle of conical spindles being, however, preserved.
The shaft is provided with loose and fast pulley and coupling
lever.

Fig. 92 illustrates a similar machine with ring-oiling.

Fig. 93 represents a belt-attachment combined with a double
grinding lathe, as constructed by the firm of Dr. G. Langbein·
& Co. The apparatus can be readily secured by means of
screws to the lathe, and is readily removed. It allows of

FIG. 92.

grinding-wheels, brushes, etc., being attached to both ends of
the shaft, while the belt can at the same time be used.

Electrically-driven grinding motors have been previously referred to. Fig. 94 shows a grinder of this type manufactured
by The Hanson & Van Winkle Co., Newark, N. J. It is of
the ribbed type, and is furnished in various sizes. The switch,
starting box and regulator are contained within the stand,
with the operating handles extending through a suitable opening. An important feature of this machine is the ability of

the operator to regulate the speed of the wheels, running them at the speeds most suitable for the work in hand. This regulation of the speed is accomplished by the simple movement of a handle, the speed remaining practically constant at any point.

A smaller type of the same machine is very suitable for use by manufacturers, jewelers, dentists, instrument makers, etc.

FIG. 93.

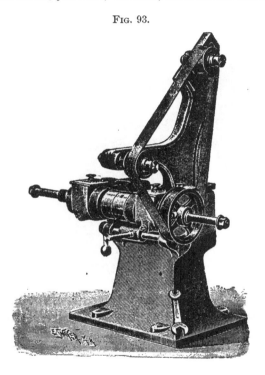

These grinding motors, as well as the polishing motors to be described later on, have the advantage of occupying no more space than that is usually required by a belt-driven lathe, while the full motive power is applied, without loss, directly to the grinding wheels. They also possess the advantage of being portable, and in a few moments' time can be moved to any part of the factory that may be best suited for the purpose required, making it possible to take the motor to the work when desired, instead of bringing the work to the motor.

Execution of grinding and brushing.   Grinding is executed by pressing the surfaces to be ground against the face of the wheel, moving the objects constantly to and fro.   The operation requires a certain manual skill, since, without good reason, no more should be ground away on one place than on

FIG. 94.

another.   Special care and skill are required for grinding large round surfaces.

If the objects are to be treated with the fine wheel, fine grinding is succeeded by brushing with oil and emery by means of circular brushes formed of bristles set in disks of wood.   Genuine bristles being at present very expensive, vegetable fiber, so-called *fibers*, has been successfully substituted for them, the wooden wheel being replaced by an iron case, in the bell-shaped cheeks of which the fiber-bundles are secured by means of strong nuts.   Before use it is advisable to saturate the fiber-bundles with oil in order to deprive them of their brittleness, and thus improve their lasting quality.

The grinding lathe (Fig. 95) is provided with a tampico brush, this fiber being particularly adapted for rough, quick work. It can, of course, just as well be placed upon the conical spindles of double machines. The iron case is provided with a conical hole corresponding exactly to the conical spindle, the large frictional surface preventing the turning of the brush upon the spindle, or its running off.

FIG. 95.

FIG. 96.

In regard to grinding the various metals, the procedure, according to the hardness of the metal, is as follows:

Iron and steel articles are first ground upon the roughing wheel, then fine-ground upon the medium wheel, and finally upon the fine wheel, or brushed with emery with the circular brush. Very rough iron surfaces may first be ground upon solid emery wheels before being worked upon the roughing wheel.

For depressed surfaces which cannot be reached with the large emery wheels, small walrus-hide wheels coated with glue and emery are placed upon the point of the spindle of a polishing lathe.

Brass and copper castings are first ground upon roughing wheels which have lost part of their sharpness and will no longer attack iron. They are then ground fine upon the medium wheel, and finally polished upon cloth or felt wheels (bobs). (See below under "*Polishing.*")

Sheets of brass, German silver and copper, as furnished by rolling-mills, are only brushed with emery and then polished with Vienna lime or rouge upon bobs.

Zinc castings, as, for instance, those produced in lamp factories, are first thoroughly brushed by means of circular brushes and emery, and then polished upon cloth bobs.

FIG. 97.   FIG. 98.

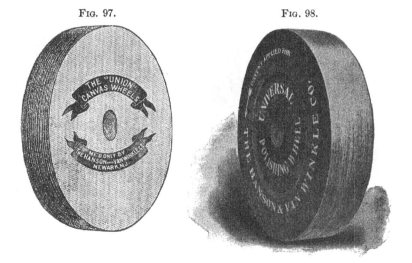

Sheet zinc is only polished with Vienna lime and oil upon cloth bobs secured to the spindle shown in Fig. 101.

Polishing.—As will be seen from the foregoing, polishing serves for making the articles ready, *i. e.*, the final luster is imparted to them upon soft polishing wheels with the use of fine polishing powder. The polishing wheels or bobs of fine felt, shirting, or cloth, are secured to the polishing lathe, and, according to the hardness of the metal to be polished, make

2000 to 2500 revolutions per minute. A foot-lathe, such as is shown in Fig. 96, makes generally not over 1800 revolutions per minute. Cloth bobs are made by placing pieces of cloth one upon another in the manner described under "Nickeling of sheet zinc," cutting out the center so as to correspond to the diameter of the spindle, and securing the disks of cloth by means of nuts between two wooden cheeks upon the spindle of the polishing lathe. In place of cloth bobs, solid wheels of felt or wooden wheels covered with a layer of felt may be used, especially for polishing smooth objects without depressions, the fineness and softness of the felt depending on the degree of polish to be imparted and the hardness of the metal to be manipulated.

An excellent polishing-wheel is the Union canvas wheel, made by the Hanson & Van Winkle Co., of Newark, N. J. It is shown in Fig. 97. It is not glued, but by a special process the weight is reduced, the elasticity and flexibility are increased, and a cloth face is obtained, which combined with the glue, presents a surface that will hold emery better than any other wheel. Being of a flexible nature, it easily adjusts itself to the irregularities of the work. No special skill is required to use it, and there is less tendency to "gouge" the work or spoil design. The wheel will do more work with one setting-up than any other. It is durable and easily kept in balance.

Fig. 98 shows the universal polishing wheel made by the Hanson & Van Winkle Co. This wheel is superior in every way. It is practically universal and decidedly economical. It can be used for "roughing," "fining", or "greasing." It retains its shape—does not rag out, and will stay in balance almost indefinitely. It is resilient and remains so until nearly worn out. A finer grade of emery can be used, and the emery can be washed off and a new face, fine, smooth, and glazed, can be obtained for another setting up. The wheel will not burn the surface of the metal and in consequence a better "color" after plating is obtained. It is made in three grades; soft, medium and hard.

Another wheel of great flexibility and elasticity is the *wal-rine wheel* manufactured by the same firm.   On account of its flexibility and elasticity, combined with its hardness, it is recommended for hard grinding, and the fact that its face can be turned to any shape—at the same time preserving all the characteristics of hide wheels—will place it before sea-horse or walrus on its merits.   One advantage of this wheel is its pliability, which allows it to adjust itself to any inequalities in the coat of emery, and consequently it wears evenly, and being lighter than any other serviceable wheel, it is much less liable to injure lathe bearings.

Double polishing lathes according to American patterns, Figs. 99 and 100, are used for polishing objects of not too large dimensions.   These polishing lathes are manufactured

Fig. 99.

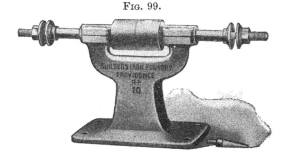

in several sizes, the largest capable of using wheels 15 inches diameter and 5 inches face.   Fig. 99 shows a 10-inch polishing head to be screwed to a bench.

Fig. 100 illustrates a 14-inch ring-oiling polishing machine. The head is constructed so as to give plenty of room between the wheels and the column, thus making the machine of special advantage where awkward pieces are to be handled, as bicycle, chandelier, and similar classes of work.   The bearings provided are peculiar to the machine, the boxes being so constructed that the spindle has four bearings, thus affording a good support and making the machine a stiff and durable one.

The polishing lathe shown in Fig. 101 serves chiefly for

FIG. 100.

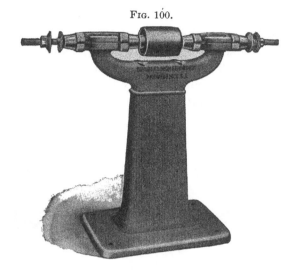

polishing large sheets, the latter being placed upon a smooth,

FIG. 101.

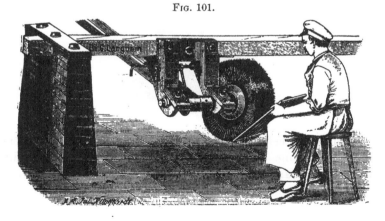

wooden support which rests upon the knees of the workman,
as will be described later on in speaking of nickeling sheet-zinc.

14

Fig. 102 shows an independent spindle-polishing and buff-ing lathe manufactured by The Hanson & Van Winkle Co., Newark, N. J.   In designing this lathe further demands for economy, *viz.*, power, shafting and belting have been antici-pated.   Countershafts, loose pulleys and incidental belting are dispensed with.   To accomplish this, a single driving pulley and double friction cones, securely housed in the lathe head, as shown in the illustration, are used on the lathe.

FIG. 102.

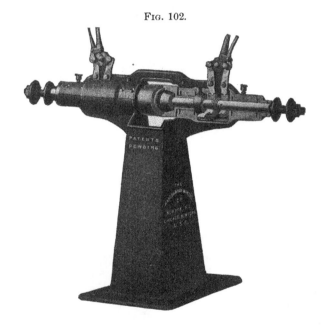

The forged steel clutches are operated by small hand levers above the casing.   A study of the illustration will show the great strength and large provision for wear in the double fric-tion cones, as well as the spindle bearings, and the means for oiling.

The lathe spindle is of large diameter, carefully ground and polished ; it runs smoothly and noiselessly and without end play.   Throwing off the clutch brings a brake into action and.

stops the spindle instantly, while the reverse motion releases the brake and starts the spindle immediately. This is done at either end without interference or waiting; as a result there is no lost time for either operator, which means a saving of from one to three hours per day.

The pedestal flares at the back so that the latter can be belted from below, a method now used in many shops, and always to be preferred, as it gives a room entirely free of belts.

Electrically-driven polishing and buffing lathes are now in frequent use. The high speed at which emery and polishing

FIG. 103.

wheels are run necessitates tight belts, heated bearings, and the dirt carried by the belt. All this is overcome in these machines. They can be placed so as to secure the best light, the speed is constant, and no power is used in driving the countershaft when not in use. The machines are furnished with and without stand.

Fig. 103 shows a type of polisher manufactured by the Hanson & Van Winkle Co. It has all the good points of the grinder (Fig. 94) manufactured by the same firm.

The belt-strapping attachment or endless-belt machine shown in Fig. 104 is made by the above-mentioned firm.   It is simple in construction and easily operated.   It can be used to great advantage by manufacturers of bicycles and bicycle parts, brass cocks, and other plumbers' fittings, gas fixtures, grate and fender work, while for cutlery it seems almost indispensable.   The attachment complete consists of a 12-inch and 6-inch diameter flanged pulley, $2\frac{1}{2}$ inches between flanges, with standard and adjusting arms.

No shop is now complete without one or more flexible shafts for grinding, polishing and buffing.   It will in many ways be

Fig. 104.

found a profitable and economical device.   For cleaning and grinding heavy castings, for polishing and buffing all metal and glass, it is an indispensable tool where power is or can be used to advantage.   These shafts are made in standard sizes, from $\frac{1}{4}$-inch diameter core, suitable for very light work, to $1\frac{1}{2}$-core, capable of driving a 3-inch drill in iron or steel.

Polishing materials.—According to the hardness of the material to be polished, rouge (ferric oxide, colcothar) tripoli, Vienna lime, etc., in the state of an impalpable powder, and generally mixed with oil, or sometimes with alcohol, are used as polishing agents.   For hard metals, an impalpable rouge of great hardness (No. F of commerce), is employed, for softer metals, a softer rouge (No. FFF), or Vienna lime, tripoli, etc.

It is of advantage to melt the rouge with stearine and a small quantity of tallow, and cast the mixture in moulds with the aid of strong pressure. The sticks thus formed are sufficiently greasy to render the use of oil superfluous. In order to impregnate the surface of the polishing bob with the polishing material, hold one of the sticks for a second against the revolving wheel, and then polish the objects by pressing them against the wheel, diligently moving them to and fro. The polishing bob must not be too heavily impregnated with rouge, since a surplus of the latter smears instead of cutting well.

In polishing with Vienna lime it is expedient to moisten the objects to be polished with stearine oil, and saturate the polishing wheels by pressing a piece of Vienna lime against them. However, this causes a great deal of dust, which not only incommodes the workman, but is also injurious to the respiratory organs. It is therefore recommended to remove the dust by means of an exhauster.

Another process of polishing, called *burnishing*, is executed by means of tools usually made of steel for the first or *grounding* process, or of a very hard stone, such as agate or bloodstone, for finishing. Burnishing is applied to the final polishing of deposits of the noble metals, and will be referred to later on.

B. Mechanical Treatment During and After Electro-plating.

In this connection *scratch-brushing* the deposits will be first considered, the object of this operation being, on the one hand to promote the regular formation of certain deposits, and, on the other, to affect the physical properties of the deposits, and finally to ascertain whether the deposit adheres to the basis-metal.

If it is noticed by the irregular formation of the deposit that the basis-metal has not been cleaned with sufficient care by the preparatory scratch-brushing, the object has to be taken from the bath and the defective places again scratch-brushed with the application of water and sand, or pumice stone, when the object is again pickled and replaced in the bath.

On the other hand, the deposits always form more or less porous, they having, so to say, a net-like structure, though it may not be visible to the naked eye. By scratch-brushing, the meshes of the net are made closer by particles of metals being forced into them by the brush, and the deposit is thus rendered capable of receiving additional layers of metal. Furthermore, by scratch-brushing, dull deposits acquire a certain luster, which is enhanced by the subsequent polishing process. Finally, by an unsparing application of the scratch-brush, it will be best seen whether the union of the deposit with the basis-metal is sufficiently intimate to stand, without becoming detached, the subsequent mechanical treatment in polishing.

According to the object in view, and the hardness of the deposit to be manipulated, scratch-brushes of steel or brass wire are chosen. For *nickel*, which, as a rule, requires scratch-brushing least, and chiefly only for the production of very thick deposits, steel wire of 0.2 millimeter thickness is taken ; ·for deposits of *copper*, *brass*, and *zinc*, brass wire of 0.2 millimeter; for *silver*, brass wire of 0.15 millimeter; and for *gold*, brass wire of 0.07 to 0.1 millimeter. Scratch-brushing is seldom done dry. The tool as well as the pieces should be constantly kept wet with liquids, especially such as produce a lather in brushing, for instance, water and vinegar, or sour wine, or solutions of cream of tartar or alum, when it is desired to brighten a gold deposit which is too dark. However, the liquid most generally used is a decoction of licorice-root, of horse-chestnut, of marshmallow, of soap-wort, or of the bark of Panama-wood, all of which, being slightly mucilaginous, allow of a gentle scouring with the scratch-brush, with the production of an abundant lather. A good adjunct for scratch-brushing is a shallow wooden tub containing the liquid employed, with a board laid across it nearly level with the edges, which, however, project a little above. This board serves as a rest for the pieces.

The hand scratch-brush, when operating upon small objects, is held by the workman in the same manner as a paint brush, and is moved over the object with a back and forward motion imparted by the wrist only, the forearm resting on the edge of the tub. For larger objects, the workman holds his extended fingers close to the lower part of the scratch-brush, so as to give the wires a certain support, and, with raised elbow, strikes the pieces repeatedly, at the same time giving the tool a sliding motion. When a hollow is met with, which cannot be scoured longitudinally, a twisting motion is imparted to the tool.

Scratch-brushing by means of circular scratch-brushes is effected in a lathe. The lathe-brush is mounted upon a spindle and is provided above with a small reservoir to contain the lubricating fluid, a small pipe with a tap serving to conduct the solution from this to a point immediately above the revolving brush. The top of the brush revolves towards the operator, who presents the object to be scratch-brushed to the bottom. The brush is surrounded by a wooden cage or screen to prevent splashing. To protect the operator against the water projected by the rapid motion, there is fixed to the top of the frame a small inclined board, which reaches a little lower than the axis of the brush without touching it. This board receives the projected liquid, and lets it fall into a zinc trough, which forms the bottom of the box. Through an outlet provided in one of the angles of the trough, a rubber tube conveys the waste liquid to a reservoir below. After scratch-brushing every trace of the lubricating liquid must be washed away before placing or replacing the objects in the bath.

Drying.—The finished plated objects are first rinsed in clean water to remove the solutions constituting the bath adhering to them. They are next immersed in hot water, where they remain until they have acquired the temperature of the water, and are then quickly dried in the manner described on p. 163.

A very good method of freeing nickeled objects from all
moisture which may have collected in the pores is to immerse
them for about ten minutes in *boiling* linseed oil, and, after
allowing them to drain off, to remove the adhering oil by
rubbing with sawdust.   According to some electro-platers, the
deposit of nickel thus treated loses its brittleness and will stand
bending several times, for instance, wire, sheets, etc., without
breaking.   Experiments made by Dr. George Langbein did
not confirm these statements, but the security against rust of
nickeled-iron objects is found to be considerably enhanced by
boiling in linseed oil.

Production of high luster.—When dry, the plated objects are
highly polished, this being effected by means of polishing
bobs of fine felt, cloth or flannel, with the use of polishing
compositions, vienna lime, etc., or by burnishing with tools of
steel and of agate or blood-stone.

Nickel deposits are almost without exception polished upon
cloth or felt bobs with rouge or Vienna lime.

Copper and brass deposits are polished with fine flannel bobs,
the polishing powder being applied very sparingly.   *Deposits
of tin* are generally only scratch-brushed, it being impossible
to impart great luster to this metal by polishing with bobs.
After drying, the deposit is polished with whiting.   *Deposits
of gold and silver* as well as of *platinum* are polished by burnishing, the steel burnisher being used for the *grounding* process, and an agate or blood-stone burnisher for finishing.   The
operation of burnishing is carried on as follows: Keep the tool
continually moistened with soap-suds.   Take hold of the tool
very near to the end, and lean very hard with it on those
parts which are to be burnished, causing it to glide by a backward and forward motion without taking it off the piece.
When it is requisite that the hand should pass over a large
surface at once without losing its point of support on the work
bench, be careful in taking hold of the burnisher to place it
just underneath the little finger.   By these means the work is
done more quickly, and the tool is more solidly fixed in the

hand. The burnishers are of various shapes to suit the requirements of different kinds of work, the first rough burnishing being often done by instruments with comparatively sharp edges, while the finishing operations are accomplished with rounded ones. Fig. 105 illustrates the most common forms of burnishers of steel and agate. Both must be free from cracks and highly polished. To keep them free from blemishes they are from time to time polished by vigorously rubbing them with fine tin putty, rouge, or calcined alum upon

FIG. 105.

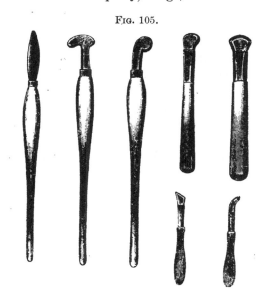

a strip of leather fastened upon a piece of wood which is placed in a convenient position upon the work bench.

Cleansing the polished objects.—The objects to which high luster has been given by means of Vienna lime and oil, or rouge, have to be freed from adhering polishing dirt. With flat, smooth objects, this is effected by wiping with a flannel rag and Vienna lime, and with those having depressions or matted surfaces, by brushing with a soft brush and soap water, and then drying in sawdust.

Cleansing is very much facilitated by brushing the polished
-articles upon a small cloth or flannel bob.

Chemical Treatment.

While it is the aim of the mechanical treatment to prepare,
·on the one hand, a pure metallic surface, and on the other, a
·smooth one, the chemical preparation of the objects serves the
purpose of facilitating the mechanical treatment by softening
·and dissolving the impurities of the surface, and of freeing the
mechanically treated objects from adhering oil, grease, dirt,
·etc., so as to bring them into the state of *absolute purity* required for the electro-plating process.

Pickling and dipping. The composition of the pickling
·liquor varies according to the nature of the metal to be
·treated.

Cast-iron and wrought-iron articles are pickled in a mixture
of 1 part by weight of sulphuric acid of 66° Bé. and 15 parts
by weight of water.*

To cleanse badly rusted iron articles without attacking the
iron itself, it is recommended to pickle them in a concentrated
solution of chloride of tin, which, however, should not contain
·too much free acid, otherwise the iron is attacked.   Bucher
·recommends a pickle composed as follows: Dissolve $3\frac{1}{2}$ ozs.
·of chloride of tin in 1 quart of water, and $1\frac{1}{2}$ drachms of tartaric acid in 1 quart of water.   Pour the former solution into
the latter, and add 20 cubic centimeters of indigo solution
·diluted with 2 quarts of water.   The object of the addition of
indigo is not intelligible.

An excellent pickle for iron ·is also obtained by mixing 10
·quarts of water with 28 ozs. of concentrated sulphuric acid, dis-
·solving 2 ozs. of zinc in the mixture, and adding 12 ozs. of
nitric acid.   This mixture makes the iron objects bright, while
·in dilute sulphuric or hydrochloric acid they become black.

Col. J. H. Hausjosten † recommends the following pickle as

---

* The acid should be poured into the water, not the water into the acid.

† Metal Industry, March, 1913.

giving good results oh cast-iron: Sulphuric acid 3 parts,
hydrofluoric acid 1 part, water 3 to 4 parts.

"The length of time required to pickle the work in this
pickle depends on the amount of scale on it, the size of the
castings, and the strength or weakness of the pickle. Small
or medium sized castings may be left in it 15 or 20 minutes,
while larger pieces and pieces with a large, smooth surface,
should be kept in longer, but the length of time required for
any class of work may be determined by the operator if he
will watch the results that the first few batches will give.
He should increase or decrease the strength of the pickle by
the amount of water or acid needed, and the condition of his
work demands.   After the castings are pickled they should be
rinsed in hot water containing about ½ pound of lime for
every 10 gallons of water.   The lime water serves a two-fold
purpose in that it neutralizes the acid and dries the castings.
The pickle tank and hot water tank should be side by side and
covered with a hood made of ½ inch boards, high enough
above the tank so that they will not interfere with the work
of the operator, but not so high that it will not carry off the
fumes and steam caused by the acid and lime water.   The
top of the hood should slant downward over the tanks from a
pipe or stack leading upward, so as to aid the natural draft
that will carry the steam and fumes away.   If the hood ends
so that the fumes will be guided into a smokestack or chimney, no exhaust will be necessary, as the natural draft will
carry them away.

"Wood boxes, with the nails driven well into the wood and
of convenient size, with holes bored in the sides and bottom,
and iron wires of sufficient strength and long enough so that
the ends will remain above the solution when attached to the
box as a handle, make good baskets for small work.   Larger
pieces with holes in them may be strung on iron wire, and the
ends looped together, so as to be convenient to take out.
Pieces of the same kind may be reversed ; that is, put face to
face or back to back, so that they will not nest and thus pre-

vent the acid from attacking them all over.   Large quantities of work may be quickly handled in this way, and, by making the tank of proper length, the operator may begin to put in work at one end and work toward the other, and by the time the tank is filled the work put in first is nearly ready to come out.   It is better to have the tank run to size in length rather than in breadth or depth.   A tank, 30 inches wide, 18 inches deep, and 5 or 6 feet long, will be large enough to pickle the work for 20 to 30 polishers.   Acid should be added to the pickle as it weakens, and when it becomes too old it should be thrown away and a new one made up.   Lime should be added to the lime water every day or so.   Both solutions may be kept in working order for a very long time by simply keeping them up to the required strength.   No set rule can be laid down to go by in adding acid, or how long the work should remain in the pickle.   This must be ascertained by the operator, and a little experience will quickly tell him what to do and when to do it.

"The scratch-brushing may also be made a factor in turning out good work, but is important only in a relative way. If it is improperly done, particles of sand may remain in the places where the polishing wheel will not reach, and in the finished piece will show up as black spots.   The cost of scratch-brushing is so little, however, that it will be well worth while to pay some attention to it, and if the pickling is properly done, it is only necessary to brush the loose sand out of the background.   An objection to pickling that has been raised is that the acid will ooze out of the work after plating and discolor it.   This may occur if the work is pickled too much, but a little care will overcome it, or will not allow it to happen at all.   All time and attention given to this part of the preparatory process will be amply repaid, as the result cannot be otherwise than a lower cost in polishing."

For many cases pickling may advantageously be effected in the electrolytic way.   Suspend the articles in a weak acid bath (hydrochloric or sulphuric acid), connect them with the positive pole of a source of current, and suspend opposite to them a sheet of metal (copper or brass), which is connected with the negative pole.

According to a patent of the Vereinigten Elektrischen Gesellschaften, Vienna and Buda-Pest, a 20-per cent. alkaline solution of common salt or of Glauber's salt is used for electrolytic pickling, the metal to be pickled serving as anode. By the passage of the current, the electrolyte is broken up into sodium-ions, which are separated on the cathodes, and SO and chlorine-ions, which are separated on the anodes, the pickling effect being produced by the latter ions.

This electrolyte may also be used for freeing sheet metal, for instance, sheet-iron, which is to be zincked, from grease, and at the same time pickling it. For this purpose the sheet-iron is suspended to the anode and object-rods of the bath, and a current allowed to enter through the anodes, the sheets serving as cathodes being thereby freed from grease by the secondarily formed caustic soda. When this has been done, the direction of the current is reversed, the former cathodes becoming now the anodes, and the former anodes, the cathodes. The first having been previously freed from grease are now pickled by the separated anions, while the latter are freed from grease. When pickling is finished, the sheets which have served last as anodes, are taken from the bath and replaced by fresh sheets, and the direction of the current having been changed, the operation is repeated, a. continuous process of freeing from grease and pickling being thus possible. With about 90 ampères and 4 volts per square meter, pickling, according to the "Metallarbeiter," requires about half an hour.

To render possible the removal in the electrolytic way of the layer of hard solder remaining after soldering bicycle frames and thus to allow of perfect enameling, the following method is, according to Burgess,* generally adopted in this country. The parts to be soldered together were simply dipped in hard

* Electro-chemical Industry, 1904, No. 1.

solder whereby a thin layer of hard solder remained upon the parts thus treated. To remove this by filing proved expensive, and to dissolve by cyanides and solutions of double chromates required much time, and was imperfect. With the assistance of the electric current and the use of a suitable electrolyte the hard solder is completely and rapidly removed without attacking the steel. A suitable electrolyte for the purpose is a 5-per cent. sodium nitrate solution. In consequence of electrolysis some sodium nitrate is formed, and this makes the iron or steel passive, *i. e.*, deprives it of the power to be attacked by the electrolyte, while the hard solder is completely dissolved. It must, however, be borne in mind that after several days' electrolysis the steel does no longer remain passive, but is perceptibly attacked by the electrolytic pickle, this being due to the fact that the electrolyte becomes alkaline and then contains free ammonia. The latter must therefore be every day neutralized by the addition of dilute sulphuric acid. The sodium nitrate used for the preparation of the electrolyte should not contain much chloride, as otherwise the iron is attacked.

The correct progress of the operation is recognized by the rapid solution of the hard solder, a brown layer remaining behind, which can readily be rubbed off. If, however, a thick, greenish, firmly adhering slime forms on the anodes, the electrolyte has become alkaline, and when a brownish foam and precipitate appear upon the surface of the electrolyte, the alkalinity has become so great that the steel is also attacked. The hard solder electrolytically dissolved from the steel frame is precipitated as copper and zinc hydroxide by the caustic soda secondarily formed on the cathode. This precipitate has from time to time to be removed from the bath. The most suitable current-density is 0.8 to 1.2 ampères with 3 to 5 volts.

The duration of pickling depends on the more or less thick layer of scale, etc., which is to be removed or softened. The process may be considerably assisted and the time shortened by frequent scouring with sand or pumice. The pickled articles are rinsed in cold water, then immersed in hot water, and dried in sawdust. In order to neutralize the acid remaining in the pores, it is advisable to make the rinsing water alkaline by the addition of caustic potash or soda, etc.

Zinc objects are only pickled when they show a thick layer of oxide, in which case pickling is also effected in dilute sulphuric or hydrochloric acid, and brushing with fine pumice. A very useful pickle for zinc consists of sulphuric acid 100 parts by weight, nitric acid 100, and common salt 1. The zinc objects are immersed in the mixture for one second, and then quickly rinsed off in water which should be frequently changed.

Copper, and its alloys *brass, bronze, tombac* and *German silver*, are cleansed and brightened by dipping in a mixture of nitric acid, sulphuric acid, and lampblack, a suitable pickle consisting of sulphuric acid of 66° Bé., 50 parts by weight, nitric acid of 36° Bé., 100, common salt 1, and lampblack 1. In order to remove the *brown* coating, due to cuprous oxide, the objects are first pickled in dilute sulphuric acid, and then dipped for a few seconds, with constant agitation, in the above-mentioned pickle until they show a bright appearance. They are then immediately rinsed in water to check any further action of the pickle.

If objects of copper or its alloys are not to be subjected, after pickling, to further mechanical treatment, or are to be at once placed in the electro-plating bath, it is best to execute the pickling process in two operations by treating them in a *preliminary pickle* and brightening them in the *bright-dipping bath*. The *preliminary pickle* consists of nitric acid of 36° Bé., 200 parts by weight, common salt 1, lampblack 2. In this preliminary pickle the articles are allowed to remain until all impurities are removed, when they are rinsed in a large volume of water, dipped in boiling water, so that they quickly dry, and plunged into the *bright-dipping bath*, which consists of nitric acid of 40° Bé., 75 parts by weight, sulphuric acid of 66° Bé., 100, and common salt 1. It is not advisable to bring the objects which have passed through the preliminary pickle and rinsing water directly, while still moist, into the bright-dipping bath, since for the production of a beautiful, pure luster the introduction of water into the bright-dipping bath must be absolutely avoided.

Hence the objects treated in the preliminary pickle should first be dried by heating in hot water, shaking the latter off.

Potassium cyanide, dissolved in ten times its weight of water, is often used instead of the acid pickle for brass, especially when it is essential that the original polish upon the objects should not be destroyed, as in the preparation of articles for nickel-plating. The objects should remain in this liquid longer than in the acid pickle, because the metallic oxides are far less soluble in this than in the latter. In all cases the final cleaning in water must be observed.

All acid pickles used for different kinds of work should be kept distinct from each other, so that one metal may not be dipped into a solution containing a more electro-negative metal, which would deposit upon it by chemical exchange.

The pickled objects must not be unnecessarily exposed to the air, and should be transferred as quickly as possible from the pickle to the wash-waters, and then to the electro-plating bath, or, if this is not feasible, kept under pure water. Pickled objects which are not to be plated are carefully washed in water, which should be frequently changed, then rinsed, drawn through a solution of tartar, and dried by dipping in hot water and rubbing with sawdust.

Places soldered with soft solder, as well as parts of iron, become black by pickling, and have to be brightened by scouring with pumice, or by scratch-brushing.

Mat-dipping. Objects of brass or other alloys of copper are frequently to be given a dead surface so that after plating they show a beautiful mat luster. Very fine effects may by this means be obtained, especially in the bronze-ware industry. Matting may be effected in various ways. Every bright dip acts as a mat dip if the objects are exposed to its effect for a longer time and at a higher temperature. Matting is, how-ever, made more effective by adding zinc sulphate to the dip, the matting being the more pronounced, the more zinc sulphate has been added.

A good mat dip is prepared by pouring a solution of 0.35 oz. of zinc sulphate in $3\frac{1}{2}$ ozs. of water in a cold mixture of $6\frac{1}{2}$ lbs. of nitric acid of 36° Bé., 4.4 lbs. of sulphuric acid of 66° Bé., and $\frac{1}{8}$ oz. of common salt. According to the shade of mat desired, the objects are allowed to remain in the dip for 2 to 10 minutes. The objects, which on coming from the mat-dip show a faded, earthy appearance, are rapidly drawn through a clean bright dip, whereby they acquire the mat luster, and are then quickly rinsed in a large volume of water.

For the production of a mat-grained surface by pickling, the following mixture may be recommended : Saturated solution of potassium dichromate 1 part by volume, and concentrated hydrochloric acid 2 parts by volume. In this mixture the brass articles are allowed to remain several hours. They are then rapidly drawn through the bright-dipping bath and rinsed in a larger volume of water frequently renewed.

A delicate matted surface may be produced by electrolytic pickling or etching. The process is the same as described above under iron.

Other methods of matting will be given under "Gilding."

Generally speaking, it may be said that less depends on the composition of the pickle than on quick and skilful manipulation ; and as good results have always been obtained with the above-mentioned mixture, there is no reason for repeating the innumerable receipts given for pickles. The main points are to have the acid mixture as free from water as possible, further to develop hyponitric acid which is effected by the reduction of nitric acid in consequence of the addition of organic substances (lampblack, sawdust, etc.), and of chlorine, which is formed by the action of the sulphuric acid upon the common salt. The volume of the dipping bath should not be *too small*, since in pickling the acid mixture becomes heated and the increased

15

temperature shows a very rapid, frequently not controllable, action, so that a corrosion of small articles may readily take place. It is therefore necessary to allow the acid mixture, after its preparation, to thoroughly cool off. Pour the sulphuric acid into the nitric acid (*never the reverse !*) and allow the mixture, which thereby becomes strongly heated, to cool off to at least the ordinary temperature.

In order to be sure of the uniform action of the pickle upon all parts, it is, in all cases, advisable previous to pickling to

Fig. 106.

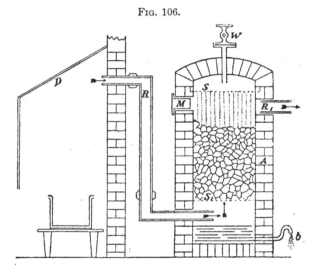

free the articles from grease by one of the methods given later on.

In pickling abundant vapors are evolved which have an injurious effect upon the health of the workmen, and corrode metallic articles exposed to them. The operation should, therefore, be conducted in the open air, or under a well-drawing vapor-flue.

In large establishments it may happen that the quantity of escaping acid vapors is so large as to become a nuisance to the neighborhood, which the proprietors may be ordered by the authorities to abate. The evil is best remedied by a small absorbing plant, as follows:

Connect the highest point of the vapor-flue D (Fig. 106) by a wide clay pipe R with a brick reservoir, A, laid in cement, so that R enters A a few centimeters above the level of the fluid, kept constantly at the same height by the discharge pipe b. Above, the reservoir is closed by an arch through which the water conduit W is introduced. Below the sieve S, which is made of wood and coated with lacquer, a wide clay pipe R leads to the chimney of the steam boiler; or the suction pipe of an injector is introduced in this place, into which the air from the vapor-flue is sucked through the reservoir and allowed to escape into the open air or into a chimney. Through the man-hole M, the sieve-bottom S of the reservoir is filled with large pieces of chalk or limestone. The manner of operating is now as follows: A thin jet of water falls upon S, where it is distributed and runs over the layer of chalk. The air of the pickling room saturated with acid vapor moves upward in consequence of the draught of the chimney of the steam boiler, the injector, or the ventilator, and yields its content of acid to the layer of chalk, while the neutral solution of calcium nitrate and calcium chloride, which is thus formed, runs off through b.

The absorption of the acid vapors may, of course, be effected by apparatus of different construction, but the one above described may be recommended as being simple, cheap, and effective.

The considerable consumption of acid for pickling purposes in large establishments makes it desirable to regain the acid and metal contained in the exhausted dipping baths. The following process has proved very successful for this purpose: Mix the old dipping baths with $\frac{1}{4}$ their volume of concentrated sulphuric acid, and bring the mixture into a nitric acid distilling apparatus. Distil the nitric acid off at a moderate temperature, condense it in cooled clay coils, and collect it in glass balloons. To the residue in the still add water, precipitate from the blue solution, which contains sulphate of copper and zinc, the copper with zinc waste, and add zinc until evolution of hydrogen no longer takes place.   Filter off the precipitated copper through a linen bag, wash, and dry.   The fluid running off, which contains zinc sulphate, is evaporated to crystallization and yields quite pure zinc sulphate, which may be sold to dye-works, or for the manufacture of zinc-white.

According to local conditions, for instance, if the zinc sulphate cannot be profitably sold in the neighborhood, or zinc waste cannot be obtained, it may be more advantageous to omit the regaining of zinc from the dipping baths.   In this case the fluid which is obtained by mixing the contents of the still with water is compounded with milk of lime until it shows a slightly acid reaction.   The gypsum formed is allowed to settle, and after-bringing the supernatant clear fluid into another reservoir, the copper is precipitated by the introduction of old iron.   The first rinsing waters in which the pickled objects are washed are treated in the same manner.   The precipitated copper is washed and dried.

Removal of grease and cleansing.   These two operations must be executed with most painstaking exactness because on them chiefly depends the success of the electro-plating process. Their object is to remove every trace of impurity, be it due to the touching with the hands or to the manipulation in grinding and polishing, and to get rid of the layer of oxide which is formed in removing the grease with lyes and other agents.

According to the preparatory treatment of the articles, the removal of grease is a more or less complicated operation. Large quantities of oily or greasy matter should be removed by washing with benzine or petroleum, it being advisable to execute this operation immediately after grinding and polishing, so that the oil used in these operations has no chance of hardening, as is frequently the case with articles preparatively polished with Vienna lime and stearine oil.   Instead of cleaning with benzine or petroleum, the articles, as far as their nature allows, may be boiled in a hot lye consisting of 1 part of caustic potash or soda in 10 of water, until all the grease is saponified, when the dirt, consisting of grinding powder, can be readily removed by brushing.   In place of solution of caustic alkalies, hot solution of soda or potash may be used, but its action is much slower and offers no advantages.   Objects of tin, lead and Britannia must be left in contact with the lye for a short time only, as otherwise they are attacked by it.

The articles thus freed from the larger portion of grease are first rinsed in water, and then, for the removal of the last traces of grease, are brushed with a bristle brush and a mixture of water, quicklime and whiting until, when rinsed in water, all portions appear equally moistened and no dry spots are visible.

The lime mixture or paste is prepared by slaking freshly-burnt lime, free from sand, with water to an impalpable powder, mixing 1 part of this with 1 part of fine whiting, and adding water, stirring constantly, until a paste of the consistency of syrup is formed.

The shape of many objects presents certain difficulties in the removal of grease, as the deeper portions cannot be reached with the brush, as, for instance, in skates, which often are to be nickeled in a finished state.   In this case the objects are drawn in succession through three different benzine vessels. In the first benzine most of the grease is dissolved, the rest in the second, while the third serves for rinsing off.   When the benzine in the first vessel contains too much grease, it is emptied and filled with fresh benzine, and then serves as the third vessel, while that which was formerly the second becomes the first, and the third the second.   After rinsing in the third benzine vessel, the objects are plunged in hot water, then for a few seconds dipped in thin milk of lime, and finally thoroughly rinsed in water.   It is recommended not to omit the treatment with milk of lime of objects freed from grease with benzine.

Electro-chemical cleaning.   It has been found that alkaline substances, such as sodium carbonate, potassium carbonate, potassium hydroxide and sodium hydroxide in solution, in varying degrees of concentration and with small proportions of potassium cyanide added, will with a sufficiently strong electric current of from 4 to 8 volts, and at a temperature nearly boiling, develop sufficient hydrogen to remove entirely all organic substances from the surface of the metal, thereby leaving it chemically clean.

This method has brought into the market several new combinations which are sold under the name of electro-chemical cleaning salts, and have given very satisfactory results.

In a paper read at the convention of the American Brass Founders' Association, at Toronto, Canada, 1908, Charles H. Proctor, in speaking of electro-chemical cleaning baths and their application, says : *

" The action of an electro-cleanser is similar to the action of an electro-plating bath. The only difference as far as the development of gases is concerned, is that no metal being in solution and the anode being insoluble, no metal is deposited. But with a strong current a copious evolution of oxyhydrogen gas is developed upon the articles, which attacks the organic matter upon the surface, practically lifting it off and by rapid evolution of the gases carries it to the surface. The small quantity of potassium cyanide contained in solution absorbs the slight oxidation that might be upon the surface, and by the combined action produces a surface clean enough, after washing in clear water, for any deposits.

" The arrangement of an electro-cleaning bath is very simple. Prepare a wrought-iron tank of proportions best adapted to the amount of work to be cleansed. This should be heated with steam coils of iron. Across the top of the tank an insulated frame should be constructed. Upon this frame place three conducting poles, as on the regular plating bath. To the two outside poles the positive current should be carried direct. This can best be accomplished with at least $\frac{1}{2}$-inch copper-wire flexible cables. To the center pole the negative current is connected with cables of the same

* The Metal Industry, June, 1908.

dimensions ; no rheostats are necessary. The stronger the current the greater the evolution of gases and the quicker the cleansing operation is accomplished.

"Although direct contact can be made with the positive current to the tank itself, in practice better results have been obtained with anodes of sheet-iron not more than 6 inches wide and of a length in proportion to the depth of the tank.

"The electro-cleaning solution should consist (for ordinary purposes) of 3 to 4 ozs. caustic potash to each gallon of water, and to every 100 gallons of solution 8 ozs. cyanide of potassium. This can be varied according to conditions. It is advisable to add at least ¼ lb. cyanide each week. Where the articles, such as iron or steel, contain much oil or grease upon the surface, the density of the solution can be increased. For articles of brass, copper or bronze that have been polished, use a solution of carbonate of soda in the proportion of 2 ozs. soda and ½ oz. caustic potash to each gallon of water, with the addition of 4 ozs. of cyanide to every 100 gallons of solution. If much organic matter is upon the surface of the articles to be cleansed, it is advisable where an air pressure can be obtained from an ordinary blower, to arrange a pipe so that the current of air can be deflected upon the surface of the solution, thus keeping the center of the solution clear of the insoluble substances that arise to the surface. When the cleanser is at rest, as much of this matter should be removed as possible.

"It should be the aim of the operator to use the same methods of avoiding all unnecessary contamination as he would in electro-depositing baths. It is obvious even to those who have not practiced this method of cleansing metallic articles that large quantities of work can be treated very rapidly, and this is the case especially where frames or racks are used in the plating operations. On account of the rapidity of operation and the efficiency of the bath, this method of cleansing should be a part of the labor-saving devices used in all great commercial establishments engaged in the electro-plating of metals."

To avoid subsequent touching with the hands the objects, before freeing them from grease, must of course be tied to the metallic wires (of soft copper) by which they are suspended in the electro-plating bath. In removing the grease by the wet method a layer of oxide scarcely perceptible to the eye is frequently formed upon the metals. This layer of oxide has to be removed, the liquid used for the purpose varying, of course, with the nature of the layer.

Fig. 107.   Fig. 108.

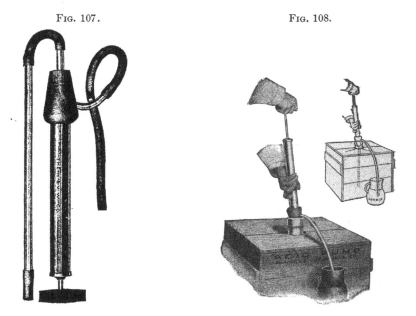

Objects of *iron* and *steel*, as well as of *zinc*, are momentarily plunged in a mixture of sulphuric acid 1 part by weight and water 20 parts, and quickly rinsed off in clean water. Highly polished objects of iron and steel, after being treated with this mixture, are best again rapidly brushed with lime paste, and, after rinsing off quickly, immediately brought into the electro-plating bath.

Copper, brass, bronze, German silver, and tombac are best

cleaned with a dilute solution of potassium cyanide, 1 part of 60-per cent. potassium cyanide in 15 to 20 of water. The objects are then quickly rinsed off and placed in the electro-plating bath.

Lead and *Britannia* may be treated with water slightly acidulated with nitric acid.

The difficult and dangerous operation of tilting heavy carboys. containing acid is overcome by the use of the Hanson & Van Winkle acid pump shown in Figs. 107 and 108. In using this pump it is not necessary for two men to handle a carboy. A workman carries the acid pitcher or receptacle to the carboy; one end of the pump tube is placed in the acid, the rubber-cork making an air-tight joint in the neck of the carboy, and the other end of the pump is carried to the pitcher. On pumping a steady flow of acid is obtained.

Electro-plating Solutions (*Electrolytes, Baths*).

Next to the proper mechanical and chemical preparations of the objects, the success of the process of electro-deposition depends on the suitable composition of the electro-plating solutions, electrolytes, or baths, and the proper current-strength which is conducted into the baths for the precipitation of the metals. In regard to the latter .the most essential conditions have already been discussed in Chap. IV., " Electro-plating Plants in General," and will be further referred to in speaking of the several electro-plating processes. * Hence, the general rules which have to be observed in the preparation of the baths will first be considered.

Solvents.—With the exception of the baths prepared with glycerin according to the patent of Marino, water is the solvent used in the preparation of all baths, and its constitution is by no means of such slight importance as is frequently supposed.

Spring and well waters often contain considerable quantities of lime, magnesia, common salt, iron, etc., the presence of which may cause various kinds of separations in the baths. On the other hand, river water is frequently impregnated to.

such an extent with organic substances that its employment without previous purification cannot be recommended. No doubt, distilled water, or in want of that rain water, is the most suitable for the preparation of baths. However, rain water collected from metal roofs should not be used, nor that running off from other roofs, it being contaminated with dust. When used, it should be caught in vessels of glass, earthenware, or wood, free from tannin, and filtered. Where river or well water has to be employed, thorough boiling and filtering before use are absolutely necessary in order to separate the carbonates of the alkaline earths held in solution. By boiling, a possible content of sulphuretted hydrogen is also driven off.

Purity of chemicals. Another important factor is the purity of the chemicals used for the baths, the premature failure of the latter being in most cases caused by the unsuitable nature of the chemicals, which also frequently gives rise to abnormal phenomena inexplicable to the operator. Chloride of zinc, for instance, may serve as an example. It is found in commerce in very varying qualities, it being prepared for dyeing purposes with about 70 per cent. actual content of chloride of zinc, for pharmaceutical purposes with about 90 per cent., and for electro-plating purposes with 98 or 99 per cent. Now it will readily be seen that if an operator who is preparing a brass bath according to a formula which calls for pure chloride of zinc uses a preparation intended for dyeing purposes, there will be a deficiency of metallic zinc in the bath, and the content of copper in the bath being too large in proportion to the zinc present, will cause reddish shades in the deposit.

Likewise, in case the operator uses potassium cyanide of low content, when the formula calls for a pure article with 98 per cent., he will not be able to effect the solution of copper or zinc salts with the quantity prescribed. Furthermore, potassium cyanide, in the preparation of which prussiate of potash containing potassium sulphate is used, will cause, by reason of the formation of potassium sulpho-cyanide, various disturbing influences (formation of bubbles in the deposit), the explanation of which is difficult to the operator, who, trusting to the purity of the chemicals, seeks elsewhere for the causes of the abnormal phenomena.

Sodium sulphite may in similar manner cause great annoyance if the suitable preparation is not used. There is a crystalized neutral salt which is employed for many gold baths, and also the bisulphite of soda in the form of powder which serves for the preparation of copper and brass baths. If the latter should be used in the preparation of gold baths, the gold would be reduced from the solution of its salts and precipitated as a brown powder.

Or, if in preparing nickel baths, a salt containing copper is used, the nickeling will never be of a pure white color, but show shades having not even a distant resemblance to the color of nickel.

The above-mentioned examples will suffice to show how careful the operator must be in the selection of the sources from which he obtains his supplies. It may be here mentioned that all the directions given in the following pages refer to chemically pure products; where products of a lower standard may be used their strength is especially given.

Concentration of baths.—For the concentration of the various baths no general rules can be laid down; neither can the determination of the density of the baths by the hydrometer be relied on. If electro-plating solutions consisted of nothing but the pure metallic salts, the specific gravity, which is indicated by the hydrometer-degrees, might serve for an estimation of their value. But such an estimation is often apt to prove deceptive, since, to decrease the resistance, the baths also require conducting salts, and by the addition of a larger quantity of them, the specific gravity of a bath may be increased to any extent without the content of the more valuable metal being greater than in a bath showing fewer hydrometer-degrees.

When the operator is acquainted with the composition of the baths, and knows how many degrees Bé. a fresh bath should show when correctly prepared, he can draw a conclusion as to the condition of the bath by changes in the specific gravity. If, for instance, a nickel bath when freshly prepared shows the standard specific gravity—70° Bé.—for nickel baths, and it shows later on 90° Bé., the greater specific gravity is due either to evaporation of water or to excessive refreshing or strengthening of the bath. Such a bath generally yields dark or spotted nickeling, the deposit is formed in a sluggish manner, and readily scales off with a stronger current. The operator in this case may recognize from the hydrometer that the cause of these phenomena is not due to a contamination of the bath, but to its over-concentration. Baths, when too concentrated, readily deposit salts in crystals on the anodes and the sides of the tanks, which should by no means take place, and there is even danger of microscopic crystals depositing upon the articles and causing holes in the deposit.

A plating bath should never be poor in metal, as otherwise it soon becomes exhausted, and besides the deposits form more slowly and with less density than in baths with a correct content of metal.

Hence in summer when the temperature of the bath, is naturally higher, they can be made more concentrated than in winter. If crystals are separated, even when a bath shows a temperature of 58° F., they should be removed and dissolved in hot water. The solution is returned to the bath and water is added to the latter until the formation of crystals ceases.

Agitation of the baths.—In order that all the strata of the bath may show an equal content of metal, it is advisable in the evening, after the day's work is done, to thoroughly stir up the solution with a wooden crutch. For practical reasons the baths are generally made one-quarter to one-third deeper than corresponds to the lengths of the objects io be plated. In consequence of this, the strata of fluid between the anodes and the objects become poorer in metal than those on the bottom, and the object of stirring up is to restore the same concentration to all portions of the bath.

While stirring up the bath, it is also advisable to see whether any metallic articles have become detached from the slings and dropped to the bottom of the vat. Such articles must be taken out, since they are dissolved by some baths, the latter being thereby spoiled. This examination must be especially thorough with nickel baths.

The strata of fluid which come in contact with the anodes become, by the absorption of metal, specifically heavier than the other strata and sink to the bottom of the tank, while, on the other hand, the strata of fluid which yield metal to the articles become specifically lighter and rise to the top. A partial compensation, of course, takes place by diffusion, but not a complete one, and from this cause arise several annoyances. The heavier and more saturated fluid offering greater resistance to the current, the anodes are attacked chiefly on the upper portions where the specifically lighter layer of fluid is; practically this is proved by the appearance of the anodes, which, at first square, after being for some time used, assume the shape shown in Fig. 109.

Fig. 109.

On the other hand, the portions of the cathodes (objects) which come in contact, near the surface, with strata of fluid poor in metal, acquire a deposit of less thickness than the lower portions which dip into the bath where it is richer in metal. Now if the bath also contains free acid, and if there is a considerable difference in the specific gravity of the lower and upper strata of fluid, the electrode, which touches both strata, produces a current, the effect of which is that metal dissolves from the upper portions and deposits upon the lower. This explains the phenomenon that a deposit on the upper portions of the objects may be redissolved, even when a current, which, however, must be very weak, is conducted into the bath from an external source.

Many authors, therefore, go so far as to demand that during the electro-plating process the baths should be kept in constant agitation by mechanical means. This, however, is scarcely necessary, because a homogeneity of the solution is to a certain extent effected by the agitation of the fluid in suspending and taking out the objects. Hence as long as objects are put in and taken out, an agitation naturally takes place in which all the strata of fluid between the objects and anodes take part, while only the deepest strata, which do not come into contact with the objects and the anodes, remain in a state of stagnation.

Constant agitation of the plating solution is of advantage in silvering, and in galvanoplastic reproduction in the acid copper bath, in which the articles have to remain four to five, and eight to ten hours. With constant agitation of the bath it is possible to work with a greater electro-motive force, whereby the deposits are finished in a shorter time; and in silvering, the formation of current-streaks is, to a certain extent, avoided; and in galvanoplastic reproduction, the formation of so-called blooms. In nickeling, with constant agitation of the bath, heavier deposits can, without doubt, be obtained in a shorter time and without premature deadening of the deposit.

Henry Sand * draws attention to the fact that, according to all known experience, the greater current-densities which are permissible in the electrolysis of given metal salt solutions, depend solely on the rapidity of the renewal of the fluid on the electrode. In his opinion it is very probable that, in the deposition of metals, the cathode potential in itself is independent of the current-density, further that the quality of the metal deposits is but slightly influenced by the current-density, and that the greater variations in the nature of the deposits with different current-densities is almost exclusively due to local changes in concentration.

These changes in concentration—the impoverishment in metal ions of the electrolyte on the place where it comes in contact with the cathode—is according to Daneel † caused, on

* Zeitschrift für Elektrochemie, 1904, S. 452.

† Zeitschrift für Elektrochemie, IX, 763.

the one hand, by the separation of metal ions on the cathode; further, in complex salt solutions, by the accumulation of ions of the salt which with the metal salt forms the complex combination, and by the conveyance by the current of the metal in those complex salt solutions which form complex anions containing the metal to be separated, and migrate away from the cathode to the anode.

The impoverishment in metal ions would proceed still more rapidly but for the counter-action of certain forces. First of all, diffusion has to be taken into consideration; it causes the entrance of strata of fluid richer in metal into those poorer in metal, and it is greatest on the sides where impoverishment has progressed furthest. Further, fresh ions are constantly supplied by dissociation, and by solutions of the simple as well as the complex salts which contain the metal in the cation, impoverishment is counteracted by the conveyance of metal by the current.

When working with low current-densities in an electrolyte, even when the latter is at rest, the slighter concentration of ions appearing on the cathode is to a certain extent obviated by diffusion of fluid richer in metal; hence under such conditions good deposits are obtained without agitation of the electrolyte.

The case, however, is different when in the same electrolyte depositions are made with high current-densities, serviceable deposits being only obtained so long as sufficient concentration of metal ions is present on the points of contact of the electrolyte with the cathode. However, as diffusion is no longer sufficiently great to replace the content of metal separated, hydrogen ions are next separated, together with the metal ions supplied by diffusion, which causes, almost without exception, the formation of spongy deposits. Hence in rapid electrolysis for which high current-densities are employed, the local exhaustion of the layers on the cathode has to be prevented by vigorous agitation of the electrolyte.

Constant agitation effects also the more rapid removal of

the hydrogen-bubbles which form on the articles, but the same end is attained without complicated contrivances by the operator accustoming himself to strike the object-rod a slight blow with the finger each time he suspends an object.

Temperature of the baths. The temperature required for the electro-plating solutions has already been referred to on p. 120, where also the means have been given for bringing baths which have cooled down too much to the proper degree of temperature. Baths which are to be used cold should under no circumstances show a temperature below 59° F., it being best to maintain them at between 64.5° and 68° F.

The warmer a bath is, the less its specific resistance and the greater its conducting power, because the salts dissolved in the bath are more dissociated than when the electrolyte is cold. In warm solutions the hydrogen-bubbles separated by the cathodes escape much more rapidly than in cold electrolytes, and consequently there is less chance of hydrogen occlusion which, according to general opinion, is the principal cause of the deposit peeling off.

Boiling the baths.—Boiling is required in the preparation of many baths, if, after cooling, they are to yield good and certain results. The kettles and boiling-pans used for the purpose are of various shapes, hemispherical or with flat bottom, and are made of different materials, those of enameled iron, or, for small baths, of porcelain or earthenware, being best. The enamel of the iron kettles must be of a composition which is not attacked by the bath. Notwithstanding their enamel these vessels become gradually impregnated with the solutions they have held, and it is risky to employ them for different kinds of baths. Thus, an enameled kettle which has been used for silvering will not be suitable, even after the most thorough washing, for a gold bath, as the gilding will certainly be white or green, according to the quantity of silver retained by the vessel. The use of metal vessels should be avoided. Copper and brass baths may, however, be boiled in strong copper kettles, though they are somewhat attacked. A copper kettle, after being freed from grease and scoured bright, may be provided with a thick deposit of nickel by filling it with a nickel bath, connecting it with the negative pole of a strong battery or dynamo machine, and suspending it in a number of nickel anodes connected with the positive pole.   Such nickeled kettles may be used for boiling nickel baths, but enameled kettles or large dishes of nickel-sheet, or vessels lined with lead, deserve the preference.   Generally speaking, nickel baths do not require actual boiling, but the nickel salts and certain conducting salts which constitute the baths, dissolve with difficulty in cold water, and hence solution is effected in hot water.

When, for the preparation of nickel baths, nickel salts soluble with difficulty have to be dissolved with the assistance of heat, and no suitable vessel is available for the purpose, solution may be effected as follows: Bring pure water in a bright copper kettle to the boiling-point.   Pour the hot water into a clean wooden bucket holding from 8 to 10 quarts, and add the quantity of nickel salt corresponding to the quantity of water. Stir with a wooden crutch until solution is complete.   Repeat the operation until all the salt required is dissolved.

For very large baths this process would, however, require too much time, and it is, therefore, advisable to use a large round or oval wooden tank, or a tank lined with pure sheet lead.   The contents of the tank are heated by means of a lead coil through which steam is introduced.

Working the baths with the current.   In case boiling of large quantities of fluid is not feasible, the same object may be attained by working the bath for some time with the electric current.   The anode rods are hung full of anodes and, a few plates of the same kind of metal having been secured to the object-rods, a current of medium strength is conducted into the bath until an object, from time to time suspended in the bath, is properly covered with a deposit.   This process is frequently used with great success for large brass baths.

Objections have been made to the process because it is sometimes carried too far, the solution becoming thereby demetal-

16

lized (?). However, there is no reason for objecting to a process because some operators carry it out in a bungling manner,
and in our opinion a bath which cannot stand working for
several days with the current without becoming poor in metal
is not of the proper composition.

Filtering the baths. Should the solutions after their preparation, and, if necessary, boiling, not be perfectly clear, they
have to be filtered. For large. baths this is best effected by
means of felt filter bags, and for smaller baths, especially those
of the noble metals, with filtering paper. This removal of
the particles held in suspension is necessary to prevent their
deposition upon the objects, which might cause small holes in
the deposit, as well as roughness and other defects. It is still
better to allow the baths to clarify by standing quietly, and
to draw off the clear solution by means of a siphon. The
turbid residue is then filtered.

Prevention of impurities. To secure *lasting* qualities to the
baths they must be carefully protected from every possible
contamination. When not in use for plating they should be
covered to keep out dust. The objects before being placed in
the baths should be free from adhering scouring material or
dipping fluid, which otherwise might, in time, spoil the bath.
The cleansing of the anode and object rods by means of sandpaper, or emery-paper, should never be done over the bath,
so as to avoid the danger of the latter being contaminated by
the oxides of the metal constituting the rods falling into it.
When a visible layer of dust has collected upon the bath, it
must be removed, as otherwise particles of dust might deposit
upon the objects and prevent an intimate union of the deposit
with the basis-metal. With large baths the removal of the
layer of dust is readily effected by drawing a large piece of
filtering or tissue paper over the surface, and repeating the
operation with fresh sheets of clean paper until all the dust is
removed. Small baths should be filtered.

The choice of anodes is also an important factor for keeping
the baths in good condition, as well as for obtaining good

results. The anodes should always consist of the metal which is deposited from the solution; and the metal used for them must be *pure* and free from all admixtures. To replace as much as possible the metal withdrawn from the bath by the electro-plating process, the anodes must be soluble; and it is wrong if, for instance, nickel baths are charged with insoluble anodes of carbon; or for smaller baths, of sheet platinum, provided the chemical composition of the bath does not in part demand insoluble anodes.   Insoluble anodes cause a steady and rapid declination in the content of metal, an excessive formation of acid in the bath, and, by the detachment of particles of carbon, a contamination of the solution. Further particulars in regard to anodes will be given in discussing the separate baths.

Absorption of the deposit.   When upon a pure metallic surface another metal is electro-deposited, the first portion of the deposit penetrates into the *basis-metal,* thus forming an alloy. This may be readily proved by repeating Gore's experiments. If a thick layer of copper be precipitated upon a platinum sheet, and then heated to a dark-red heat, the deposit can be entirely peeled off.   By then heating the platinum sheet with nitric acid, and thoroughly washing with water, it appears, after drying, entirely white and pure.   By reheating the sheet, the surface becomes again blackened by cupric oxide, and by frequently repeating the same operation, a fresh film of cupric oxide will always be obtained.

This penetration of the deposit into the basis-metal, however, does not merely take place during electro-plating, but also later on ; and it may frequently be observed that, for instance, zinc objects only slightly coppered or brassed, after some time become again white.   Since this also happens when the deposits are protected by a coat of lacquer against atmospheric influences, the only explanation of the phenomenon can be that the deposit is absorbed by the basis-metal, which is also confirmed by analysis.   This fact must be taken into consideration if durable deposits are to be produced.

Effect of the current-density. A greater or smaller current-density used in operating is not without influence upon the electrolytes. The greater the current-density is, the more metal will be deposited on the cathode, while on the anode the oxidation of the metal and its solution into acid residues cannot take place in the same measure, the result being further decompositions, oxygen gas being liberated. In con sequence of this there appear also greater differences in con centration than when depositing with less current-density and, as regards concentration, the electrolyte will remain most constant when working with a smaller current-density.

For depositions upon strongly profiled objects a medium current-strength will prove most suitable. Daneel * pictures the process as follows: If a deposit is made upon an article with elevations and depressions, the current lines will crowd together towards the elevated points, *i. e.*, those nearer to the anodes, since the current always chooses the most convenient way. Now owing to the deposit an impoverishment in metal in the electrolyte takes place on the elevated portions, and the consequence is that on these portions the separation-tension' is increased. However, when the latter is increased the current-lines will turn towards more favorable portions richer in metal ions, *i. e.*, those lying deeper, until an impoverishment in metal ions there also takes place. From this it follows that with a medium current-strength the current-lines cannot permanently favor any particular portions of the cathodes, and the deposit must become uniform. Should an unequal impoverishment take place on the electrode it would be overcome by short-closed circuits formed there, which means that they cannot distribute themselves unequally over the surface of the cathode.

With very small current-densities the process would be different. When the current-density is so small that an impoverishment in metal ions on the entire electrode can be prevented by diffusion, the current-lines will permanently

* Zeitschrift für Elektrochemie, IX, 763.

crowd together upon the more elevated portions, and the deposit will grow there, while less metal deposits in the depressions.

In practice it has been found that a great distance of the anodes from the profiled cathodes, thorough agitation of the electrolyte, not too great a conductivity of the latter, and a normal current-density, are the best auxiliaries for obtaining deposits of uniform thickness, and where these do not suffice recourse may be had for depressions to the hand anode (see later on).

Current output. In the theoretical part it has already been stated that the separation-products of electrolysis are frequently subject to secondary decomposition. Thus, the hydrogen escaping on the cathodes means a loss of electrolyzing effect of the current, and it is obvious that the quality of the deposit obtained does thereby not come up to the values which, according to the laws of Faraday, should be obtained. The quantities of deposit produced in practice referred to the theoretically calculated quantities, are called the current-output.

Reaction of the baths. A distinction is made between *acid, alkaline, neutral,* and *potassium cyanide baths.* The reaction of a bath is determined by means of suitable reagent papers. Thus, blue litmus paper, for instance, indicates the acid nature of a bath when by being dipped in it, it acquires a red color. All acids redden blue litmus paper, though many metallic salts of a perfectly neutral character produce the same change in color, and it is therefore necessary to make additional tests. Tropæolin paper, for instance, which possesses a yellow color, is only changed by mineral acids, the yellow color being converted into violet. A bath, which reddens blue litmus paper and colors tropæolin paper violet, contains without doubt, a free inorganic acid, while a bath which only reddens blue litmus paper, but does not change tropæolin paper does not contain a free inorganic acid, but an organic acid (citric, tartaric acids), and with a content of certain inorganic salts may also be free from organic acid.

An alkaline reaction of the bath is indicated by red litmus paper acquiring a blue color, while neither blue or red litmus paper is changed when the electrolyte is neutral.

In a normal state, potassium cyanide baths always show an alkaline reaction; deviations from this condition will be later on referred to.

The *general qualifications* which an electro-plating bath should possess are as follows:

1. It should possess good conducting power.

2. It must exert a sufficient dissolving effect upon the anodes.

3. It must reduce the metal in abundance and in a reguline state.

4. It must not be chemically decomposed by the metals to be plated, hence not by simple immersion; the adherence of the deposit to the basis-metal being in this case impaired.

5. It must not be essentially decomposed by air and light.

CHAPTER VI.

DEPOSITION OF NICKEL AND COBALT.

1. DEPOSITION OF NICKEL (Ni = 58.68 PARTS BY WEIGHT).

ALTHOUGH nickel-plating is of comparatively recent origin, it shall be first described, since chiefly by reason of the development of the dynamo-electrical machine it has steadily grown in popularity and become an industry of great magnitude and importance. The great popularity which nickel-plating enjoys is due to the excellent properties of the nickel itself—the almost silvery whiteness of the metal, its cheapness as compared with silver, and the hardness of the electro-deposited metal, which gives the coating great power to resist wear and abrasion ; its capability of taking a high polish ; the fact that it is not blackened by the action of sulphurous vapors which rapidly tarnish silver, and finally the fact that it exhibits but little tendency to oxidize even in the presence of moisture.

Properties of nickel.—Pure nickel is a lustrous, silvery-white metal with a slight steel-gray tinge. Its specific gravity varies from 8.3 (cast nickel plates) to 9.3 (wrought or rolled plates). It is slightly magnetic at the ordinary temperature, but loses this property when heated to 680° F. It melts at about the same temperature as iron, but is more fusible when combined with carbon.

The metal is soluble in dilute nitric acid, concentrated nitric acid rendering it passive, *i. e.*, insoluble. In hydrochloric and sulphuric acids it dissolves very slowly, especially when in a compact state.

Certain articles, for instance, hot fats, strongly attack nickel, while vinegar, beer, mustard, tea and other infusions produce stains ; hence, the nickeling of culinary utensils or the use of

(247)

nickel-plated sheet iron for that purpose cannot be recommended.

Nickel salts. The first requisite in preparing nickel baths is the use of absolutely pure chemicals, and in choosing the nickel salts to be especially careful that they are free from salts of iron, copper and other metals. Furthermore, it is not indifferent what kind of nickel salt is used, whether nickel chloride, nickel sulphate, the double sulphate of nickel and ammonium, etc., but the choice of the salt depends chiefly on the nature of the metal which is to be nickeled. There are a large number of general directions for nickel baths, of which nickel chloride, ammonio-nickel chloride, nickel nitrate, etc., form the active constituents, and yet it would be a grave mistake to use these salts for nickeling iron, because the liberated acid if not immediately and completely fixed by the anodes in dissolving, imparts to the iron objects a great tendency to the formation of rust. Iron objects nickeled in such a bath, to be sure, come out faultless, but in a short time, even if stored in a dry place, portions of the nickel layer will be observed to peel off, and by closely examining them it will be seen that under the deposit a layer of rust has formed which actually tears the nickel off. The use of nickel sulphates or of the salts with organic acids is, therefore, considered best. It might be objected that the liberated sulphuric acid produces in like manner a formation of rust upon the iron objects; but according to long experience and many thorough examinations, such is not the case, the tendency to the formation of rust being only imparted by the use of the chloride and nitrate.

Of the nickel salts with organic acids, the citrate and tartrate have been frequently employed. Nickel citrate in watery solution is not particularly well dissociated, requires a greater electro-motive force and is quite indifferent towards variations in the latter, this being the chief reason for its use in nickeling sharp ground instruments. Nickel lactate, according to Jordis's patent,[*] yields, to be sure, beautiful, lustrous deposits

* Jordis, Elektrolyse wässriger Metallsalzlösungen, S. 78.

ın thin layers, but is not serviceable for heavy nickeling, since, as the inventor himself admits, deposits in thicker layers tear. Of materially greater advantage is the use of the combinations of nickel with the ethyl sulphates * and their derivatives. According to experiments made in Dr. Langbein's laboratory, the ethyl sulphate solutions of metals are very energetically dissociated and permit the production of very thick deposits without peeling off or tearing, such as cannot be obtained in the cold bath in any other nickel solution; they are distinguished by great homogeneity and toughness.

Conducting salts. To decrease the resistance of the nickel solutions, conducting salts are added to them, which are also partially decomposed by the current. Like the use of nickel chloride in nickeling iron, an addition of ammonium chloride, which is much liked, cannot be recommended, though the subsequent easy deposition of nickel with a comparatively weak current invites its employment.

For copper and its alloys, zinc, etc., the chlorine combinations may be used, but for nickeling iron they must be avoided as the source of future evils.

The use of sodium acetate, barium oxalate, ammonium nitrate, ammonium-alum, etc., we consider unsuitable, and partially injurious, and are of the opinion that with few exceptions, which will be referred to later on, potassium, sodium, ammonia or magnesia are best for bases of the conducting salts.

The effect of the ions separated from the different conduct ing salts varies very much; the potassium ion acts different from the sodium ion, and the latter different from the magnesium ion, and an idea of this difference in action of the various ions can be formed by preparing, according to formula VIII, one nickel bath with potassium citrate and another with sodium citrate. While the bath prepared with the potassium

* German patent, No. 134,736.

salt works quite well in the deeper portions on zinc, that prepared with the sodium salt is far less effective, and several
other proofs derived from practice could be mentioned. An
attempt to explain these facts must at present be abstained
from as this suggestion cannot yet be proved by experiment so
that no objection to it could be raised.

Other additions.—Some other additions to the nickeling bath
which are claimed to effect a pure silver-white deposit have
been recommended by various experts. Thus, the presence
of small quantities of organic acid has been proposed ; for instance, boric acid by Weston, benzoic acid by Powell, and
citric acid or acetic acid by others. The presence of small
quantities of *free acid* effects without doubt the reduction of a
whiter nickel than is the case with a neutral or alkaline solution. Hence a *slightly* acid reaction of the nickeling bath, due
to the presence of citric acid, etc., with the exclusion of the
strong acids of the metalloids, can be highly recommended.
The quantity of free acid, however, must not be too large, as
this would cause the deposit to peel off.

Boric acid recommended by Weston as an addition to nickeling and all other baths, has a favorable effect upon the pure
white reduction of the nickel, especially in nickeling rough
castings, *i. e.*, surfaces not ground. Weston claims that boric
acid prevents the formation of basic nickel combinations on
the objects, and that it makes the deposit of nickel more adherent, softer, and more flexible. Whether with a correct
current-strength, basic nickel salts, to which the yellowish
tone of the nickeling is said to be due, are separated on the
cathodé, is not yet proved, and would seem more than doubtful. The action of the boric acid has not yet been scientifically explained, but numerous experiments have shown that
the deposition of nickel from nickel solution containing boric
acid is neither more adherent nor softer and more flexible
than that from a solution containing small quantities of a free
organic acid. Just the reverse, the deposition is harder and
more brittle in the presence of boric acid, and different results

.may very likely be due to the employment of varying current-densities.

In view of the fact that in the electrolysis of watery solutions, water also takes part in the processes enacted on the electrodes, and that the hydrogen appearing on the cathode promotes the formation of spongy, pulverulent and dull deposits, Marino * wants to substitute glycerin for water. Since many metallic salts dissolve only to a slight degree in glycerin, the content of metal in a glycerin bath is very low, and the resistance of the cold bath so great that enormous voltages are needed for the separation of metals, nickel, for instance, requiring more than 20 volts, and with an electro-motive force of 3 to 4 volts deposits can only be produced when the baths have been heated to quite a high temperature. However, according to Foerster and Langbein's experiments, the deposits do not possess the good qualities claimed by the patent, and cannot be forced to such thickness as, for instance, is with the greatest ease attained in nickel ethyl-sulphate baths.

The owners of the Marino patent have apparently themselves recognized the disadvantages of the glycerin electrolyte, and have applied for a patent, according to which 15 to 50 per cent. of glycerin is to be added to solutions of metallic salts in water. The glycerin is claimed to act as depolarizer, and allow of the production of lustrous nickel deposits of great homogeneousness. However, the correctness of these statements may be doubted, since by experiments made in this direction it was found impossible to produce a better technical effect with such an addition of glycerin than without it, in properly-prepared baths.

It may here again be emphasized that the compositions of the electrolyte must vary according to the results desired, and that it is impossible to attain with one electrolyte all the possible properties of the deposit.

Moreover, the addition of glycerin to aqueous electrolytes

* German patent, No. 104111.

has for a long time been known through the English patents 5300 and 22855, and it might be supposed that if such an addition were of special advantage it would have long ago come into general use.

Effect of current-density. A slighter current-density always and under all conditions causes the deposition of a *harder* and *more brittle* nickel than a current of medium strength, while with too great a current-density, the metal is separated in pulverulent form. However, as will be shown later on, nickel can also be deposited with a high current-density provided the baths are of proper composition, and agitated.

Electro-motive force. The electro-motive force given for all the baths is valid for the normal temperature of 59° to 64.4° F. and an electrode-distance of 10 cm. It has previously been mentioned, that with an increasing temperature of the electrolyte its specific resistance becomes less, but grows as the temperature decreases, less electro-motive force being required in the first case, and more in the latter.

The greater the electrode-distance in a bath is, the higher the electro-motive force which is required. A statement in figures of the changes in electro-motive force for the various electrode-distances will not be given because, on the one hand, the necessary electro-motive force is readily determined by a practical experiment, and, on the other, a calculation in advance of the electro-motive force might lead to wrong conclusions in so far as the specific resistance of the electrolyte is subject to change, and the value of the electro-motive force of the counter current varies very much for the objects suspended as cathodes, according to the nature of the basis-metal of which they are composed. For this reason a statement of the specific resistances of the bath prepared according to the directions given below is also omitted, because such statements would be valid only for freshly-prepared baths.

Reaction of nickel baths. All nickel baths work best when they possess a neutral or slightly acid reaction. Hence blue litmus-paper should be only slightly reddened, and red congo paper must not be changed at all. Baths prepared with boric acid form an exception, as they may show quite a strong acid reaction. An alkaline reaction of nickel baths is absolutely detrimental, such baths depositing the metal dull and with a yellowish color, and do not yield thick deposits.

· *Formulas for nickel baths.* I. The most simple nickel bath consists of a solution of 8 parts by weight of pure nickel ammonium sulphate in 100 parts by weight of distilled water.

Electro-motive force at 10 cm. electrode-distance, 3.0 volts.

Current-density, 0.3 ampère.

The solution is prepared by boiling the salt with the corresponding quantity of water, using in summer 10 parts of nickel salt to 100 of water, but in winter only 8 parts, to prevent the nickel salt from crystallizing out. This bath, which is frequently used, possesses, however, a considerable degree of resistance to conduction, and hence requires a strong current for the deposition of the nickel. It also requires cast-nickel anodes, since with the use of rolled anodes nickeling proceeds in a very sluggish manner. However, the cast anodes rapidly render the bath alkaline, necessitating a frequent correction of the reaction. The alkalinity is overcome by carefully adding dilute sulphuric or citric acid to neutral or slightly acid reaction.

To decrease the resistance, recourse has been had to certain *conducting salts,* and, below, the more common nickel baths will be discussed, together with their mode of preparation and action, as well as their availability for certain purposes.

II. Nickel ammonium sulphate 17 ozs., ammonium sulphate 17 ozs., distilled water 10 quarts.

Electro-motive force at 10 cm. electrode-distance, 1.8 to 2 volts.

Current-density, 0.35 ampère.

Boil the salts with the water, and, if the solution is too acid, restore its neutrality by ammonia; then gradually add solution of citric acid until blue litmus-paper is slowly but perceptibly reddened. The bath deposits rapidly, it possessing but little resistance, and all metals (zinc, lead, tin and Britannia, after previous coppering) can be nickeled in it. However, upon rough castings and iron, a pure white deposit is difficult to obtain, frequent scratch-brushing with a medium hard-steel brush being required. On account of the great content of sulphate of ammonium in the bath, the nickel deposit piles up especially on the lower portions of the objects, which, in consequence, readily become dull (*burn or over-nickel*, for which see later on), while the upper portions are not sufficiently nickeled. For this reason the objects must be frequently turned in the bath so that the lower portions come uppermost. This piling-up of the deposit also frequently prevents the latter from acquiring a uniform thickness.

III. Nickel ammonium sulphate 25½ ozs., ammonium sulphate 8 ozs., crystallized citric acid 1¾ ozs., water 10 to 12 quarts.

Electro-motive force at 10 cm. electrode-distance, 2.0 to 2.2 volts.

Current-density, 0.34 ampère.

The bath is prepared in the same manner as the preceding, the salts being dissolved in boiling water, and ammonia added until blue litmus paper is only slightly reddened.

This bath was formerly in general use in this country and is to some extent at present employed, especially for nickeling ground articles. It has the drawback of requiring very careful regulation of the current to avoid peeling off. According to experiments made by Dr. Langbein, it would be better to decrease the content of ammonium sulphate to 0.8 oz.

The reaction of this bath should be kept only very slightly acid or, still better, neutral, and it is best to use an equal number of cast and rolled nickel anodes.

If, after working for some time, the objects nickel dark, an addition of nickel sulphate is advisable, if the reaction is correct and possibly not alkaline.

IV. Nickel-ammonium sulphate 23 ozs., ammonium chloride (crystallized) 11½ ozs., water 10 to 12 quarts.

Electro-motive force at 10 cm. electrode-distance, 1.5 volts.

Current-density, 0.55 ampère.

The bath is prepared in the same manner as given for II and III. It requires exclusively rolled anodes, nickels very rapidly and quite white, but the deposit is soft and care must therefore be had in polishing upon cloth or felt bobs, the corners and edges of the objects particularly requiring careful handling. On account of the danger of peeling off a heavy deposit of nickel cannot be obtained in this bath, since, in consequence of the rapid deposition the layer of nickel condenses and absorbs hydrogen, is formed with a coarser structure and turns out less uniform and dense. These phenomena are a hindrance to a heavy deposit which, if it is to adhere, must be homogeneous and dense.

As previously mentioned, *baths with the addition of chlorides, as well as those prepared with nickel chloride and nickel nitrate, are not suitable for the solid nickeling of iron.* They are, however, well adapted to the rapid light nickeling of *cheap brass articles* on which no great demands for solidity and durability are made. To obtain nickeling of a whiter color, only 7 ozs. in place of 11¼ ozs. of ammonium chloride and 3½ ozs. of boric acid may be dissolved with the assistance of heat. The bath then requires 1.8 to 2 volts.

V. Nickel chloride (crystallized) 17½ ozs., ammonium chloride (crystallized) 17½ ozs., water 12 to 15 quarts.

Electro-motive force at 10 cm. electrode-distance, 1.75 to 2 volts; for zinc, 2.8 to 3 volts.

Current-density, 0.5 ampère.

This bath is prepared by dissolving the salts in luke-warm water and adding ammonia until the bath shows a very slightly acid, or a neutral, reaction. The bath deposits readily and is especially liked for nickeling zinc castings.

All the drawbacks of the preceding bath as regards the nickeling of iron apply also to this bath, only to a still greater extent. Rolled nickel anodes have to be exclusively used.

Nickel Baths Containing Boric Acid.

VI. Weston recommends the following composition for nickel baths: Nickel chloride $17\frac{1}{2}$ ozs., boric acid 7 ozs., water 20 quarts; or nickel-ammonium sulphate 35 ozs., boric acid $17\frac{1}{2}$ ozs., water 25 to 30 quarts. Both solutions are said to be improved by adding caustic potash or caustic soda so long as the precipitate formed by the addition dissolves.

These compositions, however, cannot be recommended, because the baths work faultlessly for a comparatively short time only. All kinds of disturbing phenomena very soon made their appearance, the deposit being no longer white but blackish, and the baths soon failing entirely. Kaselowsky's formula yields similar results. This bath is prepared by dissolving, with the assistance of heat, $35\frac{1}{4}$ ozs. of nickel-ammonium sulphate and $17\frac{3}{8}$ ozs. of boric acid in 20 quarts of water. This bath also generally fails after two or three months' use. The cause of this has to be primarily sought for in the fact that baths prepared with boric acid require, according to their composition, a definite proportion between rolled and cast nickel anodes. If rolled anodes are exclusively used, free sulphuric acid is soon formed, which causes energetic evolution of hydrogen on the articles, but prevents a vigorous deposit and imparts to the latter a tendency to peel off. The same thing happens when a nickel salt not entirely neutral has been used in the preparation of the bath. If, on the other hand, cast nickel anodes alone are employed, the bath soon becomes alkaline, with turbidity and the formation of slime, and the deposit turns out gray and dull before it possesses sufficient thickness.

From the foregoing it will be readily understood that the nickel salt used must be neutral, and that the proportion of rolled to cast anodes must be so chosen that the free sulphuric acid formed on the cast anodes is neutralized, but that the acidity of the bath dependent on the free boric acid is constantly maintained.

A recent author argues in support of Haber's proposition that the effect produced by the use of mixed anodes, *i. e.*, rolled and cast anodes, might be attained by regulating the anode current-density by the use of definite dimensions of the anodes in such a way that the electrolyte, as regards its composition, remains constant. For practical purposes this would only be feasible without trouble, if approximately the same object-surface is always present in the bath, otherwise the maintenance of the adequate anode current-density must be managed by taking out or suspending anodes according to the varying object-surfaces. However, this is far more troublesome, and the use of mixed anodes is decidedly to be preferred, it having been shown in the Galvanic Institute of Dr. George Langbein, that by this means the reaction of a bath can for years be kept constant even with considerable variations in the size of the object-surface

Such a bath containing boric acid may advantageously be prepared as follows:

VII. Nickel-ammonium sulphate 21 ozs., chemically pure nickel carbonate 1¾ ozs., chemically pure boric acid (crystallized) 10½ ozs., water 10 quarts.

Electro-motive force at 10 cm. electrode-distance, 2.25 to 2.5 volts.

Current-density, 0.35 ampère.

Boil the nickel-ammonium sulphate and the nickel carbonate * in the water until the evolution of bubbles of carbonic acid ceases and blue litmus-paper is no longer reddened. After allowing sufficient time for settling, decant the solution from any undissolved nickel carbonate, and add the boric acid. Boil the whole a few minutes longer, and allow to cool. If the nickel salt contains no free acid, boiling with the nickel carbonate may be omitted. The solution shows a *strongly acid* reaction, which must not be removed by alkaline additions.

The proportion of cast to rolled anodes used in this bath is

* In place of nickel carbonate, nickel hydrate may as well be used.

17

dependent on the quality of the anodes. The use of readily-soluble cast anodes requires the suspension in the bath of more rolled anodes than when cast anodes dissolving with difficulty are employed, since the surfaces of the latter, in consequence of rapid cooling, are not readily attacked. The proportion has likewise to be changed, with the use of soft- or hard-rolled anodes. Hence the proper proportion will have to be established by frequently testing the reaction of the bath. For this purpose the following rules may be laid down : Blue litmus-paper must always be perceptibly and intensely reddened, but congo-paper should not change its red color, for if the latter turns blue it is an indication of the presence of free sulphuric acid in the bath, which has to be neutralized by the careful addition of solution of soda or potash until a fresh piece of congo-paper dipped in the bath remains red. Ammonia cannot be recommended for neutralizing free sulphuric acid in this bath. Red litmus paper must retain its color, for if it turns blue, the bath has become alkaline, and fresh boric acid has to be dissolved in the previously heated bath until a fresh piece of blue litmus paper acquires an intense red color, or pure dilute sulphuric acid has to be added to the bath, stirring constantly, until blue litmus paper is reddened, avoiding, however, an excess which is indicated by red congo paper turning blue.

This bath is equally well adapted for nickeling ground objects, as well as for rough castings, the latter acquiring a pure white coating of nickel if thoroughly scratch-brushed, and the bath shows a normal acid reaction.

Below are given a few other formulæ for nickel baths which may be advantageously used for *special purposes*, but not for nickeling all kinds of metals with equally good results.

VIII. Nickel sulphate $10\frac{1}{2}$ ozs., potassium citrate 7 ozs., ammonium chloride 7 ozs., water 10 to 12 quarts.

For copper and copper-alloys : Electro-motive force at 10 cm. electrode-distance, 1.5 to 1.7 volts.

Current-density, 0.45 to 0.5 ampère.

For zinc: Electro-motive force at 10 cm. electrode-distance, 2 to 2.5 volts.

Current density, 0.8 to 1 ampère.

To prepare the bath dissolve 10½ ounces of nickel sulphate and 3½ ounces of pure crystallized citric acid in the water; neutralize accurately with caustic potash, and then add the ammonium chloride. This bath is especially adapted for the rapid nickeling of *polished, slightly coppered zinc* articles, for instance, tops, candlesticks, mountings, etc. Deposition is effected with a very feeble current, without the formation of black streaks, such as are otherwise apt to appear in nickeling with a weak current. The deposit itself is dull and somewhat gray, but acquires a very fine polish and pure white color by slight manipulation upon the polishing wheels. With a stronger current the bath is also suitable for the direct nickeling of zinc articles; it must, however, be kept strictly neutral. The bath works with rolled anodes, and when it has become alkaline, requires a correction of the reaction by citric acid.

IX. Nickel phosphate, 6⅓ ozs., sodium pyrophosphate, 26½ ozs., water, 10 quarts.

For copper and its alloys: Electro-motive force, at 10 cm. electrode-distance, 3.5 volts.

Current density, 0.5 ampère.

For the preparation of the nickel phosphate dissolve 12 ozs. of nickel sulphate in 3 quarts of warm water and 10 ozs. of sodium phosphate in another 3 quarts of warm water. Mix the two solutions, stirring constantly, and filter off the precipitated nickel phosphate.

Dissolve the sodium pyrophosphate in 8 quarts of warm water, add the nickel phosphate, which soon dissolves by thorough stirring, and make up the bath to 10 quarts by adding water.

This bath yields a *dark nickeling* particularly upon sheet zinc and zinc castings, and may be advantageously used for decorative purposes where darker tones of nickel are required. For zinc, work with 3.8 volts and 0.55 ampère.

For the same purpose a nickel solution compounded with a large quantity of ammonia, hence an ammoniacal nickel solution has been recommended. However, experiments with this solution always yielded lighter tones than bath IX. Special advantages cannot be claimed for this so-called dark nickeling since in arsenic and antimony we have more effective and more reliable expedients.

Black nickeling. Black deposits of nickel are frequently used particularly for decorative purposes. For the production of such deposits general directions may be given as follows: 1. A strong bath has to be used. 2. Apply a weak current. For the production of a uniform black deposit the current-strength should not exceed 1 volt. From the manner in which the deposit commences to form, it can readily be recognized whether the current is of suitable strength. The first film of deposit upon the object is iridescent, *i. e.*, shows rainbow colors and does not extend over the entire surface. The deposit next acquires a bluish tone until finally a black coating is formed. If the deposit acquires immediately in the commencement of the operation a uniform color, the current is too strong. The deposit should form slowly; it should, as mentioned above, at first be iridescent and the black deposit appear only after one or two minutes. It is of sufficient thickness as soon as the desired color appears. Very thick deposits are apt to peel off, they being more or less brittle. With the use of a weak current 30 to 60 minutes will be required for the production of a deposit of sufficient thickness. 4. A large number of nickel anodes should be used. Old anodes are to be preferred, they yielding nickel more readily than new ones. 5. Any acid which may be present in the bath should be neutralized by the addition of nickel carbonate, a neutral bath yielding a better deposit than one even only slightly acid.

A black nickel bath of the following composition yields a very uniform black deposit: Water $4\frac{1}{2}$ quarts, double sulphate of nickel and ammonium 10 ozs., ammonium sulphocyanate 1 oz., zinc sulphate 1 oz. ·This bath practically contains exclusively double nickel salts which dissolve in water. If in cold weather crystallizing takes place, the bath has to be heated. It is best to keep the temperature of the bath at from 70° to 100° F.; at a higher temperature the deposit readily acquires a gray color. At a temperature of below 59° F. some nickel salts readily crystallize out, and besides the bath does not work well.

A gray color of the deposit is an indication of too strong a current, this being also the case when streaks are formed upon the object. With the use of an old bath it may happen that it becomes acid and it will be noticed that a black coating is not produced, even by reducing the current-strength. The bath then very likely contains free acid, and the best means of neutralizing it is the addition of nickel carbonate. In case the latter is not available, ammonia may be used. Test the bath with litmus paper. If before adding the nickel carbonate or ammonia, blue limus paper when dipped into the bath turns red, the bath is acid. Add nickel carbonate until no more of it dissolves, although a small excess is no disadvantage. On the other hand, with the use of ammonia care must be had that no more than required for neutralization is added. When the limit is reached at which the color of either blue or red litmus paper is no longer changed, no more ammonia should be added.

Black nickel deposits frequently come from the bath with the proper black color and otherwise without defect, but when rinsed and dried have a brown tone. This can be removed by immersion in chloride of iron solution. The latter does not attack the black nickel deposit, provided the objects are not left too long in it, a moment's immersion being sufficient, after which they are rinsed and dried. The chloride of iron bath is composed of: Chloride of iron 8½ ozs., hydrochloric acid 18 drachms, water 4½ quarts.

Black nickel deposits when exposed to the influence of atmospheric air gradually acquire a brown color which, how-

ever, is only superficial and can be wiped off.   To prevent
such tarnishing a coat of lacquer is as a rule applied to the
nickeled object.

Black nickel deposits are much used as a priming in the
application of mat black lacquers to automobile parts and
for other purposes, where as durable a coating as possible is
desired.   If the lacquer is applied without previously giving
the brass a suitable black nickel deposit, every tiny scratch or
peeling-off becomes perceptible, which is prevented by the
black nickel deposit underneath the lacquer.

A black nickel bath of the following composition has been
recommended by Blauet : Water 95 gallons, nickel-ammonium
sulphate 12 ozs., potassium sulphocyanide 2¾ ozs., copper
carbonate 2 ozs., arsenious acid 2 ozs.

Dissolve the nickel salt in the water and add the potassium
sulphocyanide.   Then dissolve, at about 176° F., the copper
carbonate by treatment with ammonium carbonate or potassium cyanide, and add the solution, while lukewarm, to the
bath.   Finally add the arsenious acid.   If in time a gray
sediment is formed, some potassium sulphocyanide and copper
carbonate have to be added.

X. A fairly good nickel bath for many purposes is obtained
from a solution of nickel-ammonium sulphate 22½ ozs., magnesium sulphate 11¼ ozs., water 10 to 12 quarts.

For iron and copper alloys: Electro-motive force at 10 cm.
electrode-distance, 4 volts.

Current-density, 0.2 ampère.

This bath deposits with ease, and a heavy coating can be
produced on *iron* without fear of the disagreeable consequences of bath IV.   Even *zinc* may be directly nickeled in
it with a comparatively feeble current.   The deposit, however, turns out rather soft, with a yellowish tinge, and the
bath does not remain constant, but fails after working at the
utmost three or four months, even cast anodes being but little
attacked.

For the production of very thick deposits a bath composed

of nickel sulphate 17.63 ozs. and sodium citrate 17.63 ozs. dissolved in 2⅖ gallons of water has also been recommended. Pfanhauser has changed these proportions to nickel sulphate 14.11 ozs. and 12.34 ozs. sodium citrate dissolved in 2⅖ gallons of water. This bath is said to be available chiefly for the production of nickel clichés and thick deposits. It has, however, the drawback of all nickel baths prepared with large quantities of organic combinations of requiring a high electromotive force and of readily becoming mouldy. It can, however be highly recommended for nickeling articles with sharp edges and points, for instance, *knives, scissors, etc.*, it being quite indifferent towards changes in the current proportions, so that even with a higher than the normal electro-motive force and a greater current-density the objects do not readily over-nickel. The deposit is very soft, and hence in grinding such nickeled instruments peeling-off of the deposit takes place more rarely than with objects nickeled in baths of different composition. *Electro-motive force* at 10 cm. electrode-distance, 3 volts ; *current-density*, 0.35 ampère.

According to an English formula, 17.63 ozs. nickel sulphate, 9 ozs. 5½ drachms tartaric acid and 2.4 ozs. caustic potash are dissolved in 2⅖ gallons of water. The results with this bath were only moderate.

It has not been deemed necessary to give additional formulas for nickel baths, because no better results were obtained from other receipts which have been published and which have been thoroughly tested, than from those given above. In most cases success with them fell far short of expectation.

Some authors have recommended for nickeling a solution of nickel cyanide in potassium cyanide, but experiments failed to obtain a proper deposition of nickel.

The addition of carbon disulphide to nickel baths, which has been recommended by Bruce, is not advisable. According to Bruce, such an addition prevents the nickel deposit from becoming dull when reaching a certain thickness, but repeated experiments made strictly in accordance with the directions given did not confirm this statement.

The general remark may here be added that *freshly pre-pared* nickel baths mostly work correctly from the start, though it may sometimes happen that the articles first nickeled come from the bath with a somewhat darker tone.   In this case suspend a few strips of iron or brass-sheet to the object-rod; allow the bath to work for one or two hours, when nickeling will proceed faultlessly.   If, however, such should not be the case, ascertain by a test with the hydrometer whether the specific gravity of the bath is too high.   If the deposit does not turn out light, even after dilution, it is very likely that the nickel salt contains more than traces of copper, or, with black-streaked nickeling, zinc.

It has also been observed that the deposit frequently peels off when, for the purpose of neutralization, additions have been made to the nickel baths.   This phenomenon disappears in a few days, but it demonstrates that, instead of correcting the reaction of the bath by the addition of acids or alkalies, it should be done by increasing the rolled anodes in case the bath shows a tendency to become alkaline, or to increase the cast anodes in case the bath becomes too acid.

A few words may here be said in regard to what may be termed a *nickel bath without nickel salt.*   It simply consists of a 15 to 20 per cent. solution of ammonium chloride, which transfers the nickel from the anodes to the articles.   Cast anodes are almost exclusively used for the purpose, and deposition may be effected with quite a feeble current.   Before the solution acquires the capacity of depositing, quite a strong current has to be conducted through the bath until the com-mencement of a proper reduction of nickel.   This bath is only suitable for *coloring* very cheap articles, it being impos-sible to produce solid nickeling with it.   It is here mentioned because it may serve as a representative of a series of other electro-plating baths in which the transfer of the metal is effected by ammonium chloride solution without the use of metallic salts, for instance, iron, zinc, cobalt, etc.

Prepared nickel salts.   As previously mentioned, there is a large number of receipts for nickel baths, some of them being entirely unsuitable, while others are only available for certain purposes. Hence, it is impossible, even for the skilled operator, to separate the good receipts from the bad ones, if he is not qualified to do so by many years' experience and a thorough knowledge of chemistry. The choice is still more difficult for the beginner and layman, and it is recommended to them to get their supply of suitable baths from well-known dealers in electro-plating supplies.

By *prepared nickel salts* are understood preparations which, in addition to the most suitable nickel salt, contain the required conducting salt for the decrease of the resistance, and further such additions as promote a pure white separation of nickel, and are necessary for the continuously good working of the bath.

Correction of the reaction of nickel baths. When after long use a nickel bath has become alkaline, which is readily determined by a test with litmus paper, this defect can in a few minutes be overcome by the addition of an acid, and according to the composition of the bath, its neutrality or slightly acid reaction can be restored by citric, acetic, sulphuric, boric acids, etc. The use of hydrochloric acid for this purpose, which has been recommended, is not advisable. In most cases it will be best to employ dilute sulphuric acid, provided an excess of it be avoided, which is recognized by red congo-paper turning blue.

When a bath contains too much free acid, the latter may be removed by an addition of ammonia, ammonium carbonate, potash or nickel carbonate, the choice of the agent to be used depending on the composition of the bath.

Thick deposits in hot nickel baths. Nickel baths, more or less highly heated, have for years been used for nickeling, the purpose being, on the one hand, the production in a shorter time of a thick deposit, and on the other, it was expected that the product thus obtained would become especially dense in consequence of the contraction in cooling.

The results obtained in heated baths were, however, unsatisfactory, since, if the current was not carefully regulated, the deposit peeled off readily, and the polished nickeling became dull on exposure to the air.

The unsatisfactory results might primarily have been due to an unsuitable composition of the electrolytes. Foersters's *experiments have shown that almost perfectly smooth deposits of 0.5 to 1 millimeter thickness may be obtained in absolutely neutral nickel solutions with a high content of nickel—1 oz. or more per quart—if kept at a temperature of 122° to 194° F., for instance, in solutions containing 5 ozs. nickel sulphate per quart, at 167° to 176° F. Electro-motive force, at an electrode-distance of 4 cm., 1.3 volts; current-density, 2 to 2.5 ampères.

Exhaustive experiments made by Dr. George Langbein led to the result that deposits of great thickness may also be produced in slightly acidulated nickel baths of suitable composition, at a temperature kept constant at from 185° to 194° F. In a bath which contained 12.34 ozs. of nickel sulphate and 6.34 ozs. of sodium sulphate or magnesium sulphate per quart, and which was slightly acidulated with acetic acid, deposits of 0.5 millimeter thickness were in 12 hours obtained, the current-density amounting to 4 ampères.

For nickeling flat objects the current-density may, however, be materially increased, one of up to 8 ampères or more being permissible. By reason of the rapidity with which thick deposits can be produced in hot baths of the above-mentioned composition, the term *quick nickeling* has been applied to this process.

Independent of Dr. Langbein, Dr. Kugel discovered that thick deposits of nickel can be obtained in a hot nickel bath of nickel sulphate and magnesium sulphate very slightly acidulated with sulphuric acid.†

* Zeitschrift für Elektrochemie, 1897 to 1898, p. 160.
† German patent, 117054.

While, in order to avoid the formation of roughness and bud-like excrescences, Foerster found agitation of the electrolytes of the composition mentioned by him of advantage, Dr. Langbein obtained smoother deposits when the electrolyte was not mechanically agitated, and the fluid was only slowly mixed through by heating with a steam coil.

Upon flat objects, for instance, sheets, very uniform deposits 1 millimeter or more in thickness are very rapidly obtained, as well as upon round objects, if care be taken to have, by the use of anodes of the same shape, a uniform anode-distance from all object surfaces. However, the production of such deposits of entirely uniform thickness upon articles with high relief has thus far not been successfully accomplished by Dr. Langbein with the use of the above-mentioned electrolytes.

Thick deposits in cold nickel baths. By the use of an electrolyte which contains nickel ethyl sulphate (German patent, No. 134736) or the ethyl sulphates of the alkalies or alkaline earths, deposits of any desired thickness can be produced if the bath be constantly agitated by mechanical means or the introduction of hydrogen. Agitation by blowing in air is not permissible on account of oxidation of the ethyl-sulphate combinations by the oxygen of the air.

Continued experiments with such ethyl-sulphate combinations by Dr. G. Langbein & Co. resulted in finding formulas for prepared nickel salts from the solutions of which thick deposits of nickel capable of being polished can in a few minutes be obtained in the cold way. The formulas for these different prepared nickel salts will not be given, as they are protected by patents. The salts are known in commerce as Mars, Lipsia, Germania and Neptune.

In an electrolyte of given composition, which has to be constantly kept slightly acid with acetic acid, nickeling may for weeks be carried on at the ordinary temperature without any peeling-off of the deposit being noticed, and, in this respect, this bath surpasses all other known baths. In the course of six weeks, Dr. Langbein has produced upon gutta-

percha matrices, galvanoplastic nickel deposits 6 millimeters
in thickness, the metal proving thoroughly homogeneous and
firmly united throughout its entire thickness.

Coehn and Siemens * found that from electrolytes which
contain nickel salts and magnesium salts, weighable quantities
of magnesium are under certain conditions separated together
with the nickel, and they succeeded in depositing alloys containing approximately 90 per cent. of nickel and 10 per cent.
of magnesium. According to the above-mentioned authors,
the behavior of the nickel-magnesium alloys in the electrolytic
separation differs essentially from that of nickel, they showing
especially no tendency towards peeling off.

Nickel anodes. Either *cast* or *rolled* nickel plates are used
as anodes, they, of course, having to be made of the purest
quality of nickel. Every impurity of the anode passes into the
bath, and jeopardizes, if not at first, then finally, its successful
working. Rolled anodes dissolve with difficulty and cast
anodes, as a rule, with ease. If the latter dissolve only with
difficulty they fail in their object of replacing the nickel metal
withdrawn from the bath by the nickeling process.

As regards solubility, electrolytically produced nickel anodes
stand between rolled and cast anodes.

The anodes should not be too thin, otherwise they increase
the resistance. For small baths rolled anodes 2 to 3 millimeters thick are generally used. For larger baths, it is better
to use plates 3 to 10 millimeters thick, while the thickness of
cast anodes may vary from 3 to 10 millimeters, according to
their size.

Attention may here be called to the elliptic anodes, Fig.
110, patented by the Hanson & Van Winkle Co., Newark,
N. J. The great advantage claimed in the use of these
elliptic anodes over the old style flat plate is the uniformity
of deposit as disintegration takes place from all sides of the
anode ; consequently the molecules are distributed uniformly

---

* Zeitschrift für Elektrochemie, 1902, S. 591.

throughout the solution, and not only hasten the deposit, but give a heavier deposit in a given time. Another important feature in these anodes is the fact that they wear down evenly to a small, narrow strip, and when worn down to such a point that it seems desirable to put in more nickel, the old ones which take up practically no room in the tank, can remain until entirely consumed, and as a result there is practically no scrap nickel to dispose of at half price. Fig. 111 shows the small loss in the use of the elliptic anode. The weight of the orginal plate was 16 pounds. Percentage of waste only 5 per cent.

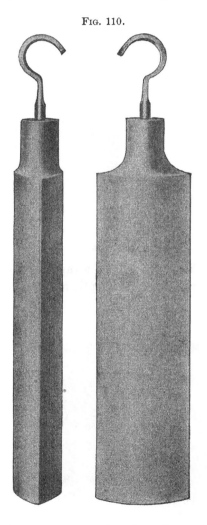

FIG. 110.

Fig. 112 shows the original shape of the flat plate still largely used and the character of the wear. The top part of the scrap-plate with its two ears is almost as heavy as the same section of the original plate. The original weight of the plate was 13⅝ lbs. Waste 2 lbs. Percentage of waste 14.6 per cent. Another form of the old-style plate is shown in Fig. 113. The original weight was 17½ lbs.   Weight of scrap 4⅞ lbs.   Percentage of waste 27.4 per cent.   The examples shown in the illustrations were

FIG. 111.

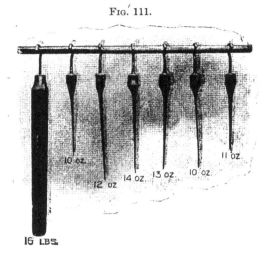

taken from a lot of scrap returned to the manufacturers.   The scrap from the elliptic anode came from a large stove concern

FIG. 112.   FIG. 113.

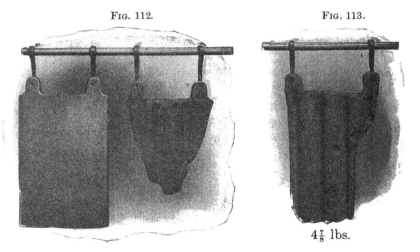

4⅞ lbs.

and the flat scrap also from a stove manufacturer.   Elliptic anodes are furnished in all commercial metals.

The use of insoluble anodes of retort carbon or platinum, either by themselves or in conjunction with nickel anodes, as frequently recommended by theorists, is not advisable. The harder and the less porous the nickel anode is, the less it is attacked in the bath and the less it fulfills the object of keeping constant the metallic content of the solution. On the other hand, the softer and the more porous the anode is, the more readily it dissolves, because it conducts the current better and presents more points of attack to the bath ; and the more it is dissolved, the more metal is conveyed to the bath. With the sole use of rolled anodes and working with a feeble current, free acid is formed in the bath; on the other hand, by working with cast anodes alone, the bath readily becomes alkaline. Now it would appear that the possibility of a bath also becoming alkaline even with the sole use of rolled anodes, especially when working with a strong current, has led to the proposition of suspending in the bath, besides the nickel anodes, a sufficient number of insoluble anodes in order to effect a constant neutrality of the bath. It would lead too far to go into the theory of the secondary decompositions which take place in a nickel bath to prove that, though neutrality is obtained, it can only be done at the expense of the metallic content of the bath. Hence this impracticable proposition will here be overthrown by practical reasons, it only requiring to be demonstrated that in baths becoming alkaline the content of nickel also decreases steadily though slowly. This fact in itself shows that in order to save the occasional slight labor of neutralizing the bath, the decrease of the metallic content should not be accelerated by the use of insoluble anodes.

For larger baths the use of expensive platinum anodes as insoluble anodes need not be taken into consideration, because for large surfaces of objects correspondingly large surfaces of platinum anodes would have to be present, as otherwise the resistance of thin platinum sheets would be considerable. But such an expensive arrangement would be justifiable only if actual advantages were obtained, which is not the case, because, though the platinum does absolutely not dissolve, the deficiency of metallic nickel in the bath caused by such anodes must in some manner be replaced.

The insoluble anodes of *gas-carbon*, which have frequently been proposed, are attacked by the bath. Particles of carbon become constantly detached, and floating upon the bath, deposit themselves upon the objects and cause the layer of nickel to peel off. Furthermore, by the use of nickel anodes in conjunction with carbon anodes, the current, on account of the greater resistance of the latter, is forced to preferably take its course through the metallic anodes, in consequence of which the articles opposite the nickel anodes are more thickly nickeled than those under the influence of the carbon anodes. With larger objects this inequality in the thickness of the deposit is again a hindrance to obtaining layers of good and uniform thickness, such as are required for solid nickeling. Since the current preferably seeks its compensation through these separate metallic anodes, they are more vigorously attacked than when nickel plates only are suspended in the bath.

With nickel baths which contain a considerable amount of ammonium chloride, the use of a few carbon anodes along with the rolled nickel anodes may be permissible, since these baths strongly attack even the rolled anodes, and thereby convey to the bath sufficient quantities of fresh nickel. Such baths containing ammonium chloride, as a rule, become very rapidly alkaline, so that frequent neutralization becomes inconvenient. However, in this case, it is advisable to place the carbon anodes in small linen bags which retain any particles of carbon becoming detached, the latter being thus prevented from depositing upon the articles in the bath.

According to long practical experience, the best plan is to use rolled and cast anodes together in a bath which does not contain chlorides, and to apportion the anode surface so that an anode-rod, about $\frac{2}{3}$ of its length, is fitted with anodes. If, for instance, a tank is 120 centimeters long in the clear and 50 centimeters deep, the width of the nickel anodes laid alongside one another should be about 80 centimeters, and their length about ⅔ of the depth of the tank, hence 30 centimeters. For each anode-rod, 8 anodes, each 30 centimeters long and 10 centimeters wide, would, therefore, be required.

The *proportion of cast to rolled anodes* depends on the composition of the bath, but it may be laid down as a rule that baths with greater resistance require more cast anodes, and baths with less resistance more rolled anodes. Baths with the greatest resistance, for instance, that prepared according to formula I, require only cast anodes, while baths with the smallest resistance, for instance, those containing ammonium chloride, may to advantage work only with rolled anodes; baths with medium resistance require mixed anodes.

The *proper proportion* has been established when, after working for some time, the original reaction of the bath remains as constant as possible. When the bath is observed to become alkaline, the number of rolled anodes should be increased, but when the content of acid increases they should be decreased, and the number of cast anodes increased.

Cast anodes, especially those not cast very hot, have, to be sure, the disadvantage of becoming brittle, and crumbling before they are entirely consumed. Nickel anodes cast in iron moulds are so hard on their surfaces as to resist the action of the bath, and dissolve only with difficulty, so that the content of metal of the bath is only incompletely replenished. Anodes cast in sand moulds, and slowly cooled, are porous and consequently dissolve readily, but by reason of their porosity their interior portions are also attacked. If such an anode be broken, it will be found that the interior contains a black powder (nickel oxide) which novices sometimes believe to be carbon. In fact cases have been heard of that customers have complained that the anodes furnished them were not nickel anodes at all, but simply carbon plates coated with a layer of nickel.

The cast anodes suspended to the ends of the conducting

18

rods are especially strongly attacked, and, therefore, when rolled and cast anodes are used together, it is best to suspend the latter more towards the center, and the former on the ends of the rods.

These drawbacks, however, are not sufficient to prevent the use of a combination of cast and rolled anodes when required by the composition of the bath. The brittle remnants of the anodes are thoroughly washed in hot water, dried, and sold.

Rolled nickel anodes are less liable to corrosion, and may be used up to the thickness of a sheet of paper before they fall to pieces. It is, however, best to replace them by fresh anodes before they become too thin, since with the decrease in thickness their resistance increases.

It is best to allow the anodes to remain quietly in the bath, even when the latter is not in use, they being in this case not attacked. By frequently removing and replacing them they are subject to concussion, in consequence of which they crumble much more quickly than when remaining quietly in the bath.

In the morning, before nickeling is commenced, the anodes will frequently show a *reddish* tinge, which is generally ascribed to a content of copper in the bath or in the anodes. This reddish coloration also appears when an analysis shows the anodes, as well as the bath, to be absolutely free from copper. It is very likely due to a small content of cobalt, from which nickel anodes can never be entirely freed. It would seem that by the action of a feeble current, cobaltous hydrate is formed, which, however, immediately disappears on conducting a strong current through the bath.

Pfanhauser is of the opinion that this reddish tinge is due to a separation of copper. In fact, even the purest brands of anodes contain traces of copper, but, on the other hand, the nickel salts are at present furnished mostly entirely free from copper, and a nickel bath would have to be worked for a long time before a content of copper would be transferred to it from the anodes. An experiment showed that a bath prepared with nickel salt absolutely free from copper produced a slight red film upon a new anode without the current having been in action ; a bright steel-sheet served as anode. This does not indicate the separation of copper, as its derivation would in this case be inexplicable.

The anodes are supported by pure nickel wire 0.11 to 0.19 inch thick, or by strips of nickel sheet riveted on.

It has previously been mentioned that the anodes in baths at rest are frequently more strongly attached at the upper than at the lower portions, because specifically lighter layers of fluid are present on top and heavier ones below, and the current takes the road where there is the least resistance. This disproportionate solution of the anodes may, however, also be noticed in baths which are agitated, and consequently in which no layers of different specific gravities are present. The lower and side edges will be found more corroded than the middle portions of the anodes, and the backs. opposite to which no objects are suspended appear also strongly attacked. These observations render plausible Pfanhauser's supposition that the current does not in all places migrate directly and in straight lines from the anodes to the cathodes, but that, as with the magnetic lines of force, this migration takes place in curves; especially when the anode-surface is small in proportion to the cathode-surface. Pfanhauser has applied the term *scattering of current lines* to this migration of the current in curves, and has noticed that it grows with the electrode-distance, and decreases as the electro-surfaces are increased.

Execution of nickeling. Next to the correct composition of the bath and the proper selection of the anodes, the success of the nickeling process depends on the articles having been carefully freed from grease and cleaned, and on the correct current-strength.

The mechanical preparation of the objects has been discussed on page 188 *et seq.*

The directions for the removal of grease, etc., given on p.

228, also apply to objects to be nickeled. In executing the operations, it should always be borne in mind that though dirty, greasy parts become coated with nickel, the deposit immediately peels off by polishing, because an intimate union of the deposit with the basis-metal is effected with only perfectly clean surfaces. Touching the cleansed articles with the *dry* hand or with dirty hands must be strictly avoided ; but, if large and heavy objects have to be handled, the hands should first be freed from grease by brushing with lime and rinsing in water, and be kept wet.

As previously mentioned, the cleansed objects must not be exposed to the air, but immediately placed in the bath, or, if this is not practicable, be kept under clean water.

While copper and its alloys (brass, bronze, tombac, German silver, etc.), as well as iron and steel, are directly nickeled, zinc, tin, Britania and lead are generally first coppered or brassed.

With a suitable composition of the nickel bath and some experience, the last-mentioned metals may also be directly nickeled ; but, as a rule, previous coppering or brassing is preferable, the certainty and beauty of the result being thereby considerably enhanced.

Security against rust.—By many operators it is preferred to copper iron and steel articles previous to nickeling, it being claimed that by so doing better protection against rust is secured. However, comparative experiments have shown that with the thin coat of copper which, as a rule, is applied, this claim is scarcely tenable, and the conclusion has been reached that a thick deposit of nickel obtained from a bath of suitable composition protects the iron from rust just as well and as long as if it had previously been slightly coppered. It cannot be denied that previous coppering of iron articles has the advantage that in case the articles have not been thoroughly cleansed, the deposit of nickel is less liable to peel off, because the alkaline copper bath completes the removal of grease ; but with objects carefully cleansed according to the directions given on page 228, previous coppering is not necessary.

The case, however, is different if the copper deposit is produced in order to act as a cementing agent for two nickel deposits. If, for instance, parts which have previously been nickeled, and from which the old deposit cannot be removed by mechanical means, are to be re-nickeled, coppering is required, because the new deposit of nickel adheres very badly to the old. Where articles are to be protected as much as possible from rust, coppering is advisable, but the best success is attained by a method different from the one generally pursued. In nickeling, for instance, parts of bicycles which are exposed to all kinds of atmospheric influences, they are first provided with a thick deposit of nickel, then with a thick coat of copper, and finally again nickeled, they thus being twice nickeled. It has previously been mentioned that every deposit is formed net-like, the meshes of the net being larger or smaller, according to the nature of the metal deposited. If now thick layers of two different metals are deposited one on the top of the other, the net-lines of one deposit do not converge into those of the previous deposit, but are deposited between them, thus consolidating the net. It will now be readily understood that by the subsequent polishing the further consolidation of the deposits will be far more complete than when the basis-metal receives but one deposit, which is to be consolidated by polishing. It is a remarkable fact that the porosity of the nickel deposit varies if the article is nickeled in several baths of different composition. Thus denser deposits may be obtained by suspending the articles in two or three baths, which proves that the different resistances of the respective baths of one and the same metal exert an influence upon the greater or slighter density of the net.

However, under certain conditions, even iron and steel objects doubly nickeled in the above-described manner do not offer a sure guarantee against rusting of the basis-metal, and to absolutely prevent the latter, the following means may be adopted :

The objects are provided with an electro-deposit of zinc. This deposit is scratch-brushed, coppered in the copper cyanide bath, rinsed in water, and finally nickeled, at first with a strong current, which is after a few minutes reduced to the normal current-density. It is recommended to polish the objects thus treated with circular brushes, and not use polishing wheels which may cause them to become heated, because by such heating blisters are readily formed.

Another plan is as follows : The objects are first coppered in the copper cyanide bath. The thickness of this deposit is then increased to 0.15 or 0.2 millimeter in the acid copper bath (see Galvanoplasty). It is then polished and nickeled. Or, if there is sufficient time, a very thick deposit of nickel is directly produced upon the object with the use of a cold ethyl sulphate nickel bath, or a hot quick nickeling bath (see pp. 266 et seq.).

The objects should never be suspended in the bath without current, since the baths, with few exceptions, exert a chemical action upon many metals, which is injurious to the electroplating process, and especially with nickel baths it is necessary to connect the anode-rods and object-rods before suspending the objects.

Over-nickeling. An error is frequently committed in nickeling with too strong a current, the consequence being that the deposit on the lower portions of the objects soon becomes dull and gray-black, while the upper portions are not sufficiently nickeled. This phenomenon is due to the reduction of nickel with a coarse grain in consequence of too powerful a current, and is called *burning* or *over-nickeling.* A further consequence of nickeling with too strong a current is that the deposit readily peels off after it reaches a certain thickness. This phenomenon is due to the hydrogen being condensed and retained by the deposit, which is thereby prevented from acquiring greater thickness.

Especially do those objects suspended on the ends of the rods nickel with great ease. This evil can be avoided by hanging on both ends of the rods a strip of copper-sheet about 0.39 inch wide, and of a length corresponding to the depth of the bath.

Normal deposition. The following criteria may serve for judging whether the nickeling progresses with a correct current-strength: In two, or at the utmost three, minutes, all portions of the objects must be perceptibly coated with nickel, but without a violent evolution of gas on the objects. Small gas bubbles rising without violence and with a certain regularity are an indication of the operation progressing with regularity. If, after two or three minutes, the objects show no deposit, the current is too *weak*, and in most cases the objects will have acquired dark, discolored tones. In such case, either a stronger current must be introduced by means of the rheostat, or, if the entire volume of current generated already passes into the bath, the object-surface has to be decreased, or, if this is not desired, the battery must be strengthened by adding more elements, or by fresh filling, etc.

If, on the other hand, a violent evolution of gas appears on the objects, and the latter are well covered in a few seconds, and the at first white and lustrous nickeling changes in a few minutes to a dull gray, the current is *too strong*, and must be weakened either by the rheostat, or by uncoupling a few elements, or diminishing the anode-surface. or finally by suspending more objects in the bath.

These criteria also apply to nickeling with the dynamo.

The most suitable current-density for nickeling varies very much, as will be seen from the preceding explanations. For the ordinary cold electrolysis it varies for copper, brass, iron, and steel from 0.3 to 1.5 ampères, while zinc, previously coppered, requires 1 to 1.2 ampères. In the hot nickel bath the current-density may be up to 5 and more ampères.

In nickeling zinc objects greater current-density and higher electro-motive force are required. If the current is not of sufficient strength, black streaks and stains are formed, zinc is dissolved, and the nickel bath spoiled. These evils are frequently complained of by nickel-platers who have not a clear perception of the prevailing conditions (see polarization-current.) A vigorous evolution of gas must take place on the zinc objects, otherwise a serviceable deposit will not be obtained.

In most cases the electro-plater will in a few days learn correctly to judge the proper current-strength by the phenomena presented by the objects, and if he closely follows the directions given but few failures will result. It may here be again repeated that the use of a voltmeter and ammeter, as well as of a rheostat, greatly facilitates a correct estimate of the proper current-strength, and these instruments should for the sake of economy never be omitted in fitting up an electro-plating plant.

It is in every case advisable first to cover the objects, i. e., to effect the first deposit of nickel, with the use of a strong current, in order to withdraw the metals from the action of the solution. The current is then reduced to a suitable strength and nickeling finished with this current. With a current thus regulated, the objects may be allowed to remain in the bath for hours, and even for days. It is further possible to nickel by weight and attain deposits of considerable thickness.

If very thick deposits of nickel are to be produced in the ordinary bath, the objects must be frequently turned, as the lower portions are more heavily nickeled than the upper; further, as soon as the deposit acquires a dull bluish luster, it has to be thoroughly scratch-brushed, in doing which, however, the objects must not be allowed to become dry. After scratch-brushing it is advisable to cleanse the deposit once more with the lime-brush, and after rinsing replace the objects in the bath. If burnt places cannot be brightened and smoothed with the scratch-brush, the desired end is readily attained with the assistance of emery paper or pumice.

For *solid* nickeling it suffices for most articles, with a normal current to allow them to remain in the bath until a mat bluish shine appears; this is an indication that the deposit has acquired considerable thickness, provided the bath has

not been alkaline.   In alkaline baths this dull deposit is frequently formed before the deposit has attained considerable
thickness and this may cause errors, if the reaction of the bath
is not frequently controlled.

If the mat appearing objects are permitted to remain longer
in the bath without scratch-brushing, the mat bluish tone
soon passes into a mat gray, and all the metal deposited in
this form must be polished away in order to obtain a bright
luster.

Whether the deposit of nickel is sufficiently heavy for all
ordinary demands is, according to Fontaine, shown by rubbing a nickeled corner or edge of the object rapidly and with
energetic pressure upon a piece of planed soft wood until it
becomes hot.   The nickeling should bear this friction.   This
test can be recommended as perfectly reliable.

Faulty arrangement of anodes.   If the objects, after having
been suspended for some time in the bath, are only partially
nickeled, it is very likely due to the defective arrangement
of the anodes.   This occurs chiefly with large round objects
and with articles having deep depressions (cups, vases, etc.).
It is, of course, supposed that the wires to which the objects
are suspended in the bath have a sufficiently large cross-
section to carry the current required for nickeling the entire
surface of the object.

For flat objects suspending them between two rows of anodes
suffices.   Round objects with a large diameter should be quite
surrounded with anodes, and be as nearly as possible equidistant from them.   This arrangement should especially not
be neglected where a heavy and uniform deposit of nickel is
to be applied to round or half-round surfaces, for instance,
large half-round stereotype plates for revolving presses.

The arrangement of two object-rods between two anode-rods
is permissible only for small and thin articles such as safety-
pins, crochet needles, lead-pencil holders, etc.   For articles
with larger surfaces it is decidedly objectionable, because the
sides of the articles turned towards the anodes acquire a

thicker deposit than the inside surfaces, and the thickness of the deposit decreases with the distance from the anodes.

Nickeling of cavities and profiled objects. While for smooth articles the most suitable distance of the anodes from the objects is $3\frac{3}{4}$ to $5\frac{3}{4}$ inches, for objects with depressions and cavities it must be larger, if it is not preferred to make use of the methods described later on. However, a deposit of a uniform thickness cannot be obtained by this means, because the portions nearer to the anodes will acquire a thicker deposit than the cavities ; hence the use of a small hand anode, which is connected by means of a thin, flexible wire with the anode-rod, and introduced into the depressions and cavities, is to be preferred. This, of course, renders it necessary for a workman to stand alongside the bath and execute the operation by hand ; but as the small anode can be brought within a few millimeters of the surface of the article, and at this distance slowly moved around it, a correspondingly thick deposit is in a short time formed.

At any rate baths in which objects with depressions and profiled articles are to be nickeled must possess greater resistance than baths for nickeling flat articles, and it is inexplicable why a bath with a large content of ammonium chloride and consequently slight conducting resistance can be recommended, as has been done, for nickeling hollow articles. When baths containing ammonium chloride are used for nickeling articles with deep cavities the portions nearest to the anodes will frequently be found overnickeled—burnt—before the deepest portions are at all covered with nickel, and if the operator waits until the deposit upon the latter portions has acquired the desired thickness, the deposit already peels off from the former portions, and frequently before that time. By comparative experiments in nickeling the inside of brass tubes, 15 millimeters in diameter, it was found that in a bath with great resistance, as well as in one with slight resistance, nickeling was equally well effected. However, the phenomenon of peeling off, above referred to, appeared in the bath which contained ammonium chloride when the ends of the 120-millimeter long tubes turned away from the anode were still so slightly nickeled that the basis-metal showed through. On the other hand, in the bath without ammonium chloride the end of the tube turned towards the anode, to be sure, became mat, but did not peel off in polishing, and nickeling in the interior of the tube had progressed well to the opposite end, the basis-metal there being well covered.

In nickeling lamp-feet of cast-zinc, the use of the hand-anode can scarcely be avoided if the depressed portions also are to be provided with a uniformly good deposit. Moreover, zinc articles form an exception to the general rule in so far as by reason of the highly positive properties of zinc, the resistance of the bath may be slighter than the baths for nickeling copper and its alloys, as well as iron and steel.

Besides the above-mentioned general rules for nickeling, which also hold good for other electro-plating purposes, the following may be given:

In suspending the objects in the bath, rub the metallic hooks or wires, with which they are secured to the rods, a few times to and fro upon the rod, in order to be sure that the place of contact is purely metallic. It is also well to acquire the habit of striking the rod a gentle blow with the finger every time when suspending an object, the gas-bubbles settling on the articles becoming thereby detached and rising to the surface. It is further advisable, before securing the objects to the object-rod, to move them up and down several times; so to say, shake them beneath the fluid, whereby, on the one hand, the layers poorer in metal are mixed with those richer in metal, and, on the other, any dust which may float upon the bath and settle on the objects is removed.

The objects suspended in the bath should not touch one another, nor one surface cover another, and thus withdraw it from the direct action of the anode. In the first case stains will readily form on the places of contact, and, in the latter, the covered surface acquires only a slight deposit. That the objects must not touch the anodes need scarcely be mentioned.

Objects with depressions and cavities should be suspended in the bath so that the air in the depressions and cavities can escape, which is effected by turning the depression upwards, or, if there are several depressions on opposite sides, by turning the articles about after being introduced into the bath. Air-bubbles remaining in the hollows prevent contact with the solution, no deposit being formed on such places.

Polarization. It remains to say a few words in regard to the so-called polarization phenomena. In the theoretical part of this work, it has been shown that by dipping two plates of different metals in a fluid a *counter or polarization current* is generated, which is the stronger the greater the difference in the potentials of the two metals in the solution is. If the anodes in a nickel bath consist of nickel and the objects of copper, the counter-current will be slight. It becomes, however, greater when iron objects are suspended in the bath, and still greater with zinc surfaces which are to be nickeled, because with the solutions here in question, zinc possesses towards nickel an essentially higher potential. Now, since the counter-current flows in a direction opposite to that of the current introduced at the bath, the latter is weakened, and the more so the stronger the counter-current is. This explains why iron requires a stronger current for nickeling than copper-alloys, and zinc a stronger one than iron.

Now it may happen that the counter-current becomes so strong as to entirely check the effect of the main current, and even to reverse the latter, the consequence being that, in the first case, the formation of the deposit is interrupted, and, in the latter, that the deposit is again destroyed, and the metals of which the articles consist dissolve and contaminate and spoil the bath. To avoid this, a main current must be conducted into the bath, which, by its sufficiently, large electro-motive force can overcome the counter-current, and the consequences of the reversal of the current can be prevented by using the galvanometer and observing the deflection of its needle, which (according to p. 143) in proper time indicates the appearance of a reversed current. Now if a nickel-plater has only a slight current at his disposal, it follows from the above explanation that, before nickeling the more electro-positive metals, such as iron, tin, zinc, it is best first to copper them, and thereby overcome the action of these metallic surfaces as regards the formation of the counter-current.

It happens comparatively seldom that the counter-current becomes so strong as to destroy the deposits formed, because for nickeling powerful Bunsen cells with two acids, or a dynamo with at least 4 volts' impressed electro-motive force, are generally used. It is, however, well to acquaint the oper-ator with all possible contingencies, and to explain the reason why the articles are preferably covered with a strong current. Sprague recommends an initial current of 5 volts' electro-motive force, but in most cases one of 3.5 volts suffices for nickeling iron and copper alloys.

Stripping defective nickel. Defective nickeling must, as a rule, be completely removed before the objects can be nick-eled, since the second deposit does not adhere to the previous one, but frequently peels off in polishing or by slightly bend-ing the object. The reasons for this behavior are : 1. Like iron, nickel readily oxidizes on the surface, but this oxidation is not so heavy as to be perceptible. Previous to nickeling this oxide has not been completely removed and in the case of quite old plated objects the nickel has had a chance to oxidize. Nickel, however, adheres firmly only to metallic nickel and not to the oxide ; hence the second deposit peels off. 2. In case the deposit is comparatively new and has not been exposed for some time to the action of atmospheric air, the peeling off of nickel deposited upon nickel is, as a rule, caused by the polishing material remaining upon the surface. Vienna lime and similar agents which contain paraffin and other mineral fats and wax are much used for polishing nickel. These substances partially penetrate into the pores of the deposited nickel or remain upon the surface. By the ordinary means of cleaning the mineral fats or wax are not removed, the consequence being that the second deposit of nickel does not adhere. With the use of animal fats, which readily saponify, as polishing agents, the case is not so bad, but even under these conditions, the nickel has a tendency to peel off. It must be borne in mind that as previously mentioned, all electrolytically produced deposits are composed of a net-work of very minute crystals, the deposit being thus of a porous nature. In polishing larger or smaller quantities of the polishing agent penetrate into these pores, and their complete removal is a very difficult matter.

For the removal of the nickel coating the following stripping acid, which may be used either cold or tepid, has been recommended: Sulphuric acid of 66° Bé., 4 lbs.; nitric acid of 40° Bé., 1 lb.; water about 1 pint. First put the water in a stoneware jar and cautiously add, a little at a time, the sulphuric acid, since considerable heat is generated when this acid is mixed with water. When the entire quantity of sulphuric acid has been added, pour in the nitric acid, when the bath is ready for use. In making up the stripping bath, the proportions of the acids may be varied, but the foregoing will be found to answer every purpose. An addition of 8 ozs. of potassium nitrate to the bath has also been recommended.

When stripping nickel-plated articles in the above bath it is necessary to watch the operation attentively, since some articles are very lightly coated and a momentary dip is frequently sufficient to deprive them of their nickel. Other articles which have been thoroughly well nickeled, but require from some accidental cause to be stripped and re-nickeled, will need immersion for several minutes; indeed well nickeled articles may occupy nearly half an hour in stripping before the underlying surface is entirely bare. The operation of stripping should be conducted in the open air, or in a fire-place, so that the acid fumes, which are very pernicious, can escape freely. The articles should be attached to a stout copper wire, and after a few moments' immersion should be removed from the bath to see how the operation progresses, it being absolutely necessary that the work should not remain in the stripping solution one instant after the nickel is removed. The object is then transferred to a large volume of cold water, and after washing twice or three times in fresh water is ready for the subsequent stages of the process. When stripping has been properly effected, the underlying metal exhibits a bright, smooth surface, giving little evidence of the mixture having acted upon it.

Many platers, however, prefer to remove the nickel coating mechanically by brushing with emery. From depressions it is as much as possible removed with the brush, after which the object is freed from grease and pickled, and coppered before nickeling. In this case the layer of copper serves for cementing together the old and new deposits, and there will be no danger of the new deposit peeling off in polishing.

It has also been proposed to strip by electrolysis by making the object the anode in an old nickel bath, Attention is equally necessary in conducting this process to guard against any attack upon the basis-metal ; but since it is impossible to prevent all action, no bath which is to be afterward employed for depositing the metal should be used for this purpose, as it will become gradually charged with impurities. A 10 per cent. solution of sulphuric acid in water may be equally readily adapted to the electrolytic stripping.

Many nickel-plated iron and steel objects are so cheap that it does not pay to strip the nickel from them, and it is best to throw them on the scrap pile. In some cases, however, for instance, surgical instruments, fire-arms, fine cutlery and other more expensive articles, it is frequently desirable to remove the old nickel deposit. To be sure, nitric acid would remove the nickel, but it also attacks iron and steel and causes pitting. For stripping such articles by electrolysis the following bath has been recommended : Water 1 lb., potassium cyanide 1.8 ozs., yellow prussiate of potash 0.5 oz. The iron or steel object to be stripped is suspended as anode in the bath, which should be used at a temperature of 122° F. A sheet of iron or steel serves as cathode. For stripping a thin deposit only a few hours are required, but a whole day for thick deposits. However, the operation requires no special attention, as the iron or steel surface is not attacked and there is no danger of pitting. The current-strength should be the ·same as usually employed for nickel-plating.

As a remedy against the yellowish tone of the nickeling, Pfanhauser recommends suspending the nickeled articles, immediately after taking them from the nickel bath, as anodes in a nickel bath acidulated with citric or hydrochloric acid, a piece of sheet nickel serving as the cathode, and to allow the current to act for a few seconds. It is claimed that thereby the basic nickel salts separated together with the nickel, and to which, according to Pfanhauser, the yellowish tinge is due, are dissolved, and the nickeling will show a pure white tone.

As nickel anodes contain, as a rule, iron, a minute quantity of this metal is deposited together with the nickel, and the latter is inclined to form a mat surface or to tarnish. If the ·objects are to be polished this does not matter, but if they are not to be polished slight mat stains frequently appear upon the surface after drying. Such stains can be removed by the use of a bath of dilute hydrochloric acid (2 parts water, 1 part acid). After thoroughly rinsing the object in water, immerse it for a moment in the acid bath, and then rinse again carefully. Now, without drying, draw the object through a soap-bath and rinse again. Since the soap solution leaves a thin film of oil upon the nickel surface not much water will adhere to it, and it will quickly dry. It will be found that the mat ·spots have disappeared or the stains are scarcely perceptible.

Defective nickeling. The following is a brief resumé of the principal defects which may occur in nickeling, as well as the means of avoiding them :

1. The articles do not become coated with nickel, but acquire discolored, generally darker, tones. *Reason :* The ·current is either too feeble to effect the reduction of nickel, and the coloration is due to the chemical action of the nickel solution upon the metals constituting the objects. This phe-

nomenon is frequently observed in nickeling zinc articles. *Remedy:* Increase the current or diminish the area of suspended objects; also examine whether the current actually passes into the bath, otherwise clean the places of contact.

2. A deposition of nickel takes place, but it is dark or spotted or marbled, even with a sufficiently strong current. *Reasons:* The bath is either alkaline, which has to be ascertained by testing with litmus-paper, and, if so, the slightly acid reaction of the bath has to be restored by the addition of a suitable acid; *or,* the bath is too concentrated, in which case a separation of crystals will be observed—this is remedied by diluting with water; *or,* the nickel solution is very poor in metal, which can be remedied by the addition of nickel salt; it should also be tested as to the admixture of copper, the production of dark tones being frequently due to this—in this case the bath is allowed to work for some time, and if the content of copper is inconsiderable a white deposit will soon be obtained; *or,* the cleaning and pickling of the articles have not been thoroughly done, which is remedied by again cleaning them; *or,* the conducting power of the bath is insufficient, which is remedied by the addition of a suitable conducting salt.

When freshly prepared baths yield dark nickeling, it can generally be remedied by working the bath two or three hours, if it is not over-concentrated and the cause, as above mentioned, has to be looked for in a small content of copper in the nickel salt.

3. A yellowish tinge of the nickeling. *Reason:* Alkalinity of the bath. *Remedy:* See under 2; *or,* with cast-iron, an insufficient metallic surface, which is remedied by repeating the scratch-brushing; *or,* unsuitable composition of the bath.

4. The objects rapidly acquire a white deposit of nickel, but the color soon changes to a dull gray-black, especially on the lower edges and corners. *Reason:* Too strong a current. *Remedies:* Regulating the current, or suspending more objects, or uncoupling elements. Frequent turning of the articles.

19

5. The nickeling is white, but readily peels off by scratching with the finger-nail, or by the action of the polishing wheel. *Reasons:* The current is too strong, which is remedied as under 4; *or,* the bath is too acid—this is remedied by the addition of ammonia, potassium carbonate, or nickel carbonate, according to the composition of the bath; *or,* freshly prepared nickel bath or freshly made additions, this being remedied by working the bath and by very careful regulation of the current in nickeling during the first days; *or,* insufficient cleaning and pickling, which is remedied by thorough cleaning after removing the defective deposit, or, if it cannot be entirely removed, coppering.

6. Though nickeling may proceed in a regular manner, some places remain free from deposit. *Reasons:* Either the surfaces of some of the objects touch one another; *or,* are stained by having been touched with dirty fingers; *or,* air bubbles are inclosed in cavities. *Remedy:* Removal of the causes.

7. The deposit appears with small holes. *Reason:* A deposit of particles of dust upon the objects. *Remedy:* Remove the dust from the surface. When there is a general turbidity of the bath in consequence of alkalinity, add the most suitable acid, and boil and filter the bath; *or,* insufficient removal of gas bubbles from the objects. *Remedy:* Shake the object-rods by blows with the finger.

8. Deposition takes place promptly upon the portions of the objects next to the anodes, while deeper portions remain free from nickel or become black. *Reason:* Too slight a distance of the objects from the anodes. *Remedy:* Increasing the distance; with large depressions, treatment with the hand-anode.

Refreshing nickel baths.—According to their composition, the amount of work performed, and the anodes used, the baths will in a shorter or longer time require certain additions in order to keep their action constant. By "refreshing" is not understood the small addition of acid or alkali from time to time required for restoring the original reaction of the baths, but additions intended to increase the metallic content and the diminished conductivity.

The metallic content is increased by boiling the bath with some of the nickel salt used in its preparation, while the conductivity is improved by adding, at the same time, so much conducting salt as is necessary to restore the electro-motive force originally required. Nothing definite can, of course, be said in regard to the quantity of such additions, it being advisable to observe their effect on a small portion of the bath, so as to be sure not to spoil the entire bath.

Nickel baths bear, as a rule, refreshing several times, but as in the course of time they take up impurities, even when the greatest care is exercised, it is best to refresh them at the utmost twice, and then to renew them entirely.

The *treatment of the articles after nickeling*, as well as after all electro-plating processes, has already been described, and it is only necessary here to refer again to the fact, that with articles of iron and steel, immersion in boiling water before drying in sawdust is absolutely necessary, and subsequent drying in a drying chamber is also a great safeguard as regards stability and protection against rust.

Nickel deposits are polished upon felt wheels or bobs of cloth, muslin or flannel, with the use of Vienna lime, rouge, Victor white polish, etc. (See " Polishing," p. 216). To give the objects the highest luster possible, it is advisable finally to polish them upon a woolen brush with dry Vienna lime.

Sharp edges, corners and raised portions should be held only with slight pressure against the polishing wheels, they being more strongly attacked by them than flat surfaces. The latter can stand a stronger pressure without fear of cutting through the deposit, provided the latter is of sufficient thickness and hardness.

Knife blades and surgical instruments with sharp edges require special care in polishing, which will later on be referred to.

Cleansing polished objects. After polishing, the nickeled

objects, especially those with depressions, have to be freed
from polishing dirt by brushing with hot soap-water, or dilute
hot caustic lye, or benzine, then rinsed in hot water and dried.

Calculation of the nickeling operation. Many inquiries regarding the mode of calculating the price to be charged for
nickeling objects give rise to the following remarks: If the
same article with the same definite surface is always to be
nickeled, the calculation is quite simple. From the current-
strength and the time required for nickeling, the weight of
the nickel-deposit can be readily determined by keeping in view
that 1 ampère theoretically deposits in 1 hour 1.1 gramme
of nickel, or about 1 gramme if the current output be taken
into consideration. The value of the ascertained weight has
to be determined by taking the cost of the anodes as the basis,
and from this is calculated the constant price of the separate
piece. To this has to be added the wages for grinding, polishing and nickeling, as well as the amount of power required,
which, according to the motors in use, has to be established
by a special calculation; further, the materials used for grinding, polishing, freeing from grease, etc., and a certain profit.
However, in most cases it is scarcely possible to make such
detailed calculations in electro-plating establishments in which
the most diverse objects have to be nickeled, because, on the
one hand, the determination of the surface of the separate
objects would be difficult and time-consuming, and on the
other, it would be very troublesome, in consequence of the
change of the object-surfaces in the bath, to keep an accurate
account of the current-strength and time required for the
separate objects.

To attain the object, it has in practice proved the simplest
plan to take as a basis the wages paid to the grinder and
polisher, and multiply them by 4, in order to obtain the selling price of the work furnished. The selling value thus determined includes all expenses and a fair profit. Somewhat more
will have to be allowed for particularly complicated objects
which require assistance with the hand-anode. This mode of

calculation has on the whole been found to answer for solid, heavy nickeling. For light nickeling—coloring white in the nickel bath—the selling value might be too high. An extra charge will of course have to be made for repairing articles which are received for nickeling.

When objects already ground and polished are sent in to be nickeled, the above-mentioned mode of calculation is of course not applicable. In that case it has to be calculated how much a charge of a bath must bring, in order to cover expenses and a certain profit, and from that the approximate selling value of the nickeling work may be determined.

Nickeling small and cheap objects in large quantities. This is

FIG. 114.

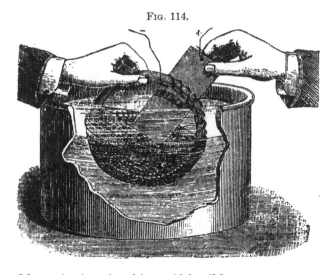

effected by stringing the objects, if feasible, upon a copper wire, and placing a large glass bead between every two objects, to prevent the surfaces from sticking together in the bath. Such objects being generally only slightly nickeled, it suffices to allow them to remain for a few minutes only in the bath with a strong current, it being advisable to diligently shake the bundles in order to effect a change of position of the objects and prevent their touching one another, notwithstanding the glass bead placed between them.

Very small objects, such as rivets, pins, etc., which cannot be strung upon wire, are nickeled in dipping baskets of stoneware or wire. To the bottom of the dipping basket is secured a copper or brass wire, which is connected with the object-rod, and the articles, not too many at a time, are then placed in the basket. During the operation the articles must be constantly shaken, and as nickel baths, as a rule, do not conduct sufficiently well to properly nickel the objects in the basket, it is advisable to hold with one hand an anode, connected by a flexible wire with the anode-rod, in the basket, while the other hand holds the basket (Fig. 114) and constantly shakes

FIG. 115.

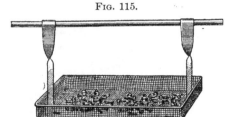

and turns it. For nickeling in the dipping basket it is further advisable to heat the nickel bath.

In place of a stoneware dipping basket, a basket tray of brass wire, Fig. 115, to which are soldered two copper wires for suspending it to the object-rod, may preferably be used. From the soldered places a few copper wires extend to the bottom of the basket. To prevent the basket from becoming covered with nickel it is coated with asphalt varnish. At a distance of about $2\frac{1}{2}$ to 3 inches below the basket an anode is arranged in horizontal position, while with one hand a hand-anode is held over the small articles in the basket. By this arrangement a thicker deposit is more rapidly obtained, especially if, with the other hand, the articles are constantly stirred by means of a glass or wooden rod.

Warren has described a solution of nickel and one of cobalt which can be decomposed in a simple cell apparatus. With the nickel solution, which was prepared by dissolving 100 parts bv weight of nickel chloride in as little water as possible and mixing with a concentrated solution of 500 parts of Rochelle salts, no satisfactory results could be obtained. The cobalt solution however yielded good results, and would seem to be suitable for electro-plating small objects in large quantities. It will be further referred to under " Deposition of Cobalt."

In the last few years a number of contrivances for electro-plating small articles in large quantities have been patented, the articles to be plated being, as a rule, contained in a revolving perforated drum. The drums of some of the contrivances are constructed of non-conducting material so that the articles receive the current through copper or other metallic strips, which are secured in the inside walls of the drums, and are brought in various ways in contact with the source of current. In other contrivances, for instance, the apparatus of Smith & Deakin, metallic pins capable of being turned around the shaft, which is in contact with the negative pole of the source of current, reach to the layer of articles in the drum, and effect the re-transmission of the current. Since in the contrivances mentioned the anodes are placed outside of the drum, and the latter acts as a diaphragm with great resistance, a very high electro-motive force is required for the production of the deposit, independent of the fact that the articles being in constant motion already require an essentially higher electro-motive force.

In another class of apparatus, the six or eight-cornered drum is constructed of the same metal which is to be deposited. Every metal plate forming one side is insulated from the next plate. The plates which, while the drum is revolving, occupy the lowest position and upon which the articles for the time being rest, are brought into contact with the negative pole of the source of current by a commutator of special construction, while the positive current is carried to the plates occupying a higher position, they thus acting as anodes. In this type of apparatus the high resistance due to the arrangement of the anodes on the outside is overcome, but the commutator with the sliding contact constitutes a very sensitive part of the construction.

Fig. 116 shows a mechanical electro-plating apparatus patented and manufactured by The Hanson and Van Winkle Co., Newark, N. J. The apparatus complete consists of an

FIG. 116.

outer wooden tank for containing the solution, a perforated revolving plating barrel, made of wood or celluloid in which to hold and tumble the work while deposition is going on, and necessary rods and connections. The size of the perforations required in the plating barrel depends on the class and shape of the work to be plated. The perforations should be as large as possible without allowing the work to slip through or catch in them. The barrel is entirely submerged, thus permitting a much larger quantity of work in each batch.

The drive is from the outside, thus avoiding the use of belts running in the solution. The barrel is removable at any time without throwing off the belt or interfering with the drive. For raising and lowering the plating barrel a lifting device is very convenient. Fig. 117 shows a hand-wheel lifting device. In operation it raises and lowers the plating barrel in a perpendicular direction, and when the barrel is suspended above the tank for a few seconds will allow the

FIG. 117.

solution to drip directly back into the tank. This reduces the loss of solution to a minimum and overcomes the difficulty of a wet and sloppy floor.

In connection with this apparatus the use of patent curved elliptic anodes, as shown in the illustration, is recommended. The anode is curved to fit the periphery of the revolving barrel, and when an anode is hung on each side of the tank, the barrel holding the work is equidistant at all times from the

-anode ; hence a regular and even deposit is obtained. These anodes are cast in all metals with square copper hooks attached.

The above-described mechanical electro-plating devices are equally well adapted for zincking articles in large quantities, such as screws, nails, rivets, etc., as well as for brassing, coppering, etc.

Nickeling sheet-zinc. The nickeling of sheet-zinc has been surrounded with a great deal of mystery by those engaged in its manufacture, which may, perhaps, be excusable on the ground that there is scarcely another branch of the electro-plating industry in which experience had to be acquired at the sacrifice of so much money and time as in this. Nevertheless, the nickeling of sheet-zinc makes no greater demand on the intelligence of the operator than any other electro-plating process, it requiring only an accurate consideration of the relations of the electric behavior of zinc towards nickel ; consequently, a knowledge of the strength of the counter-current and of the chemical behavior of zinc towards the nickel solution, which may readily dissolve the zinc ; further, a correct estimation of the proper current-strength required for a determined zinc surface, as well as of the proper anode surface, and the most suitable composition and treatment of the nickel baths.　·

With due observation of these conditions, the nickeling of sheet-zinc is accomplished as readily as that of other metals ; and the suggestions to first cover the sheets in a bath with a strong current, and finish nickeling with a weaker current, or to amalgamate the zinc before nickeling, need not be considered.

Below the conditions required for nickeling sheet-zinc, and the execution of the process itself, together with the preliminary and final polishing of the sheets, will be found fully described.

The preliminary grinding or polishing is effected upon broad cloth wheels (buffs) formed of separate pieces of cloth. The polishing lathes run with their points in movable bearings secured in a hanging cast-iron frame by a set screw and safety keys, or preferably as shown in Fig. 101, since with this construction an injury to the grinder by the lathe jumping out is impossible.

The bobs, when new, have on an average a diameter of 12 to 16 inches, and a width of $5\frac{3}{4}$ to 8 inches. The principal point in the construction of these bobs is uniform weight on all sides, quiet running and the possibility of a good polish without great exertion depending on this. Bobs not well balanced run unsteadily and jump, thereby producing fine scratches upon the sheet. The bobs are constructed as follows: A square piece of cloth if folded fourfold and the closed point cut off with a pair of scissors, so that on unfolding the cloth, the hole produced by the cut is exactly in the center of the cloth disk. According to the diameter of the spindle more or less is cut away, but in every case just sufficient for the piece of cloth to be conveniently pushed upon the spindle. The latter which is provided with a pulley and a hoop against which the pieces of cloth fix themselves, as well as with a nut and screw for securing them, is vertically fastened in a vise, and the separate pieces of cloth are pushed upon it so that the second piece placed in position forms an angle of about 30° (Fig. 118) with the first, the operation being thus continued until the bob has the desired width. Next a small, but very strong iron disk is laid upon the cloth bob, and the separate pieces are pressed together as firmly as possible with the screw. The spindle is then placed in the bearings, and after adjusting the belt upon the pulley the bob is revolved, a sharp knife being held against it to remove the projecting corners. In polishing sheet-zinc the bobs make 2,200 to 2,500 revolutions per minute, according to whether finely rolled or rougher sheets are to be polished.

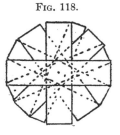

FIG. 118.

For the purpose of preparatory polishing, the operator

places the sheet upon a support of hard wood of the same size and form as the sheet, and grasps the two corners of the sheet nearest to his body, together with the support, with the hands, applying with the balls of the hands the necessary pressure to hold the sheet upon the support. The lower half of the sheet, that furthest from the body, rests upon the knees of the operator, and with them he presses the sheet against the polishing wheel, constantly moving at the same time, and at not too slow a rate, the knees from the right to the left, then from the left to the right, and so on. Previous to polishing, a streak of oil about two inches wide is applied by means of a brush to the center of the sheet in the visual line of the operator, and the revolving bob is impregnated with Vienna lime by holding a large piece of it against it, when polishing of the lower portion of the sheet begins. When about $\frac{3}{5}$ of the surface has thus been polished, the sheet is turned round and the remaining portion subjected to the same process. The sheet is then closely inspected to see whether there are still dirty or dull places, and, if such be the case, it is polished once more, after moistening it with some oil and again impregnating the bob with Vienna lime. The sheet being sufficiently polished, the oil and polishing dirt are removed by dry polishing, after providing the bob with sufficient Vienna lime, so that the sheets when finished show no streaks of dirt or oil.

Sheets 50 x 50, 100 x 50, and also 150 x 50 centimeters, can in this manner be readily polished, but it is a difficult feat, mostly subject to the risk of producing bent places, to polish sheets 6 feet long upon the knees. Numerous attempts have therefore been made to construct automatic machines for con veniently polishing sheets 12 or more feet long.

Several such automatic polishing machines have been de scribed and illustrated in the fifth edition of this book, but, while they furnish a quite good polish, they have, on the one hand, the drawback that thin sheets are readily creased or wound around the polishing roll, and, on the other hand, that the sheets are with great violence thrown out by the polishing roll if this is not prevented by placing another sheet over a portion of the sheet to be polished, and passing it together with the latter under the roll. This, however, has the drawback that the covered portion of the first sheet is not polished, and has to be again passed under the polishing roll, and the place where the edge of the second sheet has rested upon the first sheet shows a mark formed by pressure, which, as a rule, is not desirable.

The polishing machine constructed according to the patent of Hille and Müller * avoids the above-mentioned drawbacks by obliquely standing polishing rolls. In Koffler's † construction two polishing rolls move in opposite directions. · The sheets are pressed against the rolls by an oscillating table so that first one and then the other portion of the table is alternately advanced towards the corresponding polishing roll.

In the construction patented by Dr. Langbein & Co., the drawbacks of throwing out and crumpling the sheets is overcome by the arrangement of two polishing bobs, which alternately stand still and revolve, however, in opposite directions. The table consists of two movable halves; while one of the halves, in an elevated position, presses the sheet carried by the transport-rolls against the revolving polishing bob, the other half is lowered and its polishing bob remains stationary. When a certain length of the sheet has been polished the second polishing bob revolving in an opposite direction is put in action, a constant stretching of the sheet being thereby effected.

Freeing zinc sheets from grease. This is best effected in two operations, first dry and then wet. For the dry process use a very soft piece of cloth and, after dipping it in Vienna lime, very finely pulverized and passed through a hair sieve, rub over the sheet in the direction of a right angle to the polishing streaks, applying a very gentle pressure. For the wet process, dip a moist piece of cloth, or a soft sponge free from sand, into

German patent, 49736.   † German patent, 89648.

a paste of impalpable Vienna lime, whiting and water, and go carefully over the sheet so that no place remains untouched. Then rinse the sheet under a powerful jet of water, best under a rose, being particularly careful to remove all the lime, going over the sheet, if necessary, with a soft, wet rag, and observing whether all parts appear evenly moistened. If such be the case, cleaning is complete, otherwise the sheet has to be once more treated with lime.

If the sheets are to be nickeled on only one side, two of them are placed together with their unpolished sides and fastened on the two upper corners with binding screws to which is soldered a copper strip about 0.39 inch wide, by which they are suspended to the conducting rods. Plating is then at once proceeded with, without allowing the sheets to remain exposed to the air longer than is absolutely necessary. Special care must be had that the lime does not dry, as this would produce stains.

With sheets 50 x 50 centimeters, two binding screws suffice for suspending the sheets to the conducting rods. With sheets 100 centimeters long, three binding screws are generally used, with sheets 150 centimeters long, five, and with lengths of 200 centimeters, six or more, so that the current required for nickeling finds a sufficient cross-section.

Some manufacturers nickel the cleansed sheet without previous coppering or brassing, and claim special advantages for such direct nickeling. This may be done with a bath of nickel sulphate and potassium citrate without, or with a greater or smaller, addition of ammonium chloride, according to the surface to be nickeled and the intensity of current at disposal. However, sheet-zinc directly nickeled does not show the warm, full tone of sheets previously coppered or brassed ; besides, direct nickeling requires a far more powerful current, so that it is not even more economical.

For the nickeling process itself, it is indifferent whether the sheets are previously coppered or brassed, but the choice between the two is controlled by a few features which must be mentioned. The nickel deposit upon brassed sheets shows a decidedly whiter tone than that upon coppered sheets, and brassing would deserve the preference if this process did not require extraordinarily great care in the proper treatment of the bath, the nickel deposit readily peeling off, generally in the bath itself, which seldom or never occurs with coppered sheet, and then may generally be considered due to insufficient cleaning or other defective manipulation.

This peeling-off of the nickel deposit may be prevented by giving due consideration to the conditions and avoiding, on the one hand, too large an excess of potassium cyanide in the brass bath, and, on the other, by regulating the current so that no pale yellow or greenish brass is precipitated. Since nickeling with a strong current requires only a few minutes for a deposit of sufficient thickness capable of bearing polishing, it is generally desired to brass the sheets at the same time, so that the operation may proceed rapidly and continuously. To do this, a very powerful current has to be conducted into the brass bath, the result being that a deposit with a larger content of zinc and a correspondingly lighter color is formed, but also with a coarser, less adherent structure, and this is the principal reason why the nickel deposit, together with the brass deposit, peels off. To avoid this, the brassing must be done with a current so regulated that the deposit precipitates uniformly, adheres firmly, and is not porous; the correct progress of the operation is recognized by the color being more like tombac, and not pale yellow or greenish. When brassing has to be done quickly the content of copper in the brass bath must be increased to such an extent that a powerful current produces a deposit of the above-mentioned color, and, hence, too large an excess of potassium cyanide must be strictly avoided.

It will be seen that brassing requires a certain attention which is not necessary in coppering, and therefore the latter is to be preferred.

For coppering, one of the baths, formulas III to VII, given under " Deposition of Copper " can be used, to which, for this special purpose, more potassium cyanide may be added. The sheets should remain in this bath no longer than required to uniformly coat them with a beautiful red layer of copper, and under no circumstances must they be allowed to remain until the coppering commences to become dull or even discolored. They should come from the bath with a full, or at least half, luster.

When taken from the copper bath the sheets are thoroughly rinsed in a large water reservoir, the contents of which must be frequently renewed, care being had to remove any copper solution adhering to the unpolished sides which are not to be nickeled, since that would soon spoil the nickel bath. The sheets are then immediately brought into the nickel bath, it being best to suspend two, three, or four of them at the same time, to prevent one from being more thickly nickeled than the other, and take them out the same way. In suspending the sheets in the bath, care should be had to bring them as soon as possible in contact with the conducting rod, a neglect of this rule being apt to produce blackish streaks and stains.

The tanks used for nickeling sheet-zinc are generally about 7 feet long in the clear, $1\frac{1}{4}$ feet wide, and $2\frac{1}{4}$ to $2\frac{1}{2}$ feet deep. In such tanks sheets $6\frac{1}{2}$ feet long and $1\frac{1}{2}$ feet wide can be conveniently nickeled.

With the use of a nickel bath according to formula VIII, p. 258, for nickeling sheet-zinc, the most suitable electro-motive force is 3.5 volts and 1 ampère current-density per square decimeter, in order to obtain in three minutes an effective deposit. After working for some time this bath also requires a stronger electro-motive force.

If zinc is to be nickeled in baths conducting with greater difficulty, for instance, in a simple solution of nickel-ammonium sulphate without the addition of conducting salts, or in baths containing boric acid, 1.2 to 1.5 ampères and 7 volts must be allowed for 1 square decimeter, if nickeling is to be effected in the above-mentioned space of time.

For nickeling sheet-zinc, rolled anodes are, as a rule, only

used, except when working with baths containing ʋoric acid.
The anode surface must at least be equal to that of the zinc
surface. The distance between the anodes and the sheets
should be from 3 to $3\frac{3}{4}$ inches, and when the current-strength
is somewhat scant the distance may be reduced to $2\frac{1}{2}$ inches.
The nickel anodes have to be taken from the bath once daily
and scoured bright with scratch-brushes and sand. For the
rest, all the rules given for nickel anodes are valid.

Baths used for nickeling sheet-zinc soon become alkaline in
consequence of the powerful current used, which is shown by
red litmus-paper turning blue. The alkalinity also manifests
itself by the bath becoming turbid and the nickeling not turning out pure white. The slightly acid reaction required is restored by citric acid solution. The appearance of the dreaded
black streaks and *stains* is due either to the current itself being
too weak, or to its having been weakened by an extremely
great resistance of the nickel bath ; also to an insufficient metallic surface of the anodes, which may be either too small or
not sufficiently metallic on account of tarnishing ; and finally
to an excessive alkalinity of the bath, or insufficient contact
of the hooks with the connecting rods.

The metallic content of the bath must from time to time be
strengthened by the addition of nickel salt, and the bath
filtered at certain intervals. When the conductivity abates,
it has to be restored by the addition of conducting salts.

When the sheets have been sufficiently nickeled, they are
allowed to drain off, then plunged into hot water, and, after
removing the binding screws, dried by gentle rubbing with
fine sawdust free from sand and passed through a fine sieve
to separate pieces of wood. In all manipulations, the un-
nickeled sides are placed together, while a piece of paper of the
size and form of the sheets is laid between the nickeled sides.

The nickeled sheets are finally polished, which is effected
by placing them upon supports and pressing against the
revolving bob as previously described, the sheets being, however, only moderately moistened with oil, and not too much

20

Vienna lime applied to the bob. Polishing is done first in one direction and then in another, at a right angle to the first. After polishing, the sheets are finally cleansed with a piece of soft cloth and impalpable Vienna lime, when they should show a pure white lustrous nickeling, free from cracks and stains, and bear bending and rebending several times without the deposit of nickel breaking or peeling off.

Nickeling tin-plate.—For handsome and durable nickeling, tin-plate also requires previous coppering. Deposition is effected with a less powerful current than for sheet-zinc. Freeing from grease is done in the same manner as above described.

For preparatory polishing of tin-plate, the use of a polishing compound free from lime and grease is recommended, since a good polish on tin cannot be obtained with Vienna lime and oil. Nickeled tin-plate may be polished with Vienna lime and stearine oil.

It may be here mentioned as a remarkable fact that freshly nickeled tin-plate will stand every kind of manipulation, such as stamping, edging, pressing, etc., but after having been stored for a few months, the layer of nickel frequently peels of by these operations.

Nickeling copper and brass sheets.—The treatment of these sheets differs from that of sheet-zinc in that the rough sheets are first brushed with emery and then polished with the bob.

After treating the sheets with hot caustic lye or lime-paste, they are pickled by brushing them over with a solution of 1 part of potassium cyanide in 20 parts of water. They are then thoroughly and rapidly rinsed, and immediately brought into the bath. To avoid peeling off, the current-density should not exceed 0.4 ampère.

Nickeling sheet-iron and sheet-steel.—Only the best quality of sheet should be used for this purpose. After rolling, the sheets are freed from scales by pickling, then passed through the fine rolls, and finally again pickled. If the nickeled sheets are not to exhibit a high degree of polish, it suffices to brush them before nickeling with a large, broad fiber brush (p. 204) and emery No. 00. But for a high luster, such as is generally demanded, the sheets have first to be ground. For fine-grinding the pickled sheets, broad, massive wood rolls, turned and directly glued with emery are used. These wheels are 10 to 12 inches in diameter, and 2 to 4 or more inches long, according to the size of the sheets. For the first grinding, the wheels are coated with glue and rolled in emery No. 100 to 120, according to the condition of the sheets, while emery No. 00 is applied to the wheels used for the fine grinding. The grinding is succeeded by brushing, as described on page 205.

After preparing a sufficiently smooth surface, the sheets are at once rubbed with a rag moistened with petroleum, or, if preferred, with a rag and pulverized Vienna lime. They are then scoured wet in the manner described for sheet-zinc. The scouring material must be liberally applied, especially if the sheets are to be directly nickeled without previous coppering, the latter being, however, quite advisable. After rinsing off the lime-paste, the sheets are without loss of time brought into the nickel bath.

For nickeling, a bath free from chlorine should by all means be used in order to protect the sheets from rusting. The current-density should be 0.4 ampère, with which the sheets acquire in $\frac{3}{4}$ hour a deposit of sufficient thickness. With the use of cold, quick nickeling baths the same thickness of the deposit may be obtained in 15 minutes. It is not advisable to attempt to obtain a heavy deposit in a shorter time, because it would lack density which, by reason of greater protection against rust, is the principal requisite for nickeled sheet-iron.

After nickeling, the sheets are rinsed in clean water, then plunged into hot water, and dried by rubbing with warm sawdust. After this operation, it is recommended to thoroughly dry the sheets in an oven heated to between 176° to 212° F., to expel any moisture from the pores, and then to polish them with Vienna lime and oil, or with rouge.

Nickeling wire. Nickeling of wire of iron, brass or copper is scarcely ever done on a large scale.   It is, however, believed that the nickeling of iron and steel wires—for instance, piano-strings—might be of advantage to prevent rust, or at least to retard the commencement of oxidation as long as possible.

To nickel single wires cut into determined lengths, according to the general rules already given, is simple enough ; but this method cannot be pursued with wire several hundred yards long, rolled in coils, as it occurs in commerce.   Nickeling the wire in coils, however, cannot be done, as only the upper windings exposed to the anodes would acquire a coat of nickel.   Hence it becomes necessary to unwind the coil, and for continuous working pass the wire at a slow rate through the cleansing and pickling baths, as well as the nickel bath, and hot water reservoir, as shown in Fig. 119, in cross-section, and in Fig. 120, in ground plan.

The unwinding of the wire is effected by a slowly revolving shaft, upon which the nickeled wire again coils itself; but in the illustration the shaft is omitted.   In Fig. 120 four wires run over the four rolls a, mounted upon a common shaft, to the rolls b upon the bottom of the tank A, whereby they come in contact with a thickly-fluid lime-paste in the vat, and are freed from grease.   From the rolls b the wires run through the wooden cheeks i, lined with felt, which retain the excess of lime-paste, and allow it to fall back into the tank.   The wires then pass over the roll c to the roll d.   Between these two rolls is the rose g, which throws a powerful jet of water upon the wires, thereby freeing them from adhering lime-paste.   The roll d, as well as its axis, is of brass, and to the latter is connected the negative pole of the battery or dynamo, so that by carrying the wires over the roll d, negative electricity is conducted to them.   From the roll d, the wires run over the roll-bench s (Fig. 119) to the tank C, which contains the nickel solution, so that they are subjected to the action of the anodes arranged in this tank on both sides of the wires.   The wires then pass over the roll e, are rinsed under the rose h, and run finally through a hot-water reservoir and sawdust (these two apparatuses are not shown in the illustration), to be again wound in coils.   In case a high polish is required, the nick-

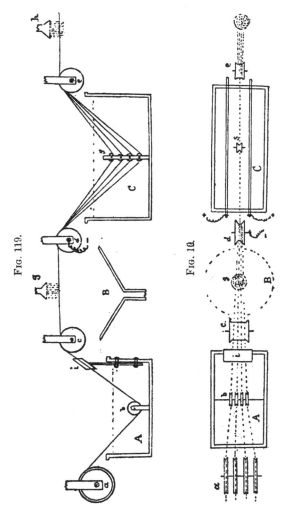

eled wires may be run under pressure through leather cheeks dusted with Vienna lime.

Nickeling knife-blades, sharp surgical instruments, etc. Considerable trouble is frequently experienced in nickeling sharp-edged instruments, the edges and points being spoiled either by the deposit of nickel or in polishing. And yet such instruments can be readily nickeled in such a manner that the edges remain in as good condition as before.

If new instruments which have never been used are to be nickeled, no special preparation is required, it being only necessary to free them at once from grease and bring them into the bath. But instruments which have been used or, by bad treatment have become partly or entirely covered with rust, must be first freed from rust by chemical or mechanical treatment, and then polished. The marks left by the stone or emery wheel are effaced by means of the circular brush, this treatment being necessary to obtain perfect nickeling. But, in brushing, the edges are rendered dull if special precautionary measures are not used. For instance, the edge of a knife-blade must never come in contact with the brush. This is prevented by firmly pressing the blade flat upon a soft support of felt or cloth, so that the edge sinks somewhat into the support, without, however, cutting into it. The edge is then held downward, and thus together with the support brought against the revolving brush. In this manner the blades may be vigorously brushed without fear of spoiling the edges.

The treatment for giving them a high polish after nickeling is the same. Freeing from grease may be done in the usual manner with lime-paste; but must also be effected upon a soft support, the same as in polishing. After thorough rinsing in clean water, the separate pieces, *without* being previously coppered, are brought directly into the nickel bath, the composition of which must, of course, be suitable for nickeling steel articles. The instruments are first coated with the use of a strong current, so that deposition takes place slowly and with great uniformity.

In suspending the articles in the bath, care should be had that neither a point nor an edge is turned towards the anodes.

It is best to use a bath with anodes on one side only, and to suspend the blades with their backs towards the anodes. If, for any reason, the instruments are to be suspended between two rows of anodes, the edges should be uppermost, as near as possible to the level of the bath ; but they should never hang deep or downwards.

These precautionary measures may be omitted by using for nickeling such articles with sharp edges, the bath consisting of nickel sulphate and sodium citrate, which has been previously mentioned. In this bath, the edges and points of the instruments do not burn as readily as in other nickel baths, and the deposited nickel being soft, it does not show a tendency to peeling off when, after nickeling, the edges of the instruments are sharpened.

The plated instruments are given a finer luster by polishing, but during this operation they must always be exposed upon a soft support, as above described, to the action of a felt wheel, or, still better, of a cloth bob.

In nickeling *skates* it is advisable to suspend them so that the runners hang upwards and that the running surfaces are level with the surface of the bath, because if the deposit upon the running surfaces is too thick, it peels off readily when injured by grains of sand upon the ice.

Nickeling of soft alloys of lead and tin, with or without addition of antimony, as are used for siphon-heads, etc., is effected, in case the objects have already a high luster, by freeing them from grease with whiting and a small quantity of Vienna lime, then rinsing in water, lightly coppering, or better, brassing, and finally nickeling in a bath containing chlorine.

If the objects require preparatory polishing, use a polishing compound free from lime and grease, as given under nickeling of tin-plate, rinse with benzine, immerse in hot water, and free from grease with whiting and Vienna lime. Then brass, nickel and polish. Direct nickeling without previous brassing is not advisable, waste in consequence of peeling off being frequently the result.

Nickeling printing plates (stereotypes, clichés, etc.). The advantages of nickeling stereotypes, etc., over steeling will be referred to under "Steeling," and hence only the most suitable composition of the nickel baths and the manipulation required will here be given.

The nickel baths according to formula I (page 253) and formula VII (page 257) are the most suitable for simple nickeling, because the ammonium sulphate not being present in too great an excess, as well as the presence of boric acid, causes the nickel to separate with considerable hardness. With nickeled stereotypes three times as large an edition can be printed as with plates of the same material not nickeled.

Hard nickeling. It being a well-known fact that a fused alloy of nickel with cobalt possesses greater hardness than either of the metals by themselves, experiments proved that an electro-deposited nickel-cobalt alloy exhibited the same behavior, the greatest degree of hardness being attained with an addition of cobalt varying between 25 and 30 per cent. For this deposit the term *hard nickeling* is proposed, the most suitable bath for the purpose being prepared according to the following formula :

Nickel-ammonium sulphate 21.16 ozs., cobalt-ammonium sulphate 5.29 ozs., crystallized boric acid 8.8 ozs., water 10 to 12 quarts.

To prepare the bath dissolve the constituents by boiling as given under formula VII, p. 257. In case the metal salts should contain free acids add, previous to the addition of the boric acid, a small quantity of nickel carbonate. The boric acid must not be neutralized and the bath should work with its acid reaction. Mixed anodes in the proportion of $\frac{1}{3}$ cast and $\frac{2}{3}$ rolled, are to be suspended in the bath.

The bath prepared according to formula No. II deserves the preference, it yielding a harder deposit than bath No. I.

For the rest, the treatment of the baths is the same as that given for nickel baths of similar composition (pp. 253 and

257), and the process of hard nickeling does not essentially differ from ordinary nickeling. The suspending hooks are soldered to the backs of the plates by means of the soldering-iron and a drop of tin ; or the plates are secured in holders of sheet-copper 0.11 inch thick, and ¾ to 1 inch wide, of the form shown in Fig. 121. The printing surface is freed from grease by brushing with lime-paste, rinsing in water, and then brushing with a clean brush to remove the lime from the depressions. The plates are then hung in the bath and

FIG. 121.

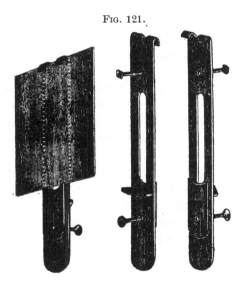

covered with a strong current. When everywhere coated with nickel, the current is weakened and the deposit allowed gradually to augment. With an average duration of nickeling of 15 to 20 minutes, with 2.8 to 3 volts, the deposit will, as a rule, be sufficiently resisting.

Stereotypes of type metal, after being freed from grease, are best lightly coppered in the acid copper bath, then rinsed and brought into the nickel bath. *Zinc etchings* are first coppered, not too slightly, in the copper cyanide bath, rinsed, and suspended in the nickel bath with a very strong current. With too weak a current, black streaks are formed, zinc is dissolved, and both the plate and bath are spoiled. With copper electros, pickling with potassium cyanide solution, after freeing from grease, must not be omitted.

The nickeled plates are rinsed in water, then plunged in hot water, and dried in sawdust, when the nickeled printing surface may be brushed over with a brush and fine whiting, it being claimed that plates thus treated take printing ink better, while the first impressions of plates not brushed with whiting are somewhat dull.

Nickel-facing is especially suitable for copper plates for color-printing, the nickel not being attacked like copper or iron by cinnabar.

Recovery of nickel from old baths. At the present price of nickel its recovery from old solutions scarcely pays. The inefficiency of the bath is in most cases due to two causes: It has either become too poor in metal or it contains foreign metallic admixtures. In the first case, the expense of evaporating, together with the further manipulations, is out of proportion to the value of the nickel recovered, and, in the second case, the reduction of the foreign metals is inconvenient and connected with expense which make it unprofitable. The recovery of nickel from old baths which have become useless, by the electric current with the use of carbon-plate anodes, as here and there recommended, is the most disastrous and expensive of all, and can only be condemned.

For *nickeling by contact and boiling*, see special chapter, " Depositions by Contact."

Deposition of nickel alloys.—From suitable solutions of the metallic salts nickel may be deposited together with copper and tin, as well as with copper and zinc. With the first combination, especially, all tones from copper-red to gold-shade may be obtained, according to which metal predominates, or according to the current-strength which is conducted into the bath, as is also the case in brassing.

A suitable bath for coating metallic articles with an alloy of nickel, copper and tin, for which the term *nickel-bronze* is proposed, is obtained by dissolving the metallic phosphates in sodium pyrophosphate solution. By mixing solution of blue vitriol with solution of sodium phosphate, cupric phosphate is precipitated, which is filtered off and washed. In the same manner nickel phosphate is prepared from a solution of nickel sulphate. These phosphates are then, each by itself, dissolved in a concentrated solution of sodium pyrophosphate, while chloride of tin is directly dissolved in sodium pyrophosphate until the turbidity, at first rapidly disappearing, disappears but slowly.

Nothing definite can be said in regard to the mixing proportions of these three solutions, because the proportions will have to be varied according to the desired color of the deposit. The operator, however, will soon find out, of which solution more has to be added to obtain the tone desired.

For depositing a nickel-copper-zinc alloy solutions of cupric sulphate (blue vitriol) and zinc oxide in potassium cyanide to which is added an ammoniacal solution of nickel carbonate, may be advantageously used. As will be seen a deposit of German silver can be obtained with the use of this solution if the latter contains the metals in the same proportions as German silver, and German silver anodes are used.

According to a French process, a deposit of German silver may be obtained as follows: Dissolve a good quality of German silver in nitric acid and add, with constant stirring, solution of potassium cyanide until all the metal is precipitated as cyanide. The precipitate is then filtered off, washed, dissolved in potassium cyanide, and the solution diluted with double the volume of water. This process, however, does not seem very feasible, since nickel separates with difficulty from its cyanide combination.

EXAMINATION OF NICKEL BATHS.

The reaction of the nickel baths have previously been briefly referred to, but the subject must here be more closely considered.

For the determination of the content of acid, a different method must be adopted according to the composition of the bath, *i. e.*, whether it has been prepared with an addition of citric acid, boric acid, etc. The reddening of blue litmus-paper simply indicates the presence of free acid in the bath, but leaves us in the dark as to which acid is present, and as to its derivation.

If, for instance, in consequence of insufficient solution of nickel, free sulphuric acid appears on the anodes, the bath becomes at the same time poorer in nickel in proportion to the increase in the content of free sulphuric acid. If we have to deal with a bath prepared from nickel-ammonium sulphate with an addition of ammonium sulphate, but without organic acids, the reddening of blue litmus-paper will at once indicate a content of free sulphuric acid, if the bath was neutral in the beginning. It is, however, quite a different matter when a bath containing boric acid is examined. In the formulæ for preparing these baths, it has been seen that before adding the boric acid, any free sulphuric acid of the nickel salt present is to be removed by treating the solution with nickel carbonate or nickel hydrate. After adding the boric acid, blue litmus-paper is strongly reddened, and this acidity due to the boric acid is to be maintained in the bath. However, in consequence of the use of too large a number of cast anodes, free sulphuric acid may form in the bath, and this, together with boric acid, cannot be recognized by blue litmus-paper, since both acids redden it. In this case red congo paper, which is not changed by boric acid, but is turned blue by sulphuric acid, has to be used. If red congo paper is colored blue, it is a sure proof that, besides boric acid, free sulphuric acid is present, which has to be neutralized for the bath to work in a correct manner.

The process is again different when a bath prepared with an addition of citric acid is to be examined. This organic acid colors certain varieties of commercial congo paper blue, just as sulphuric acid does, and hence tropaeolin paper has to be used, which is not altered by citric acid, but is colored violet by free sulphuric acid.

If a nickel bath has been prepared with the addition of organic salts, for instance, sodium citrate, ammonium tartrate or others, the formation of free sulphuric acid in the bath cannot at first be determined with reagent papers, because the sulphuric acid decomposes the organic salts, neutral sulphates being formed, and a quantity of organic acid equivalent to the sulphuric acid is liberated. For this reason the content of metal in the bath declines, though the presence of sulphuric acid cannot be established, because the sulphuric acid formed by electrolysis is not consumed for the solution of nickel on the anodes, but for the decomposition of the organic salts.

Now let us suppose the reverse, namely, that in a nickel bath prepared with the addition of one of the above-mentioned acids, free ammonia appears in consequence of the sole use of cast anodes, and of the decomposition of ammonium sulphate by a strong current. This phenomenon cannot at once be recognized, because the ammonia is first fixed by the free acid, and the bath becomes neutral or alkaline only when all the free acid which was present has been consumed for fixing the ammonia formed. With this process there will generally be connected an increase in the content of the metal, and it will be seen, without further explanation, that for the accurate determination of the processes and alterations in a nickel bath when in operation, the quantitative determination of the free acids, and as much as possible, that of the content of metal, is required.

Although it may be said that the busy electro-plater will frequently not feel inclined to familiarize himself with the methods of testing, and seldom have the necessary time for executing the determinations of the content of metal, nevertheless the methods will here be described with sufficient detail, so that those who wish to examine their baths in this respect will find the necessary instructions. To be sure, if the electro-plater himself is not a practical analytical chemist he will have to be taught by some one thoroughly conversant with the subject the management of the analytical balance, how to execute the weighings, etc. It is also advisable to procure the standard solutions required for volumetric analysis from a reliable chemical laboratory, in order to avoid the possibility of arriving at incorrect results by the use of inaccurately prepared standard solutions. For this reason directions for the preparation of standard solutions are omitted, and the methods of examination in use for our purposes will now be given.

The examinations may be made by gravimetric analysis (analysis by weight), volumetric analysis (analysis by measure), and by electrolytic analysis. The first method is based chiefly upon the precipitation in an insoluble form of the constituent to be determined, and filtering, washing, drying, and weighing the precipitate. This method requires considerable knowledge of chemistry and analytical skill, and should only be resorted to by those not versed in analysis when other more practical methods for the determination of the contents, such as volumetric and electrolytic methods, are not known.

Volumetric analysis is based upon a very different principle from that of gravimetric analysis. The constituent to be ascertained is quantitatively determined by means of a standard solution, enough of which is used until the final reaction shows that a sufficient quantity has been added. From the known content of the standard solution the constituent to be determined is then calculated. This may be explained by an example. For instance, the content of sulphuric acid in a fluid is to be determined. Measure the quantity of fluid by means of a pipette which up to a mark holds exactly 10 cubic centimeters. Allow the fluid to run into a clean beaker, dilute with about 30 cubic centimeters of water, and heat to about 122° F. Now, while constantly stirring the fluid in.

the beaker with a glass rod, add standard soda solution from a glass burette provided with a glass cork and divided into $\frac{1}{10}$ cubic centimeters until a piece of congo paper when touched with the glass rod is no longer colored blue. The addition of the standard soda solution must, of course, be effected with great care. So long as the congo paper shows a vivid blue color, a larger quantity may at one time be added, but when the colorization becomes less vivid, the solution is added drop by drop so as to be sure that the last drop is just sufficient to prevent the blue coloration which was still perceptible after the addition of the previous drop. The drop-test must, of course, be made upon a dry portion of the congo paper, which has not been previously moistened. When no blue coloration appears after the last drop has been added, it is a proof that all the sulphuric acid present has been neutralized by the standard soda solution. The number and fractions of cubic centimeters consumed are then read off on the burette, and the quantity of sulphuric acid present is calculated as follows: 1 cubic centimeter of standard soda solution neutralizes 0.049 gramme of sulphuric acid (H_2SO_4), and hence the quantity of sulphuric acid is obtained by multiplying the number of cubic centimeters of standard soda solution by 0.049. Now, since 10 cubic centimeters were measured off by the pipette and titrated, the number found is multiplied by 100, which gives the content of sulphuric acid in 1 liter of the fluid.

If, for instance, for the neutralization of 10 cubic centimeters of the fluid containing sulphuric acid, 5.4 cubic centimeters of standard soda solution were required, then the content of sulphuric acid amounts to $5.4 \times 0.049 = 0.2646$ gramme, or in 1 liter to $0.2646 \times 100 = 26.46$ grammes.

The electrolytic method of analysis is available only for the determination of such metals as can be completely separated in a coherent form from their solutions by the current. It is based upon the fact that the metallic solution contained in a platinum dish is decomposed by the current, and the metal precipitated upon the platinum dish. After washing and drying, the dish is weighed and the weight of the precipitated metal is obtained by deducting the weight of the platinum dish without precipitate, which, of course, has been ascertained before making the experiment.

The apparatus generally used for electrolytic analysis is shown in Fig. 122. The platinum dish, holding about ¼ liter, rests upon a metal ring which is secured to the rod of the stand, and is in contact with the negative pole of the source

FIG. 122.

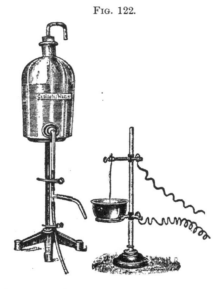

of current. Into the dish, at a distance of 1 or 2 centimeters from the bottom, dips a round platinum disk bent like the bottom, or a spiral of platinum wire, 1 millimeter thick, which serves as an anode and is secured by platinum wire in a movable support or holder. The latter is carefully insulated from the rod of the stand and connected with the positive pole of the source of current. During electrolysis the platinum dish is covered with a perforated watch-glass to prevent possible loss by the evolution of gas.

Since many precipitates have to be washed without interrupting the current, it is best to use the washing contrivance shown in the illustration to prevent the precipitated metal from being redissolved by the electrolyte. With the upper clip closed, the shorter leg of the siphon is dipped into the dish. The lower clip is then closed and the upper one opened until the short leg is filled with water. The upper clip is then closed and the lower one opened, whereby the dish is emptied. The clip of the longer leg of the siphon is then closed, the uppermost clip opened, and the dish filled up to the rim with water. The uppermost clip is then closed, the lower one opened, and the dish emptied the second time, the operation being repeated until the precipitate and dish are thoroughly washed.

Since for complete electrolytic precipitation it is essential to operate with correct electro-motive forces, it is advisable to use an accurate ammeter adjusted to 0.05 to 2.5 ampères, as well as a voltmeter.

The current for electrolysis may be supplied by cells, a thermo electric pile, a dynamo, or an accumulator, but the necessary regulating resistances must in every case be provided.

Let us now return to the examination of nickel baths. If by qualitative analysis the presence of free sulphuric acid in the bath has been established, it can be at once assumed that the content of nickel has from the first declined. Hence it will scarcely be worth while to determine by volumetric analysis the quantity of free sulphuric acid present, and to calculate from this the quantity of nickel carbonate or nickel hydrate required for neutralization. It will be only necessary to add to the bath, stirring constantly, small portions of the nickel salt rubbed up with water, until a fresh test with congo paper shows no blue coloration. The addition of a small excess of nickel carbonate or nickel hydrate is unobjectionable. Besides neutralizing the free sulphuric acid, care should at the same time be taken to prevent its further formation by increasing the number of cast-nickel anodes. The case is similar when

22

a nickel bath prepared with organic salts, for instance, with potassium citrate or sodium citrate, is to be examined. Even if it is shown by the reaction that no free sulphuric acid is present, the content of nickel, as previously mentioned, may have decreased, and the content of free organic acid increased. The latter may, however, be neutralized by the addition of nickel carbonate or nickel hydrate, and hence the determination of the content of acid by volumetric analysis is not absolutely necessary.

When, on the other hand, a nickel bath has become alkaline, the determination of the free alkali by volumetric analysis will be of little value, and it will, according to the composition of the bath, suffice to neutralize it with dilute sulphuric acid, or acidulate it with an organic acid. Since, however, baths which have become alkaline possess a higher content of nickel than the normal bath, an electrolytic determination of the nickel may be of use in order to calculate accurately the quantity of water which has to be added to reduce the content of nickel to the normal quantity.

If the bath has been prepared with nickel-ammonium sulphate with additions of ammonium sulphate, or boric acid, or if it contains only very small quantities of organic acids, it can be directly electrolyzed.

Bring by means of the pipette exactly 20 cubic centimeters of the bath into the platinum dish, add 4 grammes of ammonium sulphate and 35 to 40 cubic centimeters of ammonia of 0.96 specific gravity and electrolyze with a current-density = 0.6 ampère until no dark coloration appears after adding a drop of ammonium sulphate to a few cubic centimeters of the electrolyte. Rinse the dish, together with the precipitate, with water, remove the water by rinsing with absolute alcohol, rinse the dish with pure ether and dry at 212° F. in an air-bath. The weight of the precipitate of metallic nickel obtained by weighing the platinum dish gives the content of nickel ammonium sulphate in grammes per liter of bath by multiplying by 335. From the increase in the content of nickel ammonium sulphate shown by the analysis, it can be readily calculated how much water has to be added to the bath to reduce it to the original content.

If a nickel bath contains large quantities of organic acids, precipitate 20 cubic centimeters of the bath with sodium sulphide solution, filter and wash the precipitate, dissolve it in nitric acid, and evaporate the solution with pure sulphuric acid upon the water-bath to drive off the nitric acid. The residue is treated as above described.

2. Deposition of Cobalt.

Properties of cobalt. Cobalt (Co = 58.97 parts by weight) has nearly the same color as nickel, with a slightly reddish tinge; its specific gravity is 8.7. It is exceedingly hard, highly malleable and ductile, and capable of taking a polish. It is slightly magnetic, and preserves this property even when alloyed with mercury. It is rapidly dissolved by nitric acid, and slowly by dilute sulphuric and hydrochloric acids.

For plating with cobalt, the baths given under " Nickeling " may be used by substituting for the nickel salt a corresponding quantity of cobalt salt. By observing the rules given for nickeling, the operation proceeds with ease. Anodes of metallic cobalt are to be used in place of nickel anodes.

Nickel being cheaper and its color somewhat whiter, electroplating with cobalt is but little practiced. On account of the greater solubility of cobalt in dilute sulphuric acid, it is, however, under all circumstances, to be preferred for facing valuable *copper plates* for printing.

According to the more or less careful adjustment of such plates in the press, the facing in some places is more or less attacked, and it may be desired to remove the coating and make a fresh deposit. For this purpose Gaiffe has proposed the use of cobalt in place of nickel, because the former dissolves slowly but completely in dilute sulphuric acid. He recommends a solution of 1 part of chloride of cobalt in 10 of water. The solution is to be neutralized with aqua ammonia, and the plates are to be electro-plated with the use of a moderate current.

Cobalt precipitated from its chloride solution, however, does not yield a hard coating, and hence the following bath is recommended for the purpose: Double sulphate of cobalt and ammonium 21 ozs., crystallized boric acid 10½ ozs., water 10 quarts.

The bath is prepared in the same manner as No. VII, given under "Deposition of Nickel." It requires an electro-motive force of 2.5 to 2.75 volts; current-density, 0.4 ampère.

Prof. Sylvanus Thompson's solutions for the electro-deposition of cobalt, patented by him in 1887, yield very satisfactory results:

I. Double sulphate of cobalt and ammonium 16 ozs., magnesium sulphate 8 ozs., ammonium sulphate 8 ozs., citric acid 1 oz., water 1⅕ gallons.

II. Cobalt sulphate 8 ozs., magnesium sulphate 4 ozs., ammonium sulphate 4 ozs., water 1⅕ gallons. It is best to use the solutions warm, at about 95° F.

To determine whether copper, and how much of it, is dissolved in stripping the cobalt deposit from cobalted copper plates, a copper plate with a surface of 7¾ square inches was coated with 7.71 grains of cobalt and placed in dilute sulphuric acid (1 part acid of 66° Bè. to 12.5 parts of water). After the acid had acted for 14 hours, the cobalt deposit was partially dissolved, and had partially collected in laminæ upon the bottom of the vessel, the copper plate being entirely freed. On weighing the copper plate it was shown that it had lost about 0.0063 per cent., this loss being apparently chiefly from the back of the plate, the engraved side exhibiting no trace of corrosion. The experiment proved that there is no danger of destroying the copper plate by stripping the cobalt deposit with dilute sulphuric acid, provided the operation is executed with due care and attention.

Warren has described a cobalt solution which can be decomposed in a single-cell apparatus, and for this reason would seem suitable for electro-plating small articles in quantities. For the preparation of this bath, dissolve 3½ ounces of chloride of cobalt in as little water as possible, and compound the solution with concentrated solution of Rochelle salt until the voluminous precipitate at first formed is almost entirely redissolved, and then filter. Bring the bath into a vessel and place the latter in a clay cup filled with concentrated solution of chloride of ammonium or of common salt, and containing a zinc cylinder. Connect the objects to be plated to the zinc by a copper wire, and allow them to dip in the cobalt solution With a closed circuit the objects become gradually coated with a lustrous cobalt deposit which, after 2 hours, is sufficiently heavy to bear vigorous polishing with the bob. Coating zinc in the same manner was not successful.

CHAPTER VII.

DEPOSITION OF COPPER, BRASS AND BRONZE.

1. DEPOSITION OF COPPER.

Properties of copper. Copper (Cu = 63.57 parts by weight) has a characteristic red color, and possesses strong luster. It is very tenacious, may be rolled to thin laminæ, and readily drawn into fine wire. The specific gravity of wrought copper is 8.95, and of cast, 8.92. Copper fuses more readily than gold, but with greater difficulty than silver.

In a humid atmosphere containing carbonic acid, copper becomes gradually coated with a green deposit of basic carbonate. When slightly heated it acquires a red coating of cuprous oxide, and when strongly heated a black coating of cupric oxide with some cuprous oxide. Copper is most readily attacked by nitric acid, but is slowly dissolved when immersed in heated hydrochloric or sulphuric acid. With exclusion of the air, it is not dissolved by dilute sulphuric or hydrochloric acid, and but slightly with admission of the air. Liquid ammonia causes a rapid oxidation of copper in the air and the formation of a blue solution. An excess of potassium cyanide dissolves copper. Sulphuretted hydrogen blackens bright copper.

Copper baths. The composition of these baths depends on the purpose they are to serve, and below are mentioned the most approved baths, with the exception of the acid copper bath used for plastic deposits of copper, which will be discussed later on under " Copper Galvanoplasty "

In most cases the more electro-positive metals, zinc, iron, tin, etc., are to be coppered either as preparation for the succeeding processes of nickeling, silvering, or gilding, or to pro-

(326)

tect them against oxidation, or for the purpose of decoration. The above-mentioned electro-positive metals, however, decompose acid copper solutions and separate from them pulverulent copper, while an equivalent portion of zinc, iron, tin, etc., is dissolved. For this reason, such solutions cannot be used for coating these metals, and alkaline copper baths are exclusively employed, which may be arranged under two groups—those containing potassium cyanide, and those without it.

Copper cyanide baths are prepared by dissolving cupric salts, for instance, cupric acetate (verdigris), cupric sulphate (blue vitriol), or cuprous compounds, such as cuprous oxide, in potassium cyanide, the salts being thereby converted into potassium-copper cyanide, which is the effective constituent of all copper cyanide baths.

By compounding a solution of a cupric salt with potassium cyanide, cupric cyanide is formed, which is very unstable, rapidly changing by exposure into cupro-cupric cyanide and cyanogen gas. To avoid this loss of cyanogen, sulphites are added, which, according to one view, effect a reduction of the cupric salts to cuprous salts, which dissolve in potassium cyanide without cyanogen being liberated, while, according to another, hydrocyanic acid is produced from cyanogen gas and sodium sulphite, water being decomposed, and forms with the sodium carbonate present, sodium cyanide, the latter becoming again active in converting the cupro-cupric cyanide into the soluble double salt. It is possible that both the reactions mentioned above partly appear together.

By using from the start cuprous oxide, the latter can without loss of cyanogen be converted into potassium-copper cyanide.

In accordance with this, there will be given in the formulas for the preparation of potassium-copper cyanide baths in which cupric salts are used, larger or smaller additions of bisulphite of soda and alkaline carbonates, the former serving the purpose of decreasing the loss of cyanogen, and the latter being intended to fix any free acid formed.

Stockmeier was the first to take the trouble of calculating the combinations formed after. the conversion of the separate constituents of the copper baths, and Jordis has adopted the same course, in order to obtain a standard formula. Both these authors recommend not to produce the copper combination actually subjected to electrolysis in the bath by repeated conversions of salts, but to prepare this combination by itself, and to dissolve it direct in water in order to obtain the finished copper bath. It has been shown in practice that this method is quite practicable provided certain points are taken into consideration. However, the rational composition of a bath with potassium cyanide, which has been produced by conversion from the salts, cannot be judged according to whether the separate salts were present in stoichoimetrical proportions for the smooth conversion into new combinations without receiving in the bath an excess from one or the other substance. In practice it has long been known that an excess of sodium bisulphite has a very beneficial effect upon the separation of a lustrous copper, and prevents it from rapidly turning into a dull earthy gray, this effect being very likely due to the prussic acid liberated by the sulphurous acid. This one example may prove that standard formulas erected upon theoretical maxims should be accepted with due caution by the practical electro-plater, and that, under certain conditions, additions will have to be made to baths prepared according to such standard formulas if the result is to be as good as that from baths prepared according to older formulas.

Hossauer prepares a copper bath by dissolving 3½ ozs. of copper cyanide in a solution of 17½ ozs. of 70 per cent. potassium cyanide in 3 quarts of water, boiling, filtering and diluting with 7 quarts of water, to a 10-quart bath. This bath works very well when heated to between 113° and 122° F., but when used cold requires a very strong current.

Roseleur has recommended the use of copper acetate (verdigris) for copper baths, and suitable compositions, slightly modified, are as follows:

Copper baths for iron and steel articles.—I. *To be used at the ordinary temperature.* Water 10 quarts, bisulphite of soda in powder 7 ozs., crystallized carbonate of soda 14 ozs., neutral copper acetate 7 ozs., 75-per cent. potassium cyanide 7 ozs., ammonia 4.4 ozs.

II. *For hot coppering (at between 140° and 158° F.).* Water 10 quarts, bisulphite of soda in powder 2¾ ozs., crystallized carbonate of soda 7 ozs., neutral copper acetate 7 ozs., 75 per cent. cyanide of potassium 6¾ ozs., ammonia 4 ozs.

The baths are best prepared as follows: Dissolve the bisulphite and carbonate of soda in one-half of the water, the potassium cyanide in the other half, and mix the copper salt with the ammonia; then pour the blue ammoniacal copper solution into the solution of the soda salts, and finally add the potassium cyanide solution; the bath will then be clear and colorless. Boiling, though not absolutely necessary, is of advantage, after which the solution is to be filtered.

According to thorough investigations made, the excess of carbonate of soda in formula I serves no special purpose, but on the contrary, in many cases, is directly detrimental; neither is the use of ammonia of any special advantage, and it may just as well, or rather better, be omitted. Further, the use of separate baths for cold and warm coppering is at least questionable. It is believed that a single bath suffices for both cases, heating having been found of special advantage only for rapid and thick coppering, or for obtaining particular shades which are produced with difficulty in the cold bath, but without trouble in the heated bath.

It should be borne in mind that potassium cyanide solutions are still more rapidly decomposed when heated than when used cold, and consequently the consumption of potassium cyanide in heated baths is considerably greater than in cold baths. Recourse to heating should, therefore, only be had when the intended result cannot by any other means be obtained.

The following formula may be highly recommended, a.

copper bath composed according to it always yielding good and sure results.

III. Water 10 quarts, crystallized carbonate of soda 8½ ozs., crystallized bisulphite of soda 7 ozs., neutral copper acetate 7 ozs., 98 or 99 per cent. potassium cyanide 8½ ozs.

Electro-motive force at 10 cm. electrode-distance, 3 volts.

Current-density, 0.35 ampère.

The bath is prepared as follows: Dissolve in 7 quarts of warm water the carbonate of soda, *gradually* add the bisulphite of soda to prevent violent effervescence, and then add, with vigorous stirring, the copper acetate in small portions. Dissolve the potassium cyanide in 3 quarts of cold water, and mix both solutions when the first is cold. By thorough stirring with a clean wooden stick, a clear solution is quickly obtained, which is allowed to settle and siphoned off clear. If after the addition of the potassium cyanide the bath should not become colorless, or at least wine-yellow, add a small quantity more of potassium cyanide.

When conversion is complete, the bath contains potassium-copper cyanide, potassium acetate, sodium acetate, sodium sulphate, and potassium cyanide in excess in addition to sodium bisulphite.

For certain purposes, for instance, for the production of a very close, thick deposit, as required for cast-iron door knobs, etc., it is advisable to double the content of metal. For the preparation of such a copper bath it is only necessary to dissolve double the quantities of the salts given in formula III in 10 quarts of water.

Stockmeier recommends the following copper bath:

III*a*. Water 10 quarts, neutral bisulphite of soda 8½ ozs., 98- or 99-percent. potassium cyanide 7 ozs., crystallized carbonate of soda 6 ozs., crystallized copper acetate 7 ozs.

Dissolve the first mentioned three salts together in half the quantity of the water, and the acetate of copper in the other half, and pour the last solution into the first, stirring constantly. It is recommended to add to this bath 77 to 123 grains of bisulphite of soda per quart.

In preparing copper baths, the copper acetate prescribed in the preceding formulæ may be replaced by the carbonate or sulphate, the substitution of the latter, after its previous conversion into carbonate, being of special advantage in order not to thicken the bath by the potassium sulphate formed by eciprocal decomposition. The following formula is especially suitable for the use of sulphate of copper (blue vitriol):

IV. Blue vitriol . . . . . . $10\frac{1}{2}$ ozs.
    Crystallized carbonate of soda $10\frac{1}{4}$ ozs.

    Water . . . . 10 quarts.
    Pulverized bisulphite of soda .   7   ozs.
    Crystallized carbonate of soda. $8\frac{1}{2}$ ozs.
    98 to 99 per cent. potassium cyanide $8\frac{1}{4}$ ozs.

Electro-motive force at 10 cm. electrode-distance, about 3 volts.

Current-density, 0.35 ampère.

First dissolve the $10\frac{1}{2}$ ozs. of blue vitriol and the $10\frac{1}{2}$ ozs. of crystallized carbonate of soda, each by itself, in hot water, and mix the two solutions. Allow the precipitate of carbonate of copper to settle, and pour off the supernatant clear fluid. Then pour upon the precipitate 5 quarts of water, add the bisulphite of soda, next the carbonate of soda, and mix this solution with the solution of the potassium cyanide in 5 quarts of water. The fluid rapidly becomes clear and colorless, when it is boiled and filtered.

V. Water 15 quarts, cupron (cuprous oxide) $3\frac{1}{2}$ ozs., 99 per cent. potassium cyanide $10\frac{1}{2}$ ozs., bisulphite of soda $10\frac{1}{2}$ ozs.

Electro-motive force at 10 cm. electrode-distance, 2.8 volts.

Current-density, 0.3 ampère.

For the preparation of the bath, dissolve the potassium cyanide in about 3 quarts of the water (cold), stir in gradually the cupron, then add the solution of the bisulphite of soda in 3 quarts of the water, and with the remaining 9 quarts of water make up the bath to 15 quarts.

As previously mentioned, an addition of sulphites to the

cuprous oxide solution is not required since no cyanogen escapes. However, in dissolving cuprous oxide in potassium cyanide there is formed, in addition to potassium-copper cyanide, potassium hydroxide (caustic potash) the presence of which in the bath is for various reasons not desirable. A sufficient quantity of bisulphite of soda to convert the caustic potash into neutral potassium sulphite is therefore added, while the corresponding portion of bisulphite is converted into neutral sodium bisulphite. A sufficient excess of bisulphite of soda for the exertion of the above-mentioned favorable effects upon the coppering process remains behind.

Dr. Langbein has introduced in the copper-plating industry the cupro-cupric sulphite. It dissolves in potassium cyanide without noticeable formation of cyanogen, since it contains more than the sufficient quantity of sulphurous acid required for the reduction of the portion of cupric oxide present. Suitable formulas for copper baths with cupro-cupric sulphite are:

VI. Water 10 quarts, 99 per cent. potassium cyanide $8\frac{1}{2}$ ozs., ammonium soda $1\frac{2}{5}$ ozs., cupro-cupric sulphite $4\frac{1}{4}$ ozs., or,

VIa. Water 10 quarts, 60 per cent. potassium cyanide 14 ozs., cupro-cupric sulphite $4\frac{1}{4}$ ozs.

Dissolve the salts in the order given, stirring constantly, and then add the remaining 5 quarts of water.

The deposits obtained in these baths are of a beautiful warm color, very adherent and dense.

Pfanhauser recommends the following bath, in which separately prepared crystallized potassium-copper cyanide, in addition to suitable conducting salts, which are wanting in Hossauer's formula, is used:

VII. Water 10 quarts, ammonia-soda $3\frac{1}{2}$ ozs., anhydrous sodium sulphite 7 ozs., crystallized potassium-copper cyanide $10\frac{1}{2}$ ozs., potassium cyanide 0.35 ozs.

For the preparation of the bath the salts are to be dissolved in a suitable quantity of water, stirring constantly.

Electro-motive force at an electrode distance of 15 cm., 2.7 volts for iron, 3.2 volts for zinc. *Current-density* 0.3 ampère.

For *small zinc objects* which are to be coppered in a basket, baths III, IV, and V, may be used, or those up to and including VII, which are to be heated and compounded with a small additional quantity of potassium cyanide. For the same purpose, Roseleur recommends the following bath:

VIII. Water 10 quarts, neutral crystallized sodium sulphite 1¾ ozs., neutral copper acetate 8 ozs., 75-per cent. potassium cyanide 12¼ ozs., ammonia 3 ozs.

The bath is prepared in the same manner as the baths given under formulas I to III.

Prepared coppering salts. Combinations under various names (Schering: triple metal salts; Langbein Co.: double metal salts) are now brought into commerce, and are quite convenient for the preparation of copper baths (and brass baths) in so far that only one weighing of the substance is required.

As regards their composition these preparations are combinations of potassium-copper cyanide (relatively potassium-zinc cyanide) with alkaline sulphites, such as, for instance, are formed by dissolving cupro-cupric sulphite in potassium cyanide and subsequent evaporation to dryness. As the copper content in Langbein Co's. double copper salt amounts to from 20 to 21 per cent., copper baths with any desired content of metal can in a very simple manner be prepared. Thus, for instance, a copper bath with 92.59 grains of copper per quart is obtained by dissolving 6.6 lbs. of double copper salt in 100 quarts of water; and a copper bath with 138.88 grains of copper per quart, by dissolving 9.9 lbs. of double copper salt in 100 quarts of water. It may be added that there is very seldom occasion to exceed 138.88 grains of copper per liter.

As these preparations can contain but a small content of free potassium cyanide to prevent them from absorbing moisture when stored, it is advisable to dissolve in the baths prepared from them 46 to 78 grains of 99-per cent. potassium cyanide. It has also proved of advantage to add certain conducting salts, for instance, neutral sodium sulphite, in order to effect a better anodal solution of the copper.

A copper salt brought into commerce by the Hanson & Van Winkle Co. of Newark, N. J., under the name of *ruby oxide,* may to advantage be used in making copper solutions, it being claimed to give a much deeper red deposit than is possible to obtain with a carbonate solution.

Copper baths without potassium cyanide. Of the many directions for the preparation of these baths, only a few need here be mentioned.

For *coppering zinc objects,* Roseleur recommends the following bath :

IX. Water 10 quarts, tartar, free from lime, 6.7 ozs., crystallized carbonate of soda 15 ozs., blue vitriol 6.7 ozs., caustic soda lye of 16° Bé., $\frac{3}{4}$ lb.

To prepare this bath, dissolve the tartar and the crystallized carbonate of soda in $\frac{2}{3}$ of the water, and the blue vitriol in the remaining $\frac{1}{3}$, and mix both solutions. Filter off the precipitate, dissolve it in the caustic soda lye, and add this solution to the other.

This bath works very well, and may be recommended to electro-platers who copper zinc exclusively ; though even for this purpose the baths prepared according to III, IV and V answer equally well.

Weill obtains a deposit of copper in a bath consisting of a solution of blue vitriol in an alkaline solution of tartrate of potassium or sodium. Such a bath is composed as follows

X. Water 10 quarts, potassium sodium tartrate (Rochelle salt) 53 ozs., blue vitriol $10\frac{1}{2}$ ozs., 60 per cent. caustic soda 28 ozs.

The chief purpose of the large content of caustic soda is to keep the tartrate of copper, which is almost insoluble in water, in solution. According to Weill, the coppering may be executed in three different ways, as follows:

The iron articles tied to zinc wires, or in contact with zinc strips, are brought into the bath ; the coppering thus taking place by contact. *Or,* porous clay cups are placed in the bath containing the articles ; these clay cups are filled with soda lye in which zinc plates connected with the object-rods are allowed to dip, the arrangement in this case forming a. cell with which, by the solution of the zinc in the soda lye, a current is produced, which affects the decomposition of the copper solution and the deposition. When saturated with zinc the soda lye becomes ineffective, and, according to Weill, it may be regenerated by the addition of sodium sulphite, which separates the dissolved zinc as zinc sulphide. *The third method of coppering* consists in the use of the current of a battery or of a dynamo machine, in which case copper anodes have, of course, to be employed. According to the method used, coppering is effected in a shorter or longer time. In contact-coppering at least six hours were required for the production of a tolerably heavy deposit, and with the use of a current generated by an external source, no other advantage of this bath over potassium-copper cyanide baths could be noticed than that being free from potassium cyanide, it is not poisonous: However, the danger in the use of copper cyanide baths is generally overestimated by the layman, there being actually none, if proper care be observed.

Another copper bath, recommended by Walenn, consists of a solution of equal parts of neutral tartrate of ammonia and potassium cyanide, in which 3 to 5 per cent. of copper (in the form of blue vitriol or moist cupric hydrate) is dissolved. The bath is to be heated to about 140° F.

Gauduin's copper bath consists of a solution of oxalate of copper with oxalate of ammonia and free oxalic acid. Fontaine asserts that the bath works well when heated to between 140° and 150° F.

Tanks for potassium-copper cyanide baths. Copper baths containing cyanide cannot be brought into pitched tanks, tanks of stoneware or enameled iron being used for smaller baths, and for larger ones, basins of brick set in cement, or iron reservoirs lined with cement. Wooden tanks lined with celluloid are also useful. For large baths containing potassium cyanide wooden tanks lined with lead can be used without disadvantage, since a slight coating of cyanide of lead which may be formed upon the lead is insoluble in potassium cyanide, and even if a small quantity of cyanide of lead would be dissolved in the bath by the presence of organic acids, a separation of lead besides copper upon the cathodes does not take place.

Copper anodes. For this purpose it is best to use annealed sheets of pure copper about 0.11 to 0.19 inch thick. They should first be for some time pickled in dilute sulphuric acid, and then scratch-brushed, in order to give them a pure metallic surface.

The anode-surface should be as large as possible, at least as large as the object-surface suspended in the bath.

In all baths containing cyanide the anodes become in a comparatively short time coated with a greenish slime which consists of basic cuprous cyanide, and is mostly soluble in excess of potassium cyanide. When a very thick formation of such slime takes place, potassium cyanide is wanting and has to be added.

In addition to this coat of cuprous cyanide, there may also be formed upon the anode a brown film of paracyanide which adheres very tenaciously and cannot be removed with potassium cyanide, but has to be taken off by scratch-brushing or scouring with pumice. When coppering with small anode-surfaces, and potassium cyanide is at the same time wanting, the coat may become so thick that no current passes into the bath, and consequently no deposit is formed. This feature may even appear when the bath contains a sufficient excess of potassium cyanide, because the cyanide formed in abundance on the anode by high current-densities cannot with sufficient rapidity be dissolved by the potassium cyanide. In this case, recourse must also be had to scouring the anodes, and then increasing the anode-surfaces. The anodes are best suspended to the anode-rods by means of copper bands riveted on.

Execution of copper-plating.—The general rules given under nickeling, as regards the suitable composition of the bath, correct selection of anodes, careful scouring and pickling of the objects, and proper current-strength also apply to copper-plating.

In copper baths containing cyanide, too large an excess of potassium cyanide, to be sure, produces an evolution of hydrogen-bubbles on the objects, but it yields either no deposit at all or only a slight one, which readily peels off. If this phenomenon is noticed after an addition of potassium cyanide is made, the excess has to be removed by adding a copper salt, best, cupro-cupric cyanide. For this purpose triturate the latter with a small quantity of the copper bath in a small porcelain mortar to a thinly-fluid paste, and add the latter to the bath, stirring vigorously for some time. After each addition, see whether an object suspended in the bath becomes rapidly and properly coppered and, if such be not the case, repeat the addition of cupro-cupric cyanide until the bath works in a faultless and correct manner.

However, deposition may also fail by reason of an insufficient addition of potassium cyanide. This is recognized by the heavy formation of froth on the anodes and the appearance of a pale blue color in the fluid, though this may also be caused by the content of metal in the bath being too small. While in the first case, the simple addition of 15 to 30 grains of potassium cyanide per quart will cause the bath to deposit in the proper manner, in the second case, solution of copper cyanide in potassium cyanide is required in order to increase the content of metal, and it is advisable to add at the same time a small quantity of carbonate of soda and of bisulphite of soda. In place of preparing a solution of copper cyanide in potassium cyanide, it is recommended to use crystallized copper cyanide, which can be obtained from manufacturers of chemicals, and dissolve it in hot water.

Many platers are of the opinion that the articles to be copper-plated do not require very careful cleaning and pickling before plating, because this is supposed to be sufficiently effected by the baths themselves, by those containing potas-

22

sium cyanide, as well as by those with alkaline organic combinations. This opinion, however, is wrong.   It is true the
potassium cyanide dissolves a layer of oxide, but not, or at
least very incompletely, any grease present upon the articles,
and hence it is advisable to free objects intended for coppering
as thoroughly from grease as those to be nickeled.

The preliminary scouring and pickling of the articles to be
coppered are executed according to the directions given on p.
223.   The same precautions referred. to under " Deposition of
Nickel " have to be used in suspending the objects in the bath,
and the directions given there for the suitable arrangement of
the anodes, etc., also apply to coppering.   However, as a copper bath conducts better than a nickel bath, the distances
between the anodes and the objects may, if necessary, be somewhat greater.

With a proper arrangement of the anodes and correct regulation of the current, the objects should be entirely coated
with copper in a few minutes after being suspended in the
bath.   In five to ten minutes the objects are taken from the
bath and brushed with a scratch-brush of not too hard brass
wires, whereby the deposit should everywhere show itself to
be durable and adherent.   Defective places are thoroughly
scratch-brushed, scoured, and pickled ; the objects are then returned to the bath.   For solid and heavy plating, the objects
remain in the bath until the original luster and red tone of
the coppering disappear and pass into a dull, discolored
brown.   At this stage the objects are again scratch-brushed
until they show luster and the red copper color, and in doing
this it is of advantage to moisten them with tartar water.
They are then again returned to the bath, where they remain
until the dull, discolored tone reappears.   They are then
taken out, scratch-brushed bright, rinsed in several clean
waters, plunged into hot water, and finally dried, first in sawdust and then thoroughly, at a high temperature, in the dry
ing chamber.

Special attention must be paid to the thorough washing o

the coppered objects, because, if a trace of the bath containing cyanide remains in the depressions or pores, *small, dark, round stains* appear on those places, which cannot be removed, or at least only with great difficulty, they reappearing again in a short time after having been apparently removed. This formation of stains appears most frequently upon coppered (as well as brassed) iron and zinc castings, which cannot be produced without pores. To prevent the formation of these stains the following method is recommended : Since the rinsing in many waters, and even allowing the objects to lie for hours in running water, offer no guarantee that every trace of fluid containing cyanide has been removed, the objects are brought into a slightly acid bath which decomposes the fluid, a mixture of 1 part of acetic acid and 50 parts of water being well adapted for the purpose. The objects are allowed to remain in this mixture for three to five minutes, when they are rinsed off in water and dipped for a few minutes in dilute milk of lime. They are finally rinsed and dried. Coppered castings thus treated will in most cases show no stains.

O. Shultz obtained a patent for the following method for removing the hydrochloric acid from the pores, and for preventing the formation of stains: The plated objects are placed in a room which can be hermetically closed. The air is then removed from the room by the introduction of steam of a high tension and by means of an air-pump, and water is sprinkled upon the objects. By this treatment in vacuum the fluid in the pores comes to the surface, and the salt solution is removed by the water sprinkled over the articles.

After drying, the deposit of copper, if it is to show high luster, is polished upon soft wheels of fine flannel and dry Vienna lime. Commercial rouge FFF, moistened with a little alcohol, is also an excellent polishing agent for copper and all other soft metals.

As is well known, massive copper rapidly oxidizes in a humid atmosphere, and this is the case to a still greater extent with electro-deposited copper. Hence, the coppered

objects, if they are not to be further coated with a non-oxi-
dizing metal, have to be provided with a colorless, transparent
coat of lacquer.

It frequently happens that slightly coppered (as well as
slightly brassed) objects, especially of zinc, after some time,
become entirely white and show no trace of the deposit.
This is due to the deposit penetrating into the basis-metal, as
already explained. Lacquering in this case is of no avail,
the deposit also disappearing under the coat of lacquer. The
only remedy against this phenomenon is a heavier deposit.

If the coppered objects are to be coated with another metal,
drying is omitted, and after careful rinsing they are directly
brought into the respective bath, or into the quicking pickle,
if, as for instance, in silvering, quicking has to be done. In
such cases, where the copper deposit serves only as an intermediary for the reception of another metallic coating, the
objects need not be coppered as thickly, as previously described, by treating them three times in the bath. Preliminary coppering for 5 to 10 minutes suffices in all cases, which
is succeeded by scratch-brushing in order to be convinced
that the deposit adheres firmly, and that the basis-metal is
uniformly coated. The objects are then suspended in the
bath for from 5 to 10 minutes longer with a weak current.

In *coppering sheet-iron* or *sheet-zinc* which is to be nickeled,
the sheets are taken from the bath after 3 to 5 minutes, at any
rate while they still retain luster, scratch-brushing being in
this case omitted. For coppering such sheets a current-density
of 0.5 ampère and an electro-motive force of 3.5 to 4 volts is
required.

The treatment of copper baths when they become inactive,
or show other abnormal features, has already been referred to.

When, as is frequently done, a fluid prepared by dissolving
cuprous oxide in potassium cyanide is used in place of solution of crystallized potassium copper cyanide in water, for
increasing the content of metal, it must not be forgotten to
add the corresponding quantity of bisulphite of soda for the

conversion of the caustic potash formed into potassium sulphate.

All other general rules for plating baths given under "Electro-plating Solutions," Chapter V., must here also be observed.

In the course of time, copper-cyanide baths become thick in consequence of the decomposition of the potassium cyanide and the accumulation of alkaline carbonate and other products of transformation formed thereby, the additions for refreshing the baths also partly contributing thereto. While the normal specific gravity of freshly prepared baths is, according to their composition, from 5° to 7° Bé., the baths after having been in operation for several years may show 11° Bé. and more. It is frequently found that coppering in baths which have become thick is not effectual, the deposit not adhering so well and not showing the brilliant color of copper as when produced in a fresh bath. The only remedy for this is diluting the bath with water to 6° or 7° Bé., increasing the content of metal bv adding highly concentrated solution of potassium copper cyanide, and decomposing the alkaline carbonates, or at least a greater portion of them, by conducting sulphurous acid into the bath, or by dissolving bisulphite of soda in it.

Coppering small articles in quantities. If a large quantity of small articles is at one time to be coppered in dipping baskets, it is recommended to use the baths quite hot, this causing, to be sure, a considerably larger consumption of potassium cyanide than in cold baths. For the rest the process is the same as that given for nickeling small objects in quantities.

If very large quantities of small articles have continually to be coppered, one of the mechanical plating contrivances referred to in Chapter VI will do good service.

The *inlaying* of depressions of coppered art-castings with black may be done in different ways. Some blacken the ground by applying a mixture of spirit lacquer with lampblack and graphite, while others use oil of turpentine with lampblack and a few drops of copal lacquer. A very thin nigrosin lacquer mixed with finely pulverized graphite is very suitable for the purpose. When the lacquer is dry the elevated places which are to show the copper color are cleansed with a linen rag moistened with alcohol.

Electrolytically coppered articles may be inlaid black by coating them, after thorough scouring and pickling, with arsenic in one of the baths given under " Electro-deposition of Arsenic," and, after drying in hot water and sawdust, freeing the surfaces and profiles, which are to appear coppered, from the coating of arsenic by polishing upon a felt wheel. If this polishing is to be avoided, the portions which are not to be black may be coated with stopping-off varnish, and arsenic deposited upon the places left free.

For *coppering by contact and boiling*, see special chapter, " Depositions by Contact."

For *coloring, patinizing and oxidizing of copper*, see the proper chapter.

Examination of Copper Baths Containing Potassium Cyanide

In the preceding sections several characteristic indications which serve for the qualitative examination of these baths have already been given. Like all baths containing potassium cyanide, their original composition gradually suffers extensive alterations by the decomposition of the potassium cyanide, which by the carbonic acid of the air is changed to potassium carbonate and hydrogen cyanide, and spontaneously also to ammonia and potassium formate. The potassium cyanide is also split up by the current, potassium hydroxide being formed, together with decomposition of water, which by the carbonic acid of the air is gradually converted into potash, while hydrogen cyanide and hydrogen escape. Under certain conditions an oxidation of the potassium cyanide to potassium cyanite may also take place.

The excess of potassium cyanide required for the correct performance of the copper bath is therefore gradually consumed, and the bath, at first of a wine-yellow color, acquires a blue coloration, and does no longer yield a good deposit. When such is the case, the same quantity of copper which is withdrawn from the bath by the deposit is not dissolved from the anodes, and hence the determination of the content of free potassium cyanide, as well as that of the content of copper, may at times be necessary. A determination of the potassium carbonate (potash) formed in the bath and its removal, or conversion into potassium cyanide by the addition of the corresponding quantity of barium cyanide solution, which will be referred to under silver baths cannot be recommended. This determination in copper baths which contain, as is generally the case, sulphides, is troublesome, and an accumulation of potash in copper baths does not produce the same evils as in a silver bath. If, however, a copper bath, after working for years, has become thick in consequence of a large content of potash, it can be renewed without considerable expense, or, if this is not desired, it can be regenerated by diluting with water, and increasing the content of copper and of potassium cyanide.

Hence, the determination of the free potassium cyanide (*i. e.*, not fixed on copper) and that of the copper will here only be discussed.

Determination of potassium cyanide.—The best and most rapid method for this purpose is by titrating with decinormal solution of silver nitrate. Silver nitrate and potassium cyanide form finally potassium nitrate and insoluble silver cyanide. the latter, however, being redissolved to potassium silver cyanide so long as free potassium cyanide is present. Since potassium silver cyanide contains two molecules of cyanogen, one molecule of silver nitrate corresponds to two molecules of potassium cyanide, and 1 cubic centimeter of decinormal solution of silver nitrate corresponds to 0.013 gramme of potassium cyanide.

Bring, by means of a pipette, 5 cubic centimeters of the copper bath into a beaker having a capacity of about ¼ liter. Dilute with about 150 cubic centimeters of water, add one or two drops of saturated common salt solution, and then, whilst constantly stirring the fluid in the beaker, allow to flow in from the burette silver nitrate solution so long as the precipitate formed dissolves rapidly.

When solution becomes sluggish, add, stirring constantly, silver nitrate solution drop by drop, waiting after the addition of each drop until the fluid has again become clear. When the fluid does not become clear after adding the last drop, and it shows a slight turbidity, no more free potassium cyanide is present. By multiplying the cubic centimeters of decinormal solution of silver nitrate used by 2.6 the content of potassium cyanide per liter of bath is found.

Suppose, for instance, for 5 cubic centimeters of bath, 2.2 cubic centimeters of silver solution have been used, then 1 liter of the bath contains $2.2 \times 2.6 = 5.72$ grammes of free potassium cyanide, because 1 cubic centimeter of silver solution corresponds to 0.013 grammes of potassium cyanide, therefore 2.2 cubic centimeters $= 2.2 \times 0.013 = 0.0286$ grammes, from which results by calculation

$$5 : 0.0286 = 100 : x$$
$$x = 5.72 \text{ grammes.}$$

If now the initial content of free potassium cyanide in the freshly prepared bath has been determined, a later determination will show the deficiency of it which has come about. It must, however, be taken into consideration that the potassium formate formed by the decomposition of the potassium cyanide may, up to a certain degree, apparently fill the role of the potassium cyanide, in so far as it decreases the conducting resistance of the bath, but it does not contribute to the solution of the anodes. Hence, if the established deficiency of potassium cyanide would be replaced by equally large quantities of the salt, there would be danger of too much of it getting into the bath, and the latter would conduct too readily, which would result in the deposit precipitating too rapidly and turning out less adherent.

Hence it is evident that analytical methods alone are not sufficient for maintaining entirely constant baths containing potassium cyanide, and practical experience and a good faculty of observation are required if the results of analysis are to be utilized for the correction of the baths.   The potassium formate can neither be removed from the bath nor can it be quantitatively determined, and since its action in the bath is not accurately known, it can only be stated from practical experience, that under normal conditions only about 60 per cent. of the deficiency of free potassium cyanide in a copper bath should be replaced by pure potassium cyanide.

Determination of copper.   This may be effected by electrolytic or volumetric analysis.

For the determination of copper by electrolysis, measure off by means of the pipette, 10 cubic centimeters of the copper bath, and allow the fluid to run into a porcelain dish having a capacity of 150 to 200 cubic centimeters.   Add 10 cubic centimeters of pure, strong hydrochloric acid, cover the dish with a watch-glass and heat upon the water-bath.   When evolution of gas ceases, carefully remove the watch-glass, rinse off adhering drops with a small quantity of distilled water into the dish, and evaporate the contents of the latter nearly to dryness.   Now add about 1 cubic centimeter of strong nitric acid, swing the dish to and fro so that all portions of the residue are moistened by the acid, heat for a short time, and then add 32 cubic centimeters of pure dilute sulphuric acid (1 part acid, 2 parts water), with which the contents of each dish are heated, until every trace of odor of hydrochloric and nitric acids has disappeared.   Now pour the copper solution into the platinum dish serving for electrolysis, rinse the porcelain dish with distilled water, adding the wash-water to the contents of the platinum dish, fill the latter up to within 1 centimeter of the rim with water, add 2 cubic centimeters of pure concentrated nitric acid, and electrolyze with a current-strength of ND 100 = 1 ampère, *i. e.,* 1 ampère for 100 square centimeters surface of the platinum dish which serves as cathode.

The copper separates with a bright red color, adhering firmly to the platinum dish, which is connected with the negative pole of the source of current. That the separation of copper is finished is recognized by a narrow strip of platinum sheet, when suspended in the platinum dish, showing in 15 minutes no trace of coppering; or by a few drops of the solution when brought together with a drop of yellow prussiate of potash solution, producing no red coloration.

When by one of the above-mentioned means the complete separation of the copper has been ascertained, the platinum dish is washed, without interruption of the current, the water removed by rinsing the dish with absolute alcohol, and the latter removed by rinsing with ether. Dry for a short time in an air-bath at 212° F., and weigh the dish together with the precipitate of copper. By deducting the weight of the dish, the weight of the copper is obtained, and, since 10 cubic centimeters of the bath were electrolyzed, the weight of the copper multiplied by 100 gives the contents of copper in grammes in 1 liter of copper bath.

The volumetric determination of copper is based upon the principle that solution of sulphate or chloride of copper forms with potassium iodide, copper iodide, whilst free iodine is at the same time formed, one atom of liberated iodine corresponding to one molecule of copper salt. This free iodine is determined by titration with a solution of sodium hyposulphite of known content, and the content of copper is calculated from the number of cubic centimeters of the solution used. For the recognition of the final reaction, the blue coloration, which originates when starch solution combines with free iodine, is utilized. There are required a decinormal iodine solution which contains per liter exactly 12.7 grammes of re-sublimated iodine dissolved in potassium iodide, and a decinormal solution of sodium hyposulphite, of which 10 cubic centimeters diluted with water and compounded with a small quantity of starch solution must exactly use 10 cubic centimeters of iodine solution to give a permanent blue coloration by the formation of iodine-starch.

The mode of operation is as follows : Heat in a porcelain dish 10 cubic centimeters of the copper bath with 10 cubic centimeters of strong hydrochloric acid, evaporate nearly to dryness with 1 cubic centimeter of strong nitric acid and 2 cubic centimeters of hydrochloric acid, and heat upon the water-bath until the nitric acid is entirely removed. The residue is dissolved in water with the addition of a small quantity of dilute hydrochloric acid. The clear solution is brought into a measuring flask holding 100 cubic centimeters, the dish is rinsed with water, the free acid neutralized by the addition of dilute soda lye until a precipitate of bluish copper hydrate commences to separate, which after vigorous shaking does not disappear. Now add, drop by drop, hydrochloric acid until the precipitate just dissolves, fill the flask up to the 100-centi meter mark with water, and mix by shaking. Of this solution bring by means of the pipette, 10 cubic centimeters into a glass of 100 cubic centimeters capacity, and provided with a glass stopper, add 10 cubic centimeters of a 10 per cent. potassium iodide solution, dilute with a small quantity of water, close the glass with the stopper, and let it stand for 10 minutes. Now add from a burette, decinormal solution of sodium hyposulphite until the iodine solution has become colorless, and then add a few cubic centimeters more. Next bring into the flask a few drops of starch solution, and then add from another burette, decinormal iodine solution until a blue coloration is just perceptible. By deducting the cubic centimeters of iodine solution used from the cubic centimeters of sodium hyposulphite solution, it will be known how many cubic centimeters of the latter solution have been used for fixing the iodine liberated by the reciprocal action between copper solution and potassium cyanide solution. Since 1 cubic centimeter of a sodium hyposulphite solution, which is equivalent to the decinormal iodine solution, corresponds to 0.0063 gramme of copper, therefore, as 1 cubic centimeter of the bath has been titrated, the number of cubic centimeters found has to be multiplied by 6.3 to find the content of copper per liter of copper bath.

Suppose to 10 cubic centimeters of the copper solution mixed with potassium iodide has been added 2.8 cubic centimeters of sodium hyposulphite solution, and for titrating back the excess 0.7 cubic centimeters of iodine solution had been used up to the appearance of the blue coloration, then 2.8 — 0.7 = 2.1 centimeters have been used, which multiplied by 6.3 gives 13.23 grammes as the content of copper per liter of bath.

If now a deficiency of copper has been established by one or the other method, the original content of copper can be readily restored by the addition of crystallized potassium-copper cyanide. This salt, when pure, contains about 30 per cent. copper.

Suppose, when first prepared, the bath contained 15 grammes of copper per liter, and it has been shown by analysis that it now contains only 13.3 grammes, then a deficiency of 1.7 grammes of copper has to be made up. Since 100 grammes of potassium-copper cyanide contain 30 grammes of copper, then

$$3 : 100 = 1.7 : x$$
$$x = 3.57$$

and hence 3.57 grammes of potassium-copper cyanide per liter have to be dissolved in the bath. It is advisable to electrolytically determine in the previously described manner, after first destroying the cyanogen combinations, the content of copper in the potassium-copper cyanide to be used for strengthening the bath, so that in case the salt shows a smaller content of copper, the proper quantity of it may be added.

2. DEPOSITION OF BRASS.

Brass is an alloy of copper and zinc, whose color depends on the quantitative proportions of both metals. The alloys known as *yellow brass, red brass (similor, tombac)*, consist essentially of copper and zinc, while those known as *bell metal, gun metal*, and the *bronzes of the ancients* are composed of copper and tin. *Modern bronzes* contain copper, zinc and tin.

The behavior of brass towards acids is nearly the same as that of copper. It oxidizes, however, less rapidly in the air, is harder than copper, malleable, and can be rolled and drawn into wire.

Brass baths. In accordance with the plan pursued in this work only the most approved formulas, the greater portion of which has been practically tested, will be given. More recent propositions which cannot be recognized as improvements over the older directions, will be critically commented upon. There is a large number of receipts for brass baths which show such remarkable difference in the proportions of the two metals that, on more closely examining them, even the layman can, at the first glance, discover the doubtful result. Thus, for instance, Russell and Woolrich recommend a bath in which the quantity of copper salt to zinc salt is in the proportion of 10 : 1. Other authors give the following proportions: Copper 1 to zinc 8 (Heeren); copper 1 to zinc 2 (Salzède, Kruel); copper 2 to zinc 1 (Newton). These examples will show the difference of opinion regarding the suitable composition of brass baths.

An electro-plater understanding all the conditions and the effect of the current-strength might possibly obtain a deposit of brass even from baths which show such abnormal proportions of mixture as those made up according to the directions by Russell, Heeren, and others, but how many conditions have thereby to be taken into consideration will be shown later on. We share the opinion of Roseleur that a brass bath containing copper and zinc salts in nearly equal proportions is the most suitable and least subject to disturbances. A brass bath is to be considered as a mixture of solutions of copper cyanide and zinc cyanide, or of other copper-zinc salts, in the most suitable solvent. Now, since a solution of copper cyanide requires a different current-strength from one of zinc salt, it will be seen that according to the greater or smaller current-strength, now more of the one, and now more of the other, metal is deposited, which, of course, influences the color of the deposit.

Hence the proper regulation of the current is the chief condition for obtaining beautiful deposits, let the bath be composed as it may.

For all baths containing more than one metal in solution, it may be laid down as a rule that the less positive metal is first deposited. In a brass bath, copper is the negative, and zinc the positive metal; and hence a *weaker* current deposits more copper, in consequence of which the deposit becomes redder, while, *vice versa, a more powerful* current decomposes, in addition to the copper solution, also a larger quantity of zinc solution, and reduces zinc, the color produced being more pale yellow to greenish. By bearing this in mind it is not difficult to obtain any desired shades within certain limits.

I. *Brass bath according to Roseleur.*—Blue vitriol and zinc sulphate (white vitriol), of each 5¼ ounces, and crystallized carbonate of soda 15¾ ounces. Crystallized carbonate of soda and bisulphite of soda in powder, of each 7 ounces, 98 per cent. potassium cyanide 8¾ ounces, arsenious acid 30¾ grains, water 10 quarts.

Electro-motive force at 10 cm. electrode-distance, 2.6 to 2.8 volts.

Current-density, 0.32 ampère.

The bath is prepared as follows: In 5 quarts of warm water dissolve the blue vitriol and the zinc sulphate; and in the other 5 quarts the 15¾ ounces of carbonate of soda; then mix both solutions, stirring constantly. A precipitate of carbonate of copper and carbonate of zinc is formed, which is allowed quietly to settle for 10 to 12 hours, when the supernatant clear fluid is carefully poured off, so that nothing of the precipitate is lost. Washing the precipitate is not necessary. The clear fluid poured off is of no value and is thrown away. Now add to the precipitate so much water that the resulting fluid amounts to about 6 quarts, and dissolve in it, with constant stirring, the carbonate and bisulphite of soda, adding these salts, however, not at once, but gradually, in small portions, to avoid foaming over by the escaping carbonic acid. Dissolve the potassium cyanide in 4 quarts of cold water and add this solution, with the exception of about ½ pint in which the arsenious acid is dissolved with the assistance of heat, to the first solutions, and finally add the solution of arsenious acid in the ½ pint of water retained, when the bath should be clear and colorless. If after continued stirring, particles of the precipitate remain undissolved, carefully add somewhat more potassium cyanide until solution is complete.

The addition of a small quantity of arsenious acid is claimed to make the brassing brighter; but the above-mentioned proportion of 30¾ grains for a 10-quart bath must not be exceeded, as otherwise the color of the deposit would be too light and show a gray tone.

II. Crystallized carbonate of soda 10½ ounces, pulverized bisulphite of soda 7 ounces, neutral copper acetate 4.4 ounces, pulverized chloride of zinc 4.4 ounces, 98-per cent. potassium cyanide 14.11 ounces, arsenious acid 30¾ grains, water 10 quarts.

Electro-motive force at 10 cm. electrode-distance 2.6 to 2.8 volts.

Current density 0.32 ampére.

The preparation of this bath is more simple than that of the preceding.

Dissolve the carbonate and bisulphite of soda in 4 quarts of water, then mix the acetate of copper and chloride of zinc with 2 quarts of water, and gradually add this mixture to the solution of the soda salts. Next dissolve the potassium cyanide in 4 quarts of water, and add this solution to the first, retaining, however, a small portion of it, in which dissolve the arsenious acid with the assistance of heat. Finally add the arsenious acid solution, when the bath will become clear. If, however, the solution should not be clear and colorless, or at least wine-yellow, after adding the potassium cyanide, an additional small quantity of the latter may be used, avoiding, however, a considerable excess.

For brassing *iron* in this bath, the quantity of carbonate of soda may be increased up to 35 ozs. for a 10-quart bath. This is also permissible, when in plating zinc articles with a heavy deposit of brass, frequent scratch-brushing is to be avoided. It would seem that a large content of carbonate of soda in the bath retards to a considerable extent the brass color from changing into a discolored brown, though the brilliancy of the deposit appears to suffer somewhat. When boiled for 1 to 2 hours, or worked through with the current for 10 to 12 hours, the bath prepared according to formula II. works very well.

Cupro-cupric sulphide and cuprous oxide may also be advantageously used for the preparation of brass baths. Suitable formulas for them are as follows:

IIa. Pure crystallized zinc sulphate (zinc vitriol or white vitriol) 5½ ozs., crystallized carbonate of soda 7 ozs., pulverized bisulphite of soda 4¼ ozs., ammonium-soda 5¼ ozs., 99 per cent. potassium cyanide 10½ ozs., cupro-cupric sulphide 3⅛ ozs., water 10 quarts.

Electro-motive force at 10 cm. electrode-distance, 2.8 volts.

Current-density, 0.5 ampère.

The bath is prepared as follows: Dissolve the zinc sulphate in 5 quarts of water and the crystallized carbonate of soda in 4 quarts of warm water, and mix the two solutions. When the precipitate of zinc carbonate, which is formed, has completely settled, siphon off the supernatant fluid as much as possible, and throw the lye away.

Dissolve the bisulphite of soda, the ammonium-soda, and the potassium cyanide in 5 quarts of water, add the cupro-cupric sulphide, stirring constantly, and when solution is complete, add the precipitate of zinc carbonate.

This bath yields beautiful pale yellow deposits of a warm brass tone.

IIb. Potassium cyanide 10½ ozs., cuprous oxide 3 ozs., zinc chloride 2¾ ozs., bisulphite of soda 7 ozs., water 10 quarts.

Dissolve the potassium cyanide in 5 quarts of water, add the cuprous oxide and stir until solution is complete. Dissolve the bisulphite of soda and the zinc chloride in the other 5 quarts of water, and mix the two solutions, stirring vigorously.

If metallic cyanides are to be used for the preparation of brass baths, the following formula may be recommended :

III. Crystallized carbonate of soda 10½ ozs., pulverized bisulphite of soda 7 ozs., copper cyanide and zinc cyanide of each 3½ ozs., water 10 quarts, and enough 98 per cent. potassium cyanide to render the solution clear.

To prepare the bath dissolve the carbonate and bisulphite of soda in 2 or 3 quarts of water, rub in a porcelain mortar the copper cyanide and zinc cyanide with a quart of water to a thin paste, add this paste to the solution of the soda salts, and finally add, with vigorous stirring, concentrated potassium cyanide solution until the metallic cyanides are dissolved. Dilute the volume to 10 quarts, and, for the rest, proceed as given for formulas I and II.

Brass baths may in a still more simple manner be prepared by using the double cyanides, potassium-cupric cyanide and potassium-zinc cyanide.

IIIa. Potassium-cupric cyanide (crystallized) 5¼ ozs., potassium-zinc cyanide (crystallized) 5¾ ozs., crystallized bisulphite of soda 8¾ ozs., 98 per cent. potassium cyanide 11¼ drachms, water 10 quarts.

Electro-motive force at 10 cm. electrode-distance, 3 volts.

Current-density, 0.3 ampère.

The bath is prepared by simply dissolving the salts in warm water of about 122° F.

For *brassing zinc* exclusively, Roseleur recommends the following bath :

IV. Dissolve 9¾ ozs. of crystallized bisulphite of soda and 14 ozs. of 70 per cent. potassium cyanide in 8 quarts of water, and add to this solution one of 4¾ ozs. each of neutral copper acetate and crystallized chloride of zinc, 5½ ozs. of aqua ammonia of 0.910 specific gravity, and 2 quarts of water.

For *brassing wrought-iron, cast-iron and steel*, Gore highly recommends the following composition

23

IV*a*. Dissolve 35¼ ozs. of crystallized carbonate of soda, 7 ozs. of pulverized bisulphite of soda, 13¼ ozs. of 98 per cent. potassium cyanide in 8 quarts of water; then add, stirring constantly, a solution of fused chloride of zinc 3½ ozs., and neutral copper acetate 4¼ ozs., in 2 quarts of water. Boil and filter. The bath works well, best with an electro-motive force of 3.75 volts, and also takes readily on cast-iron.

A solution for transferring any copper-zinc alloy which serves as anode, is composed, according to Hess, as follows:

V. Sodium bicarbonate, 14¾ ozs., crystallized ammonium chloride 9½ ozs., 98 per cent. potassium cyanide 2½ ozs., water 10 quarts.

Cast metal plates are to be used as anodes. Transfer begins after a current of medium strength has for a few hours passed through the bath.

This bath is also well adapted for the deposition of tombac with the use of tombac anodes. Most suitable electro-motive force, 3 to 3.5 volts.

Fresh brass baths work, as a rule, more irregularly than any other baths containing cyanide, the deposit being either too red or too green or gray, while frequently one side of the object is coated quite well, and the other not at all. To force the bath to work correctly it must be thoroughly boiled, the water which is lost by evaporation being replaced by the addition of distilled water or pure rain water. If boiling is to be avoided, the bath, as previously mentioned, is worked through for hours, and even for days, with the current, until an object suspended in it is correctly brassed.

Prepared brass salts. Regarding these salts, we refer to what has been said in reference to coppering salts, p. 332. For a 100-quart brass bath, with 92.59 grains of brass, use 6.6 lbs. of double brass salt and 7 ozs. of 98 to 99 per cent. potassium cyanide, and for a 100-quart brass bath with 138.88 grains of brass per quart 9.9 lbs. of brass double salt and 7 ozs. of 98 to 99 per cent. potassium cyanide.

As for copper baths prepared with double salts, an addition

of 30 to 46 grains of potassium cyanide per liter and of a suitable conducting salt (neutral sodium sulphite) may be recommended.

Tanks for brass baths. What has been said on this subject under tanks for copper cyanide baths (p. 335) applies also to brass baths.

Brass anodes. Sheets of brass, annealed and pickled bright, not rolled too hard and of as nearly as possible the same composition and color as the deposit is to have, are used as anodes.

Cast anodes have the advantage of being more readily soluble, and therefore keep the content of metal in the bath more constant than rolled brass anodes. However, the latter answer very well if care is taken to from time to time increase the content of metal in the bath. The anode-surface in the bath should be as large as possible, since with slight anode current-densities the formation of slime on the anodes is less than when the contrary is the case.

Execution of brassing. As previously mentioned, the color of the deposits depends on the quantitative proportions of the two metals deposited, a weaker current depositing predominantly copper, and a stronger current more zinc. Hence by the use of a rheostat it is in the power of the operator to effect within the limits which are given by the resistances of the rheostat, deposits of brass alloys of a redder or more pale yellow to greenish color, according to whether the resistance is increased or decreased.

However, according to the composition of the brass bath, and especially with baths which have for a long time been in use, a determined color of the alloy to be deposited cannot be produced with the assistance of the rheostat. In such case the content of metal in the bath which is required and lacking for the production of a determined color, must be augmented by the addition of solution of the respective metallic salts, or in the form of double salt.

Suppose a bath which originally contained copper and zinc salts in equal proportions has been long in daily use. Now, since brass contains more copper than zinc the deposit will be richer in copper, and it is evident that more of the latter will be withdrawn from the bath than of zinc, and finally a limit will be reached when the bath with a current suitable for the decomposition of the solution will deposit a greenish or gray brass, and with a weaker current produce no deposit whatever. The only remedy in such a case is the addition of sufficient solution of copper cyanide in potassium cyanide, so that, even with quite a powerful current, a deposit of a beautiful brass color is produced, the shades of which can then again be controlled with the assistance of the rheostat. Instead of dissolving copper cyanide in potassium cyanide, it is better to directly use crystallized potassium-copper cyanide. However, it must not be forgotten that every addition of a metallic salt momentarily irritates the brass bath, making it, so to say, sick, and to confine this feature to the narrowest limit, an addition of carbonate and bisulphite of soda, or only of neutral bisulphite of soda, should at the same time be made, and the bath be worked through with the current as previously described, until a test shows that it works in a regular manner. The effect of the addition cannot be controlled in any other way, and more might be added than is required and desirable. If, however, the quantity of the addition is shown not to be sufficient, the operation is continued till the object is attained.

As in the copper bath, an abundant formation of slime on the anodes indicates the want of potassium cyanide in the bath. In this case the evolution of gas-bubbles on the objects is very slight, and the deposit forms slowly. This is remedied by an addition of potassium cyanide. However, in brass baths containing the standard excess of free potassium cyanide, the formation of slime has also a disturbing effect, when the baths are for a longer time worked without interruption. The solution by the potassium cyanide of the bath of the metallic cyanides formed on the anodes takes place more slowly than their formation. If the operation of the bath is for some time interrupted, gradual solution takes place and the greater part

of the slime on the anodes disappears.   However, with an un-
interrupted use of the bath, the layer of slime frequently increases to such an extent that the current cannot flow through
the anodes into the bath.   In such a case, the further increase
of the content of potassium cyanide would not be advisable,
since it would exceed the limit admissible for a dense deposit;
frequent mechanical cleaning of the anodes is then the best
remedy.   A particularly dense slime on the anodes is yielded
by baths which contain small quantities of conducting salts,
such, for instance, as are prepared from the double and triple
salts.   By giving such baths additions of sulphites or chlorides
a less compact slime is formed which is favorable for solution
in potassium cyanide.   This effect is without doubt due to the
anions of the conducting salts appearing on the anodes.

The sluggish formation of the deposit, however, may also be
due to a want of metallic salts.   In this case not only potassium
cyanide, but also solution of copper cyanide and zinc cyanide
in potassium cyanide, has to be added.   For this purpose prepare a concentrated solution of potassium cyanide in water,
and a solution of equal parts of blue vitriol and zinc sulphate
in water.   From the latter, precipitate the copper and zinc as
carbonates with a solution of carbonate of soda as given in
formula I.   After allowing the precipitate to settle, pour off
the clear supernatant fluid, and add to the precipitate, stirring
vigorously, of the potassium cyanide solution until it is dissolved; if heating takes place thereby, add from time to time
a little cold water.   Add this solution with a small excess of
potassium cyanide, and the addition of carbonate or bisulphite
of soda, to the bath, and boil the latter or work it through
with the current.   A more simple method is to procure copper cyanide and zinc cyanide, or concentrated solutions of
these combinations, from a dealer in such articles.   In the
first case, rub in a mortar equal parts of zinc cyanide and
copper cyanide with water to a thinly-fluid paste.   Pour this
paste into a potassium cyanide solution, containing about 7
ozs. of potassium cyanide to the quart, as long as the metallic

cyanides dissolve quite rapidly by stirring. When solution takes place but slowly, stop the addition of paste. A still more simple way is to buy crystallized potassium-copper cyanide and potassium-zinc cyanide, dissolve these salts in suitable quantitative proportions, and add the solution to the bath.

When a brass bath contains too great an excess of potassium cyanide, a very vigorous evolution of gas takes place on the objects, but the deposit is formed slowly or not at all; besides, the deposit formed has a tendency to peel off in scratch-brushing. In this case the injurious excess has to be removed, which is effected by pouring, whilst stirring vigorously, a quantity of the above-mentioned thinly-fluid paste of zinc cyanide and of copper cyanide into the bath. The addition of metallic cyanides is continued only so long as they dissolve with rapidity, so as not to fix all the free potassium cyanide of the bath.

When a brass bath has not been used for some time a white film frequently forms on its surface. This is best removed by pushing it by means of a piece of twisted paper into one corner of the bath, and lifting it off with a shallow dish. It may then be dissolved, eventually by heating, in a small quantity of potassium cyanide and the solution added to the bath.

To avoid unnecessary repetition we refer, as regards the production of thick deposits, scratch-brushing and polishing of the plated articles, to what has been said under " Execution of Coppering," the directions given there being also valid for brassing.

The deposition of several metals from a common solution is not an easy task, and requires attention and experience. If, however, the directions given in this chapter are followed, the operator will be able to conduct, after short experience, the brassing process with the same success as one in which but one metal is deposited.

Special attention must be paid to frequent thorough mixing of the contents of the bath, so that the fluids are renewed on the cathodes, because otherwise, by reason of the fluid becoming poor in metal, the deposit would show different compositions, and consequently other colors.

For the production of deposits of brass which are to show a tone resembling gold, it has been recommended to add to the brass bath an aluminium salt, such as aluminium chloride, aluminium sulphate, etc. Such baths have been offered to laymen as aluminium-bronze baths, with the assurance that the deposit obtained from them consists of an alloy of aluminium and brass, and possesses the same power of resisting atmospheric influences as aluminium-bronze. Although these commendations are evidently misleading, because, notwithstanding, innumerable receipts for the production of electro-deposits of aluminium, the reduction of this metal from solutions of its salts has thus far not been successfully accomplished. Deposits produced in such brass baths, compounded with aluminium salts, were subjected to examination, and in no case was it possible to establish even the slightest trace of aluminium in the deposit. However, notwithstanding that a reduction of aluminium does not take place, the influence of an addition of an aluminium salt to brass baths, as regards the result of the brass tone, cannot be denied, but up to the present time it has been impossible to find an explanation of this fact. If a brass bath, prepared according to the formulas given above, be compounded with 35 to 40 grains of aluminium chloride per quart of bath, the resulting deposit shows a warmer, more sad brass tone than that yielded by a bath without such an addition, and if the bath is somewhat rich in copper, the color of the deposit almost resembles red gold. However, greater power of resisting atmospheric influence could not be noticed, neither can such be expected, since the deposit consists solely of zinc and copper.

It remains to be mentioned that in brassing, the distance of the objects to be plated from the anodes is of considerable importance. If objects with deep depressions or high reliefs are suspended in the brass bath, it will be found that, with the

customary distance of $3\frac{3}{4}$ to $5\frac{3}{4}$ inches from the anodes, the
brassing of the portions in relief nearest to the anodes will turn
out a lighter color than that of the depressed portions, which
will show a redder deposit, the reason for this being that the
current acts more strongly upon the portions in relief, and consequently deposits more zinc than the weaker current which
strikes the depressions. To equalize this difference the objects
have to be correspondingly further removed from the anodes,
with lamp-feet up to $9\frac{3}{4}$ inches, and even more, when a deposit
of the same color will be everywhere formed.

The brassing of unground *iron castings* is especially troublesome, and in order to obtain a beautiful and clean deposit the
preliminary scratch-brushing has to be executed with special
care; but even then the color of the brass deposit will sometimes be found to possess a disagreeable gray tone. This is
very likely largely due to the quality of the iron itself, and it
is advisable first to give the casting a thin coat of nickel or
or tin, upon which a deposit of brass of the usual brilliancy
can be produced. In baths serving for brassing iron articles,
a large excess of potassium cyanide must be avoided. It is,
however, an advantage to increase the content of carbonate of
soda.

Inlaying of brassed objects with black is done in the same
manner as described under " Deposition of Copper."

Brassing by contact will be referred to in the chapter " Depositions by Contact."

For oxidizing, patinizing and coloring of brass, see special
chapter.

Examination of brass baths. The characteristic indications
by which a deficiency, and too large an excess of potassium
cyanide in the bath, as well as an insufficient content of metal,
may be recognized, have already been discussed, and it is here
only necessary to refer to the quantitative determination of the
separate constituents.

Free potassium · cyanide and the content of copper are determined in the same manner as described under copper baths

containing potassium cyanide. Hence only the *determination of zinc* has here to be considered. For making this deter-mination it is necessary to destroy the cyanide combinations, and entirely to remove the copper. For this purpose bring by means of the pipette 10 cubic centimeters of the brass bath into a porcelain dish, and proceed in the same manner as given on page 345 for the determination of copper by electrolysis. Dissolve the evaporated residue in the dish in water, adding a few drops of pure hydrochloric acid. Then bring the solution into a capacious beaker, dilute with water to about 250 cubic centimeters, and heat to boiling. Now add about 10 cubic centimeters of pure dilute sulphuric acid (1 : 10) and stirring constantly, mix with a solution of 2.5 grammes of crystallized sodium. Copper sulphide is separated while the sulphurous acid escapes. Cover the beaker with a watch-glass, let it stand for 15 minutes, and then filter off the precipitate. Wash the filter thoroughly with sulphuretted hydrogen water, and evaporate the filtrate together with the wash waters to about 100 to 150 cubic centimeters. The solution contains all the zinc, and can be at once titrated (see below).

For the determination of zinc by electrolysis, heat the solution to boiling, mix it with solution of sodium carbonate in excess, and after the precipitate of basic zinc carbonate has settled, filter it off. Dissolve the precipitate in the filter with pure dilute sulphuric acid, bring the filtrate, together with the waters used for thoroughly washing the filter, into a clean beaker, and neutralize accurately with sodium carbonate.

Now bring into the platinum dish, previously coppered, 5 grammes of potassium oxalate and 2 grammes of potassium sulphate dissolved in a small quantity of water, fill the platinum dish up to within 1 centimeter from the rim with distilled water, and electrolyze with a current-density of ND 100 = 0.5 ampère. The dull, bluish-white deposit of zinc is treated with water, then with alcohol and ether, dried in the exsiccator over sulphuric acid and weighed. The determined weight of the zinc deposit multiplied by 100 gives the content of zinc in grammes per liter of brass bath.

For the volumetric determination of the zinc, about 100 to 150 ·cubic centimeters of the zinc solution resulting after the precipitation of the copper are used. The determination is based upon the principle that potassium ferrocyanide solution precipitates the zinc from the solution, and that complete precipitation is indicated by an excess of potassium ferrocyanide, yielding a brown coloration with uranium acetate. If now the content of the potassium ferrocyanide solution is known, the quantity of it used gives the content of zinc. It is best to use a solution which contains per liter 32.45 grammes of pure crystallized potassium ferrocyanide, every cubic centimeter of ·this solution corresponding to 0.01 gramme of zinc. Add, ·stirring constantly, from a burette potassium ferrocyanide ·solution to the zinc solution in a beaker until a drop of the ·fluid brought upon a strip of filtering paper previously satu- ·rated with uranium acetate solution and again dried just ·shows the commencement of a brown coloration.

Since 10 cubic centimeters of the brass bath were used for the determination, the number of cubic centimeters of potassium ferrocyanide solution consumed gives the quantity of zinc in grammes per liter of brass bath. Suppose 6 cubic centimeters of solution have been consumed, they would correspond to 0.06 gramme zinc (0.01 × 6). Hence since in 10 ·cubic centimeters of bath 0.06 gramme of zinc is present, the bath contains 6 grammes (0.06 × 100) of zinc per liter.

If thus a deficiency of zinc in the bath, due to long-continued ·working, has been determined the initial content can be readily ·restored by the addition of pure potassium zinc cyanide. The latter contains 26 per cent. of zinc, and the quantity required to be added is determined in the same manner as with a copper bath (see p. 348).

Deposits of tombac, i. e., deposits having the color of tombac, are obtained by increasing the content of copper in the brass baths.

The following formula gives a tombac bath which works well :

Crystallized potassium copper cyanide 7 ozs., crystallized potassium-zinc cyanide 3½ ozs., crystallized neutral bisulphite of soda 8¾ ozs., potassium cyanide ¾ oz., water 10 quarts.

Electro-motive force at 10 cm. electrode distance, 3 volts.

Current-density, 0.3 ampère.

It is of special advantage for the deposition of tombac to heat the bath to between 86° and 95° F., the resulting deposits being of a more uniform color than at the ordinary temperature.

For tombac deposits the transferring solution, according to Hess (see "Deposition of Brass," Formula V), may also be employed, tombac sheets being used as anodes.

Deposits of bronze. Plating of metallic objects with bronze, *i. e.,* a copper-tin alloy, or an alloy of copper, tin and zinc, is but seldom practiced, the bronze tone being in most cases imitated by a brass deposit with a somewhat larger content of copper.

For coating *wrought and cast-iron* with bronze, Gountier recommends the following solution :

Yellow prussiate of potash 10¼ ozs., cuprous chloride 5¼ ozs., stannous chloride (tin salt) 14 ozs., sodium hyposulphite 14 ozs., water 10 quarts.

According to Ruolz, a bronze bath is prepared as follows : Dissolve at 122° to 140° F., copper cyanide 2.11 ozs., and oxide of tin 0.7 ozs., in 10 quarts of potassium cyanide solution of 4° Bé. The solution is to be filtered.

Elsner prepares a bronze bath by dissolving 21 ozs. of blue vitriol in 10 quarts of water, and adding a solution of 2½ ozs. of chloride of tin in potash lye.

Salzède recommends the following bath, which is to be used at between 86° and 95° F. : Potassium cyanide 3½ ozs., carbonate of potash 35¼ ozs., stannous chloride (tin salt) 0.42 oz., cuprous chloride ½ oz., water 10 quarts.

Weill and Newton claim to obtain beautiful bronze deposits from solutions of the double tartrate of copper and potash and the double tartrate of the protoxide of tin and potash, with caustic potash, but fail to state the proportions.

The above formulæ are here given with all reserve, since experiments with them failed to give satisfactory results. With Gountier's, Ruolz's and Elsner's baths no deposit was obtained, but only a strong evolution of hydrogen, while even with a strong current, Salzède's bath did not yield a bronze deposit, but simply one of tin.

The following method of preparing a bronze bath may be recommended: Prepare, each by itself, solutions of phosphate of copper and stannous chloride (tin salt) in sodium pyrophosphate. From a blue vitriol solution precipitate, with sodium phosphate, phosphate of copper, allow the latter to settle, and after pouring off the clear supernatant fluid, bring it to solution by concentrated solution of sodium pyrophosphate. On the other hand, add to a saturated solution of sodium pyrophosphate, solution of tin salt, as long as the milky precipitate formed dissolves. Of these two metallic solutions, add to a solution of sodium pyrophosphate, which contains about $1\frac{3}{4}$ ozs. of the salt to the quart, until the precipitate appears quickly and of the desired color. For anodes, use cast bronze plates, which dissolve well in the bath. Some sodium phosphate has from time to time to be added to the bath, and if the color becomes too light, solution of copper, and if too dark, solution of tin.

For nickel-bronze, see p. 315.

CHAPTER VII.

DEPOSITION OF SILVER.

Silver (Ag = 107.88 parts by weight) *and its properties.*—
Pure silver is the whitest of all known metals. It takes a fine
polish, is softer and less tenacious than copper, but harder and
more tenacious than gold. It is very malleable and ductile,
and can be made into exceedingly thin leaves and fine wire.
Its specific gravity is 10.48 to 10.55, according to whether it is
cast or hammered. It melts at about 1832° F. It is unacted
upon by the air, but in the atmosphere of towns it gradually
becomes coated with a film of silver sulphide. It is rapidly
dissolved by nitric acid, nitrogen dioxide being evolved.
Hydrochloric acid has but little action upon it even at a boiling heat; when heated with concentrated sulphuric acid it
yields sulphur dioxide and silver sulphate.

Chlorine acts upon silver at the ordinary temperature.
Silver has great affinity for sulphur, and readily fuses with it
to silver sulphide. Sulphuretted hydrogen blackens silver,
brown-black silver sulphide being formed (tarnishing of silver
in rooms in which gas is burned). Such tarnishing is most
readily removed by potassium cyanide solution.

Concentrated sulphuric acid combines at a boiling point with
silver to silver sulphate, sulphurous acid escaping. Nitric acid
readily dissolves silver at a gentle heat, and at a higher temperature with considerable violence, silver nitrate (lunar caustic) being formed, while nitrogen dioxide escapes. Watery
chromic acid converts silver into red silver chromate, and this
conversion is made use of as a test for silvering. By touching silver or genuine silver-plating with a drop of a solution
obtained by dissolving potassium dichromate in nitric acid of
1.2 specific gravity, a red stain is formed.

(365)

Electro-plating with silver was, of all electro-metallurgical processes, the first which was carried on on a large scale and has reached enormous proportions.   Large quantities of silver are annually consumed for this purpose, and it is to be regretted that no accurate statistics regarding this consumption are available.

Silver baths.   The longer an electro-plating process has been carried on, the greater, as a rule, the number of existing formulæ for baths will be; but silver baths are an exception to this rule.   If it is taken into consideration that silver-plating has been practically carried on for more than seventy years, the number of formulæ might be expected to be at least equal to those for nickel-plating, which is of much more recent origin. Such, however, is not the case, and chiefly for the reason that the attempts to improve the silver baths, which were made either with a view to banish the poisonous potassium cyanide from the silver-plating industry, or otherwise to advance the plating process, could absolutely show no better results than the baths used by the first silver-platers.   However, that attempts to make such improvements have not been entirely abandoned is shown by Zinin's proposition to substitute solution of silver iodide in potassium iodide for a solution containing potassium cyanide, or by Jordis' proposition to use silver lactate baths.   While the bath according to Zinin yields bad results as compared with the old baths containing potassium cyanide, quite good silvering is obtained with a bath according to Jordis, but independent of the fact that the bath contains no potassium cyanide, no special advantages could be established.   Hence there is no good reason for including other directions for silver baths in this work, which is primarily intended for practical use, and only formulæ for the most approved baths will be given.

However, before describing the preparation of the baths, a few words may be said in regard to the old dispute, whether it is preferable to use silver cyanide or silver chloride.   Without touching upon all the arguments advanced, it may be asserted, by reason of conscientious comparative experiments, that the results are the same, and that the life of the bath is also the same, whether one or the other salt has been used in its orginal preparation. From the chemical view-point, preference had to be given to silver cyanide, but in practice, baths prepared with silver chloride were found to possess ceitain advantages, and theory has furnished an explanation of them.

One of these advantages is found in the fact that by reason of the potassium chloride formed, the resistance of a bath prepared with silver chloride is considerably less than that of a silver cyanide bath, and, with the same current-density, the latter, therefore, requires a greater electro-motive force.

Theoretically, preference has further to be given to silver chloride, because a portion of the potassium chloride formed is dissociated, and potassium-ions and chlorine-ions thus get into the solution. These potassium-ions augment the potassium-ions which are present as a result of the dissociation of the potassium cyanide and silver cyanide and thus increase the efficiency. On the other hand, in addition to cyanogen-ions, chlorine-ions are separated on the anodes and augment the supply of silver-ions in the electrolytes. Furthermore, the presence of potassium chloride facilitates the conversion of the silver cyanide formed on the anodes into potassium silver cyanide, according to the following equation :

$$2AgCy \quad + \quad KCl \quad = \quad KAgCy_2 \quad + \quad AgCl$$

Silver cyanide.　Potassium chloride.　Potassium silver cyanide.　Silver chloride.

While, according to this, preference may be given to silver chloride for the preparation of silver baths, and the more so as it is more easily prepared than silver cyanide, yet for refresh ing the baths, which becomes from time to time necessary to increase their content of silver, recourse must be had to silver cyanide, because with the use of silver chloride for this purpose, the baths would become thick in consequence of a con-

stant supply of further quantities of potassium chloride, and
the silver separate with a coarse structure.

Silver-bath for a heavy deposit of silver (silvering by weight).

I. 98 to 99 per cent. potassium cyanide 14 ozs., fine silver
as silver chloride 8¾ ozs., distilled water 10 quarts.

Electro-motive force at 10 cm. electrode-distance, 0.75 volt.

Current density, 0.3 ampère.

I*a*. 98 to 99 per cent. potassium cyanide 8¾ ozs., fine silver
as silver cyanide 8¾ ozs., distilled water 10 quarts.

Electro-motive force at 10 cm. electrode-distance, 1 volt.

Current density, 0.3 ampère.

Preparation of bath 1. with silver chloride.   Dissolve 14 ozs.
of chemically pure nitrate of silver, best the crystallized, and
not the fused article, in 5 quarts of water, and add to the
solution pure hydrochloric acid, or common salt solution, with
vigorous stirring or shaking, until a sample of the fluid filtered
through a paper filter forms no longer a white caseous precipitate of silver chloride when compounded with a drop of
hydrochloric acid.   These, as well as the succeeding operations, until the silver chloride is ready, have to be performed
in a darkened room, as silver chloride is partially decomposed
by light.   Now separate the precipitate of silver chloride from
the solution by filtering, using best a large bag of close felt,
and wash the precipitate in the felt bag with fresh water.
Continue the washing until blue litmus paper is no longer
reddened by the wash-water, if hydrochloric acid was used
for precipitating, or, if common salt solution was used, until
a small quantity of the wash-water, on being mixed with a
drop of lunar caustic solution, produces only a slight milky
turbidity and no precipitate.   Now bring the washed silver
chloride in portions from the felt bag into a porcelain mortar,
rub it with water to a thin paste, and pour the latter into the
potassium cyanide solution consisting of 14 ozs. of 98 per
cent. potassium cyanide in 5 quarts of water, in which, by
vigorous stirring, the silver chloride gradually dissolves.   All
the precipitated silver chloride having been brought into

solution, dilute with water to 10 quarts of fluid, and boil the bath, if possible, for an hour, replacing the water lost by evaporation. A small quantity of black sediment containing silver thereby separates, from which the colorless fluid is filtered off. This sediment is added to the silver residues, and is worked together with them for the recovery of the silver by one of the methods to be described later on.

Preparation of bath Ia with silver cyanide. Dissolve 14 ounces of chemically pure crystallized nitrate of silver in 5 quarts of water, and precipitate the silver with prussic acid, adding the latter until no more precipitate is produced by the addition of a few drops of prussic acid to a filtered sample of the fluid. Now filter, wash, and proceed for the rest exactly as stated for the bath with silver chloride, except that only 8¾ ounces of potassium cyanide are taken for dissolving the silver cyanide. In working with prussic acid avoid inhaling the vapor which escapes from the liquid prussic acid, especially in the warm season of the year; and be careful the acid does not come in contact with cuts on the hands. It is one of the most rapidly-acting poisons.

Silver cyanide may also be prepared as follows: Dissolve 14 ounces of chemically pure crystallized nitrate of silver in 5 quarts of water, and add moderately concentrated potassium cyanide solution until no more precipitate is formed, avoiding, however, an excess of the precipitating agent, as it would again dissolve a portion of the silver cyanide. The precipitated silver cyanide is filtered off, washed and dissolved in potassium cyanide, as above described.

The preparation of the silver bath according to the above formulæ is more conveniently effected by using pure crystallized potassium-silver nitrate in the following proportions:

I*b*. 98-per cent. potassium cyanide, 6⅓ to 7 ozs.; crystallized potassium-silver cyanide, 17½ ozs.; distilled water, 10 quarts.

Electro-motive force and *current-density* as for I*a*.

The salts are simply dissolved in the cold water.

24

The baths prepared according to formulæ I, Ia or Ib serve chiefly for the production of a heavy deposit upon German silver articles, especially table and other household utensils. Of course, they may also be used for plating other metals by weight.

Silver bath for ordinary electro-silvering. II. 98-per cent. potassium cyanide, 6¾ to 7 ounces; fine silver (as silver nitrate or chloride), 3½ ounces; distilled water, 10 quarts.

Electro-motive force, for silver chloride, at 10 cm. electrode-distance, 1.25 volts.

Current-density, 0.3 ampère.

To prepare the bath dissolve 5½ ounces of chemically pure crystallized nitrate of silver in 5 quarts of distilled water; in the other 5 quarts of water dissolve the potassium cyanide, and mix both solutions. Or if chloride of silver is to be used, precipitate the solution of 3½ ounces of the silver salt in the same manner as given for formula I; wash the precipitated chloride of silver, and dissolve it in the potassium cyanide solution.

IIa. 98-per cent. potassium cyanide 1⅖ ozs., crystallized potassium-silver cyanide 7 ozs., distilled water 10 quarts.

Dissolve the salts in the cold water.

For the preparation of silver baths double and triple silver salts are brought into commerce by some manufacturers.

Such salts require simply dissolving in water. However, they offer no special advantages, since the preparation of baths according to formulas Ib and IIa can scarcely be surpassed as regards simplicity.

Tanks for silver baths. As receptacles for silver baths, tanks of stoneware and enameled iron tanks, as well as wood tanks lead-lined and coated or lined with celluloid can only be used.

Treatment of the silver baths.—Silver anodes. Frequently the error is committed of adding too much potassium cyanide to the bath. A certain excess of it must be present, and in the formulas given, this has been taken into consideration. For dissolving the silver cyanide prepared from 14 ounces of nitrate of silver, as given in formula Ia, only about 5½ ounces

of potassium cyanide are required, and the consequence of
working with such a bath, devoid of all excess, would be that,
on the one hand, the bath would offer considerable resistance
to the current, and, on the other, that the deposit would not
be uniform and homogeneous, and the anodes would be coated
with silver cyanide.   Hence, with the use of a normal current,
about 0.35 to 0.42 oz. more of potassium cyanide is added
per quart of bath.   However, when working with a stronger
current, this excess would already be too large, and the
deposit would not adhere properly, and rise up in scratch-
brushing.   And, again, with a very weak current, the baths
can without disadvantage stand a larger excess.   As a rule,
however, the proportions between fine silver and potassium
cyanide given in the above formulæ may be considered as
normal, and with the current-densities prescribed, a deposit
of fine structure, which adheres firmly, will result.

By reason of the slight electro-motive force required for
silvering, in plating larger object-surfaces the cells are not
coupled one after the other for electro-motive force, but in
parallel.   In no case must an evolution of hydrogen be perceptible on the objects, and the current must be the more
weakened, the larger the excess of potassium cyanide in the
bath.

In the silver baths prepared according to the formulas given
above, the excess of potassium cyanide amounts to 0.35 to
0.42 oz. per quart, and is only increased in silver baths which
are to serve for the direct silvering of tin and of alloys with a
large content of nickel, as will be shown later on.   Baths for
ordinary, as well as light, silvering, with 0.35 oz. of silver per
quart, would only require an excess of 0.17 oz. of potassium
cyanide.   However, since the larger excess of 0.35 oz. is no
disadvantage, and allows of working with less electro-motive
force, it is generally preferred.   The electro-motive force given
for the separate formulas is applicable only when the anode-
surface is of the same size, or approximately so, as the object-
surface.   If, for reasons of economy, the work is carried on

with considerably smaller anode-surfaces, the electro-motive force has to be adequately increased in order to conduct into the bath a quantity of current corresponding to the normal current-density.

Whether too much, or not enough, potassium cyanide is present in the bath is indicated by the appearance of the plated objects and the properties of the deposit, as well as by the behavior of the anodes in the bath during and after silvering. It may be accepted, as a rule, that with a moderate current the object should, in the course of 10 to 15 minutes, be coated with a thin, dead-white film of silver. If this be not the case, and the film of silver shows a meager bluish-white tone, potassium cyanide is wanting. However, if, on the other hand, the dead-white deposit forms within 2 or 3 minutes, and shows a crystalline structure, or a dark tone playing into gray-black, the content of potassium cyanide in the bath is too large, provided the current is not excessively strong. If copper and brass become coated with silver without the co-operation of the current, the bath contains too much potassium cyanide.

In silver-plating, even if the objects are to be only thinly coated, insoluble platinum anodes-should never be used, but only anodes of fine silver, which are capable of maintaining the content of silver in the bath quite constant. From the behavior and appearance of the anodes, a conclusion may also be drawn as to whether the content of potassium cyanide in the bath is too large or too small. If the anodes remain silver-white during plating, it is a sure sign that the bath contains more potassium cyanide than is necessary and desirable; but, if they turn gray or blackish, and retain this color after plating, when no current is introduced into the bath for a quarter of an hour or more, potassium cyanide is wanting. On the other hand, the correct content of potassium cyanide is present, when the anodes acquire during the plating process a gray tone, which, after the interruption of the current, gradually changes back to pure white.

The use of steel sheets as anodes for silver baths in place of silver anodes, as has been proposed, cannot be approved, especially when chloride of silver has been used for the preparation of the bath, or other chlorides are present in it. The chloride-ions would bring iron into solution and this would form potassium-ferrocyanide with the potassium cyanide.

If it is shown by the process of silvering itself, or by the appearance of the articles, or of the anodes, that potassium cyanide is wanting in the bath, it should be immediately added, though never more than 30 to 37½ grains per quart of bath at one time, so as to avoid going to the other extreme. Too large a content of potassium cyanide is remedied by adding to the bath, stirring constantly, a small quantity of cyanide or chloride of silver rubbed with water to a thinly-fluid paste, whereby the excess is rendered harmless in consequence of the formation of the double salt of silver and potassium cyanide. Instead of such addition, the current, may, however, be used for correcting the excess. For this purpose suspend as many silver anodes as possible to the anode-rods, but only a single anode as an object to the object-rod, and allow the current to pass for a few hours through the bath, whereby the excess of potassium cyanide is removed or rendered harmless by the dissolving silver.

The bath can be kept quite constant by silver anodes, provided potassium cyanide be regularly added at certain intervals, and the anode-surface is equal to that of the objects to be plated. But since, on account of the expense, a relatively small anode-surface is frequently used, the content of silver in a bath continuously worked will finally become lower, and augmentation, by the addition of silver, will be required. The manner of effecting this augmentation depends on whether the baths are used for plating by weight or for lighter silvering, or whether the baths are worked without stopping from morning till evening. For replacing the deficiency in baths prepared according to formulæ I and Ia, it is advisable to use exclusively solution of silver cyanide in potassium cyanide, or of crystallized potassium-silver cyanide in water.

It has previously been mentioned that with proper treatment baths made with chloride of silver have the same duration of life as those prepared with silver cyanide. The chief feature of such proper treatment is not to use chloride of silver dissolved in potassium cyanide for augmenting the content of silver, but to employ silver cyanide instead, since by the use of the former, the bath thickens in consequence of the potassium chloride which is simultaneously introduced. The effect of such thickening is that the deposits are formed less homogeneously and with coarser structure.

A gradual thickening of the bath may also take place if potassium cyanide containing potash is used, instead of the preparation free from potash, and of 98 to 99 per cent. purity. Even pure fused potassium cyanide produces a thickening of the bath, which, however, progresses very slowly. This thickening is due to a portion of the excess of potassium cyanide being by the action of the air converted into potassium carbonate, and if the quantity of the latter exceeds $\frac{1}{4}$ oz. per quart, it has to be neutralized. For this purpose prussic acid was formerly used in order to effect a conversion of the potassium carbonate into potassium cyanide. It is, however, a well-known fact that carbonic acid decomposes the potassium cyanide, potassium carbonate and prussic acid being formed, and the addition of prussic acid would therefore appear not very suitable for attaining the object in view

It is better to use solutions of calcium cyanide or barium cyanide, and add them so long as a precipitate of calcium carbonate or barium carbonate is formed. The solutions should, however, be freshly prepared. The precipitate formed is allowed to settle. when the clear solution is siphoned off, and the residue filtered through a paper filter.

Since, as mentioned above, the proportion of excess of potassium cyanide to the content of silver undergoes changes according to the proportion of the object-surface to the anode-surface, the temperature of the bath, etc., it becomes necessary to add one or the other in order to maintain the proper proportions and the effective working of the bath.

To determine rapidly whether the bath contains silver and excess of potassium cyanide in proper proportions, the following methods may be used : Dissolve 1 gramme (15.43 grains) of chemically pure crystallized nitrate of silver in 20 grammes (0.7 oz.) of water and gradually add this solution, whilst constantly stirring with a glass rod, to 100 grammes (3.52 ozs.) of the silver bath in a beaker, so long as the precipitate of silver cyanide formed dissolves by itself. If, after adding the entire quantity of silver solution, the precipitate dissolves rapidly, too large an excess of potassium cyanide is present in the bath ; and *vice versa*, if the precipitate does not completely dissolve, after stirring, potassium cyanide is wanting.

The quantitative determination of the content of potassium cyanide and of silver will be described later on under " Examination of Silver Baths."

Agitation of silver baths. In heavy silver-plating, constant agitation of the strata of fluid is of decided advantage, grooves and blooms being otherwise readily formed upon the plated objects, especially when the baths are over-concentrated or thickened. The depressed grooves can only be explained by the fact that the strata of fluid on the cathodes having become specifically lighter by yielding metal are subject to a current towards the surface ; the lower strata richer in silver give rise to heavier deposits on the lower cathode portions, so that agitation of the bath becomes an actual necessity.

With a bath in constant agitation a greater current-density may be used, the deposits, notwithstanding the greater current-density, forming with finer structure and in a correspondingly shorter time, which is especially noteworthy for heavy silvering. To keep the articles in gentle motion while in the bath, one method is to connect the suspending rods to a frame of iron having four wheels, about 3 inches in diameter, connected to it, which slowly travel to and fro to the extent of 3 or 4 inches upon inclined rails attached to the upper edge of the tank, the motion, which is both horizontal and vertical, being given by means of an eccentric wheel driven by steam power. By another arrangement, the frame supporting the articles does not rest upon the tank, but is suspended above the bath, and receives a slow swinging motion from a small eccentric or its equivalent. In the Elkington establishment at Birming-ham the following arrangement is in use: All the suspending rods of the bath rest upon a copper mounting, which, by each revolution of an eccentric wheel, is lifted about $\frac{3}{4}$ inch, and then returned to its position. The copper mounting is con-

Fig. 123.

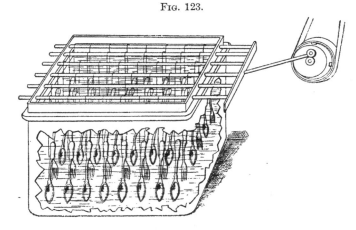

nected to the main negative wire of the dynamo-machine by a copper cable. The same object may also be attained by giving the articles a horizontal, instead of a vertical motion, as shown in Fig. 123, in which the motion is produced by an eccentric wheel on the side.

With equal, if not better, success the mechanically moved stirring apparatus, which will be described under "Copper Galvanoplasty," may be used. In this apparatus several glass rods movable around a pivot keep the bath in constant motion. Where such a stirring apparatus cannot be conveniently arranged, the motion of the bath may be produced by introducing, by means of a pump, air on the bottom of the tank.

A singular phenomenon in regard to silver baths, which has not yet been explained, may here be mentioned. A small addition of certain, and especially of organic, substances, which, however, must not be made suddenly or in too large quantities, produces a fuller and better adhering deposit of greater luster than can be produced in fresh baths. Elkington observed that an addition of a few drops of carbon disulphide to the bath made the silvering more lustrous, while others claim to have used with success solutions of iodine in chloroform, of gutta-percha in chloroform, as well as heavy hydrocarbons, tar, oils, etc.

A silver bath, as shown by experience, becomes without doubt better in the degree in which it takes up small quantities of organic substances from the air and from dust; but numerous experiments have failed to confirm Elkington's observation that the formation of the deposit or its appearance is essentially influenced by the addition of carbon disulphide or any of the above-mentioned solutions of organic origin either in very small or considerable quantities. Many baths have been entirely spoiled by an attempt to change them into bright-working baths bv the addition of such ingredients, and hence it is best to leave such experiments alone. It may, however, be stated that by the addition of a few drops of liquid ammonia, fresh silver baths accommodate themselves more rapidly to regular performance.

However, the use of carbon disulphide as an addition to the silver bath for *bright plating* is advocated by some electroplaters, and some preparations for this purpose may here be given. The carbon disulphide should not be directly added to the bath, as in that case it does not intimately mix with the bath, it settling on the bottom of the vat and the deposit would turn out defective. The following plan has been highly recommended for the ordinary silver bath prepared from chloride of silver and potassium cyanide: Add to 1 quart of the silver bath in a bottle 10 drops of carbon disulphide. Cork the bottle tightly and vigorously shake from time to time. Allow the bottle to stand over night for the fluid to settle, and then pour off the supernatant fluid. A residue of a dark color will be found on the bottom of the bottle. The fluid thus obtained should be perfectly clear, and forms the carbon disulphide solution to be added to the actual silver bath.

For the preparation of a bath for bright-plating add about ½ oz. of the carbon disulphide solution to every 45 quarts of the silver bath, and mix thoroughly. With proper treatment the deposit will be smooth and bright.

If the deposit does not show the desired surface but is still partly mat and partly white, the bath does not contain sufficient carbon disulphide solution and more has to be added to obtain satisfactory results. Since the carbon disulphide is consumed in plating, carbon disulphide solution has of course to be added from time to time to the silver bath. The want of carbon disulphide in the bath is readily recognized by the appearance of the deposit. Care must, however, be exercised in making such an addition since too much of it has an injurious effect upon the deposit. The deposit thus obtained is smooth and has a slight luster. It is considerably harder than a mat deposit but can be polished without trouble.

Another method of preparing a solution for bright-plating is as follows: Put 1 quart of ordinary silver-plating solution into a large stoppered bottle. Now add 1 pint of strong solution of cyanide, and shake well; 4 ozs. of carbon disulphide are then added, as also 2 or 3 ozs. of liquid ammonia, and the bottle again well shaken, the latter operation being repeated every two or three hours. The solution is then set aside for about 24 hours, when it will be ready for use. About 2 ozs. of the clear liquid may be added to every 20 gallons of plating solution, and well mixed by stirring. A small quantity of the brightening solution may be added to the bath every day, and the liquid then gently stirred. In course of time the disulphide solution acquires a black color; to modify this a quantity of strong cyanide solution, equal to the brightening liquor which has been removed from the bottle, should be added each time. In adding the disulphide solution to the plating bath, an excess must be avoided, otherwise the latter will be spoiled. Small doses repeated at intervals is the safer procedure, and less risky than the application of larger quantities, which may ruin the bath.

A very simple way to prepare the brightening solution is to put 2 or 3 ozs. of carbon disulphide into a bottle which holds rather more than half a gallon. Add to this about 3 pints of old silver solution and shake the bottle well for a minute or so. Then nearly fill the bottle with a strong solution of cyanide, shake well as before, and set aside for at least 24 hours. Add about 2 ozs. (not more) of the brightening liquor, without shaking the bottle, to each 20 gallons of solution in the plating vat. Even at the risk of a little loss from evaporation, it is best to add the brightening liquor to the bath the last thing in the evening, when the solution should be well stirred so as to thoroughly diffuse the added liquor. The night's repose will leave the bath in good working order for the following morning.

Yellow tone of silvering. After plating, the objects frequently show, instead of a pure white, a yellow tone, or they become yellow in the air, which is ascribed to the formation of basic silver salts in the deposit. To overcome this evil it has been proposed to allow the objects to remain in the bath for a few minutes after interrupting the current, whereby the basic salts are dissolved by the potassium cyanide of the bath ; or the same object is attained by inverting the electrodes for a few seconds, after plating, thus transforming the articles into anodes. The electric current carries away the basic salt of silver in preference to the metal. This operation should, of course, not be prolonged, otherwise the silver will be entirely removed from the objects, and will be deposited on the anodes. For the same purpose some electro-platers hold in readiness a warm solution of potassium cyanide, in which they immerse the plated articles for half a minute.

Silver alloys. It has been proposed to add to the silver baths a solution of nickelous cyanide in potassium cyanide in order to obtain a deposit of a silver-nickel alloy, which is claimed to be distinguished by its greater hardness and the property of not so readily turning dark. Numerous experiments with solutions of cyanide of silver and nickelous cyanide in potassium cyanide in all possible proportions, and with various electro-motive forces, and subsequent analysis of the deposits obtained, showed, however, only inconsiderable traces of nickel in the silver deposit, which had but a very slight influence upon the hardness and durability of the silver.

The London Metallurgical Co. endeavors to attain greater hardness and power of resistance of the silver by adding zinc-cyanide or cadmium cyanide, and has given to this process the name of *arcas* silver-plating. According to the patent, an addition of 20 to 30 per cent. of zinc or cadmium to the silver prevents the tarnishing of the plating, and besides the deposit is claimed to be lustrous and hard. For arcas silver-plating the appropriate quantity of zinc or cadmium, or a mixture of both metals, is converted into potassium-zinc cyanide or potassium-cadmium cyanide, and this solution is mixed with a corresponding quantity of solution of potassium-silver cyanide, with a small excess of potassium cyanide. Sheets of a silver-zinc or a silver-cadmium alloy are used as anodes.

This method has been favorably commented upon by Sprague, Urquart and others, and some English platers claim that for many articles, especially bicycle parts, arcas silvering may be substituted for nickeling. However these favorable opinions were not confirmed by the following experiments made by Dr. Langbein regarding the value of this process as a substitute for silver-plating instruments and articles of luxury.

A bath was prepared which contained per quart 231½ troy grains of fine silver and 77 troy grains cadmium in the form of cyanide double salts with a small excess of potassium cyanide. The most suitable tension of current for the decomposition of a pure potassium-cadmium cyanide solution which

·contained per quart 154 troy grains of cadmium with the same ·excess of potassium cyanide as the above-mentioned mixture was found to be 2 volts.

In electrolyzing the cadmium-silver bath with 0.75 volt, a uniform silver-white deposit similar to that of pure silver was at first formed. However, after two hours the deeper places ·of the objects suspended in the bath showed crystalline excrescences which felt sandy, and could be rubbed off with the fingers. After scratch-brushing the articles and again suspending them in the bath, these sandy, non-adhering metallic ·deposits were rapidly reformed. An analysis of the deposit separated from the articles showed 96.4 per cent. silver and 3.2 per cent. cadmium. This deposit could, without difficulty, be polished with the steel like a pure silver deposit, and hence its hardness would not seem greater than that of pure silver. Its capability of resisting hydrogen sulphide as compared with pure silver was scarcely greater.

In another experiment electrolysis was effected with 1.25 volts. The deposit showed from the start a coarser structure, and the formation of the sandy non-adhering deposit took place much more rapidly. But, on the other hand, the hardness of the reduced coherent metal was greater than that of pure silver, and also its power of resisting hydrogen sulphide. An analysis of the deposit showed 92.1 per cent. silver and 7.8 per cent. cadmium. In both cases the deposit was dull like that of pure silver.

With a greater electro-motive force the quantity of cadmium in the deposit increased, and the hardness of the latter became correspondingly greater. However, these deposits could not be considered serviceable for the above-mentioned purpose, because they could not be made of sufficient thickness as required for solid silver-plating of forks and spoons.

In Dr. Langbein's opinion, the decomposition-pressures of a solution of potassium-silver cyanide and of one of potassium-cadmium cyanide lie too far apart to obtain without delay ·deposits of even composition and of sufficient density and thickness.

Execution of silver-plating—A. Silver-plating by weight.—
Copper, brass, and all other copper alloys may be directly
plated after amalgamating (quicking), whilst iron, steel, nickel,
zinc, tin, lead, and Britannia are first coppered or brassed, and
then amalgamated.

The mechanical and chemical preparation of the objects for
the silver-plating process is the same as described on page 188
et seq. To obtain well-adhering deposits great care must be
exercised in freeing the objects from grease, and in pickling.
As a rule, objects to be silver-plated are ground and polished.
However, polishing must not be carried too far, since the de
posit of silver does not adhere well to highly polished surfaces;
and in case such highly polished objects are to be silvered it
is best to deprive them of their smoothness by rubbing with
pumice powder, emery, etc., or by pickling.

The treatment of copper and its alloys, German silver and
brass, which have chiefly to be considered in plating by
weight is, therefore, as follows :

1. *Freeing from grease* by hot potash or soda lye (1 part of
caustic alkali to 8 to 10 parts of water), or by brushing with
the lime-paste mentioned on page 229.

2. *Pickling* in a mixture of 1 part, by weight, of sulphuric
acid of 66° Bé. and 10 of water. This pickling is only
required for rough surfaces of castings, ground articles being
immediately after freeing from grease treated according to 3.

3. *Rubbing* with a piece of cloth dipped in fine pumice
powder or emery, after which the powder is to be removed by
washing.

4. *Pickling* in the preliminary pickle, rinsing in hot water,
and quickly drawing through the bright-dipping bath (page
223), and again thoroughly rinsing in several waters.

5. *Amalgamating (quicking)* by immersion in a solution of
mercury, called the quicking solution. This consists of a solution of 0.35 ounce of nitrate of mercury in 1 quart of water,
to which, while constantly stirring, pure nitric acid in small
portions is added until a clear fluid results. A weak solution

of potassium-mercury cyanide in water is, however, to be preferred, because the acid quicking solution mentioned above makes the metals brittle. A quicking solution for silver-plating by weight consists of: Potassium-mercury cyanide, 14 drachms to 1 oz. ; 99-per cent. potassium cyanide, 14 drachms; water, 1 quart. Care must be taken to bring the quicked objects into the bath as rapidly as possible, otherwise thin objects are liable to become brittle. The amalgam formed upon the surface penetrates to the interior of thin sheets if this action is not prevented by an immediate deposition of silver and the formation of silver amalgam. In the quicking solution the objects remain only long enough to acquire a uniform white coating, when—

6. They are rinsed in clean water, and gone over with a soft brush in case the quicking shows a gray instead of a white tone.

The articles are now brought into the silver bath, and secured to the object-rods by slinging wires of pure copper or, still better, of pure silver. The latter have the advantage that when by reason of a deposit of considerable thickness having been formed upon them they have become useless for suspending the articles, they can be directly converted into silver nitrate by dissolving in nitric acid, and used for the preparation of fresh baths, or for strengthening old baths.

FIG. 124.

When certain objects, for instance, forks and spoons, are to be plated, copper wires may be bent in the manner shown in Fig. 124. To prevent the deposition of silver upon the portions of the wire which do not serve for the purpose of contact, they are coated with fused ebonite mass or gutta-percha, only the loop in which the fork or spoon is hung and the upper end for suspending to the object-rod being left free. Silver wires are also better for this purpose.

In order to secure an extra heavy coating of silver on the convex surfaces of spoons and forks which, being subject to greater wear than the other parts, require extra protection, some plating establishments use a frame in which the articles supported therein by their tips are placed horizontally in a shallow silver bath and immersed just deep enough to allow the projecting convexities to dip into the bath. By this artifice these portions are given a second coating of silver of any desired thickness. This mode of procedure, which is termed "sectional" plating, accomplishes the intended purpose nicely and satisfactorily. In some establishments the silvered forks and spoons are placed between plates of gutta-percha of corresponding shape, and held together by rubber bands. In these plates the portions to be provided with an extra coating of silver are cut out. By suspending the forks and spoons thus protected in the bath, the unprotected places receive a further layer of silver, the outlines of which are later on smoothed down with burnishers. The second object may also be attained by coating the places which are to receive no further deposit with "stopping-off" varnish (see later on).

According to German patent No. 76975, sheets of celluloid or similar substances are suspended as shields between the inside portions of spoons and the anodes. By this means the deposit of silver on these portions, which are subject to less wear, is only slightly augmented, while the outside portions acquire a heavier deposit.

To attain the same object, J. Buck * suspends a frame in such a manner that every two spoons are turned with their inside portions towards each other, and anodes are arranged only on the outside portions.

When commencing the operation of silver-plating, introduce into the bath at first a somewhat more powerful current, so that the first deposition of silver takes place quite rapidly, and after 3 minutes regulate the current so that in 10 to 15 minutes the objects are coated with a thin, dull film of silver At this stage take them from the bath, and after seeing that

---

* German patent 126053 (expired).

all portions are uniformly coated, scratch-brush them with a brass brush, which should, however, not be too fine. In doing this the deposit must not raise up. If at this stage the objects stand thorough scratch-brushing, raising of the deposit in burnishing later on need not be feared.

Any places which show no deposit are vigorously scratch-brushed with the use of pulverized tartar, then again carefully cleansed by brushing with lime-paste to remove any impurities due to touching with the hands, pickled by dipping in potassium cyanide solution, again rinsed, quicked, and after careful rinsing returned to the bath. Special care must be taken not to contaminate the bath with quicking solution as this would soon spoil it.

The objects now remain in the bath at a normal current-density until the deposit has acquired a weight corresponding to the desired thickness. Knives, forks and spoons receive a deposit of 2.11 to 3.52 ozs. of silver per dozen.

Considerable difficulty is sometimes experienced in silver plating the steel blades of table knives as the silver will strip or pull off of the blade after it has been in use a short time. According to Mr. Charles H. Proctor * this difficulty is due to unsatisfactory conditions at the time of deposition. The coating of the knives with copper previous to silver plating will not improve matters; in fact this method has been discarded as a failure by all the manufacturers of silver-plated steel-knives and forks years ago. The most satisfactory method to pursue is to reduce all the silver from the surface of the knives by the aid of a strong cyanide solution and a strong reversed current of five to six volts or more. For cathodes use carbon and arrange the positive pole, upon which the knives are placed, so that the carbon cathodes will be placed on either side of the knives, as in a regular bath, so that the metal is reduced uniformly. After the silver is removed the surface is washed, dried and polished and then the knives should be

* The Metal Industry, December, 1912.

25

boiled out in any of the usual alkaline solutions of potash or soda. Then immerse them in undiluted hydrochloric acid and wash and scour on a tampico wheel, using sodium carbonate in the water to prevent rusting after scouring.

The articles are now ready for the bath. Frame up, wash in clean water, immerse in a 50 per cent. solution of hydrochloric acid and water, rewash and immerse directly in the strike solution. This strike should consist of: Potassium cyanide 8 ozs., silver chloride $\frac{1}{2}$ oz., water 1 gallon.

The voltage should be from one to one and one-half volts with the full ampèrage of the dynamo, and the immersion from fifteen to thirty seconds. The knives should then be placed in the regular silver bath. This bath should have very little free cyanide and should be run at a voltage not exceeding one and one-half. The ampèrage should be about three per dozen of knives, or four ampères per square foot of exposed surface. This is the method used by the majority of the large concerns.

Some platers use a first and second strike. In this case the first solution consists of a solution of cyanide in the proportion of six to eight ounces per gallon and one-eighth to one-quarter ounce of silver in the form of chloride. Two copper anodes are used, about three by eight inches, and two small silver anodes, about one-fourth the dimensions of the copper anodes. No deposit shows on the steel after the immersion in this strike. The knives are then immersed directly in the second strike, as before mentioned, and then into the bath. No copper should show from the copper and silver strike, and as soon as any becomes observable on the knives, more cyanide should be added to the bath. This is practically only an electric cleaner. For the deposit of silver, by following the above instructions carefully, no trouble with peeling of the deposit will be experienced.

Determination of weight. In order to control the weight of the deposit, proceed as follows : Remove one of the pans of a sensitive beam balance and substitute for it a brass rod which keeps the other pan in equilibrium. Under this rod place a vessel filled with pure water, and of sufficient diameter and depth to allow of the article suspended to the rod dipping entirely into the water without touching the sides of the vessel. Suppose now that several dozen spoons of the same size and shape are at the same time to be provided with a deposit of

Fig. 125.

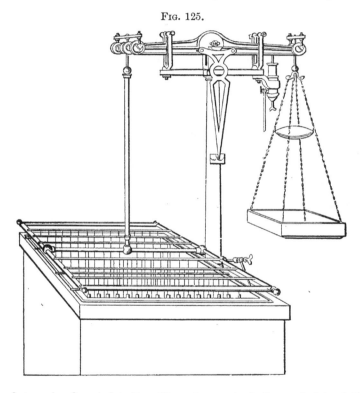

a determined weight, it suffices to control the weight of the deposit of a single spoon, and when this has acquired the necessary deposit all the other spoons will also be coated with a deposit of silver of the same thickness as the test spoon. The spoons having been quicked and carefully rinsed, one of them is suspended to the brass rod of the balance so that it dips entirely under water.   The equilibrium is then re-established by placing lead shot upon the pan of the scale, and adding the weight corresponding to the deposit the spoon is to receive. Now bring the weighed spoon together with the rest into the bath, and proceed with the silvering process in the ordinary manner.   After some time the weighed spoon is taken from the bath, rinsed in water, and hung to the brass rod of the scale.   If it does not restore the equilibrium of the latter, it is returned to the bath, and after some time again weighed, and so on until its weight corresponds to that of the lead shot and weight placed in the pan of the scale, when it is assumed that the other objects have also received their proper quantity and that the operation is complete.

A more complete weighing apparatus is the metallometric balance first used by Brandely, and later on improved by Roseleur.   The apparatus, which is shown in Fig. 125, is

FIG. 126.

designed for obtaining deposits of silver " without supervision and with constant accuracy, and which spontaneously breaks the current when the operation is terminated."   It is manufactured in various sizes, suitable for small or large operations.

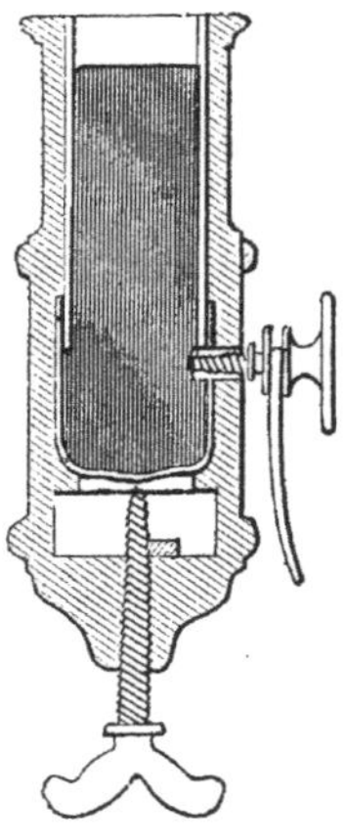

It consists of : 1. A wooden vat, the upper edge of which carries a brass winding-rod having a binding screw at one end to receive the positive conducting wire of the battery.   From this rod the anodes are suspended, which are entirely immersed in the solution, and communicate with brass cross-rods by means of platinum wire hooks.   These cross-rods are flattened at their ends so that they may not roll, and at the same time have a better contact with the " winding-rod."   2. A cast-iron column screwed at its base to the side of the vat, and which carries near the top two projecting arms of cast-iron, the extremities of which are vertical and forked, and may be opened or closed by iron clamps.

These forks are intended for sustaining the beam and preventing the knives from leaving their bearings under the influence of too violent oscillations. In the middle of the two arms are two wedge-shaped recesses of polished steel to receive the knife edges of the beam. One of the arms of the column carries at its end a horizontal ring of iron in which is fixed a heavy glass tube supporting a cup of polished iron which is insulated from the column (Fig. 126).

This cup has at its lower part a small pocket of lamb-skin or of India rubber, which by means of a screw beneath may be raised or lowered. This flexible bottom allows the operator to lower or raise at will the level of the mercury introduced afterwards into the iron cup. Another lateral screw permits connection to be made with the negative electrode. 3. A cast-iron beam carrying in the middle two sharp knife edges of the best steel hardened and polished. At each extremity there are two parallel bearings of steel separated by a notch, and intended for the knife-edges of the scale-pan that receives the weights, and those of the frame supporting the articles to be plated. One of the arms of the beam is provided with a stout platinum wire, placed immediately above and in the center of the cup of mercury. According as the beam inclines one way or the other, this wire plays in or out of the cup. 4. A scale-pan for weights, with two knife-edges of cast steel, which is attached to four chains supporting a wooden pan for the reception of weights. A smaller pan above is intended for the weights corresponding to that of the silver to be deposited. 5. The frame for supporting the articles to be plated, which is also suspended from two steel knife-edges, and the rod of which is formed of a stout brass tube attached below to the brass frame proper, which last is equal in dimensions to the opening of the vat, and supports the rods to which the articles are suspended.

It must, however, be borne in mind that the weight of a body immersed in water is less than when weighed in the air, it being as much less as the weight of the volume of fluid displaced by it. The specific gravity of silver is 10.5, hence 1 cubic centimeter of silver weighs 10.5 more than 1 cubic centimeter of water, so that 10.5 grammes of silver weighed in the air weigh only $10.5 - 1 = 9.5$ grammes when immersed in water. Since the specific gravity of the silver bath is greater than that of water—of fresh baths about 5° Bé = $1.035 - 10.5$ grammes of silver, while immersed in the silver bath, weigh only $10.5 - 1.035 = 9.465$ grammes. Hence for the determination of the weight to be placed in the scale-pan, which corresponds to the actual weight of the silver deposit, the desired weight of the deposit has to be multiplied by 9.465 and divided by 10.5, or what amounts to the same thing, multiplied by 0.901. Suppose 300 grammes of silver are to be deposited upon forks, knives and spoons, not 300 grammes, but only $300 \times 0.901 = 270.3$ grammes have to be placed in the scale-pan. The weight of the articles themselves is not taken into account as the objects are tared under the solution and remain in the same bath-fluid to the end of the process. Hence, according to the above calculation, the weight, to be placed in the scale-pan should, in round figures, be 10 per cent. less than the desired weight of silver. However, as silver also deposits on the slinging wires, it has been shown that a reduction of 4 to 5 per cent. of the weight is about the right thing.

Fig. 127 shows a metallometric balance in operation, as coupled with the rheostat, voltmeter, and the silver bath, and will be understood without further explanation. These metallometric balances must of course be very carefully constructed so as to render possible accurate weighings with a load of about 11 lbs. They are used by most large silver-plating establishments for forks, knives and spoons. They may also be employed to advantage in plating coffee-pots, tea-pots, sugar bowls, etc., but with such articles special attention has to be paid to the anode arrangement in order to obtain a deposit of uniform thickness upon all portions. It is evident that with the use of straight silver anodes the portions of round vessels nearest to the anodes will receive a thicker deposit of silver than the portions at a greater distance from them. However, this also happens in every silver bath in operation not connected with a metallometric balance, the latter indicating of course only the total weight of deposit upon all the articles in the bath, and means for the formation

Fig. 127.

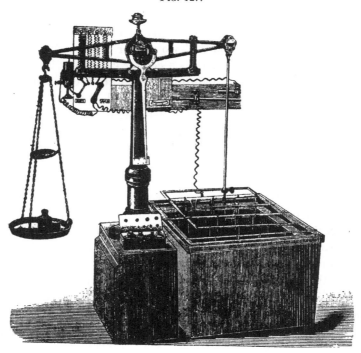

of a uniform deposit on all portions must therefore be provided. This is effected by the use of curved anodes suspended around the objects at equal distances; further by frequently hanging the position of the objects so as to bring the more remote portions nearer to the anodes.

For the determination of the weight of the deposit, Pfanhauser Jr. uses a voltametric balance,* a combination of a voltameter with a balance, the principle of which, according to Ferchland,† is similar to Edison's registering voltameter, and has for some time been practically utilized by Prof. Domalip ‡ in Prague.

A *copper voltameter* is an apparatus which allows of the determination of the quantity of current conducted in a certain time through an acid copper sulphate solution (water 35 ozs., copper sulphate 7 ozs., concentrated sulphuric acid 1¾ ozs., alcohol ¼ oz.) by the quantity of copper electrolytically separated on the cathode. Two copper anodes dip into the solution and between them is a copper cathode. The weight of the latter is exactly determined, and after the current has for an accurately measured time passed through the voltameter, is washed with water, rinsed with alcohol, dried and again weighed. From the increase in weight the quantity of current which has passed through the voltameter can be readily calculated, since one ampère deposits in one hour 1.1858 grammes of copper.

Now by placing such a copper voltameter in the current-conductor of a silver bath so that the current passes through the bath and the voltameter one after the other, the same quantity of current must pass through the bath as well as the voltameter since, according to Kirchhoff's law, the current-strength is equally great in all points of a current-circuit. According to Faraday's law, the quantities of metals deposited are proportional to their electro-chemical equivalents and hence the quantity of copper separated will be to the quantity of silver deposited as 1.858 : 4.0248 ; thus when the quantity of copper separated is known, the weight of the silver deposit can be readily calculated.

In Pfanhauser's voltameter balance the copper cathode is

---

* German patent 120843.

† Zeitschrift fur Elektrochemie, 1891, No. 71.

‡ Ahrens, Handbuch der Elektrochemie, ıı, Aufl. S. 151.

by means of a conductor suspended to the metal-beam of a smaller balance, the equilibrium being restored by placing weights or shot in the scale-pan on the other end of the beam. Now in order to determine the weight adequate to the desired deposit in the silver bath, the weight has to be multiplied by 1.1858 and divided by 4.0248. The number of grammes found is the weight the copper-deposit in the copper voltameter must attain to correspond to the desired weight of the deposit in the silver bath. The weight thus determined has to be placed in the pan of the scale. When the deposit has acquired the desired weight the current is automatically interrupted by a contrivance similar to that described when speaking of the metallometric balance.

As compared with the metallometric balance, the voltametric balance possesses advantages and disadvantages, the latter being chiefly that besides the swelling-up of the deposit in the voltameter, which has to be taken into consideration, the reduction of the weight of silver to be deposited to the weight of the copper deposit to be placed in the scale-pan may readily lead to errors if left to the ordinary workman. To be sure, this can be avoided by consulting tables which are furnished with the balance.

It must furthermore be borne in mind that the voltametric balance gives reliable results only when the current-output of the silver bath remains constant and is exactly equal to that of the copper solution in the voltameter, a third correction of the weight being otherwise required. The current-output in the voltameter is 100, but that of fresh silver baths does not reach this height. According to numerous determinations by Friessner, executed in Dr. G. Langbein & Co.'s electro-chemical laboratory, the current-output in cyanide silver baths varies according to whether the bath is at rest or in motion, the current-output with 0.3 ampère current-density being in a bath which contains per liter 25 grammes of silver as silver cyanide and 27 grammes 99-percent potassium cyanide, 99.63 per cent. without agitation of the bath, and 99.18 per cent. with agitation, hence 0.45 per cent. less in the latter case.

The increased electro-motive force when a voltametric balance is placed in the current-circuit must also not be lost
sight of. Since the voltameter solution and the silver bath
are coupled one after the other, about double the electro-
motive force is required, and this greater performance of work
increases the cost of current. However, this is not of sufficient
importance to prevent the use of voltametric balances.

The advantages of a voltametric balance consist in that it
is not necessary to place it in the vicinity of the bath. It may
be located in a special dry room where it is not exposed to the
effects of the damp atmosphere of the work-room. However,
these effects are as a rule over-estimated, since metallometric
balances with knife-edges of specially prepared steel, which
quite well resists the action of rust, working in agate bearings,
have been known to work with great accuracy after having
been in use for fifteen years. The slighter load of the beams
is also in favor of metallometric balances, so that they can be
of lighter construction.

Metallometric as well as voltametric balances have the
drawback that the current must pass through the beam and
other sensitive parts to reach the bath. To prevent corrosion
on the contacts and avoid large sparks on the sensitive portions, special protective measures are required which render
the construction both more complicated and more expensive.

The advantage of metallometric balances, namely, simple
attendance and, when due care is observed, absolute certainty
of results, may be advantageously utilized, together with the
advantages of voltametric balances, i. e., slighter load and
location at any desired distance from the bath, by the following combination devised by Neubeck.

If in front of the silver bath be placed a smaller controlling
bath of exactly the same composition as the silver bath in
operation, or in other words, is taken from the latter, the current-output of both these baths must be the same, provided
the current-density is the same. Now by using as cathodes
for this controlling bath very thin silver sheets—0.05 to 0.1

millimeter thick—with a total surface approximately equal to
that of the objects in the bath, the weight of the cathodes will,
on the other hand, be materially less than that of the articles
in the bath, for instance, of an equally large surface represented by spoons, and consequently the balance can be of
lighter construction, while, on the other hand, the current-
density in the controlling bath will be approximately equal
to that in the silver bath.

One dozen spoons weigh on an average 540 to 550 grammes
and have a surface of about 13.2 square decimeters. The same

Fig. 128.

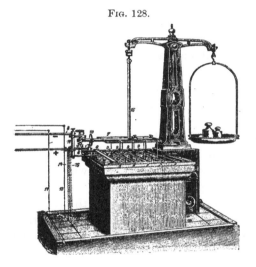

surface in silver sheet, 0.1 millimeter thick, weighs about 70
grammes, and in silver sheet, 0.05 millimeter thick, about 35
grammes. Hence the load of the ware is, in the commencement, $7\frac{1}{2}$ to 15 times less than with the metallometric balance
and when, on attaining a thickness of about $\frac{1}{2}$ millimeter, the
cathodes are reversed, always only half as great.

Now by allowing the current to pass through the controlling bath and then through the silver bath, exactly the same
quantity of silver is deposited in the former as in the latter;

thus by connecting the cathodes of the controlling bath with a balance, the quantity of metal deposited in the silver bath upon the objects can be accurately determined.

These considerations led to the construction of the *volta-metric controlling apparatus*, Fig. 128, which can be connected with any kind of cheap beam-balance. The apparatus * constructed by Dr. G. Langbein & Co. is arranged as follows:

The screw 1 secured to the anode-frame (Fig. 128) is directly connected with the anode wire. Upon the anode-frame of copper sit the conducting rods 9, 9, 9, etc., for the anodes, the latter being secured to them by means of platinum wire. To the rod 16 is secured the movable cathode from 7, to which the thin cathode sheets are suspended by means of platinum wire. The current enters the bath through the binding-post 1, the anode-rods 9, and the anodes, passes to the cathodes, and returns through the cathode-frame 7 to the source of current so long as the screw 2 fixed to the support 17 dips, by reason of the load of the balance, into the mercury vessel 3. The latter is secured to the anode-frame 7 and conductively connected with it. The celluloid disk 10 serves the purpose of protecting the bath from mercury, which may be spilled by careless handling, passing into it.

When the cathodes have attained the required weight the flow of current to the bath is interrupted by the beam of the balance tilting over. The two steel pins 4 and 5, which are insulated from the mercury vessel 3, dip thereby into the mercury cups 6 fixed below them on the movable arm of the support 17 and insulated one from the other, the short-circuit of the bell wire being thus effected. The wire 11 leads direct to the bell 13, while a second wire 12 leads through the spirals 14 to the support 17 and from there through spiral 15 to the bell. When the pins 4 and 5 dip into the mercury cups, 6, a second connection with the bell is made and the latter rings so long as the pins 4 and 5 dip into the mercury.

* German patent.

Contrary to the principle of the metallometric and the volta-metric balances, the beam and other portions of the balance are here entirely excluded from the current-circuit. Hence, any ordinary beam-balance can be used and no corrosion of sensitive portions of the balance by sparks takes place.

When the cathodes of the controlling bath have attained a thickness approximately the same as that possessed by the silver anodes of the bath in the commencement of the operation, they are used as anodes by suspending them to the anode-rods, while the anodes which have become thinner are suspended as cathodes.

With this arrangement there is to be sure, the same drawback as with the voltametric balance, namely, that by reason of the baths being coupled one after the other, double the electro-motive force than that used for a silver bath connected with a metallometric balance is required. The interest, which, however, amounts to very little, on the dead metallic silver in the controlling bath may also be called a disadvantage. On the other hand, the controlling apparatus has the advantage that two or more silver baths of the same composition can be connected with it when the same quantity of silver is to be deposited upon approximately the same object-surface in each bath.

The case is somewhat different when the baths in operation are of considerable size and furnished with many object-rods. In order to obtain the same current-density in the actual silver bath as in the controlling bath, the latter would have to be of quite large dimensions and require so much electrode-material as to cause considerable expense for providing it. In such a case the advantages presented by the same composition of the operating bath and the controlling bath have to be abandoned and the controlling bath has to be used as a copper voltameter. Since in the copper solution depositions can be made with current-densities up to 1.5 ampères per square decimeter only one-fifth part of the object-surface in the operating bath is required as cathode surface in the controlling copper bath. Of course the weight of the desired silver deposit, reduced to copper, which is found from tables furnished with the apparatus, has to be placed in the scale-pan. The tables are calculated for current-outputs of from 98 to 99.6 per cent. in $\frac{1}{10}$ per cent. for baths for silvering by weight with 25 grammes of fine silver as silver cyanide and 25 grammes of 99-percent potassium cyanide, and a current-density of 0.3 ampère per square decimeter. Hence the current-output of the bath to be used has to be determined, and the weights of copper corresponding to the silver deposit are then found in the table for the determined current-output. As the current-output is subject to change by the bath becoming gradually contaminated by foreign metals, which cannot be avoided in silvering metals soluble in potassium cyanide, for instance, zinc and copper, it will be necessary to determine the current-output at least twice a year.

As previously mentioned, the current-output is materially smaller when the bath is agitated than when it is at rest, and hence the controlling silver bath cannot be used for agitated baths because, in order to obtain accurate weighing results, agitation of the controlling bath has to be avoided. In this case the copper solution (see p. 392) has also to be used with reference to the tables for the respective current-output. However, when the controlling apparatus works with the copper solution, it still has the advantage of no current passing through sensitive portions of the balance, and thus they are not subject to wear.

Calculation of the weight of the silver deposit from the current-strength used. This can be done if the current conducted into the bath during silvering be constantly kept at the same strength, and the current-output of the bath be taken into consideration. According to the table on p. 61, one ampère deposits in 1 hour 4.0248 grammes of silver, hence after t hours with a current-strength i: $4.0248 \times i \times t$ grammes of silver will be deposited, if the current-output amounts to 100. However, the latter is, as a rule, only 98 to 99 per cent., and the value obtained is therefore to be multiplied by the fraction $\frac{98}{100}$ or $\frac{99}{100}$, in order to determine the actual weight of the deposit.

Attention must, however, be drawn to the fact that it is very difficult to keep the current-strength quite constant for a longer time, especially where numerous baths are connected to a common circuit, and for this reason a calculation, based upon the measurement of the current-strength, will very likely be in most cases impracticable.

For the calculation of the time which has been consumed for depositing a certain weight of silver when the current-strength is known, the desired weight of the deposit has to be divided by the product from current-strength times chemical equivalent of the silver, and the value found multiplied by $\frac{100}{98}$ or $\frac{100}{99}$.

If, for instance, 50 grammes of silver are to be deposited upon one dozen spoons, and the current-strength be 3.2 ampères and the current-output 99 per cent., the time is found from the following calculation :

$$\frac{50 \times 100}{3.2 \times 4.0248 \times 99} = \frac{5000}{1275.05} = 3.92 \text{ hours} = 3 \text{ hours } 55 \text{ minutes.}$$

If the current-strength is to be calculated, which is required to produce in a given time a determined thickness of deposit, the product from the desired thickness in millimeters, times the object-surface in square decimeters, times the specific gravity of the metal to be deposited, times 1000, has to be divided by the product from time, times electro-chemical equivalent, times current-output. Suppose, for instance, the desired thickness of the deposit is to be 0.1 millimeter, the object-surface 1.5 square decimeter, the specific gravity 10.5, the time 4 hours, the electro-chemical equivalent 4.025, and the current-output 98 per cent., then the current-strength required is:

$$\frac{0.1 \times 1.5 \times 10.0 \times 100}{4 \times 4.0248 \times 98} = \frac{1575}{1577.72} = 0.99 \text{ ampère.}$$

When the articles have received a deposit of the required weight, they are treated for the prevention of subsequent yellowing according to one of the methods given on p. 379, then scratch-brushed bright with the use of decoction of soap-root, plunged in hot water and dried in sawdust.

Mat silver.—Articles which are to retain the beautiful crystalline dead white with which they come from the bath are, without touching them with the fingers or knocking them against the sides of the vessel, rinsed thoroughly in clean water, plunged into very hot, distilled water, and then suspended free to dry. Immediately after drying they are to be provided with a thin coat of transparent lacquer to protect the dead-white coating, which readily turns yellow, and, moreover, is very sensitive.

Frosting silver is affected by means of scratch-brushes. They take different forms, according to the kind of work to be frosted. They are made of several strengths, that is, the wires of them are especially prepared of several thicknesses, and when a very fine satin finish is required, a brush of very fine wire is taken, and so on. A brush with wires thicker and thicker in proportion is taken as a more extended roughness is desired. These wire brushes are fixed upon a horizontal spindle in the lathe. Frosting requires great speed to do the work nicely. The wires of the scratch-brush must be even on the surface, all of the same length, and always kept straight at the points, otherwise the frosting will not be regular. Sometimes the little hand scratch-brushes are employed for coarser work; four of them are taken, and firmly secured in four corresponding grooves in a circular chuck, which screws into the lathe. The ends of the four little brushes are repeatedly cut off as occasion requires in order to present a straight surface for continual contact with the work, without which it would not present a uniform appearance.

According to Gee, the following mixture may also be used for frosting silver: Sulphuric acid 1 oz., water 1 oz., saltpetre 2 dwts. Add the sulphuric acid to the water and afterwards put in the saltpetre in a state of fine powder. The mixture is used in the boiling state and takes a few minutes to accomplish the desired object.

Polishing the deposits.—The silvered articles having been scratch-brushed, must finally be polished, which may be effected upon a fine felt wheel with the use of rouge, but imparting high luster by burnishing is to be preferred, the deposit being first treated with the steel burnisher, and then with the stone burnisher, as explained on p. 216.

In some establishments in which plated table-ware in large quantity is turned out, ingeniously-devised burnishing machines driven by power are in use, by which much of the manual labor is saved. The knife, spoon, etc., each supported by its tips in a suitable holder, are very slowly rotated, while the burnishing-tool moves quickly over the surface, performing the work rapidly and satisfactorily.

When burnishing is completed, the surface is wiped off longitudinally with an old, soft calico rag. Sawdust, hard cloth, and tissue paper produce streaks.

B. Ordinary silver-plating.—Objects which are to receive a deposit of less thickness have to undergo exactly the same operations described under plating by weight, the only difference being that for quicking a weaker solution (15 to 31 grains of nitrate of mercury to 1 quart of water), or very dilute solution of potassium-mercury cyanide (77 grains of potassium-mercury cyanide and 77 grains of potassium cyanide to 1 quart of water) is used, and that the objects remain a shorter time in the bath.

Direct silvering of Britannia, tin, German silver. As previously mentioned, iron, steel, zinc and tin should first be coppered or brassed. However, tin and Britannia may also be directly plated. but the bath must be rich in silver and contain a large excess of potassium cyanide. Further, the current should be so strong that the articles acquire a blue-gray color. They are then suspended in. the silver bath of normal composition, and plating is finished with a normal current.

26

The same process is also suitable for plating articles of German silver rich in nickel. In polishing such articles it is frequently observed that the deposit rises, but by plating in the above-mentioned preparatory bath, and finishing in the normal bath, the deposit will very well bear polishing with the steel.

For silver-plating *Britannia ware* and *articles of tin,* Gore recommends the following process. Boil the articles in caustic potash solution, scratch-brush them, and plate them preparatively with a strong current and the use of large anode-surfaces in a hot silver bath (194° F.), and then finish deposition to the desired thickness in the ordinary cold silver bath.

According to an Australian patent, the following process is claimed to yield good results in *directly silver-plating iron and steel :* The article to be plated having first been dipped in hot dilute hydrochloric acid is brought into solution of mercury nitrate and then connected with the zinc pole of a Bunsen element. It becomes quickly coated with a layer of mercury, when it is taken out, washed and brought into an ordinary silver bath. When covered with a layer of silver of sufficient thickness, it is heated to 572° F., the mercury evaporating at this temperature. It is claimed that silver deposited in this manner adheres more firmly than by any other process, but it is doubtful whether for solid silver-plating this method can replace previous coppering.

Stopping off. If certain parts of a metallic article are not to receive a deposit, as for instance, when a contrast is to be effected by depositing different metals upon the same object, these parts are covered, or "stopped-off," with a varnish. Stopping-off varnish is prepared by dissolving asphalt or dammar with an addition of mastic in oil of turpentine. Apply with a brush, and after thoroughly drying the articles in the drying-chamber, place them for an hour in very cold water, whereby the varnish hardens completely. After plating, the varnish is removed, best with benzine, the articles plunged in hot water and dried in sawdust.

For a varnish that will resist the solvent power of the hot alkaline gilding liquid, Gore recommends the following composition : Translucent rosin 10 parts, yellow beeswax 6, extra-fine red sealing-wax 4, finest polishing rouge 3.

Quick-drying, stopping-off varnishes, which harden immediately at the ordinary temperature and resist cyanide baths, are now found in commerce.

Special applications of electro-silvering.—It remains to mention a few special applications of electro-silvering as well as processes of decorating with silver by electrical and chemical means.

Silvering of fine copper wire is effected in an apparatus, which is described and illustrated in Chapter IX, "Deposition of Gold," where further details will be found. Luster is imparted to the silvered wire by drawing through a draw-plate.

Incrustations with silver (and gold, and other metals).—By incrusting is understood the inlaying of depressions, produced by engraving or etching upon a metallic body, with silver, gold and other metals, such as Japanese incrustations, which are made by mechanically pressing the silver or gold into the depressions. Such incrustations, however, can also be produced by electro-deposition, the process being as follows: The design which is to be incrusted upon a metal is executed with a pigment of white-lead and glue-water or gum-water. The portion not covered by the design is then coated with stopping-off varnish. The article is next placed in dilute nitric acid, whereby the pigment is first dissolved, and next the surface etched, which is allowed to progress to a certain depth. Etching being finished, the article is washed in an abundance of water and immediately brought into a silver or gold bath, in which, by the action of the current, the exposed places are filled up with metal. This being done, the stopping-off varnish is removed with benzine, the surface ground smooth, and polished. In this manner one article may be incrusted with several metals; for instance, brass may be incrusted with copper, silver and gold, and by oxidizing or coloring portions of the copper beautiful effects can be produced. The principal requisites for these incrustations are manual skill and much patience. Expensive apparatus is not required, every skilled electro-plater being able to execute the work.

Imitation of niel or nielled silvering. By nielling is understood the inlaying of designs produced either by engraving or stamping, with a black mixture of metallic sulphides. The nielling powder is prepared by melting silver 20 parts by weight, copper 90 parts and lead 150 parts. To the liquid metallic mass add 26½ ozs. of sulphur and ¾ oz. of ammonium chloride, quickly cover the crucible and continue heating until the excess of sulphur is volatilized. Then pour the contents of the crucible into another crucible, the bottom of which is covered about ⅛ inch deep with flowers of sulphur, cover the crucible and allow the mixture to cool. When cold bring the contents once more to the fusing point, and pour the fused mass in a thin stream into a bucket filled with water, whereby granulated metal is formed, which can be readily reduced in a mortar to a fine powder. This powder is mixed with ammonium chloride and gum-water to a thin paste. This paste is brought into the designs produced by engraving or stamping, and after drying burnt-in in a muffle. When cold, any roughness is removed by grinding, and after polishing a sharp, black design in white silver is obtained.

To imitate niel by electro-deposition, the design is executed upon the surface with a pigment consisting of white lead and glue- or gum-water. The portions which are to remain free are coated with stopping-off varnish, and the design is uncovered by etching with very dilute nitric acid. The article is then brought as the anode into dilute solution of ammonium sulphide, while a small sheet of platinum connected to the negative pole is dipped into the solution. Sulphide of silver being formed, the design becomes rapidly black-gray, and after removing the stopping-off varnish with benzine, stands out in sharp contrast from the white silver.

Upon *brass*, nielling may be imitated by silvering the

article and then engraving the design, by which the silver is removed and the brass uncovered. The article is then brought into the black bright-dip, by which the uncovered brass is colored black while the silvered portions remain unchanged. If portions in relief are to be made black, the silvering is removed by grinding, the article dipped into cream of tartar solution, and then brought into the black bright-dip. This process is largely employed by manufacturers of buttons when silvered buttons are to be supplied with the name of the firm and the quality number in black.

Old (antique) silvering.—To give silvered articles an antique appearance coat them with a thin paste of 6 parts graphite, 1 red ochre and sufficient spirits of turpentine. After drying, gentle rubbing with a soft brush removes the excess of powder, and the reliefs are set off (discharged) by means of a rag dipped into alcohol.

A tone resembling antique silvering is also obtained by brushing the silvered articles with a soft brush moistened with very dilute alcoholic solution of chloride of platinum.

In order to impart the old silver tinge to small articles, such as buttons, rings, etc., they are agitated in the above-mentioned paste, and then " tumbled " with a large quantity of dry sawdust until the desired shade is obtained.

With the use of the electric current and carbon anodes, antique silver may be produced as follows : Bring the silvered articles, previously thoroughly freed from grease, into the silver bath at an electro-motive force of 4 to 5 volts, and allow them to remain for a few minutes until they become covered with a uniform blue-gray deposit. They are then thoroughly rinsed in water, and the raised portions rubbed with very fine pumice, to lay bare the silver. If surfaces are to appear in antique silver, the deposit is only sufficiently removed with pumice for the silver to shine through, and the surface to show the proper antique-silver tone.

Oxidized silvering. This term is incorrect, as silver oxide does not form the coloring film or at least only to a very slight extent, the coloration being due to the formation of silver sulphide or silver chloride upon the objects.   This process of coloring silver is frequently employed to obtain decorative contrasts.   Solution of pentasulphide of potassium (liver of sulphur of the shops) is generally used for the purpose.   Dissolve liver of sulphur 1 oz., and ammonium carbonate 2 ozs., in 1 gallon of water heated to 176° F.   Immerse the objects in the solution and allow them to remain until they have acquired the desired dark tone.   Immediately after immersion the articles become pale gray, then darker, and finally deep black blue.   For coloring in this manner, the silvering should not be too thin.   For objects with a very thick deposit of silver, solution of double the strength may be used.   Very slightly silvered objects cannot be colored in this manner, as the bath would remove the silvering, or under the most favorable conditions produce only a gray color.   If the operation is not successful, and the objects come from the bath stained or otherwise, dip them in warm potassium cyanide solution, which rapidly dissolves the silver sulphide formed.

A bath which produces the same effect as potassium sulphide solution may be cheaply prepared as follows : Pour 1 quart of water over 13 ozs. of unslaked lime and 22½ ozs. flowers of sulphur.   The mixture becomes quickly heated and thick.   Dilute it with 1 quart hot water and boil half an hour.   The resulting liquor is now ready for use and is best employed very hot.   If, whilst boiling the solution, 1¾ ozs. of antimonious sulphide or 1¼ ozs. of arsenious sulphide be added, an agreeable bluish-gray coloration is obtained which, with the use of antimonious sulphide passes later on into a beautiful gray-brown.

A beautiful brown tone is imparted to silver objects by immersion in the following solution : Copper sulphate 10 ozs., saltpetre 5 ozs., ammonium chloride 10 ozs.

Another process is as follows : Place the objects in a porcelain dish, cover them with ammonium sulphide, and heat gradually. When the objects have acquired a blue-black color, take them from the dish, place them in soap-water and, so long as they remain in the latter, rub them with a soft brush.

A *yellow color* is imparted to silvered articles by immersion in a hot saturated solution of copper chloride, rinsing and drying.

For silvering by contact, boiling and friction, see special chapter " Depositions by Contact."

Stripping silvered articles.—When a silvering operation has failed, or the silver is to be stripped from old silvered articles, different methods have to be used according to the nature of the basis-metal. Silvered *iron articles* are treated as the anode in potassium cyanide solution in water (1 : 20), the iron not being attacked by potassium cyanide. As cathode suspend in the solution a few silver anodes or a copper-sheet rubbed with an oily rag ; the silver precipitates upon the copper sheet but does not adhere to it. Articles, the basis of which is *copper*, are best stripped by immersion in a mixture of equal parts of anhydrous (fuming) sulphuric acid and nitric acid of 40° Bé. This mixture makes the copper passive, it not being attacked while the silver is dissolved. Care must, however, be had not to introduce any water into the acids, nor to let them stand without being hermetically closed, since by absorbing moisture from the air they become dilute, and may then exert a dissolving effect upon the copper. The fuming sulphuric acid may also be highly heated in a shallow pan of enameled cast iron. Then at the moment of using it pinches of dry and pulverized nitrate of potassium (saltpeter) are thrown into it, and the article, held with copper tongs, is plunged into the liquid. The silver is rapidly dissolved, while the copper or its alloys is but slightly corroded. According to the rapidity of the progress of solution, fresh additions of saltpeter are made. All the silver has been dissolved when, after rinsing in water and dipping the articles into the cleansing acids, they present no brown or black spots, that is to say, when they behave like new. In this hot acid stripping proceeds more quickly than in the cold acid mixture, but the latter acts more uniformly.

Determination of silver-plating.—By applying to genuine silver-plating a drop of nitric acid of 1.2 specific gravity, in which red chromate of potash has been dissolved to saturation, a red stain of chromate of silver is formed. According to Gräger, this method may also be used, to a certain extent for the recognition of any other white metal which may be mistaken for silver. A drop of the mixture applied to *German silver* becomes brown, no red stain appearing after rinsing with water; upon *Britannia* the drop produces a black stain; *zinc* is etched without a colored spot remaining behind ; upon *amalgamated* metals a brownish precipitate is formed, which does not adhere and is washed away by water; upon *tin* the drop also acquires a brownish color, and by diluting with water a yellow precipitate is formed ; upon *lead* a beautiful yellow precipitate is formed.

Custom-house officers in Germany are directed by law to use the following process for the determination of genuine silver-plating : Wash a place on the article with ether or alcohol, dry with blotting paper, and apply to the spot thus cleansed a drop of a 1 to 2 per cent. solution of crystallized bisulphite of soda prepared by boiling 1.05 ozs. of sodium sulphite and 2.36 drachms of flowers of sulphur with 0.88 oz. of water until the sulphur is dissolved, and diluting to 1 quart of fluid. Allow the drop to remain upon the article about ten minutes and then rinse off with water. Upon silver articles, a full, round, steel-gray spot is produced. Other white metals and alloys, with the exception of amalgamated copper, do not show this phenomenon, there appearing at the utmost a dark ring at the edge of the drop. Amalgamated copper is more quickly colored, and acquires a more dead-black color than silver.

Examination of silver baths.

For the quantitative examination of silver baths, the determination of the content of free potassium cyanide and of metallic silver as well as of the potassium carbonate which is formed by the action of air, etc., upon the potassium cyanide, has to be taken into consideration.

Regarding the determination of the free potassium cyanide, the reader is referred to the method given under " Examination of copper baths containing potassium cyanide," and what has been said there in reference to replacing the deficiency also applies here.

The potassium carbonate which is formed in constantly increasing quantities in the bath, is best removed by the addition of barium cyanide solution, whereby, in consequence of reciprocal decomposition, potassium cyanide is formed, while barium carbonate in an insoluble state is separated.

The determination of the potassium carbonate present in the bath is desirable, so as to be able on the one hand, to calculate the quantity of barium cyanide required for its decomposition and, on the other, to become acquainted with the quantity of free potassium cyanide formed thereby

The determination of the potassium carbonate is effected as follows: Bring by means of the pipette, 20 cubic centimeters of silver bath into a beaker, dilute with 50 cubic centimeters of water, and compound with barium nitrate solution in excess. Allow to settle for some time, then filter through not too large a paper filter, taking care that the entire precipitate reaches the filter, and wash the filter thoroughly with water until a few drops of the filtrate, when evaporated upon a platinum sheet, leave no residue. Now take the filter, together with the residue, carefully from the funnel, bring it into a beaker, and add water, as well as a carefully measured quantity of standard nitric acid, which should, however, be somewhat larger than required for dissolving the barium carbonate. While solution is being effected, keep the beaker

·covered with a watch glass, and then rinse any drops appearing upon the latter into the beaker by means of distilled water. Add to the solution, as an indicator, a few drops of methyl-orange, whereby the solution is colored red, and add, while stirring constantly, from a burette, standard soda solution until the red color of the solution passes into yellow. By now deducting the cubic centimeters of soda solution used from the cubic centimeters of standard nitric acid added to the solution of the barium carbonate, and multiplying the number of the remaining cubic centimeters of standard nitric acid by 3.45, the quantity of potassium carbonate in grammes present in 1 liter of silver bath is obtained.

Now the quantity of barium cyanide has to be calculated, which is required for the conversion of the quantity of potassium carbonate found, into potassium cyanide with the separation of barium carbonate. It is best to use a 20½ per cent. barium cyanide solution, and since 1 gramme of potassium carbonate requires for conversion 1.36 grammes of barium ·cyanide, 6.80 grammes of 20 per cent. barium cyanide solution are necessary for the purpose, and each gramme of potassium ·carbonate yields 0.942 gramme of potassium cyanide. Hence for the determination of the potassium cyanide present after the destruction of the potassium carbonate, there has to be added to the potassium cyanide found by titration, the content ·of free potassium cyanide calculated from the conversion with barium cyanide. If this shows a deficit as compared with the ·original content, it is to be made up by adding only about one-half the quantity, for the same reason as given in speaking of the copper bath, namely, because the potassium for·mate, which is at the same time formed, performs the function ·of the potassium cyanide.

To save calculation a table by Steinach and Buchner for ·the use of a 20½ per cent. barium cyanide solution is here ·given

Potassium carbonate in 1 liter of silver bath.	For 1 liter of silver bath have to be added	
	20½ per cent. barium cyanide solution.	Potassium cyanide formed thereby.
1 gramme	6.7 grammes	0.95 gramme
2 "	13.4 "	1.90 "
3 "	20.1 "	2.85 "
4 "	26.8	3.80
5 "	33.5	4.70
6 "	40.2	5.70
7	46.9	6 65
8	53.6	7.60
9	60.3	8.55
10	67.0	9.50
11	73.7	10.40
12	80.4	11.40
13	87.1	12.35
14	93.9	13.30
15	100.5	14.20

For the determination of the silver, the electrolytic method is the most simple and suitable in so far as the silver bath can be directly used for the purpose.

Bring by means of the pipette into the platinum dish 10 cubic centimeters of the silver bath, or 20 cubic centimeters if the bath is weak ; add, according to the greater or smaller excess of the potassium cyanide present, ½ to 1 gramme of potassium cyanide dissolved in water, and dilute up to 1 or 1½ centimeters from the edge of the dish. Heat, by means of a small flame, the contents of the dish to from 140° to 149° F., and maintain this temperature as nearly constant as possible. Electrolysis is effected with a current-density ND 100 = 0.08 ampère. Complete precipitation, which requires 3 to 3½ hours, is recognized by ammonium sulphide producing no dark coloration of the fluid. The dish is then washed, without interrupting the current, rinsed with alcohol and ether, dried for a short time at 212° F., and weighed. The weight of the precipitate multiplied by 100 gives the content of silver in grammes per liter of bath. If 20 cubic centimeters of silver bath have been electrolyzed, multiply only by 50.

If the analysis has shown a deficit of silver in the bath, it can be readily replaced. For strengthening the bath it is best to use pure crystallized potassium-silver cyanide, which in round numbers contains 50 per cent. of silver. Suppose the bath contains per liter 2 grammes of silver less than it should, then for each liter of bath (52 : 100 = 2 : x ; x = 3.8 grammes), 3.8 grammes of pure crystallized potassium-silver cyanide have to be added.

The more troublesome volumetric analysis may be omitted, it offering no advantage over the electrolytic method.

Recovery of silver from old silver baths, etc.—Old solutions which contain silver in the form of a silver salt are easily treated. It is sufficient to add to them, in excess, a solution of common salt, or hydrochloric acid, when all the silver will be precipitated in the state of chloride of silver, which, after washing, may be employed for the preparation of new baths.

For the recovery of silver from solutions which contain it as cyanide, the solutions may be evaporated to dryness, the residue mixed with a small quantity of calcined soda and potassium cyanide, and fused in a crucible, whereby metallic silver is formed, which, when the heat is sufficiently increased, will be found as a button upon the bottom of the crucible ; or if it is not desirable to heat to the melting-point of silver, the fritted mass is dissolved in hot water, and the solution containing the soda and cyanide quickly filtered off from the metallic silver. The evaporation of large quantities of fluid, to be sure, is inconvenient, and requires considerable time. But the reducing process above described is without doubt the most simple and least injurious.

According to the *wet method*, the bath is strongly acidulated with hydrochloric acid, provision being made for the effectual carrying-off of the hydrochloric acid liberated. Remove the precipitated chloride of silver and cyanide of copper by filtration, and, after thorough washing, transfer it to a porcelain dish and treat it, with the aid of heat, with hot hydrochloric acid, which will dissolve the cyanide of copper. The resulting chloride of silver is then reduced to the metallic state by mixing it with four times its weight of crystallized carbonate of soda, and half its weight of pulverized charcoal. The whole is made into a homogeneous paste, which is thoroughly dried, and then introduced into a strongly-heated crucible. When all the material has been introduced, the heat is raised to promote complete fusion and to facilitate the collection of the separate globules of silver into a single button at the bottom of the crucible, where it will be found after cooling. If granulated silver is wanted, pour the metal in a thin stream and from a certain height, into a large volume of water.

A very simple method is as follows : Bring the silver bath into flasks, mix the contents of the flask with zinc dust (zinc in a finely-divided state) in the proportion of about $\frac{1}{3}$ oz. per quart of bath, and shake thoroughly 5 or 6 times every day. In five days all the silver is precipitated. Decant the clear liquid from the precipitate, wash the latter several times with water, and dissolve the zinc contained in the precipitate in pure hydrochloric acid. The silver remains behind in a pulverulent form, and may be dissolved in nitric acid, and worked up into silver chloride or silver cyanide. In place of zinc, aluminium powder may be used for precipitation, the excess of aluminium being then dissolved by caustic potash, or caustic soda solution.

From acid mixtures used for stripping, the silver may be obtained as follows : Dilute the acid mixture with 10 to 20 times the quantity of water, and precipitate the silver as chloride of silver by means of hydrochloric acid. Interrupt the addition of hydrochloric acid, when a drop of it produces no more precipitate of chloride of silver in the clear fluid. The precipitated chloride of silver is filtered off, washed, and either directly dissolved in potassium cyanide, or the silver is regained as metal by fusing the chloride of silver with calcined soda and wood charcoal powder, previously thoroughly mixed.

Still more simple is the reduction of the chloride of silver by pure zinc. For this purpose suspend the chloride of silver by water, add hydrochloric acid, and place pure zinc rods or granulated zinc in the fluid. While zinc dissolves, metallic silver is separated, which is filtered off, washed and dried.

CHAPTER IX.

GOLD (Au = 197.2 parts by weight) is generally found in the metallic state. It is one of the metals possessing a yellow color. Precipitated from its solution with green vitriol (ferrous sulphate) or oxalic acid, it appears as a brown powder without luster, which on pressing with the burnisher acquires the color and luster of fused gold. Pure gold is nearly as soft as lead, but possesses considerable tenacity. In order to increase the hardness when used for articles of jewelry and for coinage, it is alloyed with silver or copper. The "fineness of gold," or its proportion in the alloy, is usually expressed by stating the number of carats present in 24 carats of the mixture. Pure gold is stated to be 24 carats "fine;" standard gold is 22 carats fine; 18 carat gold is a mixture of 18 parts of gold and 6 of alloy. Gold is the most malleable and ductile of the metals. It may be beaten out into leaves not exceeding $\frac{1}{10,000}$ of a millimeter in thickness. When beaten out into thin leaves and viewed by transmitted light, gold appears green; when very finely divided it is dark red or black. The specific gravity of fused gold is 19.35, and that of precipitated gold powder, from 19.8 to 20.2. Pure gold melts at about 2016° F., and in fusing exhibits a sea-green color. The melting-points of alloyed gold vary according to the degree of fineness. Thus, 23 carat gold melts at 2012° F.; 22 carat at 2009°; 20 carat at 2002°; 18 carat at 1995°; 15 carat at 1992°; 13 carat at 1990°; 12 carat at 1987°; 10 carat at 1982°; 9 carat at 1979°; 8 carat at 1973°; 7 carat at 1960°. The fineness of gold may be approximately estimated by means of the *touch-stone*, a basaltic stone formerly obtained

(415)

from Asia Minor, but now procured from Saxony and Bohemia. The sample of gold to be tested is drawn across the stone, and the streak of metal is treated with dilute nitric acid. From the rapidity of the action and the intensity of the green color produced—due to the solution of the copper, as compared with streaks made by alloys of known composition —the assayer is enabled to judge of the proportion of inferior metal which is present. Gold preserves its luster in the air, and is not acted upon by any of the ordinary acids. Nitric, hydrochloric, or sulphuric acid by itself does not dissolve gold, but it dissolves in an acid mixture which develops chlorine, hence in aqua regia (nitro-hydrochloric acid).

The gold found in commerce under the name of *shell-gold* or *painter's gold*, which is used in painting and for repairing smaller defects in electro-gilding, is prepared by triturating waste in the manufacture of leaf gold with water, diluted honey, or gum-water. Gold solution may also be precipitated with antimonic chloride. The resulting precipitate is triturated with barium hydrate, extracted with hydrochloric acid, and after washing, the gold powder is triturated with gum arabic solution.

Gold baths. Gold-plating may be effected in a hot or cold bath, large objects being generally plated in the latter, and smaller objects in the former. The hot bath has the advantage of requiring less current-strength, besides yielding deposits of greater density and uniformity, and of sadder, richer tones. Hot baths work with a moderate content of gold—$11\frac{1}{2}$ to $12\frac{1}{2}$ grains per quart of bath—while cold baths should contain not less than 54 grains per quart.

Baths prepared with potassium ferrocyanide are preferred by some authors, while others work with a solution of gold salt and potassium bicarbonate, and others recommend a solution of cyanide of gold in potassium cyanide. With proper treatment of the bath, good results may be obtained with either. However, the use of baths prepared with potassium ferrocy--anide cannot be recommended on account of the secondary decompositions which take place during the operation of plating, and because the baths do not dissolve the gold anodes. Below only approved formulas for the preparation of gold baths will be given.

I. *Bath for cold gilding.*—Fine gold in the form of fulminating gold 54 grains, 98 per cent. potassium cyandide 0.35 to 0.5 oz. (according to the current-strength used), water 1 quart.

Electro-motive force at 10 cm. electrode-distance, and with the use of 0.35 oz. of potassium cyanide, 1.35 volts; with the use of 0.5 oz. of potassium cyanide, 1.2 volts.

Current-density, 0.15 ampère.

To prepare this bath, dissolve 54 grains of fine gold in aqua regia in a porcelain dish heated over a gas or alcohol flame, and evaporate the solution to dryness. Continue the heating until the solution is thickly-fluid and dark brown, and on cooling congeals to a dark brown mass. Heating too strongly should be avoided, as this would cause decomposition and the auric chloride would be converted into aurous chloride, and eventually into metallic gold and chlorine, which escapes. The neutral chloride of gold formed in this manner is dissolved in 1 pint of water and ammonia added to the solution so long as a yellow-brown precipitate is formed, avoiding, however, a considerable excess of ammonia. The precipitate of fulminating gold is filtered off, washed, and dissolved in 1 quart of water containing 0.5 oz. of potassium cyanide in solution. The solution is boiled, replacing the water lost by evaporation, until the odor of ammonia which is liberated by dissolving the fulminating gold in potassium cyanide disappears, when it is filtered. Instead of dissolving the gold and preparing neutral chloride of gold by evaporating, it is more convenient to use 108 grains of chemically pure neutral chloride of gold as furnished by chemical works, and precipitate the fulminating gold from its solution.

Too large an excess of potassium cyanide yields gold deposits of an ugly, pale color. When working with a more

27

powerful current, the excess of potassium cyanide need only be slight ; with a weaker current it may be larger.

The fulminating gold must not be dried, as in this condition it is highly explosive, but should be immediately dissolved while in a moist state.

If the cost of a bath for cold gilding with such a high content of gold as given in formula I should appear too great, only 27 grains of gold per quart may be used. With a suitable electro-motive force, deposits of a beautiful, sad-yellow color are thus also obtained. Such a bath is yielded by the following formula :

I*a*. Fine gold in the form of neutral gold chloride 27 grains, 98 to 99 per cent. potassium cyanide 0.26 oz., water 1 quart.

Electro-motive force at 10 cm. electrode-distance, 2.0 volts.

Current-density, 0.15 ampère.

For cold gilding, Roseleur recommends the following bath :

II. Fine gold as neutral chloride of gold, 0.35 oz.; 98 per cent. potassium cyanide, 0.7 oz.; water, 1 quart.

Electro-motive force at 10 cm. electrode-distance, about 1.5 volts.

Current-density, 0.12 ampère.

Dissolve the gold-salt from 0.35 oz. of fine gold or about 0.7 oz. of neutral chloride of gold in ½ pint of the water, and the potassium cyanide in the other ½ pint of water, and after mixing the solutions boil for half an hour. The preparation of this bath is more simple than that of formula I, but the color of the gold deposit obtained with the latter is warmer and sadder. The high content of gold in the bath, prepared according to formula II, readily causes a red-brown gold deposit, and hence special attention has to be paid to the regulation of the current.

For those who prefer gold baths prepared with yellow prussiate of potash instead of potassium cyanide, the following formula for *cold-gilding* is given :

III. Yellow prussiate of potash (potassium ferrocyanide), 0.5 oz.; carbonate of soda, 0.5 oz.; fine gold (as chloride of gold or fulminating gold), 30.75 grains; water, 1 quart.

Electro-motive force at 10 cm. electrode-distance, 2 volts.

Current-density, 0.15 ampère.

To prepare the bath, heat the solutions of the yellow prussiate of potash and of the carbonate of soda in the water to the boiling-point, add the gold-salt, and boil ¼ hour, or with use of freshly-precipitated fulminating gold, until the odor of ammonia disappears. After cooling, the solution is mixed with a quantity of distilled water, corresponding to the water lost by evaporation, and filtered. This bath gives a beautiful bright gilding upon all metals, even upon iron and steel.

The yellow prussiate of potash baths are deservedly popular for decorative gilding, when gold deposits of different colors are to be produced upon an object. Certain portions have then to be covered with stopping-off varnish, the latter being less attacked by this bath than by one containing an excess of potassium cyanide.

This bath is especially suitable for the so-called clock gilding. The articles are first provided with a heavy deposit of copper in the alkaline copper bath, then matt-coppered in the acid copper bath, next drawn through the bright-pickling bath, thoroughly rinsed; and finally gilded in the bath heated to about 122° F.

Gold baths for hot gilding.—IV. Fine gold (as fulminating gold) 15.4 grains, 98 per cent. potassium cyanide 77 grains, water 1 quart.

Electro-motive force at 10 cm. electrode-distance, 1.0 volt.

Current-density, 0.1 ampère.

This bath is prepared in the same manner as that according to formula I, from 15.4 grains of fine gold, which is converted into neutral chloride of gold by dissolving in aqua regia and evaporating; or dissolve directly 29.32 to 30.75 grains of chemically pure neutral chloride of gold in water, precipitate the gold as fulminating gold with aqua ammonia, wash the precipitate, dissolve it in water containing the potassium cyanide, and heat until the odor of ammonia disappears, replacing the water lost by evaporation. This bath yields a beautiful sad gilding of great warmth. All that has been said in regard to the content of potassium cyanide in the bath prepared according to formula I also applies to this bath. The temperature should be between 158° and 176° F., and the current-strength 2.0 to 2.5 volts.

Roseleur recommends for hot gilding :

V. Chemically pure crystallized sodium phosphate 2.11 ozs., neutral sodium sulphite 0.35 oz., potassium cyanide 30.86 grains, fine gold (as chloride) 15.43 grains, distilled water 1 quart.

Electro-motive force at 10 cm. electrode distance 1.5 volts.

Current-density, 0.12 ampère.

If this bath is to serve for directly plating *steel,* only half the quantity of potassium cyanide is to be used, and the objects should be covered with the use of a somewhat greater electro-motive force. Increasing the content of neutral sodium sulphite to 0.5 or 0.7 oz. also appears advisable.

Dissolve in a porcelain dish, with the aid of moderate heat, the sodium phosphate and sodium sulphite, and when the solution is *cold,* add the neutral chloride of gold prepared from 15.43 grains of gold = about 30.86 grains of commercial chloride of gold, and the potassium cyanide. For use, heat the bath to between 158° and 167° F.

For the preparation of gold baths for hot and cold gilding, double gold salts and triple gold salts, as well as gold solutions, as brought into commerce by some manufacturers may also be used.

Many gold-platers prepare their gold baths with the assistance of the electric current. For this purpose prepare a solution of 3.52 ozs. potassium cyanide (98 to 99 per cent.) per quart of water and, after heating to between 122° and 140° F., conduct the current of two Bunsen cells through two sheets of gold, not too small, which are suspended as electrodes in the potassium cyanide solution. The action of the current is interrupted when the solution is so far saturated with gold that an article immersed in it and connected to the negative pole in place of the other gold sheet, is gilded with a beautiful warm tone. By weighing the sheet of gold serving as anode, the amount of gold which has passed into the solution is ascertained. According to English authorities, a good gold bath prepared according to this method should contain 3.52 ozs. of potassium cyanide and 0.7 oz. of fine gold per quart of water.

The only advantage of this mode of preparing the bath is that it excludes a possible loss of gold, which may occur in dissolving gold, evaporating the gold solution, etc., by breaking the vessel containing the solution. However, by using commercial chemically pure chloride of gold such loss is avoided, and the bath prepared according to the formulæ given yields richer tones than a gold bath produced by electrolysis. Besides, the preparation of the gold bath with the assistance of the electric current can only be considered for smaller baths, since the saturation of a larger volume of potassium cyanide solution requires considerable time, and the potassium cyanide is strongly decomposed by long heating.

Gold anodes. Management of gold baths.—It is advisable to keep the content of gold in the baths prepared according to the different formulæ as constant as possible, which is best effected by the use of fine gold anodes.

Insoluble platinum anodes are better liked in gilding than for all other electro-plating processes, partly because they are somewhat cheaper, and partly because they are recommended in most books on the subject. However, a bath which has become low in gold does not yield a beautiful gold color, and has to be frequently strengthened by the addition of chloride of gold or concentrated solution of fulminating gold in potassium cyanide, the preparation of which consumes time and causes expense, so that the use of gold anodes is the cheapest in the end, especially with the present high price of platinum.

The use of steel anodes for cold and warm cyanide gold baths, advocated by some, cannot be recommended. Every gilder knows from experience that, when the enamel of the tanks containing the gold baths becomes defective, the baths in a short time fail. The reason for this is simply that the iron on the defective places of the tank decomposes the gold bath, metallic gold being reduced. Iron, in this respect, acts like zinc, which, in a still shorter time, precipitates metallic gold from gold baths. Now, when iron anodes remain suspended in the baths, a reduction of gold takes place, while a quantity of iron equivalent to the reduced gold is dissolved, and, in the form of ferric oxide, falls to the bottom of the vat.

In hot gold baths this separation of gold proceeds still more rapidly and the content of potassium cyanide in the bath is destroyed, yellow prussiate of potash being formed. The argument made in favor of the use of steel anodes, that the old practitioners often added intentionally yellow prussiate of potash to their baths to heighten the gold tone is fallacious. A plater who works with gold baths prepared with yellow prussiate of potash cannot expect to replace the gold by the solution of the gold anodes, and when working with gold cyanide and potassium cyanide baths there is no inducement for gradually changing the bath into a yellow prussiate of potash bath by the use of steel anodes.

According to one statement, a hot gold bath with steel anodes showed, after being electrolyzed for 70 hours, scarcely a trace of iron. To ascertain the correctness of this statement by an experiment, a gold bath prepared according to formula IV was electrolyzed at 158° F., with a blue annealed steel anode weighing 12.092 grammes. During the first two hours only a moderate yellow-reddish bloom of iron salt was perceptible on the anode, which became detached from the latter and fell to the bottom of the beaker. The bloom, however, became gradually heavier, the bottom of the beaker was covered with a precipitate of a yellow-brown color, the previously colorless bath acquiring a yellow color and after electrolyzing for five hours, the blue color of the anode had largely disappeared. The anode weighed now 11.832 grammes, and had consequently lost 2.2 per cent. After again suspending it in the bath it was more rapidly attacked in consequence of the destruction of the blue annealing color, which retarded corrosion. After five more hours the anode weighed 11.105 grammes, the loss being therefore 8.16 per cent. The bath now showed a deep yellow color, and the precipitate on the bottom of the beaker had increased, while small, lighter flakes of ferric hydrate spun around in the bath and attached themselves to the anode. Electrolysis was now discontinued, since the last mentioned phenomena proved the uselessness of steel anodes for the reasons given under "Deposition of Nickel and Cobalt."

As regards the advantage claimed for the use of steel anodes, that a large anode-surface corresponding to the object-surface can be rendered effective without taxing too severely the pocket-book of the gilder, it may be said that the same object can in a more rational manner be attained by employing carbon anodes, which to prevent contamination of the bath by particles of carbon, are placed in linen bags. Crosses and balls of unusually large dimensions for church towers have frequently been gilded in Dr. Geo. Langbein & Co.'s establishment, for which a large anode-surface was required in order to obtain a uniformly heavy deposit, and in such cases carbon anodes of the best quality of retort graphite were used. These anodes, to be sure, become saturated with gold bath, and for that reason cannot be used for other baths. When not required for some time, they are kept in a vessel filled with clean water, and the latter is added to the bath to replace that lost by evaporation.

The employment as anodes of platinum strips or platinum wire may, perhaps, be advocated for coloring the deposit, i. e., for the purpose of obtaining certain tones of color when gilding in the hot bath. By allowing the platinum anode to dip only slightly in the bath a pale gilding is obtained, because the current thereby becomes weaker ; by immersing the anode deeper the color becomes more yellow, and by immersing it entirely the tone becomes more reddish.

However, instead of producing these effects of the current-strength by the anode, which requires the constant presence of the operator, it is better to obtain the coloration by means of the rheostat. By placing the switch upon " strong," a reddish-gold tone is obtained, and by placing it upon " weak," a paler gold tone, while the beautiful gold-yellow lies in the middle between the two extremes. However, since even with the use of gold anodes the content of gold in the bath is not entirely restored, the bath has after some time to be strengthened, which is effected by a solution of fulminating gold or chloride of gold in potassium cyanide, according to the composition of the bath.

The excess of potassium cyanide must not be too large, otherwise the gilding will be pale ; but, on the other hand, it must not be too small, since in this case quite a strong current would have to be used to effect a normal deposition of gold, which, besides, would not be dense and homogeneous. Too small a content of potassium cyanide is indicated by the gold anodes showing dark streaks.

As in the silvering baths, the excess of potassium cyanide in the gold baths is also partially converted into potassium carbonate by the action of air, heat, etc., and it is, therefore, advisable from time to time to add a small quantity of potassium cyanide.

The presence of larger quantities of organic substances which may get into the bath by dust or some other way, shows itself, as a rule, by a brownish coloration. Such baths rarely yield a beautiful gold color, but deposit gold of a dark tone.

Unsightly and spotted deposits are also caused by a contamination of gold baths with compounds of lime which reach the bath by the use of water containing much lime, or by insufficient removal of lime paste after freeing the objects from grease.

Tanks for gold baths. Gold baths for cold gilding are kept in tanks of stoneware or enameled iron, or small baths in glass tanks, which, to protect them against breaking, are placed in a wooden box. Baths for hot gilding require enameled iron tanks in which they can be heated by a direct fire, or better, by placing in hot water (water bath), or by steam. For small gold baths for hot gilding, a porcelain dish resting upon a short-legged iron tripod may be used (Fig. 129). Beneath the iron tripod is a gas burner supplied with gas by means of a flexible India-rubber tube connected to an ordinary gas burner. Across the porcelain dish are placed two glass rods, around which the pole-wires are wrapped.

In heating larger baths in enameled tanks over a direct.

FIG. 129.

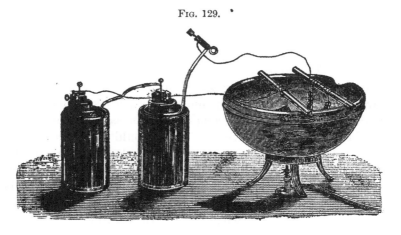

fire it may happen that on the places most exposed to the heat the enamel may blister and peel off; it is, however, better to heat the baths in a water or steam bath. For this purpose have made a box of stout iron or zinc sheet about $\frac{3}{4}$ inch wider and longer, and about 4 inches deeper than the enameled tank containing the gold bath. To keep the level of the water constant, the box is to be provided with a water inlet- and overflow-pipe. In this box place the tank so that its edges rest upon those of the box, and make the joints tight with tow. The water-bath is then heated over a gas flame or upon a hearth, the water lost by evaporation being constantly replaced, so that the enameled tank is always to half its height surrounded by hot water. For heating by steam the arrangement is the same, only a valve for the introduction, and a pipe for the discharge, of steam, are substituted for the water inlet- and overflow-pipe.

Execution of gold-plating.—Most suitable current-density, 0.15 to 0.2 ampère. Like all other electro-plating operations, it is advisable to effect gold-plating with an external source of current, that is, to use a battery or other source of current separated from the bath, and to couple the apparatuses as previously described and illustrated by Figs. 44 and 45.

To be sure, there are still gilders who gild without a battery or separate external source of current and obtain good results, the process being, as a rule, employed only in gilding small articles. The apparatus used for this purpose consists of a glass vessel containing the gold solution compounded with a large excess of potassium cyanide, and a porous clay cup filled with very dilute sulphuric acid or common salt solution, which is placed in the glass vessel. Care should be taken to have the fluids in both vessels at the same level. Immerse in the clay cup an amalgamated zinc cylinder or zinc plate, to which a copper wire is soldered. Outside the cup this copper wire is bent downwards, and the article to be gilded, which dips in the gold solution, is fastened to it. In working with this apparatus there is always a loss of gold, since the gold solution penetrates through the porous cup, and on coming in contact with the zinc is reduced by it, the gold being separated as black powder upon the zinc. In cleaning the apparatus this black slime has to be carefully collected and worked for fine gold.

For the sake of greater solidity, only articles of silver and copper and its alloys should be directly gilded, while all other metals are best first brassed or coppered. Cleaning from grease and pickling is done in the same manner, as described on page 228. The preparation of the articles for gilding differs from

that for silvering only in that the surfaces, which later on are
to appear with high luster, are not artificially roughened with
emery, pumice, or by pickling, because, on the one hand, the
gold deposit seldom needs to be made extravagantly heavy,
and the rough surface formed would require more laborious
polishing with the burnishers; and, on the other, the gold deposits adhere quite well to highly-polished surfaces, provided
the current-strength is correctly regulated, and the bath
accurately composed according to one of the formulæ given.
Quicking the articles before gilding, which is recommended
by some authors, is not necessary.

The current-strength must, under no circumstances, be so
great that a decomposition of water, and consequent evolution
of hydrogen on the objects, takes place, since otherwise the
gold would not deposit in a reguline and coherent form, but as
a brown powder. By regulating the current-strength so that
it just suffices for the decomposition of the bath, and avoiding
a considerable surplus, a very dense and uniform deposit is
formed; and by allowing the object to remain long enough in
the bath, a beautiful, mat gold deposit can be obtained in all
the baths prepared according to the formulæ given. It may,
however, be mentioned that this mode of mat gilding is the
most expensive, since it requires a very heavy deposit, and it
will, therefore, be better to matten the surface previous to
gilding, according to a process to be described later on.

Constant agitation of the objects in the baths, or of the latter itself, is of great advantage for obtaining good gilding. It
is evident that by reason of the small amount of metal in the
gold baths, especially in warm ones, the strata of fluid on the
cathodes become rapidly poor in metal, and if care be not
taken to replace them by strata of fluid richer in gold, disturbances in deposition will result.

For gilding with cold baths, two freshly-filled Bunsen cells
coupled for electro-motive force suffice in almost all cases,
while for hot baths one cell is, as a rule, sufficient, if the
anode surface is not too small. The more electro-positive the
metal to be gilded is, the weaker the current can and must be.

Though gold solutions are good conductors, and, therefore, the portions of the articles which do not hang directly opposite the anodes gild well, for solid plating of larger objects it is recommended to frequently change their positions, except when they are entirely surrounded by anodes.

The inner surfaces of *hollow-ware*, such as drinking-cups, milk pitchers, etc., are best plated after freeing them from grease and pickling, by filling the vessel with the gold bath and suspending a current-carrying gold anode in the center of the vessel, while the outer surface of the latter is brought in contact with the negative conducting wire. The lips of vessels are plated by placing upon them a cloth rag saturated with the gold bath and covering the rag with the gold anode.

For gold-plating in the cold bath the process is as follows: The objects, thoroughly freed from grease and pickled (and if of iron, zinc, tin, Britannia, etc., previously coppered), are suspended in the bath by copper wires, where they remain with a weak current until in about 8 or 10 minutes they appear uniformly plated. At this stage they are taken from the bath, rinsed in a pot filled with water, and the latter, after having been used for some time, is added to the bath to replace the water lost by evaporation. The articles are finally brushed with a fine brass scratch-brush and tartar solution, thoroughly rinsed, again freed from grease by brushing with lime-paste and then returned to the bath, where they remain until they have acquired a deposit of sufficient thickness.

When an article is to have a very heavy deposit, it is advisable to scratch-brush it several times with the use of tartar or its solution, or with a solution of size and water, between the intermediate coats of gold. By these means a very durable and lasting coating of gold will be secured. For gold plating by weight the same plan as given for silver-plating by weight (p. 382) is pursued.

For gold-plating with the hot bath, the operations are the same, with the exception that a weaker current is introduced into the bath and the time of the plating process shortened.

Frequent scratch-brushing also increases the solidity of the deposit and prevents its prematurely turning to a dead brown-black. Since in hot plating more gold than intended is readily deposited, it is especially advisable to place a rheostat and voltmeter in the circuit, as otherwise the operator must remain standing along-side of the bath and regulate the effect of the current by immersing the anodes more or less.

When taken from the bath, the finished gilded objects should show a deep yellow tone, which, after polishing, yields a full gold color. If the objects come from the bath with a pale gold tone, the deposit, after polishing, shows a meager, pale gold color, which is without effect. Gold deposits of a dark or brown color also do not yield a sad gold tone.

With a somewhat considerable excess of potassium cyanide, and if the objects to be plated are not rapidly brought in contact with the current-carrying object rod, hot gold baths cause the solution of some metal. Therefore when silver or silver-plated objects are constantly plated in them they yield a somewhat greenish gilding in consequence of the absorption of silver, or a reddish gilding due to the absorption of copper, if copper or coppered articles are constantly plated in them. Hence, for the production of such green or reddish color, gold-plating baths which have thus become argentiferous or cupriferous, may be advantageously used. In order to obtain a deposit of green or red gold with fresh baths, the tone-giving addition of metal must be artificially effected, as will presently be seen.

If, however, such extreme tones are not desired, the content of gold in the baths may be exhausted for preliminary plating with the use of platinum anodes, the sad gold color being then given in a freshly prepared bath.

The gold deposits are *polished* in the same manner as silver deposits, with the burnisher and red ochre, and moistening with solution of soap, decoction of flaxseed, or soap-root, etc. For less heavy gilding the articles, previous to gilding, are given high luster, and after gilding, burnished with rouge and buckskin.

Red-gilding. In order to obtain a red gold with the formulæ given, a certain addition of copper cyanide dissolved in potassium cyanide has to be made to them. The quantity of such addition cannot be well expressed by figures, since the current strength with which the articles are plated exerts considerable influence. It is best to triturate the copper cyanide in a mortar to a paste with water, and add of this paste to a moderately concentrated potassium cyanide solution as long as copper cyanide is dissolved. Of this copper solution add, gradually and in not too large portions, to the gold solution until, with the current-strength used, the gold deposit shows the desired red tone, and if fine gold anodes are used, the bath is kept constant with this content of copper by an occasional addition of the above-mentioned copper solution.

The absorption of copper in the bath may also be effected by suspending, in place of gold anodes, anodes of copper or copper-gold alloys, for instance, fourteen-carat gold, and allowing the current to circulate (suspension of a few gold anodes to the object-rod). The direct addition of copper cyanide, however, deserves the preference.

In place of preparing the solution of copper cyanide in potassium cyanide, commercial crystallized potassium-copper cyanide may be used. It is dissolved in warm water, and of the solution a sufficient quantity is gradually added to the gold bath.

For the determination of the content of copper required for the purpose of obtaining a beautiful red gold, a bath for hot gilding which contained 10.8 grains of gold per quart was compounded with a solution of copper cyanide in potassium cyanide with 1.08 grains content of copper. The tone of the gilding, which previously was pure yellow, immediately passed into a pale red gold. By the further addition of 1.08 grains of copper, a fiery red gold tone was obtained, while a third addition of 1.08 grains of copper yielded a color more approaching that of copper than of gold. These experiments show that 20 per cent. of copper of the weight of gold contained in the bath seems to be the most suitable proportion· for obtaining a beautiful red gold.

Rings, watch-chains and other objects of base metal are frequently to be plated with red gold, so as to show no perceptible sign of having been attacked by nitric acid, even after remaining in it for several hours. This may be effected by first giving the objects a deposit of a strongly yellow color by gilding in a bath containing 10.8 to 15.43 grains of gold per quart and then coloring them in the red gilding bath. This process may be called an imitation of mechanical gold plating, and is frequently made use of in the jewelry industry.

A method of gilding chains and other articles manufactured from common metal, in imitation of genuine gold articles is given by Gee as follows: A bath is prepared by dissolving a quantity of pure gold and making a solution of it in the usual manner, and then using a large copper anode instead of a gold one in the process of gilding.

The articles are gilt until they stand the acid test, when they are well burnished until they present a bright gold-like appearance. If the articles are slightly gilt as a first process and then burnished, and afterwards more thickly gilt and again burnished, much less gold is required than if the process is conducted straight to the end without any intermediate burnishing. The burnishing stops up all the pores of the metal by the adoption of this plan, and more quickly renders the articles gilt acid proof and that at the expense of much less gold. When the solution begins to gild of an inferior color it is abandoned and another one made. It produces a surface alloy of about 16 or 18 carat, and well answers the purpose for which it has been designed.

Green gilding. To obtain greenish gilding, solution of cyanide or chloride of silver in potassium cyanide has to be added to the gold bath. It is not easy to prepare greenish gilding of a pleasing color, and to obtain it the current-strength must be accurately proportioned to the object-surface, since with too weak a current silver predominates in the deposit.

the gilding then turning out whitish, while too strong a current deposits too much gold in proportion to silver, the gilding becoming yellow, but not green.

Rose-color gilding may be obtained by the addition of suitable quantities of copper and silver solutions, but such coloration requires much attention and thought.

Rose gold solution.—Probably one of the best solutions for the rose gold, sometimes also termed *old gold* is, according to Mr. Chas. H. Proctor,* made from ¼ oz. of pure 24-karat gold. dissolved in aqua regia in the usual manner, then precipitated as fulminate with ammonia (26°), and then well washed. Add the gold salt to a solution of 1 gallon of cyanide solution, standing 2 to 3° Bé, and add ½ oz. of hyposulphite of sodium to each gallon of solution so prepared. This solution will produce a good flash gold with a weak current. Cheap rose gold work is first acid copper plated for a few minutes, then relieved on the high lights, and then gilded. Gold work is run for five to ten minutes with a strong current, according to the tone required, then relieved with sodium bicarbonate instead of pumice stone. '

To produce a rose gold finish without the use of gold a number of concerns are using the Electrochroma Process, in which the articles are immersed in a special bath in the same manner as employed in plating. In a minute or two the articles are coated with a pinkish yellow surface that resembles rose gold. The surface is afterwards relieved to produce a contrast effect on the high lights.

This finish can also be imitated very successfully by the following method : All articles, except those of brass, should be previously brass-plated and then a surface similar to the brush brass finish produced. Use floated silax instead of pumice stone so that the surface will be even and not have a scratchy appearance. Then gold lacquer, using a yellowish instead of a red toned lacquer. The surface should be thor-

* Metal Industry, June, 1913.

oughly dried on the lacquer heater. The rose tone is then produced by mixing dry orange chrome and a very little finely powdered gold rouge, mixed with turpentine and a teaspoonful of turpentine varnish per pint of the mixture. This should be mixed to a thinly fluid paint and then applied to the detail work with a soft brush. The articles should then be dried for a short time by the aid of heat and allowed to become cool. The surface should be opaque without any luster when dry. Now mix up equal parts of boiled linseed oil and turpentine and use this for reducing the color from the surface. To accomplish this operation, moisten soft rags with the mixture and remove the colors from the high lights or detail work. After this is done the articles will have the appearance of true rose gold.

For the sake of completeness, a method of gilding which is a combination of fire-gilding with electro-deposition, may here be mentioned, though experiments made with it failed to show the advantages claimed for it, because it does not yield as dense a deposit as fire-gilding, nor can the volatilization of mercury be avoided, the latter operation being the most dangerous part of fire-gilding.

According to Du Fresne, the process is as follows:

The articles are first coated with mercury, with the assistance of the current, in a mercurial solution consisting of cyanide of mercury in potassium cyanide, with additions of carbonate and phosphate of soda, then gilded in an ordinary gilding-bath, next again coated with mercury, then again gilded, and so on, until a deposit of sufficient thickness is obtained. The mercury is then evaporated over glowing coals, and the articles, after scratch-brushing, are burnished.

According to another process, the articles are gilded in a bath, consisting of 98 per cent. potassium cyanide 1.2 qzs., cyanide of gold 92½ grains, cyanide of mercury 22½ grains, distilled water 1 quart, a strong current being used. When the objects are sufficiently gilded, the mercury is evaporated in the above-mentioned manner, and the objects are scratch-brushed, and finally polished.

28

Mat gilding.—As previously mentioned, a beautiful mat gold deposit may be obtained by the use of any of the formulas given, and a current correctly regulated, and allowing sufficient time for gilding. The heavy deposit of gold required for this process makes it, however, too expensive, and it is, therefore, advisable to produce mat gilding by previously matting the basis-surface, since then a thinner deposit of gold will answer very well. The process of graining will be referred to later on under "Silvering by Contact," etc.

Another method is to mat the first slight deposit by means of brass or steel-wire brushes, and then to give a second deposit of gold, which also turns out mat upon the matted surface. The character of the mat produced depends on the thickness of the wire of the brushes. Thicker wire gives a mat of a coarser grain, and thinner wire one of a finer grain.

Objects may be readily matted with the use of the sand blast, after which they are quickly drawn through the bright dipping bath, thoroughly rinsed, and brought into the gold bath.

Matting by chemical or *electro-chemical means* is effected by one of the following methods :

For this purpose the mixture of 1 volume of saturated solution of bichromate of potash and 2 volumes of concentrated hydrochloric acid, mentioned on p. 225, may be used. Brass articles are allowed to remain several hours in the mixture, and are then quickly drawn through the bright-dipping bath. Copper alloys might also be successfully matted by suspending them as anodes in a mixture of 90 parts water and 10 parts sulphuric acid, and drawing the matted articles through the bright-dipping bath.

Or, they are mat-silvered, and the gold is deposited upon the matted layer of silver. Articles gilded upon a mat silver basis, however, acquire before long an ugly appearance, since in an atmosphere containing sulphuretted hydrogen the silver turns black, even under the layer of gold and shines through.

More advantageous is the process of providing the articles

with a mat copper coating in the acid galvanoplastic bath. They are then drawn through a not too strong pickle, rinsed, and gilded. This process is used for the so-called French clock gilding, and yields a very sad, beautiful gilding. The articles consisting of zinc are first heavily coppered in a cyanide copper bath, then matted in the acid copper bath (see " Galvanoplasty "), care being taken that the slinging wire is in contact with the object-rod, which conducts the current, before the coppered zinc objects is suspended in the bath. This process of coppering zinc in the acid copper bath is, however, quite a delicate operation, and it will frequently be noticed, even with apparently very heavy coppering in the cyanide copper bath, that in suspending the articles in the acid bath, brownish-black places appear on which, by contact of the acid bath with zinc, copper in a pulverulent form is deposited. When this is observed, the articles must be immediately taken from the bath, thoroughly scratch-brushed, and again thoroughly and heavily coppered in the cyanide copper bath, before replacing them in the acid copper bath. It may be recommended to provide the coppered zinc articles with a thick deposit of nickel, and then to copper them mat in the acid bath, the percentage of unsuccessful coppering being much smaller than without previous nickeling. The mat-coppered articles are rapidly drawn through the bright-dipping bath and then gilded, the bath prepared according to formula III, and heated to about 140° F., being very suitable for the purpose.

Coloring of the gilding. It has been repeatedly mentioned that the most rational and simple process of giving certain tones of color to the gilding is by means of a stronger or weaker current. Many operators, however, cling to the old method of effecting the coloration by gilder's wax or brushing with certain mixtures, and for this reason this process, which is generally used for coloring fire-gilding, shall be briefly mentioned.

To impart to the gold-deposit a *redder* color, the gilding-wax is prepared with a greater content of copper, while for greenish gilding more zinc-salt is added. There are innumerable receipts for the preparation of gilding wax, nearly every gilder having his own receipt, which he considers superior to all others. Only two formulæ which yield good results will have to be given one (I) for *reddish* gilding and one (II) for *greenish* gilding.

I. Wax 12 parts by weight, pulverized verdigris 8, pulverized sulphate of zinc 4, copper scales 4, borax 1, pulverized bloodstone 6, copperas 2.

II. Wax 12 parts by weight, pulverized verdigris 4, pulverized sulphate of zinc 8, copper scales 2, borax 4, pulverized bloodstone 6, copperas 2.

Gilder's wax is prepared as follows: Melt the wax in an iron kettle, add to the melted mass, while constantly stirring, the other ingredients, pulverized and intimately mixed, in small portions, and stir until cold, so that the powder cannot settle on the bottom or form lumps. Finally, mould the soft mass into sticks about $\frac{1}{3}$ inch in diameter.

Gilder's wax is applied as follows: Coat the heated gilded articles uniformly with the wax, and burn off over a charcoal fire, frequently turning the articles. After the wax flame is extinguished, plunge the articles into water, scratch-brush with wine-vinegar, dry in sawdust, and polish.

To give gilded articles a beautiful, rich appearance, the following process may also be used: Mix 3 parts by weight of pulverized alum, 6 of saltpetre, 3 of sulphate of zinc, and 3 of common salt, with sufficient water to form a thinly-fluid paste. Apply this paste as uniformly as possible to the articles by means of a brush, and after drying, heat the coating upon an iron plate until it turns black; then wash in water, scratch-brush with wine-vinegar, dry and polish.

According to a French receipt, the same result is attained by mixing pulverized blue vitriol 3 parts by weight, verdigris 7, ammonium chloride 6, and saltpetre 6, with acetic acid 31; immersing the gilded articles in the mixture, or applying the latter with a brush ; then heating the objects upon a hot iron plate until they turn black, and, after cooling, pickling in concentrated sulphuric acid.

Some gilders *improve bad tones of gilding* by immersing the articles in dilute solution of nitrate of mercury until the gilding appears white. The mercury is then evaporated over a flame and the articles are scratch-brushed. Others apply a paste of pulverized borax and water, heat until the borax melts, and then quickly immerse in dilute sulphuric acid.

Incrustations with gold are produced in the same manner as incrustations with silver, described on p. 403.

Gilding of metallic wire and gauze.—Fine wire of gilded copper and brass is much used in the manufacture of metallic fringes and lace, for epaulettes and other purposes. The fine copper and brass wires being drawn through the draw-irons and wound upon spools by special machines, and hence not touched by the hands, freeing from grease may, as a rule, be omitted. The first requisite for gilding is a good winding machine, which draws the wires through the gold bath and wash-boxes, and further effects the winding of the wire upon spools. The principal demand made in the construction of such a machine is that by means of a simple manipulation a great variation in the speed with which the wire or gauze passes through the gold bath can be obtained. This is necessary in order to be able to regulate the thickness of the gilding by the quicker or slower passage of the wire. A machine well adapted for this purpose is that constructed by J. W. Spaeth, and shown in Fig. 130.

The variation in the passage of the wire is attained by the two friction-pulleys F, which sit upon a common shaft with the driving pulley R, and transmit their velocity by means of the friction-pistons KK' to the friction-pulley F', which is firmly connected to the belt-pulley R driving the spool spindle. Since by a simple device the pistons K and K' may be shifted, it is clear that the transmission of the number of revolutions from F to F' is dependent on the position of the friction-pistons K and K', and that the velocity will be the greater the shorter the distance they are from the center of friction-pulleys F and F'. In order that the friction between F, K and F' may always be sufficient for the transmission of the motion, even when the pistons are worn, four weights, G, are provided, which press the above-mentioned parts firmly against each other.

In front of each spool of this machine is inserted a small

FIG. 130.

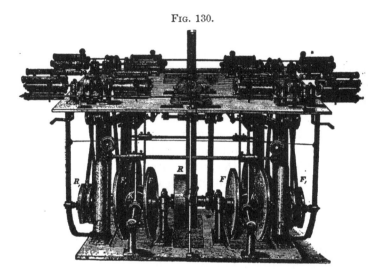

enameled iron tank which contains the gold bath, and is heated by a gas flame to about 167° F. Between this bath and the winding machine is another small tank with hot water in which the gilded wire is rinsed

The wire unwinds from a reel placed in front of the gold baths, runs over a brass drum which is connected to the negative pole of the source of current and transmits the current to the wire. The dipping of the wire into the gold bath is effected by porcelain drums, which are secured to heavy pieces of lead placed across the tanks, as shown in Fig. 131. The gilded wire being wound upon the spools of the winding machine, these spools are removed and thoroughly dried in the drying chamber. The wire is then again reeled off onto a simple reel, in doing which it is best to pass it through between two soft pieces of leather to increase its luster.

For gilding wire the most suitable gold bath is that prepared according to formula IV. The electro-motive force should be from 6 to 8 volts, in order to produce a deposit of sufficient thickness, even when the wire passes at the most rapid rate through the bath. For this reason a dynamo with a voltage of 10 volts is almost exclusively used for wire gilding.

As a rule an anode of platinum—a strip of platinum sheet—of the same length as the tank is placed upon the bottom of the latter, and connected by means of platinum wire to the positive pole of the source of current. The use of gold anodes

FIG. 131.

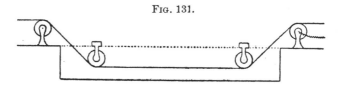

for wire gilding is not required, since the small gold baths—generally only 2 to 4 quarts—are as far as possible to be worked till exhausted, when they are replaced by fresh baths.

In place of platinum anodes, Stockmeir recommends the use of blued Bessemer steel anodes for wire gilding. In this case there can be no objection to steel anodes, because the baths are rapidly exhausted, and then go amongst the gold-residues. But, nevertheless, the use of an indestructible platinum anode would appear to deserve the preference, the baths being without doubt kept cleaner than with steel anodes.

Silver-plated wires are, as a rule, to be gilded, and since the color of the basis-metal exerts an influence upon the gilding, Stockheimer recommends brassing the silver-plated or solid silver wires previous to gilding, because a gold-deposit of less thickness than for covering the white silver, would thus be required. The proposition to gild nickeled wires, in place of silver-plated wires, because they are less subject to rapid discoloration in an atmosphere containing sulphuretted hydrogen, also deserves consideration.

Stripping gold from gilded articles.—Gilded articles of·*iron* and *steel* are best stripped by treating them as anodes in a solution of from 2 and 2¾ ozs. of 98 per cent. potassium cyanide in 1 quart of water, and suspending a copper plate greased with oil or tallow as the cathode. Gilded *silverware* is readily stripped by heating to ignition, and then immersing in dilute sulphuric acid, whereby the layer of gold cracks off, the heating and subsequent immersion in dilute sulphuric acid being repeated until all the gold is removed. Before heating and immersing in dilute sulphuric acid, the articles may first be provided with a coating of a paste of ammonia chloride, flowers of sulphur, borax and nitrate of potash which is allowed to dry. On the bottom of the vessel containing the dilute sulphuric acid, the gold will be found in laminæ and scales, which are boiled with pure sulphuric acid, washed and finally dissolved in aqua regia, and made into chloride of gold or fulminating gold.

To strip articles of *silver*, *copper* or *German silver* which will not bear heating, the solution of gold may be effected in a mixture of 1 lb. of fuming sulphuric acid, 2.64 ozs. of concentrated hydrochloric acid, and 1.3 ozs. of nitric acid of 40° Bé. Dip the articles in the warm acid mixture, and observe the progressive action of the mixture by frequently removing the articles from it. The articles to be treated must be perfectly dry before immersing in the acid mixture, and care must be had to preserve the latter from dilution with water in order to prevent the acids from acting upon the basis-metal.

The process by which scratched or rubbed rings are, so to say, electrolytically smoothed and polished, may be called a sort of stripping. For this purpose the rings are suspended as anodes in a bath consisting of: Water 1 quart, yellow prussiate of potash 1 oz., 99 per cent. potassium cyanide $\frac{7}{10}$ ozs. By conducting a current of high electro-motive force, of about 20 to 25 volts, through the bath any roughness or unevenness is in a few minutes removed, and the rings will be almost perfectly smooth when taken from the bath. A sheet of gold or platinum is used as cathode.

Determination of genuine gilding.—Objects apparent gilded are rubbed upon the touchstone, and the streak obtained is treated with pure nitric acid of 1.30 to 1.35 specific gravity. The metal contained in the streak thereby dissolves, and as far as it is not gold, disappears, while the gold remains behind. The stone should be thoroughly cleansed before each operation, and the streak should be made, not with an edge or a corner of the object to be tested, but with a broader surface. If no gold remains upon the stone, but there is nevertheless, a suspicion of the article being slightly gilded, proceed with small articles as follows: Take hold of the article with a pair of tweezers, and after washing it first with alcohol, and then with ether, and drying upon blotting paper, pour over it in a test glass, cleansed with alcohol or ether, according to the weight of the article, 0.084 to 5.64 drachms of nitric acid of 1.30 specific gravity free from chlorine. The article will be immediately dissolved, and if it has been gilded never so slightly, perceptible gold spangles will remain upon the bottom of the glass.

Examination of Gold Baths.

The determination of free potassium cyanide and of the potassium carbonate which is formed, is effected in the same manner as given under "Examination of copper baths and of silver baths."

The determination of the gold is effected by the electrolytic method. With baths poor in gold, 50 cubic centimeters are used for electrolysis, and with baths rich in gold, 25 cubic centimeters. After diluting with water to within 1 centimeter of the rim of the platinum dish, the liquid is electrolyzed for about three hours with a current-density ND 100 = 0.067 ampère, the complete separation of the gold being recognized by a platinum strip suspended over the rim of the dish and dipping into the fluid showing in fifteen minutes no trace of a separation of gold.

The dish is then washed, rinsed with alcohol, and dried at 212° F. To obtain the content of gold in grammes per liter of bath, multiply the weight of the precipitate by 20, when 50 cubic centimeters, or by 40, when 25 cubic centimeters, of the bath have been used.

The content of gold in the baths declines constantly, especially with the use of platinum and carbon anodes. For strengthening the bath neutral gold chloride dissolved in potassium cyanide is used, 2 grammes neutral gold chloride and 1.4 grammes 99 per cent. potassium cyanide dissolved in a small quantity of water or directly in the bath, being required for every gramme of gold deficit in the baths.

The determination of gold described above is suitable only for baths prepared with potassium cyanide, which contain the gold in the form of potassium-gold cyanide. The determination of gold in baths prepared with yellow prussiate of potash is more difficult and should be made by a skilled analyst.

Recovery of gold from gold baths, etc. To recover the gold from old cyanide gilding baths, evaporate the baths to dryness, mix the residue with litharge, and fuse the mixture. The gold is contained in the lead button thus obtained. The latter is then dissolved in nitric acid, whereby the gold remains behind in the form of spangles. These spangles are filtered off and dissolved in aqua regia.

The recovery of gold from gold baths may also be advantageously effected by precipitation with zinc dust according to the same process as given for the recovery of silver, p. 413. After removing the zinc by means of hydrochloric acid and washing the gold powder, the latter is dissolved in aqua regia and the chloride of gold solution evaporated to dryness. Aluminium powder is still more suitable for precipitating the gold; the excess of aluminium is dissolved by potash or soda lye.

From the *acid mixtures* serving for mat pickling gold, or for stripping, the gold is precipitated by solution of sulphate of iron (copperas) added in excess. The gold present is precipitated as a brown powder mixed with ferric oxide. This powder is filtered off and treated in a porcelain dish with hot hydrochloric acid, which dissolves the iron. The gold which remains behind is then filtered off, and, after washing, dissolved in aqua regia in order to work the solution into fulminating gold or neutral chloride of gold.

For gilding by contact, boiling and friction, see special chapter " Deposition by Contact."

CHAPTER X.

1. DEPOSITION OF PLATINUM (PT = 195.2 PARTS BY WEIGHT).

Properties of platinum.—Pure platinum is white with a gray-ish tinge. It is as soft as copper, malleable and very ductile. At a white heat it can be welded, but is fusible only with the oxyhydrogen blowpipe or by the electric current. Its specific gravity is 21.4.

Air has no oxidizing action upon platinum. It is scarcely acted upon by any single acid; prolonged boiling with concentrated sulphuric acid appears to dissolve the metal slowly. The best solvent for it is aqua regia, which forms the tetra chloride, $PtCl_4$. Chlorine, bromine, sulphur and phosphorus combine directly with platinum, and fusing saltpetre and caustic alkali attack it.

Besides, in the malleable and fused state, platinum may be obtained as a very finely divided powder, the so-called *platinum black*, which is precipitated with zinc from dilute solution of platinum chloride acidulated with hydrochloric acid.

Platinum baths.—In view of the valuable properties of platinum of oxidizing only under certain difficult conditions, of possessing an agreeable white color, and of taking a fine polish, it seems strange that greater attention has not been paid to the electro-deposition of this metal than is actually the case. The reason for this may perhaps be found in the fact that the baths formerly employed for experiments possessed serious defects, causing the operator many difficulties, and besides, allowed only of the production of thin deposits. Giving due consideration to the requirements of the process of electro-deposition of platinum, and with the use of a suit-

(444)

able bath, deposits of platinum of a certain thickness can be readily produced, and necessary conditions will be described under "Treatment of Platinum Baths."

The platinum baths formerly proposed did not yield satisfactory results, because the content of platinum was too small in some of them, while with others dense deposits could not be obtained. A more recent formula by Böttger, however, gives quite a good bath. A moderately dilute, boiling-hot solution of sodium citrate is added to platoso-ammonium chloride until an excess of the latter no longer dissolves, even after continued boiling. The following proportions have been found very suitable: Dissolve 17½ ozs. of citric acid in 2 quarts of water, and neutralize with caustic soda. To the boiling solution add, whilst constantly stirring, the platoso-ammonium chloride freshly precipitated from 2.64 ozs. of chloride of platinum, heat until solution is complete, allow to cool, and dilute with water to 5 quarts. To decrease the resistance of the bath, 0.7 or 0.8 oz. of ammonium chloride may be added; a larger addition, however, will cause the separation of dark-colored platinum.

The platoso-ammonium chloride is prepared by adding to a concentrated solution of platinic chloride, concentrated solution of ammonium chloride until a yellow precipitate is no longer formed on adding a further drop. The precipitate is filtered off and brought into the boiling solution of sodium citrate. The bath works very uniformly if the content of platinum is from time to time replenished.

"The Bright Platinum Plating Company," of London, has patented the following composition of a platinum bath: Chloride of platinum 0.98 oz., sodium phosphate 19¾ ozs., ammonium phosphate 3.05 ozs., sodium chloride 0.98 oz., and borax 0.35 oz., are dissolved, with the aid of heat, in 6 to 8 quarts of water, and the solution is boiled for 10 hours, the water lost by evaporation being constantly replaced. The results obtained with this bath were not much better than with Böttger's.

Jordis obtained useful results from a platinum lactate bath

prepared by transposition from platinic sulphate with ammonium lactate. There are, however, difficulties in obtaining platinic sulphate of uniform composition.*

Management of platinum baths. Copper and brass may be directly plated with platinum, but iron, steel and other metals are first to be coppered, otherwise they would soon decompose the platinum bath, independent of the fact that an unexceptionable deposit cannot be produced upon them without the cementing intermediary layer of copper.

Platinum baths must be used hot, and even then require an electro-motive force of 5 to 6 volts, and hence, in plating with a battery at least three, or better four, Bunsen cells must be coupled one after the other. An abundant evolution of gas must appear on the objects and anodes. The anode-surface (platinum anodes) must not be too small, and should be only at a few centimeters' distance from the objects. Since the platinum anodes do not dissolve, the content of platinum in the bath decreases constantly, and the bath must from time to time be strengthened. For this purpose, the bath, prepared according to Böttger's formula, is heated in a porcelain dish or enameled vessel to the boiling-point, a small quantity of fresh solution of sodium citrate is added and platoso-ammonium chloride introduced so long as solution takes place. A concentrated solution of platoso-ammonium chloride in sodium citrate (so-called platinum essence) may be kept on hand and a small quantity of it be at intervals added to the bath. Baths prepared according to the English method are strengthened by the addition of platinum chloride.

Execution of platinum plating. The objects, thoroughly freed from grease and, if necessary, coppered, are suspended in the bath heated to between 176° and 194° F., and this temperature must be maintained during the entire operation. The current should be of sufficient strength and the anodes placed so close to the objects that a liberal evolution of gas appears

* Jordis, Die Elektrolyse wässeriger Metallsalzlösungen, 1901.

on them.   For plating large objects, it is recommended to go round them, at a distance of 0.31 to 0.39 inch, with a hand-anode of platinum sheet which should not be too small and should be connected to the anode-rod.   When the current has vigorously acted for 8 to 10 minutes, the objects are taken from the bath, dried and polished.   However, for the production of heavy deposits—for instance, upon points of lightning rods—the deposit is vigorously brushed with a steel-wire scratch-brush or fine pumice-powder.   The objects are then once more freed from grease and returned for 10 or 15 minutes longer to the bath to receive a further deposit of platinum with a weaker current, which must, however, be strong enough to cause the escape of an abundance of gas-bubbles.   The objects are then taken out, and after immersion in hot water, dried in sawdust.   The deposit is then well burnished, first with the steel tool and finally with the stone, whereby the gray tone disappears and the deposit shows the color and luster of massive platinum sheet.   Points of lightning-rods platinized in this manner were without flaw after an exposure to atmospheric influences for more than six years.

For plating directly, without previous coppering, iron, nickel, cobalt and their alloys with platinum, the following process has been patented in Germany : * Nickel or cobalt is first electrolytically deposited upon base metals fusing with difficulty, such as, iron, nickel, cobalt, or their alloys, for instance, nickel steel, a suitable bath for this purpose being composed of nickel-ammonium sulphate 290 parts, ammonium sulphate 75, citric acid 20, distilled water 4000.   The metal is then heated in a reducing hydrogen atmosphere at 1652° to 1832° F., this operation being repeated after each electrolytical treatment in the platinum bath.   The latter is best composed of : Platinum-ammonium phosphate 25 parts, sodium phosphate 500, distilled water 4000.

The advantage claimed for this process is that the deposit

* German patent 201664.

of platinum does not peel off even when exposed to great heat, as is the case with an alloy previously coppered, and that by frequently repeating the operation the content of platinum steadily increases until the deposit finally possesses the properties of pure platinum.

Recovery of platinum from platinum solutions. From not too large baths, precipitation of the platinum with sulphuretted hydrogen is the most suitable method, and preferable to evaporating and reducing the metal from the residue. The process is as follows : Acidulate the platinum solution with hydrochloric acid ; and, after warming it, conduct sulphuretted hydrogen into it. The metal (together with any copper present) precipitates as sulphide of platinum. The precipitate is filtered off, dried, and ignited in the air, whereby metallic platinum remains behind. From larger baths the platinum may be precipitated by suspending bright sheets of iron in the acidulated bath. In both cases the precipitated platinum is treated with dilute nitric acid in order to dissolve any copper present. After filtering off and washing the pure platinum, dissolve it in aqua regia. The solution is then evaporated to dryness in the water bath, and the chloride of platinum thus obtained may be used in making a fresh bath. Precipitation by zinc sheets or zinc dust can also be recommended.

2. DEPOSITION OF PALLADIUM.

Properties of palladium. Palladium, when compact, has a white color and possesses a luster almost equal to that of silver. Its specific gravity is about 12.0 ; it is malleable and ductile, and may be fused at a white heat. In the oxyhydrogen flame it is volatilized, forming a green vapor. It is less permanent in the air than platinum. It is dissolved by nitric acid ; it is scarcely attacked, however; by hydrochloric or sulphuric acid. Hydriodic acid and free iodine coat it with the black palladium iodide.

On account of the high price of its salts, palladium has been but little used for electro-plating purposes ; nor, for the same reason, is it likely to be more extensively employed in the future.

According to M. Bertrand, the most suitable bath consists of a neutral solution of the double chloride of palladium and ammonium, which is readily decomposed by 3 Bunsen cells coupled one behind the other (therefore about 5.4 volts). A sheet of palladium is used as anode.

A solution of palladium cyanide in potassium cyanide does not yield as good results as the above bath.

Palladium is entirely constant in the air, and in color closely resembles silver. It possesses further the property of not being blackened by sulphuretted hydrogen, and for this reason it is sometimes employed for coating silver-plated metallic articles.

Palladium has also of recent years been employed for plating watch movements. According to M. Pilet, 4 milli-grammes (about $\frac{1}{17}$ grain) of palladium are sufficient to coat the works of an ordinary-sized watch. M. Pilet recommends the following bath : Water 2 quarts, chloride of palladium $5\frac{1}{2}$ drachms, phosphate of ammonia $3\frac{1}{2}$ ozs., phosphate of soda $17\frac{1}{2}$ ozs., benzoic acid $2\frac{3}{4}$ drachms.

Deposits of iridium and *rhodium* have recently been produced from baths similar in composition to those mentioned under palladium. But as these metals would be used for plating purposes only in isolated cases, it is not necessary to enter into details.

29

CHAPTER XI.

I. DEPOSITION OF TIN (Sn = 119 parts by weight).

Properties of tin. Tin is a white, highly lustrous metal. It possesses but little tenacity, but has a high degree of malleability, and tin-foil may be obtained in leaves less than $\frac{1}{30}$th of a millimeter in thickness. Tin melts at about 446° F., and evaporates at a high temperature. The fused metal shows great tendency to crystallize on congealing. By treating the surface of melted tin with a dilute acid, the crystalline structure appears in designs (*moiré métallique*), resembling the ice-flowers on frosted windows.

Tin remains quite constant even in moist air, and resists the influence of an atmosphere containing sulphuretted hydrogen. Strong hydrochloric acid quickly dissolves tin on heating, hydrogen being evolved and stannous chloride formed. Dilute sulphuric acid has but little action on the metal ; when heated with concentrated sulphuric acid, sulphur dioxide is evolved. Dilute nitric acid dissolves tin in the cold without evolution of gas ; concentrated nitric acid acts vigorously upon the metal, whereby oxide of tin, which is insoluble in the acid, is formed. Alkaline lyes dissolve the metal to sodium stannate, hydrogen being thereby evolved.

Tin baths. The bath used by Roseleur for tinning with the battery works very well. It is composed as follows :

I. Pyrophosphate of soda 3.5 ozs., tin salt (fused) 0.35 oz., water 10 quarts.

Electro-motive force at 10 cm. electrode-distance, 1.25 volts.
Current density, 0.25 ampère.

To prepare the bath dissolve the pyrophosphate of soda in

(450)

10 quarts of rain water, suspend the tin-salt in a small linen bag in the solution, and move the bag to and fro until its contents are entirely dissolved.

Objects of zinc, copper and *brass* are directly tinned in this bath. Articles of *iron* and *steel* are first coppered or preliminarily tinned by boiling in a bath given later on under tinning by contact, the deposit of tin being then augmented in bath I with the battery current. Cast-tin anodes as large as possible are used, which, however, will not keep the content of tin in the bath constant. It is therefore necessary, from time to time, to add tin-salt, which is best done by preparing a solution of 3.5 ozs. of pyrophosphate of soda in 1 quart of water and introducing into the solution tin-salt as long as the latter dissolves clear. Of this tin-essence add to the bath more or less, as may be required, and also augment the content of pyrophosphate of soda, if notwithstanding the addition of tin-salt, the deposition of tin proceeds sluggishly.

. Though the bath composed according to formula I suffices for most purposes, an alkaline tin bath, first proposed by Elsner, and later on recommended by Maistrasse, Fearn, Birgham and others, with or without addition of potassium cyanide, may be mentioned as follows:

II. Crystallized tin-salt. 0.7 ozs., water 1 quart, and potash lye of 10° Beaumè until the precipitate formed dissolves.

As seen from the formula the solution of tin-salt is compounded with potash lye of the stated concentration (or with a solution of 1 oz. of pure caustic potash in water), until the precipitate of stannous hydrate again dissolves.

Some operators recommend the addition of 0.35 oz. of potassium cyanide to the solution.

In testing Salzède's bronze bath (p. 363), it was found to yield quite a good deposit of tin directly upon *cast iron*, and it was successfully used for this purpose by omitting the cuprous chloride, and using instead 0.88 oz. of stannous chloride, so that the composition became as follows:

II*a*. 98 per cent. potassium cyanide 3.5 ozs., carbonate of potassium 35¼ ozs., stannous chloride 0.88 ozs., water 10 quarts. With 4 volts a heavy deposit was rapidly obtained.

Very good results were obtained in a hot bath (158° to 194° F.), first made public by Neubeck, which consists of:

III. 70 per cent. caustic soda 35¼ ozs., ammonium soda 35¼ ozs., fused tin-salt 7 ozs., water 10 quarts.

Electro-motive force at 10 cm. electrode-distance, and 155° F., 0.8 volt.

Current-density 1 ampère.

The chemicals are sufficiently dissolved in the water. When the bath commences to work sluggishly, about 0.35 to 0.5 oz. of fused tin-salt has to be added.

Management of tin baths.—Tin baths should not be used at a temperature below 68° F. Too strong a current causes a spongy reduction of the tin, which does not adhere well, while with a suitable current-strength quite a dense and reguline deposit is obtained. Cast-tin plates, with as large a surface as possible, are used as anodes. The choice of the tin-salt exerts some influence upon the color of the tinning. By using, for instance, crystallized tin-salt, which is always acid, in preparing the bath according to formula I, a beautiful white tinning with a bluish tinge is obtained, which, however, does not adhere so well as that produced with fused tin-salt. Again, the latter yields a somewhat dull gray layer of tin, and therefore the effects of the bath will have to be corrected by the addition of one or the other salt.

As previously mentioned, *iron* and *steel objects* are best subjected to a light preliminary tinning by boiling. However, instead of this preliminary tinning, they may first be electro-coppered and, after scratch-brushing the copper deposit, brought into the tin bath.

Process of tin-plating.—From what has been said, it will be evident that the execution of tin-plating is simple enough. After being freed from grease, and pickled, the objects are brought into the bath and plated with a weak current. For heavy deposits the objects are frequently taken from the bath and thoroughly brushed with a brass scratch-brush, not too hard, and moistened with dilute sulphuric acid (1 part acid of 66° Bé. to 25 water) and, after rinsing in water, are returned to the bath. If, with the use of too strong a current, the color of the deposit is observed to turn a dark dull gray, scratch-brushing must be repeated. When the tinning is finished the articles are brushed with a brass scratch-brush and decoction of soap-root, then dried in sawdust, and polished with fine whiting.

For tinning by contact and boiling, see special chapter, " Depositions by Contact."

2. DEPOSITION OF ZINC (Zn = 65.37 parts by weight).

Properties of Zinc. Zinc is a bluish-white metal, possessing high metallic luster. It melts at 776° F. At the ordinary temperature zinc is brittle, but it is malleable at between 212° and 300° F., and can be rolled into sheets. At 392° F. it again becomes brittle, and may be readily reduced to powder. The specific gravity of zinc varies from about 6.86 to 7.2. When strongly heated in the air, or in oxygen, it burns with a greenish-white flame, producing dense white fumes of the oxide

In moist air it becomes coated with a thin layer of basic carbonate, which protects the metal beneath from further oxidation. Pure zinc dissolves slowly in the ordinary mineral acids, but the commercial article containing foreign metals is rapidly attacked, hydrogen being evolved.

Since zinc is a very electro-positive metal and precipitates most of the heavy metals from their solutions, especially copper, silver, lead, antimony, arsenic, tin, cadmium, etc., this being the reason why in dissolving impure zinc, the admixed metals do not pass into solution so long as zinc in excess is present. Potash and soda lyes attack zinc, especially when it is in contact with a more electro-negative metal, hydrogen being evolved.

Zinc in contact with iron protects the latter from rust, and also prevents copper from dissolving when in contact with it.

Up to within a few years, objects were, as a rule, zincked by the so-called galvanizing process, and electro-plating with zinc was only used for parts which could not stand hot galvanizing, for instance, finer qualities of cast-iron objects or parts of machines, such as parts of centrifugals for sugar houses.

Further researches and practical experience in this line led to the application of zinc by the electrolytic cold process to other objects, such as sheet-iron and iron for constructive purposes. Thus, for instance, all the iron in lengths of up to 36 feet and 7 feet 6 inches projection used in the construction of the palm houses in the new botanical garden at Dahmen-Berlin were electro-plated with a thick deposit of zinc, and thorough investigations by the authorities have shown that, as regards protection from rust, the iron thus plated with zinc is at least not inferior to iron zincked by the hot process.

Electro-zincking has also proved of value for protecting the tubes of the Thornycroft boiler and several plants of this character are now in operation. Of great importance is also the electro-zincking of iron and steel network for corsets.

Electro-zincking is also of great advantage for small iron articles, for instance, screws and nuts, since the worms do not fill up with metal, as is the case in the hot process, and consequently do not require re-cutting.

While in the further manipulations, such as bending, punching, etc., of sheets, angle-iron, T-iron, pipes, etc., zincked bv the hot galvanizing process, the layer of zinc readily cracks off, the electro-deposit adheres very firmly, if the basis-metal has been properly cleansed ; and if the deposit is not of excessive thickness, which would be entirely useless, it cannot be detached by bending and beating. If it be further taken into consideration that in zincking by the hot galvanizing process much more zinc than is necessary for the protection against rust adheres to the objects, that the loss of zinc by the formation of hard zinc (an iron-zinc alloy) is considerable, and that it is quite expensive to keep the plant in repair, it will have to be admitted that in an economical aspect also, zincking with the assistance of the current presents many advantages.

Exhaustive comparative experiments regarding zincking by the hot process and by electro-deposition (cold process), have been published by Burgess.* These experiments relate to the duration of protection of the basis-metal, adhering power of the zinc coating, ductility and flexibility of the latter, uniformity of the coatings as regards strength and density, as well as resistance against mechanical wear. The results of the experiments were as follows:

The disadvantages of hot galvanizing are: Considerable consumption of heat and a material loss of zinc by oxidation on the surface of the fused zinc, by alloying with the cover of sal-ammoniac, and by the formation of hard zinc—a zinc-iron alloy which is formed at the expense of the walls of the iron tank. Burgess estimates the loss of zinc at 50 per cent. of the zinc used, and only a portion of it can be regained by a special process.

On the other hand, in electro-zincking no loss by heat is incurred, and hence such articles as steel-wire, steel-springs, etc., can be zincked, the treatment of which, at the temperature of the melted zinc, would be out of the question. Electro-zincking of certain kinds of work is now specified by the Governments of Great Britain and Germany, and the United States Government has installed at its various shipyards complete equipments for the purpose of treating articles by the electrical method.

The loss of zinc in electro-zincking is nominal and the wear of the vessels used is less than 10 per cent. per annum, while in the hot process it amounts to from 50 to 100 per cent.

By electro-deposition articles of any size may be zincked; the bath is always ready for use, and the thickness of the coating can be controlled and regulated, which in hot galvanizing

* Lead and Zinc News, 1904, viii, Nos. 8 to 10.

is possible only to a limited degree. Both processes possess the drawback of never yielding coatings of uniform thickness; the edges of hot-zincked pieces, especially of those which come last from the bath, are smeared over, *i. e.*, they are more heavily zincked than others, while, by reason of the current-density being greater on these portions, the edges of sheets zincked by electro-deposition are also more heavily zincked than parts in the center of the sheets.

The usual method of determining, by immersion in a 20 per cent. copper sulphate solution, the thickness of the coating of zinc obtained by hot-galvanizing, was found by Burgess to be quite unsuitable for judging the thickness and quality of electro-zincking. This test, known as Preece's test, consists in placing the galvanized iron in the copper solution for $\frac{1}{2}$ to 1 minute, and continuing the immersions until the test-piece shows a red deposit of copper, which is a true indication that the zinc has been penetrated and the iron exposed. In Germany it is as a rule required that hot-galvanizing must stand for at least 30 seconds constant immersion before the red copper color appears; so long as the coating of zinc is intact there is only a black coloration. On applying Preece's test to electro-zincked articles it was found that they would not stand as many immersions in the copper solution as coatings obtained by hot-galvanizing, but nevertheless they were more resisting to atmospheric influences. For testing the power of resistance of the coatings, Burgess therefore made use of dilute sulphuric acid, and found that an electro-deposited coating $\frac{1}{3}$ the weight of one produced by hot-galvanizing possesses the same power of resisting corrosion as the latter, and that for coatings of equal thickness the proportion of the resisting power is as 10 : 1. This superiority of electro-zincking has to be ascribed to the greater purity of the deposit effected by electro-deposition.

On measuring the adhesive power, Burgess ascertained quite different values from those of hot-galvanizing. With electro-deposited zinc the force required to tear the coatings from the basis-metal (iron) amounted on an average to 482 lbs. per square inch, and only to 280 lbs. for coatings obtained by hot-galvanizing. Hence the adhesive power of electro-deposited coatings is materially greater.

Attempts were made to ascertain the ductility and flexibility. of the coatings by rolling. However, no positive results were obtained, some deposits becoming thereby more or less cracked, while others remained intact. The flexibility of the deposit is without doubt affected by the reaction of the bath, and it has been observed that from very slightly acid electrolytes, with an electro-motive force of 1.5 ampères, deposits free from cracks were obtained while very brittle deposits were obtained from more strongly acid solution with the same current-density. Burgess's experiments in this respect only made sure of the fact that there is no material difference in the behavior of hot- and cold-galvanized sheets with coatings of equal thickness. In all cases the zinc, when subjected to rolling, showed a tendency to separate from the basis-metal; the zinc detached. from electro-zincked sheet, however, possessed greater strength. and was less brittle than that from hot-galvanized sheet.

On examining the detached coatings under the microscope it was further found that, contrary to the generally accepted opinion, the coating produced by hot-galvanizing was far more porous than that obtained by electro-deposition. An electro-deposit of less than 100 grammes zinc per square meter surface, was to be sure also porous, but with a thickness of 200 grammes zinc per square meter the pores had grown together.

The resistance against mechanical wear was apparently the same with the different deposits of equal thickness. Only in one case the electro-deposit proved of less value than the hot-galvanizing, namely, when electro-zincked sheets were subjected by heating and cooling to frequent and considerable changes in temperature, blisters were more frequently formed, and the zinc became detached to a greater extent than was the case with hot-galvanized sheet. This shows that electro-zincked sheets should not be used for heating pipes for high

temperatures. Blistering is less to be feared with temperatures not exceeding that of steam of 3 atmospheres.

Zinc baths. While flat articles can be readily coated with
a firmly-adhering layer of zinc of uniform thickness, the production of such a deposit upon large, shaped articles and profiled objects is attended with difficulties, because zinc baths do
not work quite well in the deeper portions. As will be seen
later on, these difficulties may be overcome, on the one hand,
by heating the baths and, on the other, by the use of anodes
with somewhat the same profile as the article to be zincked,
so that all portions of it are as nearly as possible at the same
distance from the anodes.

In plating articles with depressions, better results are obtained by depositing not pure zinc, but zinc in combination
with other metals. Of course, zinc must be largely in excess
if the deposit is to have the same effect as pure zinc in protecting the plated article from rust. By the addition of salts
of magnesium and aluminium to the zinc bath, Schaag, Dr.
Alexander and others have endeavored to deposit zinc in combination with these metals. While the possibility of depositing aluminium from aqueous solutions is doubtful, it is very
likely that in Schaag's, as well as in Dr. Alexander's patented
process neither the magnesium nor the aluminium is the
effective agent, but the tin or mercury salts which are also
added to the bath. But such additions are nothing new, since
deposits of zinc-tin alloys with or without mercury salts have
for many years been produced. The same object is attained
by an addition of tin and nickel to the zinc bath, and experiments have conclusively shown that deposits upon iron produced in such a bath protect the iron from rust as well as a
deposit produced in a bath of pure zinc, or in Dr. Alexander's
zinc baths, the patents for which are now expired. The good
effect of aluminium sulphate in zinc baths might solely be
due to the fact that the acidity of the baths is longer maintained.

In connection with the Alexander patent it may here be

stated that an important decision was rendered by Judge Cross of the Circuit Court of the United States for the District of New Jersey, in favor of the Hanson & Van Winkle Co., of Newark, N. J., and Chicago, Ill., and against the United States Electro-Galvanizing Co. of Brooklyn, owners of these patents. The decision ends as follows: "For the following, among other reasons, then, the defendant does not infringe; it does not make the alloyed coating of the patent, employs no basic salts, but rather makes and maintains throughout an acid bath ; does not use chloride of aluminium in its salts, does not use any organic substance with its salts or bath, or any equivalent thereof, and its bath is composed in part of different ingredients from the complainants, is prepared differently and under different conditions, and its ingredients, in so far as they are the same, appear in the different proportions. The bill of complaint will accordingly be dismissed, with costs."

Whatever may be said of the validity of the Alexander patents as against others, as against the salts and processes of the Hanson & Van Winkle Co., the patent is of no effect. The largest cold galvanizers of this country have been fitted up by the experts of this company.

While the protection against rust of deposits from alloy-baths is about the same as from pure zinc baths, the use of the latter, without the addition of foreign metals can nevertheless be recommended, since with a suitable composition of the bath and proper arrangement of the anodes perfect zinc deposits can in all cases be obtained.

During the last few years several investigations regarding the electrolysis of zinc have been made and numerous propositions have been advanced, but space will not permit to consider them here. According to O. Hildebrand very satisfactory results are obtained with the so-called regenerative process, a lead plate being used as anode instead of a zinc anode. Solution of zinc sulphate in water, to which is added a small quantity of sulphuric acid, is employed as electrolyte. By the use of a regenerative vat charged with zinc dust, the electrolyte is kept constantly in circulation and regenerated by coming in contact with the zinc dust in the regenerative vat.

By this method zinc coatings of good quality are obtained. The deposit is almost free from impurities, adheres firmly to the iron and is more uniform than that obtained by the hot galvanizing process. The zinc being used in the form of finely divided zinc dust the electrolyte comes in intimate contact with it and consequently is very quickly neutralized. Besides the drawbacks connected with the use of zinc anodes are avoided. As disadvantages may be mentioned the considerably greater electro-motive force required with the use of insoluble lead anodes and the consequently larger cost of current, and further, the operating expenses caused by the apparatus forcing the bath-liquor into the regenerating vats.

Dr. Szirmay and von Kollerich want to add solution of magnalium (aluminium-magnesium alloy) in sulphuric acid and dextrose to white vitriol (zinc sulphate) solution. In dissolving magnalium, aluminium sulphate and magnesium sulphate are formed. Neither aluminium or magnesium in watery solutions are reducible as metals by the current, as they oxidize at the moment of reduction, water being decomposed. The effect of the aluminium sulphate with its acid reaction is simply that the bath does not readily become alkaline, while the magnesium sulphate acts as a conducting salt, the separated magnesium-ions of it causing the secondary reduction of zinc from the sulphate solution. The addition of carbohydrates to which dextrose belongs, which became known through the English patent No. 12691, 1897, is claimed to prevent the formation of sponge, which, however, according to experiments made in Dr. Langbein's laboratory, is the case only to a limited extent. An addition of dextrose appears to have the further effect of the bath working better in the deeper portions and the deposits turning out less tufaceous. The deposits frequently come from the bath with a slight luster, this being especially the case when electrolysis is for some time continued after the addition of the dextrose.

According to Goldberg (German patent 151336) an addition of pyridine to zinc baths is claimed to effect a dense deposit of zinc of a beautiful white color and velvety appearance. On testing this process these claims were found to be correct, and furthermore such a bath works better in the depression.

Classen has patented the addition of glucosides, and claims to obtain thereby the deposition of lustrous zinc coatings.

The reason why in baths of the compositions formerly given, actually thick deposits without showing a spongy structure could not be obtained, is found in the fact that these baths contained too little metal and had an unsuitable, generally alkaline, reaction. Even when electrolysis has only been carried on for a short time, alkaline baths do not yield a coherent and purely metallic deposit of zinc, a basic zinc oxide being reduced together with the metallic zinc, which readily gives rise to the formation of sponge.

The formula for an alkaline zinc bath, namely, $3\frac{1}{2}$ ozs. of white vitriol dissolved in 1 quart of water, and adding potash lye until the precipitated zinc hydroxide is again dissolved, which was given in former editions of this work, yields quite fair results. This bath works best when, in place of potash or soda lye, ammonia is used for precipitating and dissolving the zinc hydroxide, and the bath contains a large excess of ammonia. Hence, in the above-mentioned formula, the potash lye should be replaced by ammonia and, in addition to the quantity required for the solution of the precipitate formed, enough of it should be used to impart to the bath a strong odor of ammonia. However, by reason of this odor of ammonia, the operation of such a bath becomes disagreeable, and even injurious to health.

In order to force the bath to work better in the deeper portions, mercury salts in the form of potassium-mercuric cyanide may be added to alkaline baths. It must, moreover, be borne in mind that an addition of mercury is of advantage because the anodes are thereby superficially amalgamated and kept in a purely metallic state. In alkaline zinc baths particularly, an abundant coat of zinc hydroxide is formed upon the anodes, and because this coat does not dissolve to the same extent as it is formed, it has to be frequently removed by mechanical means.

Below formulas for zinc baths which have stood the test for a long time are given.

I. Chemically pure crystallized zinc sulphate 44 lbs., pure crystallized sodium sulphate 8.8 lbs., chemically pure zinc chloride 2.2 lbs., crystallized boric acid 1.1 lbs., dissolved in water to a 100-quart bath.

Electro-motive force at 10 cm. electrode-distance and at 64.4° F., 1.1, 1.5, 1.8, 2.2, 2.4, 2.7, 3.7 volts.

Current-density at 64.4° F., 0.55, 0.75, 0.95, 1.15, 1.25, 1.45, 1.9 ampères.

Electro-motive force at 10 cm. electrode-distance and at 113° F., 0.9, 1.05, 1.25, 1.40, 1.8, 2.0, 2.3, 3.5 volts.

Current-density at 113° F., 0.7, 0.8, 1.0, 1.1, 1 4, 1 55, 1.8, 2.75 ampères.

This bath, as well as others of similar composition, will stand considerably higher current-densities if provision is made for vigorous agitation of the electrolyte. If agitation is to be avoided, an increase of the content of zinc salt and boric acid is of advantage.

To prepare the bath, dissolve the zinc sulphate, the zinc chloride and sodium sulphate (Glauber's salt) in luke-warm water. Heat a portion of this fluid to about 194° F., dissolve in it the boric acid, and mix it with the other solution. An addition of 0.8 to 1 oz. of dextrose per quart is recommended.

For the production of a good deposit of zinc it is of importance to use zinc salts free from other metals, it having been shown that a content of foreign metals, especially iron, causes disturbances.

The reaction of the bath should be kept slightly acid, so that blue litmus paper is intensely reddened, but congo paper is not perceptibly blued. The bath gradually loses its acid reaction and does not work as well, the deposit becoming

darker instead of pale gray, and inclining towards the formation of sponge.   It should then be acidulated by the addition
of pure dilute sulphuric acid.   For flat objects (sheets, etc.)
the bath may be used cold, but for profiled objects, such as
angle-iron, beams, etc., it is advisable to heat it between 104°
and 122° F.

II. Crystallized sodium citrate 5.5 lbs., chemically pure
zinc chloride 8.8 lbs., pure crystallized ammonium chloride
6.6 lbs.   Dissolve with water to a 100-quart ,bath.

Electro-motive force at 10 cm. electrode-distance and at 64.4°
F., 0.8, 1.0, 1.5, 1.8, 2.2, 3.4 volts.

Current-density at 64.4° F., 0.7, 0.9, 1.4, 1.7, 1.9, 3.0 ampères.

Electro-motive force at 10 cm. electrode-distance and 113° F.,
0.8, 1.0, 1.5, 1.75, 2.5, 3.2 volts.

Current-density at 113° F., 1.0, 1.25, 1.9, 2.3, 3.2, 4.3 ampères.

The bath is prepared by dissolving the constituents in the
water, which should not be too cold ; best luke-warm.   What
has been said under formula I in reference to the reaction
also applies to this bath.

III. Wm. Schneider * recommends a bath of the following.
composition : Water 1 gallon, sulphate of zinc 2 lbs., sulphate
of aluminium 2 ozs., glycerine ½ oz.   *Electro-motive force:* 10
ampères to 1 square foot of surface.   The work must be agitated while being coated.   The use of pure zinc anodes is
imperative if satisfactory results are to ·be obtained, and the
iron to be plated must be perfectly clean.   If this solution
is carefully attended to it will plate a good light gray, and by
the addition of a few of the various reagents of which there
are several, such as glue and dextrine, on the market, a very
bright deposit of zinc can be obtained.

Zinc anodes.   Treatment of zinc baths.   For anodes it is best
to use very pure rolled-zinc sheets 0.11 to 0.19 inch or more

* Metal Industry, No. 9, 1909.

in thickness. Strips of zinc riveted to the anodes with zinc rivets serve for suspending the anodes to the anode rods. For zincking sheet-iron, wires, etc., on a large scale, cast zinc anodes may be preferred on account of being cheaper. It must, however, be borne in mind that cast anodes readily crumble, especially when they have frequently to be cleansed mechanically by scraping and scratch-brushing for the removal of the basic zinc salts forming on them. The loss of zinc caused by the formation of basic zinc salt and by crumbling is considerable, and according to Cowper-Coles may be as large as 30 per cent. of the entire consumption of zinc. This estimate, however, appears to be excessive; to be sure the formation of a coat on the anodes as well as the crumbling of the anodes is disagreeable, but it cannot be considered a direct loss since the zinc salt as well as the detached crumbs of metal can, by dissolving in sulphuric acid, be converted into zinc sulphate, and the latter be used for strengthening the bath. The surface of the anodes should be as large as possible. Rolled anodes also become readily coated with a layer of basic zinc salt, and it is advisable from time to time to remove this layer by scratch-brushing. The coating thus removed may be dissolved in dilute sulphuric acid and added to the bath as neutral zinc solution. As previously mentioned, the zinc baths should show a perceptibly acid reaction in order to avoid as much as possible the formation of sponge, and therefore the reaction should at short intervals be tested and, if necessary, corrected by the addition of dilute sulphuric acid.

The zinc anodes should be removed from the bath when the latter is not in operation, otherwise the free acid of the bath would be neutralized by the solution of zinc and it would have to be again acidified.

Although the zinc baths, without exception, work well at a temperature of 64.4° to 68° F. upon flat articles, it is recommended, in view of the slight electro-motive force required, to keep them somewhat warmer.

For zincking strongly-profiled objects it is advisable to heat the baths to between 104° and 113° F., since at a higher temperature the deposit penetrates better into the deeper portions. Anodes with profiles similar to those of the objects are used. As shown by the current conditions given with the formulas for the baths, a fixed current-density is not obligatory in electro-zincking. For the bath, according to formula I, 1.25 to 1.5 ampères may be designated as the lowest rational current-density, at which 5.29 to 6.34 ozs. of zinc per square meter (10.76 square feet) are in one hour deposited. With heated and agitated baths, the maximum current-density may be given as about 3 ampères, with which 12.91 ozs. of zinc per square meter are in one hour deposited. However, in certain cases, this current-density may be exceeded. For baths, according to formula II, it is best to use a slighter current-density. Should it, however, be necessary to work with higher current-densities, provision has to be made for a sufficiently acid reaction and thorough agitation in order to avoid the formation of sponge. In zincking, agitation of the baths is of special value, and, if possible, should never be omitted.

Tanks for zinc baths.—For smaller baths it is best to use stoneware vessels, while for larger baths, tanks of pitch-pine, or still better, of wood lined with lead, may be employed. Zinc salt solutions gradually impair the swelling capacity of wood, and even pitch-pine tanks, most carefully built, commence in the course of time to leak. For this reason tanks of wood lined with lead, or of sheet-iron, deserve the preference. Brick tanks lined with cement may also be used, provided several coats of thinly-fluid asphalt lacquer be applied to the cement lining to prevent the latter from being attacked by the acid baths.

Heating the zinc baths is best effected by steam introduced through a hard lead (alloy of antimony and lead) coil on the bottom of the tank.

Execution of zincking.—Since the principal object of electro-zincking is to prevent rusting, embellishing the metallic objects being only in very rare cases effected, the mechanical

30

refinement of the surface by grinding is as a rule omitted, a purely metallic surface free from scale being produced in a cheaper manner.

This is done by pickling, scratch-brushing, scrubbing with sand in a drum, or by the sand-blast. The latter deserves the preference for large quantities of small articles, as well as for objects with not too large surfaces. For freeing large surfaces of sheet from scale, the use of the sand-blast is, however, too expensive on account of the great consumption of power, and, besides, takes too much time.

In electro-zincking particularly, the mode of operating depends entirely on the nature and form of the objects, and it is, therefore, advisable to discuss separately the various manipulations required for certain objects.

Zincking sheet-iron. When the sheets have been freed from grease by means of hot alkaline lyes or lime paste, they are pickled in dilute sulphuric or hydrochloric acid, and the loosened scale is removed by scouring with fire brick and sand. The use of a pickle of hydrochloric acid 2 parts, sulphuric acid of 66° Bé., 1 part, and water 17 parts is also of advantage. To what extent the electrolytic method of pickling, previously referred to, can be used to advantage for this purpose, has thus far not been practically determined. By the use of a sand-blast the cleanest and most complete results are obtained, but the expense for power has to be taken into consideration.

As in all other electro-plating processes, a purely metallic surface free from scale is an absolutely necessary condition for a well-adhering deposit of zinc. Portions of the sheets coated with scales, would come out zincked, but in the further manipulation of the sheets, the layer of zinc becomes detached, and this must be avoided.

The pickled and scoured sheets, generally in lengths of 6 feet and 1 foot wide, are secured to binding screws and brought into the zinc bath, that given under formula I being especially suitable for the purpose. Heating the bath to between 104°

and 113° F., and vigorous agitation by blowing in air, or by means of a mechanical contrivance, allows of working with a current-density of $2\frac{1}{2}$ ampères, and, if necessary, more, at which a sufficiently heavy deposit to protect the objects from rust is in 20 to 25 minutes obtained.

In working on a large scale, it is advisable to couple the baths in series and to zinc in each bath 3 sheets, each 2 x 1 meters, with a total surface of 12 square meters per bath. Hence, for zincking the sheets on both sides and working with a current-density of 2.5 ampères per square decimeter, there will be required four baths in series, each with 12 square meters of surface, a dynamo of 12 x 250 = 3000 ampères, and the electro-motive force should be 12 volts, $2\frac{1}{2}$ to 3 volts being required per bath. With such a plant working for 10 hours, 340 to 360 sheets, each 2 x 1 meters, can be zincked on both sides so as to protect them from rust, and by working day and night, 820 to 850 sheets.

When zincking is finished, the sheets are rinsed in water, then immersed in boiling water until they have acquired the temperature of the latter, and finally set up free, or hung up to dry. The hot sheets then dry in a few minutes.

The zincked sheets show a pale-gray, mat, velvety appearance, and are generally used in this state. If the zinc deposit is to be lustrous, the sheets are scratch-brushed, best dry, with steel scratch-brushes.

Zincking of pipes. The pipes are freed from scale either by means of the sand blast or by pickling and scouring. To protect the screw threads from the pickle, they are coated with tallow which, however, previous to zincking, has to be removed with hot soda lye or rubbing with benzine, and care must be had thoroughly to free them from grease with lime paste.

The pipes, best four or six pieces one above the other, are placed upon a frame which also serves for conducting the current, and, if they are of considerable diameter, it is advisable to turn them 90° when half the time for zincking has expired, in order to obtain a uniform deposit.

While in zincking straight flat sheets, the distance of the anodes from the cathodes need only be 1.96 to 2.35 inches, the distance of the zinc anodes from the pipes should be the greater, the larger the diameter of the latter is. Pipes of very large diameter are best suspended alongside each other, instead of one above the other, a row of anodes of zinc sheet suitably bent being arranged between every two pipes. Frequent turning of the pipes is required, and uniform zincking is promoted by heating the bath and by vigorous agitation. The further manipulation of the zincked pipes is similar to that given for sheets.

Zincking the insides of the pipes is a more difficult operation. To commence with, it is as a rule a difficult task to find out whether pickling and scratch-brushing has been sufficiently done and a pure metallic surface have everywhere been produced. Special directions for inside zincking depend partly on the diameter, and it can only be said in general that it is best to zinc the pipes while in a vertical or half-lying position and to provide for a constant renewal of the electrolyte in the interior.

Zincking of wrought-iron girders, T-*iron,* U-*iron,* L-*iron, etc.* The cheapest plan of freeing the objects from scale is by pickling in dilute sulphuric or hydrochloric acid and vigorous scrubbing with sand, or scratch-brushing. The use of a transportable sand blast ·is also very suitable, but this method is more expensive than the former.

For the uniform zincking of such profiled objects, the use of flat zinc anodes is not practicable, far more zinc being deposited upon the edges and surfaces next to the anodes than upon the depressed portions. Hence, in addition to heating and vigorously agitating the bath, recourse must be had to profiled anodes corresponding to the shape of the object to be zincked, the object of such profiled anodes being solely to bring all portions of the objects at as nearly an equal distance as possible from the anodes.

The profiled anodes may be made by bending zinc sheets

into the proper shape, or what is better, by riveting or screwing square cast-zinc bars to the zinc sheets, this being of advantage, for instance, in zincking girders.   Figs. 132 and 133 show the arrangement of the anodes, and require no further explanation.

In zincking profiled objects it is of advantage to add to the bath prepared according to formula I, 0.8 oz. of dextrose per quart, the bath working better in the deeper portions with

FIG. 132.   FIG. 133.

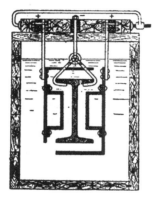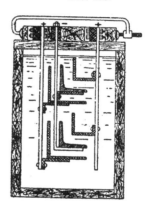

such an addition than without it.   Of still greater advantage is the pyridine-zinc bath according to Goldberg.

Zincking of wire, steel tapes, cords, etc.—Under this heading will chiefly be considered iron and steel wire which is to be protected from rust by a deposit of zinc.   As previously mentioned when speaking of nickeling wire, the latter has to be uncoiled and passed at a suitable rate of speed through the pickling solutions and the zinc bath.

Bright-drawn iron and steel wire, requiring but little preparatory work, is most suitable for zincking.   It suffices for coils of such wire, when free from rust, to push them upon a shaft of corresponding diameter, and bring the whole into a tank with hot soda lye, which is furnished with bearings for the shaft. From this tank the wire, freed from grease by the hot lye, passes through a few felt rolls or cloth cheeks supplied with thin lime paste, an additional freeing from grease being thus, for the sake of greater security, effected. The wire is then brought under a rose for the removal by water of adhering lye and lime, then slides over a metallic roll which is in contact with the negative pole of the source of current, and passes into the zinc bath. In the latter zinc anodes are arranged below and above the lengths of wires running parallel to each other. Such zinc baths for wire zincking are from 20 to 26 feet long. The average velocity with which wire 0.039 inch in diameter passes through the bath is about 20 feet per minute. For wire of less diameter the velocity may be increased to 59 feet, while for wire of greater diameter it has to be correspondingly decreased.

When the wire comes from the bath it is conducted through a tank containing boiling water, and is then reeled up. The contrivances for reeling up the wire are best driven by an electro-motor with the use of a starting resistance for regulating the turns, it being thus possible to choose and change at will the velocity of the passage of the wire in the bath.

Zincking of wires produced by rolling is not quite so simple. Such wire, which is readily recognized by its black appearance, is coated with a scale, which adheres very firmly by reason of the rolling process. Previous to zincking, such wire has to be carefully freed from scale in order to obtain a purely metallic surface. This may be accomplished in various ways, the method selected depending on the nature and properties of the material to be manipulated.

Experiments in cleansing such wire by means of the sand blast did not yield satisfactory results, the process being too slow notwithstanding the use of suitable annular blast-pipes. Hence recourse will have to be had to pickling in acids, but many kinds of wire and steel tapes stand pickling only for a very short time, as they readily become brittle. Wire which does not show this drawback is pickled until the scale is partially dissolved, or at least very much loosened. After rinsing in water, it is immersed in boiling water so that it will dry rapidly and then, for the removal of the loosened scale, passed by means of a suitable contrivance through the scratch-brushing machine. Wire which will not bear pickling at all, or only for a very short time, has to be brightened by mechanical means, either by the drawing-plate, or by conducting it over revolving, hard grindstones or emery wheels provided with insertions for holding it.

Zincking of screws, nuts, rivets, nails, tacks, etc. Such small objects are freed from grease either by means of the sand blast or in tumbling barrels with the use of wet, sharp sand. When the latter process is employed, the objects have of course to be immediately zincked to prevent rusting.

For zincking quantities of such small objects, one of the mechanical plating contrivances referred to under " Depositions of Nickel " is very suitable. When zincking is effected in baskets the position of the objects has to be frequently changed by stirring in order to insure a uniform deposit.

The bath prepared according to formula II, when heated, is very well adapted for zincking small objects in large quantities. With the use of a mechanical plating apparatus, it is possible in consequence of the constant agitation, to work with high current-densities and to deposit a correspondingly large quantity of zinc in a comparatively short time.

The further manipulation of the zincked small objects consists in washing them in baskets and drying them quickly by immersion in boiling water and shaking with heated clean sawdust, though the latter operation may be omitted.

For zincking by contact, see special chapter " Depositions by Contact."

Zinc alloys.—The production of the principal zinc alloy, brass, by the electric method, having already been mentioned, and also that of a zinc-nickel-copper alloy (German silver), it remains to give an alloy of zinc with tin, or of zinc, tin and nickel, which can be produced by the same means.

A suitable bath for depositing this alloy consists of: Chloride of zinc 6¾ drachms, crystallized stannous chloride 9 drachms, pulverized tartar 9 drachms, pyrophosphate of soda 2¾ drachms, water 1 quart. Dissolve the salt at a boiling heat, and filter the cold solution, when it is ready for use. For anodes, cast plates of equal parts of tin and zinc are used.

These deposits have no special advantages, but, on the other hand, a deposit containing zinc in large excess has the same effect of protecting iron from rust as a deposit of pure zinc.

By preparing a bath which contains as conducting salt sodium citrate, and ammonium chloride and the chlorides of the metals in the proportion of 4 zinc chloride to 1 tin chloride, a deposit is obtained, which not only is a perfect protection against rust, but also enters far better into depressions than pure zinc. By adding to the bath a small quantity of chloride of mercury, or of nickel, alloys of zinc, tin, mercury, or of zinc, tin and nickel are formed, which are distinguished from pure zinc deposits by a finer structure.

2. DEPOSITION OF LEAD (Pb = 207.10 parts by weight).

The properties of lead only interest us in so far as it is less attacked by most mineral acids than any other metal, and against the action of such agents. For decorative purposes electro-deposits of lead are scarcely used, and those as a protection against chemical influences cannot be produced of sufficient thickness for that purpose.

Lead baths.—I. Dissolve, by continued boiling, caustic potash 1.75 ozs. and finely pulverized litharge 0.17 oz. in 1 quart of water.

II. According to Watt, the following solution is used. Acetate of lead 0.17 oz., acetic acid 0.17 oz., water 1 quart.

The bath prepared according to formula I deserves the preference.

Lead baths require anodes of sheet-lead or cast-lead plates, a weak current and, in order to produce a dense deposit of some thickness, the objects have to be frequently scratch-brushed. Iron is best previously coppered. Peroxide of lead. is separated on the anodes, and they have to be frequently cleansed with a scratch-brush. The formation of peroxide of lead on the anodes is utilized for the production of the so-called Nobili's rings (electrochromy).

Metallo-chromes (*Nobili's rings, iridescent colors, electro chromy*). The reduction of .peroxide of lead upon the anodes or upon objects suspended as anodes, produces superb effects of colors. For the production of such colors, a bath is prepared by boiling for half an hour $3\frac{1}{2}$ ozs. of caustic potash, 14 drachms of litharge, and 1 quart of water. The operation is as follows : Suspend the articles, carefully freed from grease and pickled, to the anode-rods, and with a weak current introduce in the lead solution a thin platinum wire connected with the object-rod by flexible copper wire, without, however, touching the article. The latter will successively become colored with various shades—yellow, green, red, violet and blue. By the continued action of the current, these colors pass into a discolored brown, which also appears in the beginning if the current be too strong, or if the platinum wire be immersed too deep. Such unsuccessful coloration has to be removed by rapidly dipping in nitric acid, and, after rinsing in water, suspending the article in the bath. For coloring not too large surfaces, a medium-sized Bunsen cell is, as a rule, sufficient, if the platinum wire be immersed about $\frac{3}{4}$ inch.

Colors of all possible beautiful contrasts may be obtained by perpendicularly placing between the objects to be colored and the platinum wire a piece of stout parchment paper, or providing the latter with many holes or radial segments.

Another process of producing these effects of colors is as follows : Prepare a concentrated solution of acetate of lead (sugar of lead), and after filtering, pour it into a shallow porcelain dish. Then immerse a plate of polished steel in the solution, and allow it to rest upon the bottom of the dish. Now connect a small sheet of disc copper with the wire proceeding from the zinc element of a constant battery of two or three cells, the wire connected with the copper element being placed in contact with the steel plate. If now the copper disc be brought as close to the steel plate as possible without touching it, in a few moments a series of beautiful prismatic colorations will appear upon the steel surface, when the plate should be removed and rinsed in clean water. These colorations are films of lead in the form of peroxide, and the varied hues are due to the difference in thickness of the precipitated peroxide of lead, the light being reflected through them from the polished metallic surface beneath. By reflected light every prismatic color is visible, and by transmitted light a series of prismatic colors complementary to the first colors will appear occupying the place of the former series. The colors are seen to the greatest perfection by placing the plate before a window with the back to the light, and holding a piece of white paper at such an angle as to be reflected upon its surface. The colorations are not of a fugitive character, but will bear a considerable amount of friction without being removed. In proof of the lead oxide being deposited in films or layers, it may be stated that if the deposit be allowed to proceed a few seconds beyond the time when its greatest beauties are exhibited, the coloration will be less marked, and become more or less red, green or brown. If well rubbed, when dry, with the finger or fleshy part of the hand, a rich blue-colored film will be laid bare by the removal of the delicate film above it.

The plan recommended by Mr. Gassiot to obtain the metallochromes is to place over the steel plate a piece of cardboard or parchment paper cut into some regular design, and over this a rim of wood, the copper disc being placed above this. Very beautiful effects are obtained when a piece of fine copper wire is turned up in the form of a ring, star, cross or other pattern, and connected with the positive electrode, this being in fact one of the simplest and readiest methods of obtaining the colorations upon the polished metal. Metallochromy is extensively employed in Nüremberg to ornament metallic toys. It has been adopted in France for coloring bells, and in

Switzerland for coloring the hands and dials of watches. In using the lead solutions to produce metallochromes, it must be remembered that metallic lead becomes deposited upon the cathode, consequently the solutions in time become exhausted, and must therefore be renewed by the addition of the lead salt.

For the preparation of iridescent sheets, i. e., nickeled zinc sheet coated with peroxide of lead, a sheet of lead of the same size as the sheet to be made iridescent is used as object, and a current of about $2\frac{1}{2}$ volts is employed. The slime formed in the bath must from time to time be removed, as otherwise the tones of color will not turn out pure.

4. DEPOSITION OF IRON (Fe = 55.85 parts by weight) (STEELING).

The principal practical use of the electro-deposition of iron is to coat printing plates of softer metal to increase their wearing qualities. We are indebted to Böttger for calling attention to the employment of iron deposits, but notwithstanding the efforts of many scientific and practical men to improve the process, the expectation entirely to replace copper galvanoplasty for clichés by iron-galvanoplasty has not been fulfilled.

Only such baths as are suitable for steeling will here be given. Solutions for the production of thick iron deposits, and the conditions under which they can be obtained, will be referred to later on under "Galvanoplasty in Steel."

Iron (steel) baths. I. According to Varrentrapp: Pure green vitriol $4\frac{1}{2}$ ozs., ammonium chloride $3\frac{1}{2}$ ozs., water 1 quart.

Electro-motive force at 10 cm. electrode-distance, 1.0 volt.

Current-density, 0.2 ampère.

Boil the water for $\frac{1}{2}$ hour to expel all air, and, after cooling, add the green vitriol and ammonium chloride. By the action of the air, and the oxygen appearing on the anodes, this bath is readily decomposed, insoluble basic sulphate of iron being separated as a delicate powder, which has to be frequently removed from the fluid by filtering. To decrease decomposition, the double sulphate of iron and ammonium, which can be more readily obtained pure and free from oxide, may be used.

II. Ammonium chloride 3½ ozs., water 1 quart.

Electro-motive force at 10 cm. electrode distance, 1.0 volt.

Current-density, 0.2 ampère.

This neutral solution of ammonium chloride may be made into an iron bath by hanging in it iron sheets as anodes, suspending an iron or copper plate as cathode, and allowing the current to circulate until a regular separation of iron is attained, which is generally the case in 5 to 6 hours. Although a separation of hydrated oxide of iron also takes place in this bath, it does so in a less degree than in that prepared according to formula I. For the production of not too heavy a deposit of iron, some operators claim to have obtained the best results with this bath.

According to Böttger, the following bath serves for steeling :

IIa. Potassium ferrocyanide (yellow prussiate of potash) 0.35 oz., Rochelle salt 0.7 oz., distilled water 200 cubic centimeters. To this solution is added a solution of 1.69 drachms of persulphate of iron in 50 cubic centimeters of water, whereby a moderate separation of Berlin blue takes place. Then add, drop by drop, whilst stirring constantly, solution of caustic soda until the blue precipitate has disappeared. The clear, slightly yellowish solution thus obtained can be used directly for steeling.

A heavy and very hard deposit of iron is obtained in a bath of the following composition.

III. Ammonio-ferrous sulphate 1¼ ozs., crystallized citric acid 0.88 oz., water 1 quart ; sufficient ammonia for neutral or slightly acid reaction.

Elecro-motive force at 10 cm. electrode-distance, 2.0 volts.

Current-density, 0.3 ampère.

Management of iron baths. As previously mentioned, the insoluble precipitate from time to time formed in the bath has to be removed by filtration. This precipitate is, however, very delicate, and when stirred up might settle upon the objects and prevent the adherence of the deposit. It is, therefore, advisable to use for steel baths, tanks of much greater depth than corresponds to the height of the objects, whereby the stirring-up of the sediment in suspending the objects is best avoided.

With the use of steel anodes the baths may become readily acid. This can be avoided by suspending a few small linen bags filled with carbonate of magnesia in the bath. On the other hand, anodes of soft iron make the electrolyte alkaline, and when such anodes are employed, the reaction of the bath must from time to time be tested and the neutral reaction be restored by the addition of very dilute sulphuric acid, or better, citric acid.

Deposits produced in an iron bath which has become alkaline show slight hardness, form very rapidly with a mat appearance and have a tendency to peel off.

Execution of steeling.—The cleansed and pickled objects are placed in the baths according to formulæ I and II, with a current of 1.5 to 2 volts, and the anodes at a distance of 4 to $4\frac{3}{4}$ inches, after which the current is reduced to 1 volt. To produce iron deposits of any kind of thickness, the escape of the hydrogen bubbles which settle on the objects must be promoted by frequent blows with the finger upon the object-rod. When steeling is finished, the articles are thoroughly rinsed, then plunged into very hot water, and, after drying in sawdust, placed for several hours in a drying chamber heated to about 212° F., to expel all moisture from the pores.

Steeling of printing plates has the advantage over nickeling, that when the plates are worn they can be rapidly freed from the deposit by dilute sulphuric acid or very dilute nitric acid, and resteeled. It has been ascertained by experiments that the capability of resistance of steeled plates is less than that of nickeled plates, 200,000 impressions having been made with the latter without any perceptible wear.

For steeling printing plates a bath prepared according to formula II or III is very suitable.

CHAPTER XII.

1. DEPOSITION OF ANTIMONY (Sb = 120.2 PARTS BY WEIGHT).

Properties of antimony.—Electro-deposited antimony possesses a gray luster, while native, fused antimony shows a silver-white color. Antimony is hard, very brittle, and may easily be reduced to powder in a mortar. It melts at 842° F., and at a strong red heat takes fire and burns with a white flame, forming the trioxide. Its specific gravity is 6.8. It is permanent in the air at ordinary temperatures. Cold, dilute, or concentrated sulphuric acid has no effect upon antimony, but the hot concentrated acid forms sulphide of antimony. By nitric acid the metal is more or less energetically oxidized, according to the strength and temperature of the acid.

Antimony baths.—Electro-depositions of antimony are but seldom made use of in the industries, though they are very suitable for decorative contrasts. This is no doubt due to the fact that a thoroughly reliable bath yielding deposits without the appearance of drawbacks during the operation is thus far not known.

For the special study of electro-depositions of antimony we are indebted to Böttger and Gore, the latter having discovered the explosive power of deposits of antimony deposited from a solution containing chloride or hydrochloric acid.

According to Gore, a bath consisting of tartar emetic 3 ozs., tartaric acid 3 ozs., hydrochloric acid 4¼ ozs., and water, 1 quart, yields a gray, crystalline deposit of antimony. This bath requires a current of about 3 volts. The deposit possesses the property of exploding when scratched or struck with a hard object. The explosion is attended by a cloud of white

(478)

vapor, and sometimes by a flash of light, considerable heat being always evolved. This explosibility is due to a content, of antimony chloride. Böttger found 3 to 5 per cent. of chloride of antimony in the deposit, and Gore 6 per cent. A similar explosive deposit is obtained by electrolyzing a simple solution of chloride of antimony in hydrochloric acid (liquid, butter of antimony, *liquor stibii chlorati*) with the current.

A lustrous, non-explosive deposit of antimony is obtained by boiling 4.4 ozs. of carbonate of potash, 2.11 ozs. of pulverized antimony sulphide, and 1 quart of water for 1 hour, replacing the water lost by evaporation, and filtering. Use the bath boiling hot, employing cast antimony plates or platinum sheets as anodes.

An antimony bath which yields good results is composed as follows:

Schlippe's salt 1¾ ozs., water 1 quart. Dissolve the salt in the water. Electro-motive force required, 4 volts. An unpleasant feature of this bath is that during electrolyzing sulphuretted hydrogen escapes, which limits its application.

2. DEPOSITION OF ARSENIC (AS 74.96 PARTS BY WEIGHT).

Properties of arsenic.—Arsenic has a gray-white color, a strong metallic luster, is very brittle, and evaporates at a red heat. In dry air arsenic retains its luster, but soon turns dark in moist air. It is scarcely attacked by dilute hydrochloric and sulphuric acids, while concentrated sulphuric acid, as well as nitric acid, oxidizes it to arsenious acid. If caustic alkalies are fused together with arsenic, a portion of the latter is converted into alkaline arsenate.

Arsenic baths.—Arsenic solutions are extensively used in the plating room for decorative purposes in order to produce blue-gray to black tones of a certain warmth, which are very effective in combination with bright copper, brass, etc.

For coloring all kinds of metals blue-gray the following solutions are very suitable:

I. Pulverized white arsenic 1¾ ozs., crystallized pyrophosphate of soda 0.7 oz., 98 per cent. potassium cyanide 1¾ ozs., water 1 quart.

Dissolve the pyrophosphate of soda, and the potassium cyanide in the cold water, and after adding, whilst stirring, the arsenic acid, heat until the latter is dissolved. In heating, fumes containing prussic acid escape, the inhalation of which must be carefully avoided. The bath is used warm, and requires a vigorous current of at least 4 volts, so that, at the least, 3 Bunsen cells have to be coupled for electro-motive force. After suspending the objects they are first colored black-blue, the color passing with the increasing thickness of the deposit into pale blue, and finally into the true arsenic gray. Platinum sheets or carbon plates are to be used as anodes.

In place of the bath prepared according to formula I, a solution of the following composition may be used :

II. Sodium arsenate 1¾ ozs., 98 per cent. potassium cyanide 0.8 oz., water 1 quart. Boil the solution for half an hour, then filter and use it at a temperature of at least 167° to 176° F with a strong current. It yields a good deposit.

Large baths, to be used cold, must be more concentrated, and require a stronger current than hot baths.

When the baths begin to work irregularly and sluggishly, they have to be replaced by fresh solutions.

The same rules as for other electro-plating processes are to be observed in depositing arsenic and antimony.

However, attention may here be called to one feature which is frequently the cause of defective deposits. When, for instance, mountings of zinc, such as are used for book covers, jewel boxes, etc., are to be provided with a deposit of copper and arsenic, and hence are to show two colors, it is necessary to first copper them. After polishing and cleaning the coppered mountings, the places which are not to receive the blue-gray deposit of arsenic are coated with stopping-off varnish. When articles thus treated, after being again freed from grease and pickled, are brought into the arsenic bath, they frequently show ugly stains the size of a pin-head. This feature, however, does not appear when the articles before being brought into the bath are drawn through water acidulated with a small quantity of nitric acid (about $\frac{1}{2}$ oz. of nitric acid to 1 quart of water), and thoroughly rinsed in clean water.

Mr. Emmanuel Blassett, Jr.,* gives the following solutions for coloring articles black :

III. White arsenic 1 lb., potassium cyanide 2 lbs., water 6 gallons, ammonium carbonate 10 ozs.

Dissolve the white arsenic and potassium cyanide in 5 gallons of water, and the ammonium carbonate separately in 1 gallon of water, and mix the two solutions. This solution is worked cold. Steel or brass anodes may be used and a current with the tension of 1 volt is sufficient. Without the ammonium carbonate the deposit is not so black and is inclined to be steel-gray in color. It is a difficult operation to color small light pieces in this bath, possibly due to the poor conductivity of arsenic solutions. On very light articles the bath gives an iridescent color, or blue-black at the most. For this reason if a good black color is desired, the following dip should be used for small articles : White arsenic 2 ozs., potassium cyanide 5 ozs., water 1 gallon.

This dip is used hot and without the current. The solution is made up in an enamel or agate ware vessel, and brought to a boiling point by means of a gas stove. Work to be colored is fastened to wires or immersed in the solution by means of a plating or dip basket. Some platers make up a new dip every day or two, but by careful management, such as replacing the water lost by evaporation and making small additions of arsenic and cyanide, the dip may be made to last a long time.

Another arsenic bath is made up as follows :

* Metal Industry, March, 1913.

IV. White arsenic 3 lbs., potassium cyanide 4 ozs., commercial caustic soda 1½ lbs., water 5 gallons.

The ingredients are boiled together. Soft steel or brass anodes are used, and a current of about 1 volt is required.

A favorite bath with some platers is composed of:

V. White arsenic 2 lbs., copper carbonate 4 ozs., potassium cyanide 2½ lbs., water 5 gallons.

In addition to the solutions described, which are all alkaline, there are several acid arsenic baths which are frequently employed. The most simple is composed as follows:

VI. Muriatic acid 5 gallons, white arsenic 2 lbs.

Carbon anodes are often used in operating this solution, and for that reason it is often spoken of as the " carbon solution." Acid solutions are indispensable on work where a portion of the surface is stopped off with an asphalt paint or varnish. Strong alkaline solutions ordinarily will remove paint or varnish and spoil the finish, unless unusual care is exercised.

A well-known acid bath is composed as follows:

VII. Muriatic acid 1 gallon, iron filings 4 ozs., white arsenic 4 ozs.

This formula is very suitable for producing an oxidized brass effect. The coating is a little heavier, as iron is deposited simultaneously with the arsenic. The deposit is soft, and may be easily relieved for a cut through finish. In all acid baths it is best to use either carbon or brass anodes. If iron or steel anodes are used, they are attacked by the acid, producing a very concentrated solution. Under such conditions particles of undissolved iron may interfere with the operation.

Some platers prefer to use iron sulphate instead of iron filings and make up their solution as follows:

VII. Muriatic acid 5 gallons, white arsenic 1 lb., iron sulphate 1 lb.

Electro-depositions of chromium, tungsten, cadmium and bismuth have thus far not become of any practical importance, and their discussion may, therefore, with good reason, be omitted.   As regards silver-cadmium deposits the reader is referred to *Arcas-silvering* under "Deposition of Silver."

3. DEPOSITION OF ALUMINIUM (Al = 27.1 PARTS BY WEIGHT).

There is actually no reason or authority for the heading of this section, but it has been introduced because inquiries are frequently received as to whether baths for the deposition of aluminium can be furnished.

A number of receipts for the preparation of aluminium baths have been published, but in testing them nothing further could be obtained than the confirmation of the fact that the deposition of aluminium from aqueous solutions of its salts by the current upon the cathode has thus far not been feasible.

Reports have from time to time appeared in newspapers of the iron construction of tall buildings having been provided with a heavy deposit of aluminium, and even the processes used have been fully described.   It is to be regretted that such elaborate reports are even admitted into scientific journals, though the separation of romance from truth could be readily accomplished by a conscientious examination.   Unscrupulous dealers offer their customers aluminium baths, charging a high price for them, and on testing a deposit produced with such baths, it has frequently been found to consist solely of tin.   Others, who actually had faith in the value of their invention have submitted for examination objects plated with aluminium by their processes.   On testing such deposits it was found that in one case, it consisted of a thin deposit of zinc, the origin of which was due to zinckiferous aluminium salts, and in other cases, of a deposit of iron due to the same cause.

The concurrence of the above-mentioned influences and the recent rapid development of the aluminium industry explains the demand for aluminium baths by many electro-platers. However, without entering into scientific reasons, which would not be within the scope of this work, it can here only be repeated that the reduction of metallic aluminium from its solutions will very likely remain an empty dream.

4. DEPOSITION UPON ALUMINIUM

The electro-deposition of other metals upon aluminium presents many difficulties which are chiefly due to the behavior of this metal towards the plating baths. The deposits to be sure are formed, but they possess no adherence, and especially baths containing potassium cyanide yield the worst results in consequence of the effect of alkaline solutions upon the basis-metal. Since the production of aluminium has so largely increased, and a great number of articles of luxury and for practical use are now made of this metal, the need of decorating such articles by electro-plating or covering them entirely with other metals has been felt, since the color of aluminium is by no means a sympathetic one. Look into a show window where aluminium articles are exposed—nothing but gray in gray. Offended, the eye of the observer turns away, and seeks a more agreeable resting-place.

Aluminium behaves so differently from other metals towards the cleansing agents generally used, that different methods from those previously described have to be employed in preparing it for plating. Nitric acid has almost no effect on aluminium, and pickle just a little ; but, on the other hand, the metal is attacked by concentrated hydrochloric acid, dilute hydrofluoric acid, and especially by alkaline lyes. Hence, if polished articles of aluminium are to be prepared for plating, alkaline lyes will have to be avoided in freeing them from grease, it being best to use only benzine for the purpose. Unpolished articles may without hesitation be freed from grease with caustic potash or soda lye, and, for the production of a dead white surface, be for a short time pickled in dilute hydrofluoric acid, and then thoroughly rinsed in running water.

For producing an electro-deposit upon aluminium it has been considered advisable to first copper the metal, and the

Aluminium Gesellschaft of Neuhausen recommends for this purpose a solution of nitrate of copper. But the adherence of the copper proved also insufficient, because in the subsequent silvering, nickeling, etc., the deposit raised up.

The copper bath recommended by Delval, consisting of sodium pryophosphate 3 ozs., copper sulphate (blue vitriol) ¾ oz., sodium bisulphite ¾ oz., water 1 quart, also proved unreliable.

According to another patented process, plating of aluminium is claimed to be effected successfully, and without defect, by lightly coating the metal with silver amalgam by boiling in a silver bath compounded with potassium-mercury cyanide. However, this treatment did not always yield reliable results.

According to Villon, articles of aluminium are to be immersed for one hour in a bath consisting of glycerin 5¼ ozs., zinc cyanide 0.88 oz., zinc iodide 0.88 oz., and then heated to a red heat. When cold, they are washed with a hard brush and water, and brought into the gold or silver bath. The success of this process seems also questionable.

The best and most reliable process is without doubt the one patented, in 1893, by Prof. Nees. It consists in first immersing the aluminium articles previously freed from grease in caustic soda lye until the action of the lye upon the metal is recognized by gas bubbles rising to the surface. The articles without being previously rinsed are then for a few minutes immersed in a solution of 77 troy grains of chloride of mercury, rinsed, again brought into the caustic soda lye, and then, without rinsing, suspended in the silver bath. The deposit of silver thus obtained adheres very firmly, and can be scratch-brushed, and polished with the steel without raising up. It can also be directly gilded, brassed, or, after previous coppering in the potassium cyanide copper bath, provided with a heavy deposit of nickel and polished upon polishing wheels.

Burgess and Hambuechen * found it best to first zinc the

* Electro-chemical Industry, 1904, No. 3.

aluminium in an acid zinc bath containing 1 per cent. fluoric acid ; the fluoric acid acts as a solvent upon the film of oxide formed so that the deposit is effected upon a pure metallic surface.   According to these authors, the aluminium article is to be immersed in dilute fluoric acid until its surface appears slightly rough and attacked.   It is then to be rinsed in water, and for a few seconds immersed in a bath of sulphuric acid 100 parts and nitric acid 75 parts.   It is then again thoroughly rinsed in water, next brought into a zinc bath of 15° Bé., consisting of zinc sulphate and aluminium sulphate, and acidulated with 1 per cent. of fluoric acid or the equivalent quantity of potassium fluoride, and zincked for 15 to 20 minutes. For subsequent silvering or gilding the zinc deposit is first coppered in the potassium cyanide copper bath.

According to Göttig, a thin, firmly-adhering deposit of copper is first to be produced upon the aluminium by triturating blue vitriol solution with tin-powder and whiting; or a tin deposit is to be produced by applying stannous chloride-ammonium chloride solution by means of a soft brass brush.

The Mannesmann Pipe Works, Germany, produce durable electro-deposits by brushing the aluminium with solutions of sulphide of gold and sulphide of silver in balsam of sulphur * and volatile oils, and burning in the metals in a muffle, under exclusion of the air, at 840° to 930° F.   Thin layers of metal which are reduced adhere firmly to the aluminium, and are then provided with an electro-deposit desired.   According to a process patented by the same corporation, the articles are provided with a firmly adhering (?) film of zinc by immersing them in boiling solution of zinc dust in caustic soda, and are then electro-plated.

However, in view of the fact that all the methods mentioned above partly yield uncertain results, it has recently been proposed first to provide the aluminium in readily-fusible metallic salts (cupric chloride, tin salt) with a coat of these metals, and then treat it further in aqueous electrolytes.

* Solution of sulphur in linseed oil.

CHAPTER XIII.

DEPOSITION BY CONTACT, BY BOILING, AND BY FRICTION.

If a sheet of metal, for instance, copper, be brought into a solution which contains the cations of a metal of slighter solution-tension (p. 61), for example, a solution of potassium-silver cyanide in water with an excess of potassium cyanide, which has been heated to about 158° F., the following process takes place: By the osmotic pressure (p. 49) the metal-ions, in this case silver-ions, are reduced upon the copper sheet, the osmotic pressure of the solution being thereby decreased. In consequence of this decrease in the osmotic pressure. copper-ions can be forced into the solution by the solution-tension, while additional silver-ions are brought to separate upon the copper sheet. Hence, during the formation of the deposit, a solution of the metal to be coated in the case in question, copper, takes place at the same time, this process coming to a standstill when the copper sheet has been covered with a coherent coat of silver, which prevents further solution of copper in the electrolyte.

The process may also be explained in a different way namely, that by the immersion in the silver solution the copper is negatively charged, positive silver-ions being by reason of electrostatic attraction attracted and reduced on the copper.

The deposits produced in this manner are generally known as *deposits by immersion*, or when the electrolyte is highly heated, *by boiling*.

The same process takes place when metallic objects are plated by applying by means of a *brush* or by *friction* an

(487)

electrolyte which contains a metal with slighter solution tension than possessed by the metal of an object to be coated.

Since the more electro-positive metals of the old series of electro-motive force possess a greater solution-tension than the electro-negative metals, it may be briefly stated, that electro-positive metals when immersed in suitable solutions of electro-negative metals reduce the latter, and, under certain conditions, become coated with them so that a coherent deposit is formed.

From the process above described, according to which reduction only takes place till the electro-positive metal has been provided with a coherent coating of the electro-negative metal, it is plain that such deposits can be only very thin and cannot be increased by continued action of the electrolyte, except recourse be had to other means.

The process, however, is a different one when a deposit is to be produced by the *contact* of one metal with another in an electrolyte. If a copper sheet dipping in a potassium cyanide solution of potassium-silver cyanide be touched with an electro-positive metal, for instance, a zinc rod or a zinc sheet, the latter dipping also in the electrolyte, an electric current is generated which reduces silver-ions on the copper-sheet, while on the zinc sheet zinc-ions are forced into solution. However, even when the copper sheet has been covered with a coherent deposit of silver, the reduction of the latter goes on in so far as the silver which is also reduced upon the zinc, and which interrupts contact with the electrolyte, as well as prevents further migration of zinc-ions into the solution, is only from time to time removed.

The contact processes can, however, be applied only to a limited extent. On the one hand, the formation of uniformly heavy deposits upon the metallic objects is excluded, because by reason of the greater current-densities appearing at the point of contact with the contact metal, a heavier reduction of metal takes place there than on the portions further removed from the point of contact, except the latter be frequently

changed.   On the other hand, the constant increase of dis-·
solved contact-metal in the electrolyte constitutes a drawback,.
and is the cause of the electrolytes, as a rule, giving out long
before their content of metal is exhausted. Finally, the·
reduction of metal upon the contact-metal is not a desirable·
feature.

As *contact-metals*, zinc, cadmium and aluminium are chiefly.
used.   In many cases, aluminium being a highly positive
metal, considerably surpasses in its effect the first-mentioned.
metals, and possesses the advantage of not bringing into the
electrolyte, metals reducible by the current.   Furthermore,
the quantities of metal deposited upon the aluminium can be
dissolved with nitric acid without materially attacking the
contact-metal.   Darlay has recommended magnesium as a
contact-metal (German patent 127,464).   It presents, however,
no advantage, on the one hand, on account of its high price
and, on the other, by reason of the deficient results in connec-·
tion with the baths of the above-mentioned patent.

The electrolytes serving for depositions by contact must.
possess definite properties if they are to yield good results.

Since the currents generated by contact are weak, the elec-·
trolyte should possess good conductivity, so that the reduction
of metal does not take place too slowly, and it must attack—·
chemically dissolve—the contact-metal, as a current can only
be generated if such be the case.   Let us consider, for instance,.
a well-known gold bath for hot gilding by contact, which con-·
tains in 1 quart of water 77 grains of crystallized sodium phos-·
phate, $46\frac{1}{4}$ grains of caustic potash, $15\frac{1}{3}$ grains of neutral chloride of gold, and 0.56 oz. of 98 per cent. potassium cyanide.
It will be found that only a slight portion of the potassium.
cyanide is consumed for the conversion of the chloride of gold.
into potassium-gold cyanide, the greater portion of it serving
to increase the conductivity of the electrolyte.   The caustic
potash, together with the sodium phosphate, effects the alkalinity of the bath which is required for attacking and dissolving the contact-metal, whether it be zinc, cadmium, aluminium.

or magnesium. The effect of the alkaline phosphate as such is claimed to be that the deposit of metal which results not only upon the objects in contact with the contact-metal, but also upon the latter itself, does not firmly combine with it, but can be readily removed by scratch-brushing.

For increasing the conductivity of electrolytes containing potassium cyanide, a greater or smaller excess of the latter is used either by itself or in combination with chlorides, for instance, ammonium chloride or sodium chloride, nearly all known baths for contact-deposition containing these salts in varying quantities. For nickel and cobalt baths, an addition of ammonium chloride, in not too small quantity, is most suitable, it assisting materially the attack upon the contact-metal and may in some cases serve for this purpose by itself without the co-operation of an alkali.

The attack on the contact-metal is most effectually promoted by sufficient alkalinity of the electrolyte, mostly in connection with chlorides, in a few rarer cases without chlorides, and as previously mentioned, occasionally by chlorides alone without the co-operation of an alkali.

In judging the formulas for contact baths to be given later on, the effects here explained will have to serve as a basis.

Small objects in quantities are generally plated in baskets made of the contact-metal, and, as previously mentioned, the deposition of quantities of the same metal with which the objects are to be coated upon the contact-body cannot be avoided. To be sure, claim is made in a few patents to prevent deposition upon the contact-metal and to keep the contact-body free by certain additions, for instance, alkaline pyrophosphates and phosphates, but experiments failed to prove the correctness of these claims.

The useless reduction of metal is one of the many weak points of the contact-process. The bath thereby becomes rapidly poor in metal, requires frequent refreshing or regeneration, which as a rule is not so readily done, and thus in practice the contact-process becomes quite expensive. It must further be borne in mind that so soon as reduction of metal upon the contact-body takes place, the formation of a deposit upon the object ceases, this being the reason why only very thin deposits can be produced, which do not afford protection against atmospheric influences, and are not sufficiently resistant to mechanical attack.

To avoid as much as possible the drawback of metal being reduced on the wrong place, Dr. G. Langbein & Co. use, according to a method for which a patent has been applied for, baskets of contact-metal, the outsides of the latter, which do not come in contact with the objects, being insulated from the electrolyte by enameling, or coating with hard rubber, celluloid or similar materials capable of resisting the hot solution. Or, they use baskets of contact-metal the outsides of which are provided, either mechanically by rolling or welding, or electrolytically by deposition, with the same metal contained in solution in the electrolyte, the baskets being thus protected from the deposit; while, in addition, a partial regeneration of the bath is in many cases attained. With certain combinations a portion of the electro-negative metal or alloy combined with the contact-metal or fixed insulated from it, passes into solution, and partly replaces the metal which has been withdrawn from the bath and deposited upon the objects.

Further drawbacks of the contact process are, working with baths almost boiling hot, and the consequent evolution of steam which is injurious to the workmen as well as to the work-rooms.

Hence, the contact process is suitable only for coating—so to say coloring—objects in large quantities with another metal, when no demands as regards solidity of the deposit are made.

Nickeling by Contact and Boiling.

According to Franz Stolba, articles can be sufficiently nickeled in 15 minutes by boiling them, mixed with fragments of zinc in a solution of nickel sulphate. A copper kettle tinned inside is to be used. Since stains are readily formed by this process, especially when nickeling polished iron and steel articles, on the places where the metal to be nickeled comes in contact with the zinc, Stolba in later experiments omitted the zinc, and thus the contact process becomes a boiling process. The articles are to be boiled for 30 to 60 minutes in a 10 per cent. zinc chloride solution to which is added enough nickel sulphate to give the solution a deep green color.

However, Stolba's process cannot be recommended to the nickel-plater. To be sure a thin nickel deposit of a light color might be obtained upon brass articles, but that on iron objects generally turned out dark and mostly stained. The nickeling is so thin that it will not stand polishing with any kind of pressure, and the cheapness claimed for the process is quite illusive, the solution soon becoming useless by reason of the absorption of copper, iron, etc., from the metals to be nickeled.

For small articles, which are not to be nickeled with the assistance of the current, one of the following processes is to be preferred :

By boiling a solution of 8½ ozs. of nickel-ammonium sulphate and 8½ ozs. of ammonium chloride in 1 quart of water, together with clean iron filings free from grease, and introducing into the fluid copper or brass articles, the latter become coated with a thin layer of nickel capable of bearing light polishing.

In place of iron filings, it is of greater advantage to bring the objects to be coated in contact with a piece of sheet-zinc of not too small a surface, or to nickel them in an aluminium basket. The hotter the solution is, the more rapidly coating with nickel is effected, and when the bath is made slightly alkaline with ammonia, iron objects also nickel quite well in an aluminium basket.

In place of the zinc contact, Basse & Selve use an aluminium contact for nickeling (as well as for coppering and silvering). According to the patent specification, objects nickel gray and show no metallic luster when brought in a zinc basket into a boiling solution of 20 parts of nickel-ammonium sulphate, 40 parts ammonium chloride and 60 parts water, which, after the addition of a slight excess of ammonia and filtering, is rendered slightly acid with citric acid.   By substituting for the zinc basket an aluminium basket, a lustrous, more firmly adhering layer of nickel is in about two minutes obtained.

Still better results are obtained by keeping the bath slightly alkaline with ammonia or ammonium carbonate.

A. Darlay has patented in France, as well as in Germany, a process of nickeling (as well as cobalting) by aluminium or magnesium contact.   However, the object of the invention is not the aluminium contact, which has been known for a long time, nor the special kinds of baths, the compositions of which are similar to those of other known contact-baths, but the use of the aluminium or magnesium contact in connection with baths of exactly defined compositions.

These patented baths fulfill nothing further than the general conditions given in detail on p. 487 *et seq.*, and which are also fulfilled by most of the long known contact-baths as shown by the bath for gilding by contact (p. 489).   Darlay's patent is, therefore, a combination-patent, and its right of existence appears rather doubtful in view of the fact that Basse and Selve's patent has expired, and that baths of the composition of the Darlay electrolytes have long been known and used for deposition.

Darlay's baths are brought into commerce by " Electrometallurgie " under the name of *autovolt baths*, and in answer to many inquiries it may here be stated that for the reason given on p. 488, no heavier deposits can be produced with these autovolt baths, than with the contact process in general, and that this autovolt method shows the same drawbacks as all other contact processes.

In his patent specification, Darlay gives the following composition of the electrolyte which is to be used hot :

Water 1 quart, nickel chloride $7\frac{1}{2}$ drachms, sodium phosphate 8¼ ozs., ammonium chloride 11¼ drachms, ammonium carbonate and sodium carbonate each 4¾ drachms.

The sodium phosphate is claimed to effect the production of a bright attacking surface of the contact-metal, and the sodium and ammonium carbonates, the alkaline reaction and, hence the generation of the current, by dissolving the aluminium, while the ammonium chloride produces good conductivity. As regards the action of alkaline pyrophosphates, the reader is referred to p. 489.

The inventor asserts that the proportions given above have to be kept within quite narrow limits. With the exception of the nickel chloride, the quantities of sodium phosphate and of one of the other chemicals can without fear be increased 50 per cent., the results thus obtained being still better than with Darlay's formula.

The chemical process of Darlay's electrolyte consists no doubt in that a transposition takes place between the nickel chloride and the sodium phosphate, sodium chloride and nickel phosphate being formed, which are soluble in the excess of sodium phosphate, and are not precipitated by the alkaline carbonates.

Hence the bath given on p. 259 under formula IX for nickeling with an external source of current, should be suitable for contact-nickeling with aluminium, if the quantity of sodium phosphate be materially increased, the conductivity enhanced by the addition of ammonium chloride, and the solution of the aluminium promoted by adding caustic potash, caustic soda, or better, alkaline carbonates.

Cobalting by Contact and Boiling.

Cobalting by contact is readily accomplished with the use of the following bath : Crystallized cobalt sulphate 0.35 oz., crystallized ammonium chloride 0.07 oz., water 1 quart. Heat the bath to between 104° and 122° F., and immerse the previously cleansed and pickled articles in it, bringing them in contact with a bright zinc surface not too small; for small articles a zinc basket may be used.   In 3 or 4 minutes the coating is heavy enough to bear vigorous polishing.

It is a remarkable fact that with aluminium-contact no satisfactory results are obtained in this bath, the reaction of aluminium in cobalt solutions thus appearing to be different from that in nickel solutions.   What has been said in re- gard to Darlay's contact process for nickeling applies also to cobalting.

For cobalting small objects in quantities, the reader is re- ferred to Warren's process, p. 324.

Coppering by Contact and Dipping.

According to Lüdersdorff, a solution of tartrate of copper in neutral potassium tartrate serves for this purpose.   A suitable modification of this bath is as follows : Heat 10 quarts of water to 140° F., add 2 lbs. of pulverized tartar (cream of tartar) free from lime, and $10\frac{1}{2}$ ozs. of carbonate of copper.   Keep the fluid at the temperature above mentioned until the evolution of gas due to the decomposition of the carbonate of copper ceases, and then add in small portions, and with constant stir- ring, pure whiting until effervescence is no longer perceptible. Filter off the fluid from the tartrate of lime, separate and wash the precipitate, so that the filtrate, inclusive of the wash water, amounts to 10 or 12 quarts, and dissolve in it $1\frac{3}{4}$ ozs. of caustic soda and 1 oz. of 99 per cent. potassium cyanide.

With zinc-contact the bath works somewhat slowly, but more rapidly with aluminium-contact.   Zinc is coppered in this bath by simple immersion.

The bath for coppering by contact, proposed by Weill, has been given on p. 334, under formula X.   The bath is to be heated to between 185° and 194° F., and with zinc contact yields a tolerably good deposit upon small iron objects.   With aluminium-contact, iron screws as well as iron articles in quantities are quickly and nicely coppered.

According to Bacco, a copper bath in which zinc may be coppered by immersion, and iron and other metals in contact with zinc, is prepared by adding to a saturated solution of blue vitriol, potassium cyanide solution until the precipitate of cyanide of copper which is formed is again dissolved. Then add $\frac{1}{10}$ to $\frac{1}{5}$ of the volume of liquid ammonia and dilute with water to 7° Bé.

The bath is to be heated to 194° F. To the same extent as zinc passes into solution the copper bath is gradually changed to a brass bath.

Every strongly alkaline copper cyanide bath may serve for coppering by contact, provided only a small quantity of free potassium cyanide is present in the bath, and the latter is heated to 194° F.

Zinc when used as a contact-metal shows the drawback of the copper depositing quite firmly upon it, so that it has to be removed by pickling in nitric acid. Furthermore, with the use of zinc as contact-bodies, the content of free alkali has to be much larger than with aluminium contacts, and so much zinc passes, in the first case, into solution that, in place of copper deposits, brass deposits with tones of color varying according to the temperature are in a short time obtained.

According to Darlay's patent, an alkaline copper cyanide bath heated to between 185° and 194° F. is to be used, the electrolyte consisting of:

Water 1 quart, cupric sulphate 0.35 oz., potassium cyanide 0.42 oz., caustic soda 0.52 oz.

When in such formulas the quantity of potassium cyanide is given without stating its content in per cent., it would, as a rule, be understood to refer to the 98 or 99 per cent. article. However, according to experiments made with Bacco's bath, with the use of 98 per cent. potassium cyanide, the excess would be too large, and it may be supposed that Darlay's formula refers to 60 per cent. potassium cyanide.

However, in this respect, the patent specification does not agree with the facts. For instance, the content of potassium cyanide "is exactly to be adhered to" in order to prevent a deposit of copper upon the contact-body—an aluminium basket.   However, no matter whether potassium cyanide with a content of 60 per cent., or more is used, a heavy deposit of copper is always formed upon the aluminium,* and the formation of a deposit of copper upon the objects is not in the least dependent upon adhering exactly to the quantity of potassium cyanide given.

The chemical process of Darlay's formula consists therefore in the conversion of cupric sulphate and potassium cyanide to potassium-copper cyanide.   With the use of 68 per cent. potassium cyanide scarcely any free potassium cyanide is contained in the bath, while with 98 per cent. potassium cyanide, free potassium cyanide remains in the bath.   If, now "the accurately-fixed content of potassium cyanide" in Darlay's formula refers to the 60 per cent. article, we come back to Bacco's formula, in which just enough potassium cyanide is added to the cupric sulphate solution to redissolve the separated cupro-cupric cyanide, a content of free potassium cyanide being avoided.   Bacco effects alkalinity by ammonia and Darlay by caustic soda.   From this it will be seen that Darlay's formula is very similar to Bacco's, and it is doubtful whether a patent-right can be claimed on the substitution of caustic soda for ammonia.   At any rate, now that Basse and Selve's patent has expired, it is obvious that Bacco's bath with the use of aluminium-contact can be employed for coppering by contact without infringing on Darlay's patent.

The so-called *brush-coppering*, which has been recommended, may here be mentioned.   This process may be of practical advantage for coppering very large objects which by another method could only be coated with difficulty.   The deposit of copper is, of course, very thin.   The process is executed as follows: The utensils required are two vessels of sufficient size, each provided with a brush, preferably so wide that the entire surface of the object to be treated can be coated with one ap-

---

* According to experiments made by Friessner, about 90 per cent. of the metal contained in the bath was deposited upon the contact-body, and only 10 per cent. upon the objects.

32

plication. One of the vessels contains a strongly saturated solution of caustic soda, and the other a strongly saturated solution of blue vitriol. For coppering, the well-cleansed object is first uniformly coated with a brushful of the caustic soda solution, and then also with a brushful of the blue vitriol solution. A quite thick film of copper is immediately deposited upon the object. Care must be had not to have the brush too full, and not to touch the places once gone over the second time, as otherwise the layer of copper does not adhere firmly.

Many iron and steel objects, for instance, wire, springs, etc., are provided with a thin film of copper in order to give them a more pleasing appearance. For this purpose a copper solution of 10 quarts of water, $1\frac{3}{4}$ ozs. of blue vitriol, and $1\frac{3}{4}$ ozs. of pure concentrated sulphuric acid may be used. Dip the iron or steel objects, previously freed from grease and oxide, for a moment in the solution, moving them constantly to and fro ; then rinse them immediately in ample water, and dry. By keeping the articles too long in the solution the copper separates in a pulverulent form, and does not adhere.

Steel pens, needles' eyes, etc., may be coppered by diluting the copper solution just mentioned with double the quantity of water, moistening sawdust with the solution, and revolving the latter, together with the objects to be coppered, in a wooden tumbling barrel.

Brassing by Contact.

Some older authors have given formulas for baths for brassing by contact, but the results obtained are not very satisfactory.

Darlay has patented the bath given below. It is brought into commerce under the name of *autovolt brass bath*, and yields thin brass deposits of an agreeable color and good luster :

Water 1 quart, cupric sulphate 0.14 oz., zinc sulphate 0.35 oz., potassium cyanide 0.44 oz., caustic soda 0.52 oz.

On testing this formula it was found that with the use of

98 per cent. potassium cyanide the bath yielded no deposit, one being, however, obtained with the 60 per cent. article. What has been said in reference to the autovolt copper bath applies also to the brass bath.

As previously mentioned, deposits produced by contact cannot be obtained of any thickness, the contact-metal soon becoming covered with a deposit when the process comes to a standstill. Aluminium, to be sure, relinquishes the deposited metal in coherent laminæ, this being promoted by the heavy evolution of hydrogen. However, it shows also how large are the quantities of metal which are deposited upon the aluminium, and that deposition by contact is consequently connected with a waste of metallic salts, which considerably increases the cost of manufacture. For removing the deposit upon the aluminium body, mixtures of nitric and sulphuric acids have to be used, so that, in addition to the loss of metal, there is a considerable consumption of acids.

Iron objects brassed in the above-mentioned baths have, to be sure, quite a neat appearance, but soon commence to rust, and for this reason cannot serve as substitutes for objects thickly brassed by means of an external source of current.

Silvering by Contact, Immersion and Friction.

For contact-silvering of copper and brass objects the following bath may be used ·

Water 1 quart, crystallized silver nitrate 0.52 oz., 60 per cent. potassium cyanide 1.4 ozs.

The bath is to be somewhat heated, so that deposition does not take place too slowly. Zinc is very suitable for a contact-metal, but to avoid formation of stains, the contact-points have to be frequently changed.

If iron articles are to be silvered, it is recommended to add to the bath, heated to between 176° and 194° F., about 0.28 to 0.35 oz. of caustic soda, and to use an aluminium contact; for smaller objects in quantities an aluminium basket. It is of greater advantage, in all cases, first to brass or copper the iron objects.

According to Darlay's German patent, 128,318, the following baths serve for silvering by contact with aluminium :

Water 1 liter (2.11 pints), silver nitrate 30 grammes (0.7 oz.), potassium cyanide 10 grammes (0.35 oz.), caustic potash 4 grammes (0.14 oz.).

Information regarding the content in per cent. for potassium cyanide is wanting. Besides, the quantity of potassium cyanide in proportion to silver nitrate is too low, which may be due to a typographical error, and it may be supposed that the formula for a 25-liter bath, as given in the patent specification, should read 0.05 kilogramme of silver nitrate instead of 0.5 kilogramme. This, calculated to 1 liter, gives 2 grammes instead of 20 grammes of silver nitrate.

For silvering *iron and steel* in hot baths :

Water 1 quart, silver nitrate 0.44 oz., potassium cyanide 4.4 ozs., sodium phosphate 0.88 oz.

The object of the sodium phosphate here is not to prevent the adhesion of the metal deposited upon the contact metal, which is to be effected by the excess of potassium cyanide. However, in experiments made with this bath, more silver was deposited upon the contact-body than upon the objects.

Silvering by immersion. For silvering coppered or brassed objects by immersion, the following solution may be used :

Water 1 quart, silver nitrate 0.35 oz., 98 per cent. potassium cyanide 1.23 ozs.

To prepare the bath dissolve the silver salt in 1 pint of the water, then the potassium cyanide in the remaining pint of water, and mix the two solutions. The bath is heated in a porcelain or enameled iron vessel to between 176° and 194° F., and the thoroughly cleansed and pickled objects are immersed in it until uniformly coated, previous quicking being not required. The deposit is lustrous if the articles are left but a short time in the bath, but becomes dull when they remain longer. In the first case the deposit is a mere film, and, while it is somewhat thicker in the latter, it can under no circumstances be called solid. The thickness of the deposit does not increase by continued action, as much metal being dissolved as silver is deposited, and the silver deposit prevents a further dissolving effect upon the basis metal.

The bath gradually works less effectively, and finally ceases to silver, when its action may be restored by the addition of $2\frac{3}{4}$ to $5\frac{1}{2}$ drachms of potassium cyanide per quart. Should this prove ineffectual, the content of silver is nearly exhansted, and the bath is evaporated to dryness, and the residue added to the silver waste. Frequent refreshing of the bath with silver salt cannot be recommended, the silvering always turning out best in a fresh bath.

A solution of nitrate of silver in sodium sulphite is, according to Roseleur, very suitable for silvering by immersion. The solution is prepared by pouring into moderately concentrated solution of sodium phosphite, while constantly stirring, solution of a silver salt until the precipitate of silver sulphide formed begins to be dissolved with difficulty. The bath can be used cold or warm, fresh solution of silver being added when it commences to lose its effect. If, however, the bath is not capable of dissolving the silver sulphide formed, concentrated solution of sodium sulphite has to be added.

For the preparation of the solution of sodium sulphite, Roseleur recommends the following method:

Into a tall vessel of glass or porcelain (Fig. 134) introduce 5 quarts of water and 4 pounds of crystallized soda, after pouring in mercury about an inch or so deep to prevent the glass tube through which the sulphurous acid is introduced from being stopped up by crystals. The sulphurous acid is evolved by heating copper turnings with concentrated sulphuric acid, washing the gas in a Woulff bottle filled an inch or so deep with water, and introducing it into the bottle containing the soda solution, as shown in the illustration. A part of the soda becomes transformed into sodium sulphite, which dissolves, and a part is precipitated as carbonate. The latter, however, is transformed into sodium sulphite by the continuous action of sulphurous acid, and carbonic acid gas escapes with effervescence.   When all has become dissolved, the introduction of sulphurous acid should be continued until the liquid slightly reddens blue litmus paper, when it is set aside for 24 hours.   At the end of that time a certain quantity of crystals will be found upon the mercury, and the liquid above, more or less colored, constitutes the sodium sulphite of the silvering bath.   The liquid sodium sulphite thus prepared should be stirred with a glass rod, to eliminate the carbonic acid which may still remain in it.   The liquid should then be again tested with blue litmus paper, and if the latter is strongly reddened, carbonate of soda is cautiously added, little

Fig. 134.

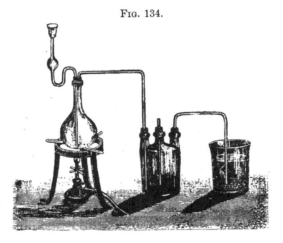

by little, in order to neutralize the excess of sulphurous acid. On the other hand, if red litmus paper becomes blue, too much alkali is present, and more sulphurous acid gas must be passed through the liquid, which is in the best condition for our work when it turns litmus paper violet or slightly red. The solution should mark from 22° to 20° Bé., and should not come in contact with iron, zinc, tin, or lead.

As will be seen, this mode of preparing the sodium sulphite solution is somewhat troublesome, and it is therefore recommended to proceed as follows: Prepare a saturated solution of commercial sodium sulphite. The solution will show an alkaline reaction, the commercial salt frequently containing some sodium carbonate. To this solution add, while stirring, solution of bisulphite of sodium saturated at 122° F., until blue litmus paper is slightly reddened. Then add to this solution concentrated solution of nitrate of silver until the flakes of silver sulphide separated begin to dissolve with difficulty.

The immersion-bath, prepared according to one or the other method, works well, the silvering produced having a beautiful luster, such as is desirable for many cheap articles. If the articles are allowed to remain for a longer time in the bath, a mat deposit is obtained. For bright silvering, the bath should always be used cold. It must further be protected as much as possible from the light, otherwise decomposition gradually takes place.

According to Dr. Ebermayer, a silver immersion-bath for bright silvering is prepared as follows: Dissolve 1.12 ozs. of nitrate of silver in water, and precipitate the solution with caustic potash. Thoroughly wash the silver oxide which is precipitated, and dissolve it in 1 quart of water which contains 3.52 ozs. of potassium cyanide in solution, and finally dilute the whole with one quart more water. For silvering, the bath is heated to the boiling-point, and the silver withdrawn may be replaced by the addition of moist silver oxide as long as complete solution takes place. When the silvering is no longer beautiful and of a pure white color, the bath is useless, and is then evaporated. Experiments with a bath prepared according to the above directions were never quite satisfactory. Better results were, however, obtained by diluting the bath with 3 to 4 quarts of water and using it without heating. It then yielded very nice, lustrous silvering.

The process of coating with a thin film, or rather whitening, with silver, small articles, such as *hooks and eyes, pins, etc.*, differs from the above-described immersion method, which effects the silvering in a few seconds, in that the articles require to be boiled for a longer time. The process is as follows:

Prepare a paste from 0.88 oz. of silver nitrate precipitated as silver chloride, cream of tartar 44 ozs. and a like quantity of common salt, by precipitating the silver nitrate with hydro- chloric acid, washing the chloride of silver and mixing it with the above-mentioned quantities of cream of tartar and common salt, and sufficient water to a paste, which is kept in a dark glass vessel to prevent the chloride of silver from being decom- posed by the light. Small articles of copper or brass are first freed from grease and pickled. Then heat in an enameled kettle 3 to 5 quarts of rain-water to the boiling-point; add 2 or 3 heaping teaspoonfuls of the above-mentioned paste, and bring the metallic objects contained in a stoneware basket into the bath, and stir them diligently with a rod of glass or wood. Before placing a fresh lot of articles in the bath, additional sil- ver paste must be added. If finally the bath acquires a green- ish color, caused by dissolved copper, it is no longer suitable for the purpose, and is then evaporated and added to the sil- ver residues.

Cold silvering with paste.—In this process an argentiferous paste, composed as given below, is rubbed, by means of the thumb, a piece of soft leather, or rag, upon the cleansed and pickled metallic surface (copper, brass, or other alloys of cop- per) until it is entirely silvered. The paste may also be rubbed in a mortar with some water to a uniformly thin-fluid mass, and applied with a brush to the surface to be silvered. By allow- ing the paste to dry naturally, or with the aid of a gentle heat, the silvering appears. The application of the paste by means of a brush is chiefly made use of for decorating with silver, articles thinly gilded by immersion. For articles not gilded, the above-mentioned rubbing-on of the stiff paste is to be preferred.

Composition of argentiferous paste.—I. Silver in the form of freshly precipitated chloride of silver,* 0.352 oz., common salt 0.35 oz., potash 0.7 oz., whiting 0.52 oz. and water a suffi- cient quantity to form the ingredients into a stiff paste.

II. Silver in the form of freshly precipitated chloride of silver * 0.35 oz., potassium cyanide 1.05 ozs., sufficient water to dissolve these two ingredients to a clear solution, and enough whiting to form the whole into a stiff paste. This paste is also excellent for polishing tarnished silver; it is, however, poisonous.

The following non-poisonous composition does excellent service: Silver in the form of chloride of silver 0.35 oz., cream of tartar 0.7 oz., common salt 0.7 oz., and sufficient water to form the mixture of the ingredients into a stiff paste.

Another composition is as follows: Chloride of silver 1 part, pearl ash 3, common salt 1½, whiting 1, and sufficient water to form a paste. Apply the latter to the metal to be silvered and rub with a piece of soft leather. When the metal is silvered, wash in water, to which a small quantity of washing soda has been added.

Graining.—In gilding parts of watches, gold is seldom directly applied upon the copper; there is generally a preliminary operation called graining, by which a grained and slightly dead appearance is given to the articles. Marks of the file are obliterated by rubbing upon a whetstone, and lastly upon an oil stone. Any oil or grease is removed by boiling the parts for a few minutes in a solution of 10 parts of caustic soda or potash in 100 of water, which should wet them entirely if all the oil has been removed. The articles being threaded upon a brass wire, cleanse them rapidly in the acid mixture for a bright luster, and dry them carefully in white wood sawdust. The pieces are fastened upon the even side of a block of cork by brass pins with flat heads. The parts are then thoroughly rubbed over with a brush entirely free from grease, and dipped into a paste of water and very fine pumice-stone powder. Move the brush in circles, in order not to rub one side more than the other; thoroughly rinse in cold water, and no particle of pumice-stone should remain upon the pieces of cork. Next place the cork and the pieces

---

* From 0.56 oz. of nitrate of silver.

in a weak mercurial solution, composed of water $2\frac{1}{2}$ gallons, nitrate or binoxide of mercury $\frac{1}{14}$ oz., sulphuric acid $\frac{1}{7}$ oz., which slightly whitens the copper. The pieces are passed ·quickly through the solution and then rinsed. This operation gives strength to the graining, which without it possesses ·no adherence.

The following preparations may be used for graining: I. ·Silver in impalpable powder 2 ozs., finely-pulverized cream of tartar 20 ozs., common salt 4 lbs. II. Silver powder 1 oz., ·cream of tartar 4 to 5 ozs., common salt 13 ozs. III. Silver powder, common salt, and cream of tartar, equal parts by weight of each. The mixture of the three ingredients must be thorough and effected at a moderate and protracted heat. The graining is the coarser the more common salt there is in the mixture, and it is the finer and more condensed as the proportion of cream of tartar is greater, but it is then more diffi-·cult to scratch-brush. The silver powder is obtained as follows: Dissolve in a glass or porcelain vessel $\frac{2}{3}$ oz. of crystallized nitrate of silver in $2\frac{1}{2}$ gallons of distilled water, and place 5 or 6 ribbands of cleansed copper, $\frac{3}{4}$ inch wide, in the solution. These ribbands should be long enough to allow of a portion of them being above the liquid. The whole is kept in a dark place, and from time to time stirred with the copper ribbands. This motion is sufficient to loosen the deposited silver, and present fresh surfaces to the action of the liquor. When no more silver deposits on the copper the operation is complete, and there remains a blue solution of nitrate of copper. The ·silver powder is washed by decantation or upon a filter until ·there remains nothing of the copper solution.

For the purpose of graining, a thin paste is made of one of the above mixtures and water, and spread by means of a spatula upon the watch parts held upon the cork. The cork itself is placed upon an earthenware dish, to which a rotating move-·ment is imparted by the left hand. An oval brush with close bristles, held in the right hand, rubs the watch parts in every ·direction, but always with a rotary motion. A new quantity

·of paste is added two or three times and rubbed in the manner
indicated. The more the brush and cork are turned, the
rounder becomes the grain, which is a good quality, and the
more paste added, the larger the grain. When the desired
grain is obtained, the pieces are washed and scratch-brushed.
The brushes employed are of brass wire, as fine as hair, and
very stiff and springy. It is necessary to anneal them upon
an even fire to different degrees; one soft or half annealed for
the first operation or uncovering the grain; one harder for
bringing up the luster; and one very soft or fully annealed,
used before gilding for removing any marks which may have
been made by the preceding tool, and for scratch-brushing
after gilding, which, like the graining, must be done by giving a rotary motion to the tool. If it happens that the same
watch part is composed of copper and steel, the latter metal
requires to be preserved against the action of the cleansing
acids and of the graining mixture by a composition called
resist. This consists in covering the pinions and other steel
parts with a fatty composition which is sufficiently hard to
resist the tearing action of the bristle and wire brushes, and
insoluble in the alkalies of the gilding bath. A good composition is: Yellow wax, 2 parts by weight; translucent rosin,
$3\frac{1}{2}$; extra-fine red sealing-wax, $1\frac{1}{2}$; polishing rouge, 1. Melt
the rosin and sealing-wax in a porcelain dish, upon a water-
bath, and afterwards add the yellow wax. When the whole
is thoroughly fluid, gradually add the rouge and stir with a
wooden or glass rod, withdraw the heat, but continue the stirring until the mixture becomes solid, otherwise all the rouge
will fall to the bottom. The flat parts to receive this resist
are slightly heated, and then covered with the mixture, which
melts and is easily spread. For covering steel pinions employ
a small gouge of copper or brass fixed to a wooden handle.
The metallic part of the gouge is heated upon an alcohol
lamp and a small quantity of resist is taken with it. The
composition soon melts, and by turning the tool around, the
steel pinion thus becomes coated. Use a scratch-brush with

long wires, and their flexibility prevents the removal of the composition. When the resist is to be removed after gilding, put the parts into warm oil or tepid turpentine, then in a very hot soap-water or alkaline solution; and, lastly, into fresh water. Scratch-brush and dry in warm, white wood saw-dust. The holes of the pinions are cleansed and polished with small pieces of very white, soft wood, the friction of which is sufficient to restore the primitive luster. The gilding of parts of copper and steel requires the greatest care, as the slightest rust destroys their future usefulness. Should some gold deposit upon the steel, it should be removed by rubbing with a piece of wood and impalpable pumice dust, tin putty, or rouge.

The gilding of the grained watch parts is effected in a bath prepared according to formula I or III, given under "Deposition of Gold."

Gilding by Contact, by Immersion, and by Friction.

For contact-gilding by touching with zinc, formulas I, II, IV and V, given in Chapter IX "Deposition of Gold" may be used, IV and V being especially suitable, if the addition of potassium cyanide is somewhat increased and the baths are sufficiently heated.

A contact gold bath prepared with yellow prussiate of potash according to the following formula also yields a good deposit.

I. Fine gold as chloride of gold 54 grains, yellow prussiate of potash 1 oz., potash 1 oz., common salt 1 oz., water 1 quart.

The bath is prepared as given for formula III under "Deposition of Gold." For use, heat it to boiling.

II. Chemically pure crystallized sodium phosphate 2.11 ozs., neutral crystallized sodium sulphite 0.35 oz., potassium cyanide 0.28 oz., fine gold (as chloride) 15.43 grains, water 1 quart.

The bath is prepared as given for formula V under "Deposition of Gold." Temperature for contact-gilding 185° to 194° F. If red gilding is to be effected in this bath a corresponding addition of potassium-copper cyanide has to be made, 7½ grains sufficing for paler red, while 15 grains have to be added for redder tones.

Gilding by contact is done the same way as silvering by contact. The points of contact must be frequently changed, since in the gold bath intense stains are still more readily formed than in the silver bath.

Gilding by immersion (without battery or contact). The following two formulas have proved very useful:

I. Crystallized sodium pyrophosphate 2.82 ozs., 12 per cent. prussic acid 4.51 drachms, crystallized chloride of gold 1.12 drachms, water 1 quart. Heat the bath to the boiling-point, and immerse the pickled objects of copper or its alloys, moving them constantly until gilded. Iron, steel, tin, and zinc should be previously coppered, coating the objects with mercury (quicking) being entirely superfluous.

All gold baths prepared with sodium pyrophosphate, when fresh, give rapid and beautiful results, but they have the disadvantage of rapidly decomposing, and consequently can seldom be completely exhausted. In this respect the following formula answers much better.

II. Crystallized sodium phosphate 2.82 drachms, chemically pure caustic potash 1.69 drachms, chloride of gold 0.56 drachm, 98 per cent. potassium cyanide 9.03 drachms, water 1 quart. Dissolve the sodium phosphate and caustic potash in ¾ of the water, and the potassium cyanide and chloride of gold in the remaining ¼, and mix both solutions. Heat the solution to the boiling-point. This bath can be almost entirely exhausted, as it is not decomposed by keeping. Should the bath become weak, add about 2¾ drachms of potassium cyanide, and use it for preliminary dipping until no more gold is reduced. To complete gilding, the objects subjected to such preliminary dipping are then immersed for a few seconds in a freshly prepared bath of the composition given above.

The bath prepared according to formula II is also very suitable for contact gilding.

The layer of gold produced by immersion is in all cases very thin, since only as much gold is deposited as corresponds to the quantity of basis-metal dissolved. For heavier gilding by this process the action of zinc or aluminium contact will have to be employed as auxiliary means.

Gilding by friction. This process is variously termed *gilding with the rag, with the thumb, with the cork.* It is chiefly employed upon silver, though sometimes also upon brass and copper. The operation is as follows: Dissolve 1.12 to 1.69 drachms of chloride of gold in as little water as possible, to which has previously been added 0.56 drachm of saltpetre. Dip in this solution small linen rags, and, after allowing them to drain off, dry them in a dark place. These rags saturated with gold solution are then charred to tinder at not too great a heat, whereby the chloride of gold is reduced, partially to protochloride and partially to finely-divided metallic gold. This tinder is then rubbed in a porcelain mortar to a fine, uniform powder.

To gild with this powder, dip into it a charred cork moistened with vinegar or salt water and rub, with not too gentle a pressure, the surface of the article to be gilded, which must be previously cleansed from adhering grease. The thumb of the hand may be used in place of the cork, but in both cases care must be had not to moisten it too much, as otherwise the powder takes badly. After gilding, the surface may be carefully burnished.

Reddish gilding by friction is obtained by adding about 8 grains of cupric nitrate to the gold solution.

For gilding by friction, a solution of chloride of gold in an excess of potassium cyanide may also be used, after thickening the solution to a paste by rubbing in whiting. The paste is applied to the previously zincked metals by means of a cork, a piece of leather or a brush. Martin and Peyraud, the originators of this method, describe the operation as follows: Articles of other metals than zinc are placed in a bath consisting of concentrated solution of ammonium chloride, in which

has been placed a quantity of granulated zinc.   The articles
are allowed to boil a few minutes, whereby they acquire a
coating of zinc.   For the preparation of the gilding composition, dissolve 11.28 drachms of chloride of gold in a like
quantity of water, and add a solution of 2.11 ozs. of potassium
cyanide in as little water as possible (about 2.8 ozs.).   Of this
solution add so much to a mixture of 3.52 ozs. of fine whiting
and 2.82 drachms of pulverized tartar that a paste is formed
which can be readily applied with a brush to the article to be
gilded.   When the article is coated, heat it to between 140°
and 158° F.   After removing the dry paste by washing, the
gilding appears and can be polished with the burnisher.

Platinizing by contact.

Though a thick deposit cannot be produced by the contact-
process, Fehling's directions may here be mentioned as suitable for giving a thin coat of platinum to fancy articles.   He
recommends a solution of 0.35 oz. of chloride of platinum and
7 ozs. of common salt in 1 quart of water, which is made alkaline by the addition of a small quantity of soda lye, and for
use heated to the boiling-point.

If larger articles are to be platinized by contact, free them
from grease, and after pickling, and if necessary, coppering,
wrap them round with zinc wire, or place them upon a bright
zinc sheet, and introduce them into the heated bath.   All the
remaining manipulations are the same as in other contact-
processes.

Tinning by Contact and by Boiling.

For *tinning by zinc-contact* in the boiling tin bath, the following solutions are suitable :

According to Gerhold : Pulverized tartar and alum, of each
3.5 ozs., fused stannous chloride 0.88 oz., rain-water 10 quarts.

According to Roseleur : Potassium pyrophosphate 7 ozs.,
crystallized stannous chloride (tin-salt) 0.38 oz., fused stannous chloride 2.8 ozs., rain-water 10 quarts.

It might be advisable to increase the content of potassium pyrophosphate, and to add about 0.7 oz. of caustic soda.

According to Roseleur by *immersion :*

Potassium pyrophosphate 5.6 ozs., fused stannous chloride 1.23 ozs., rain-water 10 quarts.

For *tinning by contact,* heat the bath to boiling and suspend the clean and pickled objects in contact with pieces of zinc, or, better, wrapped around with zinc wire spirals, care being had from time to time to shift them about to prevent staining. Large baths which cannot be readily heated are worked cold, the objects being covered with a large zinc plate. In the cold bath the formation of the tin deposit requires, of course, a longer time. By using the electric current the deposit can be made as heavy as desired. By immersion in the bath prepared according to the last formula, zinc can only be coated with a very thin film of tin.

For *tinning by contact in a cold bath,* Zilken has patented the following solution : Dissolve with the aid of heat in 100 quarts of water, tin-salt 7 to 10.5 ozs., pulverized alum 10.5 ozs., common salt 15¾ ozs., and pulverized tartar 7 ozs. The cold solution forms the tin bath. The objects to be tinned are to be wrapped round with strips of zinc. Duration of the process, 8 to 10 hours.

Darlay uses for a cold tin bath with aluminium-contact :

Water 10 quarts, stannous chloride 1.05 ozs., potassium cyanide 1.41 ozs., caustic soda 1.76 ozs.

It might be advisable to heat the bath to at least between 113° and 122° F. For a hot tin bath Darlay uses :

Water 10 quarts, stannous chloride 0.88 oz., potassium cyanide 10.58 ozs., caustic soda 0.88 oz., sodium pyrophosphate 8.8 ozs.

The contact-body cannot be kept free from deposit by the addition of potassium cyanide, and tinning is effected as well without as with the addition of potassium cyanide.

Tinning solution for iron and steel articles. . Crystallized ammonium-alum 7 ozs., crystallized stannous chloride 2.8

drachms, fused stannous chloride 2.8 drachms, rain-water 10 quarts. Dissolve the ammonium-alum in the hot water, and when dissolved add the tin-salts. The bath is to be used boiling hot and kept at its original strength by an occasional addition of tin-salt. The clean and pickled iron objects, after being immersed in the bath, become in a few seconds coated with a firmly-adhering film of tin of a dead, white color, which may be polished by scratch-brushing, or scouring with sawdust in the tumbling barrel. Tinning by boiling in the above bath is the most suitable preparation for iron and steel objects which are finally to be provided with a heavy electro-deposit of tin. To insure entire success it is recommended thoroughly to scratch-brush the objects after boiling, then to return them once more to the bath, and finally to suspend them in a bath composed according to formula I, IIa or III, given under "Deposition of Tin."

A tinning solution for small brass and copper articles (pins, eyes, hooks, etc.), consists of a boiling solution of: Pulverized tartar 3.5 ozs., stannous chloride (tin-salt) 14.11 drachms, water 10 quarts. After heating the bath to the boiling-point, immerse the objects to be tinned in a tin basket, or in contact with pieces of zinc in a stoneware basket. Frequent stirring with a tin rod shortens the process.

A tinning solution highly recommended by Roseleur consists of:

Crystallized sodium pyrophosphate 7 ozs., crystallized stannous chloride 0.7 oz., fused stannous chloride 2.82 ozs., water 10 quarts.

The solution is prepared in the same manner as the preceding one.

Another solution, given by Böttger, also yields good results: Dissolve oxide of tin by boiling with potash lye, and place the copper or brass objects to be tinned in the boiling solution in contact with tin shavings.

Elsner's bath yields equally good results. It consists of a solution of equal parts of tin-salt and common salt in rain-water. The manipulation is the same as given above.

33

A *characteristic method of tinning* by Stolba is as follows : Prepare a solution of 1.75 ozs. of tin-salt and 5.64 drachms of pulverized tartar in one quart of water. Moisten with this solution a small sponge and dip the latter into pulverulent zinc. By then rubbing the thoroughly cleansed and pickled articles with the sponge, they immediately become coated with a film of tin. To obtain uniform tinning, the sponge must be repeatedly dipped, now into the solution, and then into the zinc-powder, and the rubbing continued for a few minutes.

Zincking by Contact.

For zincking iron by contact, a concentrated solution of zinc chloride and ammonium chloride in water is very suitable. The objects are placed in the solution in contact with a large zinc surface.

Darlay (German patent 128319) gives the following bath which, with an aluminium contact is claimed to yield a useful coating of zinc :

Water 10 quarts, zinc sulphate 0.35 ozs., potassium cyanide 1 oz., caustic soda 5.29 ozs.

It may be supposed that the bath is to be heated to between 170° and 194° F., though the patent specification is silent on this point. Experiments to obtain, according to these directions, a good coating of zinc on iron did not yield satisfactory results.

To coat brass and copper with a bright layer of zinc proceed as follows : Boil for several hours commercial zinc-gray, *i. e.,* very finely-divided metallic zinc, with concentrated solution of caustic soda. Then immerse the articles to be zincked in the boiling fluid, when, by continued boiling, they will in a short time become coated with a very bright layer of zinc. When a copper article thus coated with zinc is carefully heated in an oil bath to between 248° and 284° F., the zinc alloys with the copper, forming a sort of bronze similar to tombac.

Depositions of Antimony and of Arsenic by Immersion.

A heated solution of chloride of antimony in hydrochloric acid—*liquor stibii chlorati* of commerce—deposits upon brass objects immersed in it a coating of antimony of a steel-gray color inclining to bluish.

A purer steel-gray color is obtained with the use of a hot solution of arsenious chloride in water.

CHAPTER XIV

COLORING OF METALS.

METAL coloring and bronzing is an important branch of the metal industry, its object being, on the one hand, to embellish the original metallic surface and, on the other, to protect it from the influence of atmospheric air, moisture, various gases, etc. Although, strictly speaking, these operations do not form a part of a work on the electro-deposition of metals and cannot be adequately treated within the limits of a chapter, a few methods and approved formulas will here be given since the electro-plater is frequently forced to make use of one or the other method to furnish basis-metals or electro-deposits in certain shades of colors demanded by customers.

Metal coloring may be effected by electrolytic, chemical and mechanical means. Methods of coloring electrolytically have been given under Deposition of Nickel (black nickelling), and under Deposition of Antimony and Arsenic.

Mechanical methods of coloring require the use of pigments, bronze powders, varnishes, etc. and cannot be here fully described. To the electro-plater the most important of these operations is lacquering which will be described in the next chapter.

Attention will here be given chiefly to coloring metals by chemical means.

The practice of coloring metals requires considerable talent for observation and a certain knowledge of, the behavior of metals or metallic alloys towards the chemical substances used.

Especially in coloring alloys, for instance, brass, their per-

(516)

centage composition makes a difference, and patinas can be produced upon a brass richer in zinc, which cannot be obtained upon an alloy richer in copper. Hence instructions for patinizing have to be changed in one or the other direction, and this problem cannot be readily solved without a certain chemical knowledge. The temperature of the solutions used is also of great importance, and the directions given in this respect must be accurately observed.

1. *Coloring copper.*—With the use of chemicals nearly all colors can be produced upon copper, as well as upon other metals and alloys, by first coating them electrolytically with copper and afterwards coloring the deposit. For the production of yellow and brown, alkaline sulphides are, for instance, used, for green, copper salts, for black, metallic silver, bismuth, platinum, etc.

All shades from the pale red of copper to a dark chestnut brown can be obtained by superficial oxidation of the copper. For small objects it suffices to heat them uniformly over an alcohol flame. With larger objects a more uniform result is obtained by heating them in oxidizing fluids or brushing them over with an oxidizing paste, the best results being obtained with a paste prepared, according to the darker or lighter shades desired, from 2 parts of ferric oxide and 1 part of black-lead, or 1 part each of ferric oxide and black-lead, with alcohol or water. Apply the paste as uniformly as possible with a brush, and place the object in a warm place (oven or drying chamber). The darker the color is to be the higher the temperature must be, and the longer it must act upon the object. When sufficiently heated, the dry powder is removed by brushing with a soft brush, and the manipulation repeated if the object does not show a sufficiently dark tone. Finally the object is rubbed with a soft linen rag moistened with alcohol, or brushed with a soft brush and a few drops of alcohol until completely dry, and then with a brush previously rubbed upon pure wax. The more or less dark shade produced in this manner is very warm, and resists the action of the air.

Brown color on copper.—Apply to the thoroughly cleansed object a paste made of verdigris 3 parts, ferric oxide 3, sal ammoniac 1, and sufficient vinegar and heat until the paste turns black, then wash and dry the object. By the addition of some blue vitriol to the paste the color may be darkened to chestnut-brown.

A brown layer of cuprous oxide on copper is produced as follows: After polishing the articles with pumice powder apply with a brush a paste made of verdigris 4 parts, ferric oxide 4, finely rasped horn shavings 1, and a small quantity of vinegar. Dry, heat over a coal fire, wash, and smooth with the polishing stone.

A brown color is also obtained by brushing to dryness with a hot solution of 1 part of potassium nitrate, 1 of common salt, 2 of ammonium chloride, and 1 of liquid ammonia in 95 of vinegar. A warmer tone is, however, produced by the method introduced in the Paris Mint, which is as follows: Powder and mix intimately equal parts of verdigris and sal ammoniac. Take a heaping tablespoonful of this mixture and boil it with water in a copper kettle for about twenty minutes, and then pour off the clear fluid. To give copper objects a bronze-like color with this fluid, pour part of it into a copper pan ; place the objects separately in it upon pieces of wood or glass, so that they do not touch each other, or come in contact with the copper pan, and then boil them in the liquid for a quarter of an hour. Then take the objects from the solution, rub them dry with a linen cloth, and brush them with a waxed brush.

A beautiful and uniform brown tone on copper is produced as follows: Place the articles, previously freed from grease and pickled, in a solution of 5¼ ozs. of copper sulphate, and 2¾ ozs. potassium chloride heated to 140° F. until the desired tone is produced. Then brush with a soft brass-wire brush, rinse again for a short time in the pickle, and finally wipe dry with a soft cloth.

Brown of various shades on copper is produced as follows:

Bring the objects previously cleansed and pickled into a solution of liver of sulphur 1¼ ozs., or a solution of trichloride of antimony (butter of antimony) 1¼ ozs. in water 1 quart. When the desired tone is produced, rinse the objects thoroughly in water and dry. The shade of the color may be varied by the concentration of the bath as well as by the length of time of its action. The color is finally fixed by rubbing with a rag saturated with oil varnish or by rubbing heated wax upon the object.

A beautiful brown on copper by the so-called Chinese process is produced as follows : Crystallized verdigris 2 parts, cinnabar 2, ammonium chloride 5, finely powdered alum 5, intimately mixed and made into a thin paste with water or wine vinegar. Apply this paste with a brush to the polished surfaces. Then heat uniformly over a coal fire and when cold wash carefully with water. By the addition of copper sulphate a color shading more into chestnut-brown is obtained, and by the addition of borax one shading more into yellow.

Gold-yellow on copper. Treat the objects with a hot solution diluted with water, of mercury 10 parts and zinc 1 part in hydrochloric acid, to which some pulverized tartar has been added.

According to Manduit, copper and coppered articles may be bronzed by brushing with a mixture of castor oil 20 parts, alcohol 80, soft soap 40, and water 40. This mixture produces tones from *bronze Barbédienne to antique green patina*, according to the duration of the action.. After 24 hours the article treated shows a beautiful bronze, but when the mixture is allowed to act for a greater length of time the tone is changed and several different shades of great beauty can be obtained. After rinsing, dry in hot sawdust, and lacquer with colorless spirit lacquer.

Yellowish-brown on copper is produced by boiling the objects in a saturated solution of potassium chloride and ammonium nitrate. By heating the objects after drying them, a more reddish-brown color is obtained.

Dark brown to black on copper is obtained by dissolving nitrates of bismuth, copper, silver or cupriferous silver in water and adding some nitric acid. Copper to which such a fluid has been applied is, when heated, colored *brown* with the use of bismuth, and *black* with the use of copper and silver salts. *Very dark black* is produced by placing the objects for half an hour over a vessel containing a saturated solution of liver of sulphur to which some hydrochloric acid has been added. The luster may be increased by rubbing with a woolen cloth and a waxed brush.

Red to violet shades on copper articles. According to a process patented in Germany by M. Mayer, the highly polished copper article is electrolytically provided with a thin deposit of arsenic or antimony. For the preparation of the bath, solution of an antimony or arsenic salt is poured into a ferric chloride solution till the precipitate formed redissolves. A sheet of iron may serve as anode. The articles thus treated are then heated to cherry red and again polished. It is claimed that the electro-deposit as a carrier of oxygen effects a uniform oxidation of the copper underneath, but at the same time prevents it from becoming too highly oxidized so that by heating a layer of oxide is chiefly formed. The coating thus obtained shows red to violet shades, adheres firmly and resists physical as well as chemical influences.

Copper is colored blue-black by dipping the object in a hot solution of $11\frac{1}{4}$ drachms of liver of sulphur in 1 quart of water, moving it constantly. *Blue gray* shades are obtained with more dilute solutions. It is difficult to give definite directions as to the length of time the solution should be allowed to act, since this depends on its temperature and concentration. With some experience the correct treatment, however, will soon be learned.

The so-called *cuivre-fumé* is produced by coloring the copper or coppered objects blue-black with solution of liver of sulphur, then rinsing, and finally scratch-brushing them, whereby the shade becomes somewhat lighter. From raised portions which are not to be dark, but are to show the color of copper, the coloration is removed by polishing upon a felt wheel or bob.

Black color upon copper is produced by a heated pickle of 2 parts of arsenious acid, 4 of concentrated muriatic acid, 1 of sulphuric acid of 66° Bé., and 24 of water.

Mat-black on copper.—Brush the object over with a solution of 1 part of platinum chloride in 5 of water, or dip it in the solution. A similar result is obtained by dipping the copper object in a solution·of nitrate of copper or of manganese, and drying over a coal fire. These manipulations are to be repeated until the formation of a uniform mat-black.

A solution recommended for obtaining a *deep black color* on copper and its alloys is composed as follows: Copper nitrate 100 parts, water 100 parts. The copper nitrate is dissolved in the water, and the article, if large, is painted with it; if small, it may be immersed in the solution. It is then heated over a clear coal fire and lightly rubbed. The article is next placed in, or painted, with a solution of the following composition: Potassium sulphide 10 parts, water 100, hydrochloric acid 5.

More uniform results, however, are obtained by using a solution about three times more dilute than the above, *viz.:* Copper nitrate 100 parts, water 300. Small work can be much more conveniently treated by immersion in the solution, and after draining off, or shaking off the excess of the solution, heating the work on a hot plate until the copper salt is decomposed into the black copper oxide. It would be difficult to heat large articles on a hot plate, but a closed muffle-furnace would give better results than an open coal fire. In any case heating should not be continued longer than necessary to produce the change mentioned above.

Black color on copper, coppered objects and alloys rich in copper. For this purpose Dr. Groschuff gives the following directions: Heat a suitable quantity of 5 per cent. soda lye in a vessel of glass, porcelain, stoneware or enameled iron to 212° F., add 1 per cent. powdered potassium persulphate and immerse the article previously secured to a wire; an evolution of oxygen will be perceptible. The article is moved to and fro in the hot bath till the black color desired is produced which, with smaller articles, is generally the case within five minutes. Should the evolution of oxygen cease previous to this, add 1 per cent. more of potassium persulphate.

The article presenting at first a velvety appearance is rinsed in cold water, dried with a soft towel and rubbed ; it will then be of a deep black color with mat luster.

The solution may also be used for coloring black a large number of alloys with a high percentage of copper. Generally speaking, more time is required for coloring alloys than copper.

Patina. This term is applied to the beautiful green colors antique statues and other art-works of bronze have acquired by long exposure to the action of the oxygen, carbonic acid, and moisture of the air, whereby a thin layer of copper carbonate is formed upon them. It has been sought to accelerate by chemical means the formation of the patina thus slowly produced by the action of time and the term patinizing has been applied to the production of such colors.

Artificial patina. There are numerous directions for the production of an artificial patina on metallic objects, and, in conformity with the natural principle of formation, the various artificial processes are based upon the slowest possible action of the patinizing fluid.

To avoid stains the surfaces of the metallic objects should be as bright as possible, and any adhering grease must first be removed by washing with dilute soda lye. The objects are then placed, without touching them with the bare hands, on the bench or other place where they are to be patinized

Patinizing is effected with a dilute solution applied with a brush or sponge. After allowing the first application to dry at a temperature of about 60° F., the process is several times repeated. The composition of the metal to which the patinizing fluid is to be applied, exerts an influence upon the formation of a patina of good quality, the latter being most readily formed upon bronze, while copper and brass are more difficult to patinize; alloys containing arsenic easily turn black.

Donath makes a distinction between acid and alkaline patinizing fluids. The former contain acetic acid, oxalic acid, hydrofluo-silicic acid, and the latter, ammonia, ammonium carbonate, etc. Coatings effected with acids require a longer time for their formation; they are in the beginning less crystalline and at first blue-green, later on, of the color of verdigris, but possess less resistance towards water. Coatings produced with ammoniacal fluids have a dull, earthy appearance, and a blue-green to gray-green color. Yellow-green tones are obtained by the addition of chlorides—common salt, sal ammoniac—to the solution, while copper nitrate or copper acetate yield more blue-green colorations. If a yellow-green coloration is to be changed into blue-green, only ammonium carbonate solution can subsequently be used.

Imitation of genuine green patina, as well as its rapid formation upon objects of copper, and of bronze and brass, is obtained by repeatedly brushing the objects with solution of ammonium chloride in vinegar, the action of the solution being accelerated by the addition of verdigris. A solution of 9 drachms of ammonium chloride and $2\frac{1}{4}$ drachms of potassium binoxalate (salt of sorrel) in 1 quart of vinegar acts still better. When the first coating is dry, wash the object, and repeat the manipulations, drying and washing after each application, until a *green* patina is formed. It is best to bring the articles after being brushed over with the solution into a hermetically closed box, upon the bottom of which a few shallow dishes containing very dilute sulphuric or acetic acid and a few pieces of marble are placed. Carbonic acid being thereby evolved, and the air in the box being kept sufficiently moist by the evaporation of water, the conditions required for the formation of genuine patina are thus fulfilled. If the patina is to show a more *bluish* tone, brush the objects with a solution of $4\frac{1}{4}$ ozs. of ammonium carbonate and $1\frac{1}{2}$ ozs. of ammonium chloride in 1 quart of water, to which a small quantity of gum tragacanth may be added.

A blue-green patina, much used in Paris, is produced by heating in the following solution : Water 500 grammes, corrosive sublimate 2.5 grammes, saltpetre 8.6 grammes, borax 5.6 grammes, zinc oxide 11.3 grammes, copper nitrate 22 to 22.5 grammes.

A brown patina is obtained with the following solution : Oxalic acid 3 grammes, sal ammoniac 15 grammes, distilled water 280 grammes.

The article is to be frequently brushed with the solution · this process requires considerable time.

Patina for copper and brass. The production of two fine tones of color upon copper and brass articles is due to the fact that ammonia attacks and eventually dissolves copper. The following directions are given by *La Nature :* If to objects of copper is to be given the appearance of very antique art objects recently dug up, it is only necessary to immerse them in ammonia. The effect does not show itself immediately, but only after 24 hours. A beautiful dark green coating, which adheres quite firmly, is formed. By allowing the copper object to remain for several days in the fluid the surface is more strongly attacked and the antique effect is heightened.

Another kind of patina which cannot be produced upon copper but only upon brass is obtained by immersing the object in a hot, nearly boiling, mixture of 75 cubic centimeters of ammonia, the same quantity of water and 10 grammes of potash. A uniform durable patina shows itself in half a minute. By allowing the article to remain longer in the solution the patina acquires, without being materially altered, a steely bluish-gray luster.

To produce a *steel-gray color upon copper*, immerse the clean and pickled objects in a heated solution of chloride of antimony·in hydrochloric acid. By using a strong electric current the objects may also be coated with a steel-gray deposit of arsenic in a heated arsenic bath.

For coloring copper *dark steel-gray*, a pickle consisting of 1 quart of hydrochloric acid, 0.125 quart of nitric acid, 1½ ozs. of arsenious acid, and a like quantity of iron filings is recommended.

Various colors upon massive copper.—First draw the object through a pickle composed of sulphuric acid 60 parts, hydrochloric acid 24.5, and lampblack 15.5 ; *or* of nitric acid 100 parts, hydrochloric acid 1½ and lampblack ¼. Then dissolve in a quart of water, 4½ ozs. of sodium hyposulphite, and in another quart of water, 14¼ drachms of blue vitriol, 5¼ drachms of crystallized verdigris, and 7¾ grains of sodium arsenate. Mix equal volumes of the two solutions, but no more than is actually necessary for the work in hand, and heat to between 167° and 176° F. By dipping articles of copper, brass, or nickel in the hot solution they become immediately colored with the colors mentioned below, one color passing within a few seconds into the other, and for this reason the effect must be constantly controlled by frequently taking the objects from the bath. The colors successively formed are as follows :

Upon copper :	*Upon brass :*	*Upon nickel :*
Orange,	Golden-yellow,	Yellow,
Terra-cotta,	Lemon color,	Blue,
Red (pale),	Orange,	Iridescent.
Blood-red,	Terra-cotta,	
Iridescent.	Olive-green.	

Some of these colors not being very durable, have to be protected by a coat of lacquer or paraffine. It is further necessary to diligently move the objects, so that all portions acquire the same color. The bath decomposes rapidly, and hence only sufficient for 2 or 3 hours' use should be mixed at one time.

2. *Coloring brass and bronzes.* Most of the directions given for coloring copper are also available for brass and bronzes, especially those for the production of patinas and the oxidized tones by a mixture of ferric oxide and blacklead.

Many colorations on brass are, however, effected only with difficulty, and are partially or entirely unsuccessful as, for instance, coloring black with liver of sulphur. As a pickle for the production of a

Lustrous black on brass, the following solutions may be used ·
Dissolve freshly precipitated carbonate of copper, while still moist, in strong liquid ammonia, using sufficient of the copper salts so that a small excess remains undissolved, or, in other words, that the ammonia is saturated with copper. The carbonate of copper is prepared by mixing hot solutions of equal parts of blue vitriol and of soda, filtering off and washing the precipitate.

Dilute the solution of the copper salt in ammonia with one-fourth its volume of water, add 31 to 46 grains of graphite and heat to between 95° and 104° F.

According to experiments in the laboratory of the Physikalisch-Technischen Reichsanstalt, the following proportions have proved very effective : Copper carbonate 3½ ozs., liquid ammonia 26½ ozs., and an addition of 5½ ozs. of water. Place the clean and pickled articles in this pickle until they show a full black tone, then rinse in water, immerse in hot water, and dry in sawdust. The solution soon spoils, and hence no more than required for immediate use should be prepared.

For *black pickling* in the hot way, a solution of 21 ozs. of copper nitrate in 7 ozs. of water mixed with a solution of 3¾ grains of silver nitrate in ½ oz. of water, is recommended.

Black of a beautiful luster may be produced, especially upon nickeled brass, by suspending the objects as anodes in a solution of lead acetate (sugar of lead) in caustic soda, using a slight current-density.

Black color on brass optical instruments is produced by placing the brass in a solution of platinum or chloride of gold mixed with stannous nitrate. The Japanese bronze brass with a solution of copper sulphate, alum and verdigris. Success in bronzing depends on the temperature of the alloy, the proportions of metals used in the alloy, drying, and many other small details which can be learned only by practical experience.

Steel gray on brass.—Use a mixture of 1 lb. of strong hydrochloric acid with 1 pint of water to which are added 5¼ ozs. of iron filings and a like quantity of pulverized antimony sulphide.

Hydrochloric acid compounded with white arsenic is also recommended for the purpose. The mixture is brought into a lead vessel, and the object dipped in it should be in contact with the lead of the vessel, or be wrapped around with a strip of lead.

Solution of antimony chloride produces a gray color with a bluish tinge, and a hot solution of arsenious chloride in a small quantity of water a steel gray color.

Silver color on brass. Dissolve in a well-glazed vessel 1½ ozs. cream of tartar and ½ oz. of tartar emetic in 1 quart of hot water, and add to the solution 1¾ ozs. of hydrochloric acid, 4½ ozs. of granulated, or better, pulverized tin and 1 oz. of powdered antimony. Heat the mixture to boiling and immerse the articles to be colored. After boiling at the utmost for half an hour, the articles will be provided with a beautiful, hard and durable coating.

Pale gold color on brass. Dissolve in 90 parts by weight of water, 3.6 parts by weight of caustic soda, and the same quantity of milk sugar. Boil the solution ¼ hour. Then add a solution of 3.6 parts by weight in 10 parts by weight of hot water. Use the bath at a temperature of 176° F.

Straw color, to brown, through golden yellow, and tombac color on brass may be obtained with solution of carbonate of copper in caustic soda lye. Dissolve 5.25 ozs. of caustic soda in 1 quart of water, and add 1¾ ozs. of carbonate of copper. By using the solution cold, a *dark, golden-yellow* is first formed, which finally passes through *pale brown* into *dark brown* with a green luster. Coloration is more rapidly effected by using the solution hot.

Color resembling gold on brass, according to Dr. Kayser: Dissolve 8½ drachms of sodium hyposulphite in 17 drachms of water, and add 5.64 drachms of solution of antimonious chloride (butter of antimony). Heat the mixture to boiling for some time, then filter off the red precipitate formed, and after washing it several times upon the filter with vinegar, suspend it in 2 or 3 quarts of hot water; then heat and add concentrated soda lye until solution is complete. In this hot solution dip the clean and pickled brass objects, removing them frequently to see whether they have acquired the desired coloration. By remaining too long in the bath, the articles become *gray*.

Brown color, called bronze Barbédienne, on brass. This beautiful color may be produced as follows : Dissolve by vigorous shaking in a bottle, freshly prepared arsenious sulphide in liquid ammonia, and compound the solution with antimonious sulphide (butter of antimony) until a slight permanent turbidity shows itself, and the fluid has acquired a deep yellow color. Heat the solution to 95° F., and suspend the brass objects in it. They become at first golden-yellow and then brown, but as they come from the bath with a dark dirty tone, they have to be several times scratch-brushed to bring out the color. If, after using it several times, the solution fails to work satisfactorily, add some antimonious sulphide. The solution decomposes rapidly, and should be prepared fresh every time it is to be used.

A suitable solution may also be prepared by boiling 0.88 oz. of arsenious acid and 1 oz. of potash in 1 pint of water until the acid is dissolved and, when cold, add 250 cubic centimeters of ammonium sulphide. According to the degree of dilution, brown to yellow tones are obtained.

By this method only massive brass objects can be colored brown. To *brassed zinc* and *iron*, the solution imparts brown-black tones, which, however, are also quite beautiful.

Upon massive brass, as well as upon brassed zinc and iron objects, bronze Barbédienne may be produced as follows : Mix 3 parts of red sulphide of antimony (*stibium sulfuratum aurantianum*) with 1 part of finely pulverized bloodstone, and triturate the mixture with ammonium sulphide to a not too thickly-fluid pigment. Apply this pigment to the objects with a brush, and, after allowing to dry in a drying-chamber, remove the powder by brushing with a soft brush.

In Paris *bronze articles* are colored *dead-yellow* or *clay-yellow* to *dark brown* by first brushing the pickled and thoroughly rinsed objects with dilute ammonium sulphide, and, after drying, removing the coating of separated sulphur by brushing. Dilute solution of sulphide of arsenic in ammonium is then applied, the result being a color resembling mosaic gold. The more frequently the arsenic solution is applied, the browner the color becomes. By substituting for the arsenic solution one of sulphide of antimony in ammonia or ammonium sulphide, colorations of a more reddish tone are obtained.

Dead red color on brass. Suspend the articles, previously thoroughly freed from grease, in a solution of equal parts of potassium-lead oxide and red prussiate of potash heated to 122° F. until they have acquired a sufficiently dark color.

For coloring brass articles in large quantities brown by boiling, the following solution is recommended : Water 1 quart, potassium chromate 1½ ozs., nickel sulphate 1½ ozs., potassium permanganate 4½ drachms.

Solution of blue vitriol and potassium permanganate serves the same purpose. However, after boiling, the articles must not be scratch-brushed, but after drying rubbed with vaseline.

Violet and cornflower-blue upon brass: Dissolve in 1 quart of water 4½ ozs. of sodium hyposulphite, and in another quart of water 1 oz. 3¾ drachms of crystallized lead acetate (sugar of lead), and mix the solutions. Heat the mixture to 176° F., and then immerse the cleansed and pickled articles, moving them constantly. First a gold-yellow coloration appears, which, however, soon passes into violet and blue, and if the bath be allowed to act further, into green. The action is based upon the fact that in an excess of hyposulphite of soda, solution of hyposulphite of lead is formed, which decomposes slowly and separates sulphide of lead, which precipitates upon

34

the brass objects, and, according to the thickness of the deposit, produces the various lustrous colors.

Upon the same action is based the spurious gilding of small silvered brass and tombac articles. Though this process has been known for many years, Joseph Dittrich obtained a German patent for it. He dissolves in 6½ lbs. of water, 10½ ozs. of sodium hyposulphite, and 3½ ozs. of lead acetate (sugar of lead).

Similar lustrous colors are obtained by dissolving 2.11 ozs. of pulverized tartar in 1 quart of water, and 1 oz. of chloride of tin in ½ pint of water, mixing the solution, heating, and pouring the clear mixture into a solution of 6.34 ozs. of sodium hyposulphite in 1 pint of water. Heat this mixture to 176° F., and immerse the pickled brass objects.

Ebermayer's experiments in coloring brass.—Below the results of Ebermayer's experiments are given. In testing the directions, the same results as those claimed by Ebermayer were not always obtained ; and variations are given in parentheses.

I. Blue vitriol 8 parts by weight, crystallized ammonium chloride 2, water 100, give by boiling a *greenish* color. (The color is *olive-green*, and useful for many purposes. The coloration, however, succeeds only upon massive brass, but not upon brassed zinc.)

II. Potassium chlorate 10 parts by weight, blue vitriol 10, water 1000, give by boiling a *brown-orange* to *cinnamon-brown* color. (Only a *yellow-orange* color could be obtained.)

III. By dissolving 8 parts by weight of blue vitriol in 1000 of water, and adding 100 of caustic soda until a precipitate is formed, and boiling the objects in the solution, a *gray-brown* color is obtained, which can be made darker by the addition of colcothar. (Stains are readily formed. Brassed zinc acquires a pleasant *pale-brown.*)

IV. With 50 parts by weight of caustic soda, 50 of sulphide of antimony, and 500 of water, a pale *fig-brown* color is produced. (Fig-brown could not be obtained, the shade being rather *dark olive-green.*)

V. By boiling 400 parts by weight of water, 25 of sulphide of antimony and 600 of calcined soda, and filtering the hot solution, *mineral kermes* is precipitated. By taking of this 5 parts by weight and heating with 5 of tartar, 400 of water, and 10 of sodium hyposulphite, a beautiful *steel-gray* is obtained. (The result is tolerably sure and good.)

VI. Water 400 parts by weight, potassium chlorate 20, nickel sulphide 10, give, after boiling for some time, a *brown* color, which, however, is not formed if the sheet has been pickled. (The brown color obtained is not very pronounced.)

VII. Water 250 parts by weight, potassium chlorate 5, carbonate of nickel 2, and sulphate of ammonium and nickel 5, give, after boiling for some time, a *brown-yellow color*, playing into a magnificent red. (The results obtained were only indifferent.)

VIII. Water 250 parts by weight, potassium chlorate 5, and sulphate of nickel and ammonium 10, give a beautiful *dark brown*. Upon massive brass a good dark brown is obtained. The formula, however, is not available for brassed zinc.

3. *Coloring zinc.* Direct coloring of zinc does not give, as a rule, reliable results, and it is therefore recommended to first copper or tin the zinc and color the coating thus obtained.

Black on zinc. a. Dissolve crystallized copper nitrate 2 parts and copper chloride 2 parts in acidulated water 64 parts, and add to the solution hydrochloric acid of 1.1 specific gravity 8 parts. The resulting fluid has a slightly bluish color. A sheet of zinc, previously scoured bright by means of dilute hydrochloric acid and fine sand, will, when immersed in the fluid, immediately be colored intensely black. By removing the sheet thus treated, at once from the fluid and rinsing it without loss of time in a large quantity of pure water and allowing it to dry, the black coating will adhere very firmly to the zinc.

b. Dip the object in a boiling solution of pure green vitriol 5.64 ozs. and ammonium chloride 3.17 ozs. in 2½ quarts of water.   Remove the loose black precipitate deposited upon the object by brushing, again dip the object in the hot solution and then hold it over a coal fire until the ammonium chloride evaporates.   By repeating the operation three or four times, a firmly adhering black coating is formed.

Gray, yellow, brown to black colors upon zinc.—Bring the articles into a bath which contains 6 to 8 quarts of water, $3\frac{1}{2}$ ozs. of nickel-ammonium sulphate, $3\frac{1}{2}$ ozs. of blue vitriol and $3\frac{1}{2}$ ozs. of potassium chlorate.   The bath is to be heated to 140° F.   By increasing the content of blue vitriol a dark color is obtained, and a brighter one with the use of a larger proportion of nickel salt.   The correct proportions for the determined shades will soon be learned by practice.   When colored, the articles are thoroughly rinsed, dried, without rubbing, in warm sawdust, and finally rubbed with a flannel rag moistened with linseed oil, whereby they acquire deep luster, and the coating becomes more durable.

Brown patina on zinc.—The objects are first coppered in a copper bath containing potassium cyanide, then in the acid-copper bath, rinsed, and finally suspended in a pickle consisting of a solution of 5.29 ozs. of blue vitriol and 2.82 ozs. of potassium chlorate in one quart of water at 140° F., until they show the desired brown tone.   They are then rinsed in water, scratch-brushed with a fine brass-brush, for a short time replaced in the pickle, again thoroughly rinsed in water, and dried with a soft cloth.

By suspending zinc in a nickel bath slightly acidulated with sulphuric acid, a firmly adhering *blue-black* coating is, after some time, formed without the use of a current.   This coating is useful for many purposes.   A similar result is obtained by immersing the zinc objects in a solution of 2.11 ozs. of the double sulphate of nickel and ammonium and a like quantity of crystallized ammonium chloride in 1 quart of water.   The articles become first *dark yellow*, then successively *brown*, *purple-violet* and *indigo-blue*, and stand slight scratch-brushing and polishing.

A *gray coating on zinc* is obtained by a deposit of arsenic in a heated bath composed of 2.82 ozs. of arsenious acid, 8.46 drachms of sodium pyrophosphate and $1\frac{3}{4}$ drachms of 98 per cent. potassium cyanide, and 1 quart of water. A strong current should be used so that a vigorous evolution of hydrogen is perceptible. Platinum sheets or carbon plates are used as anodes.

A sort of *bronzing* on zinc is obtained by rubbing it with a paste of pipe-clay to which has been added a solution of 1 part by weight of crystallized verdigris, 1 of tartar, and 2 of crystallized soda.

Red-brown shades on zinc. Rub with solution of copper chloride in ammonia.

Yellow-brown shades on zinc. Rub with solution of copper chloride in vinegar.

4. *Coloring iron. Browning of gun barrels.* Apply a mixture of equal parts of butter of antimony and olive oil. Allow the mixture to act for 12 to 14 hours, then remove the excess with a woolen rag and repeat the application. When the second application has acted for 12 to 24 hours, the iron or steel will be coated with a bronze-colored layer of ferric oxide with antimony, which resists the action of the air, and may be made lustrous by brushing with a waxed brush.

A patina which protects metals—iron, zinc, tin, etc.—from rust, is, according to Haswell, obtained as follows: The article, previously freed from grease and pickled, is suspended as negative electrode in a solution of $15\frac{1}{2}$ grains of ammonium molybdate and $\frac{1}{3}$ oz. of ammonium nitrate in 1 quart of water. A weak current should be used—0.2 to 0.3 ampère per $15\frac{1}{2}$ square inches.

To protect gun barrels and other articles of iron and steel from rust, they are, according to Haswell, suspended as anodes in a bath consisting of a solution of lead nitrate and sodium nitrate, into which manganous oxide has been stirred.

Lustrous black on iron. Apply solution of sulphur in turpentine prepared by boiling on the water-bath. After the evaporation of the turpentine a thin layer of sulphur remains upon the iron, which on heating immediately combines with the metal.

A lustrous black is also obtained by freeing the iron articles from grease, pickling, and after drying, coating with sulphur balsam,* and burning in at a dark-red heat. If pickling is omitted, coating with sulphur balsam and burning-in must be twice or three times repeated.

The same effect is produced by applying a mixture of three parts flowers of sulphur, and 1 part graphite with turpentine, and heating in the muffle.

According to Meritens a bright black color can be obtained on iron by making it the anode in distilled water, kept at 158° F., and using an iron plate as cathode. The method was tested as follows : A piece of bright sheet pen-steel was placed in distilled water and made the anode by connecting it with the positive pole of a plating dynamo, and a similar sheet was connected with the negative pole to form the cathode. An electromotive force of 8 volts was employed. After some time a dark stain was produced, but it lacked uniformity. The experiment was repeated with larger plates, when a good blue-black color was obtained on the anode in half a hour. On drying in sawdust the color appeared less dense, and inclined to a dark straw tint. The back of the plate was also colored, but not regularly. The face of the cathode was discolored with a grayish stain on the side opposite to the anode, but on the other side the appearance was almost identical with the black of the anode. The water became of a yellowish color.

Fresh distilled water was then boiled for a long time so as to expel all trace of oxygen absorbed from the atmosphere, and the experiment repeated as in the former cases. No perceptible change took place after the connection had been made with the dynamo for a quarter of an hour. After the interval of one hour a slight darkening occurred, but the effect

---

* Sulphur dissolved in linseed oil.

was much less than that produced in five minutes in aerated water.

The action of the liquid in coloring the steel is evidently one of oxidation, due to the dissolved oxygen, which becomes more chemically active under the influence of the electric condition, and gradually unites with the iron.

The *mat black coating upon clock cases of iron and steel*—the so-called *Swiss mat*—is not produced by the electric process, but by a slow process of oxidation, ferroso-ferric oxide being formed. The objects, previously cleaned from grease with the greatest care, are brushed over by means of a sponge or brush with a ferric chloride solution, called *ferroxydin*, allowed to dry, and then steamed. For the production of a very strong mat, the process is to be twice or three times repeated. By one operation a beautiful black with semi-luster is obtained.

Blue color on iron and steel. Immerse the article in ½ per cent. solution of red prussiate of potash mixed with an equal volume of ½ per cent. solution of ferric chloride.

Brown-black coating with bronze luster on iron. Heat the bright objects and brush them over with saturated potassium dichromate solution. When dry, heat them over a charcoal fire, and wash until the water running off shows no longer a yellow color. Repeat the operation twice or three times. A similar coating is obtained by heating the iron objects with a solution of 10 parts by weight of green vitriol and 1 part of sal ammoniac in water.

To give iron a silvery appearance with high luster.—Scour the polished and pickled iron objects with a solution prepared as follows: Heat moderately 1½ ozs. of chloride of antimony, 0.35 ozs. of pulverized arsenious acid, 2.82 ozs. of elutriated bloodstone with 1 quart of 90 per cent. alcohol upon a water-bath for half an hour. Partial solution takes place. Dip into this fluid a tuft of cotton and go over the iron portions, using slight pressure. A thin film of arsenic and antimony is thereby deposited, which is the more lustrous the more carefully the iron has previously been polished.

5. *Coloring of tin.*—*A bronze-like patina* on tin may be obtained by brushing the object with a solution of $1\frac{3}{4}$ ozs. of blue vitriol and a like quantity of green vitriol in 1 quart of water, and moistening, when dry, with a solution of $3\frac{1}{2}$ ozs. of verdigris in $10\frac{1}{2}$ ozs. of vinegar. When dry, polish the object with a soft waxed brush and some ferric oxide. The coating thus obtained being not durable, must be protected by a coating of lacquer.

Durable and very warm sepia-brown tone upon tin and its alloys.—Brush the object over with a solution of 1 part of platinum chloride in 10 of water, allow the coating to dry, then rinse in water, and, after again drying, brush with a soft brush until the desired brown luster appears.

A *dark* coloration is also obtained with ferric chloride solution.

6. *Coloring of Silver.*—See " Deposition of Silver. "

Electrochroma. The process for the production of colors on metals by electro-deposition, known under this term, is the invention of Mr. F. Arquimedas Rojaz. By this method either deposits or smuts of any desired color or texture can be produced upon any metal used as a cathode. The anodes used are of pure carbon, no metal of any sort being put into the tank containing the plating solution except the work itself. In starting to color a piece of metal, be it brass, copper, tin, lead or iron, etc., the metal is first dipped into a cleaning solution, then into a hot water bath, next into the tank containing the solution for whatever background color is desired. A current of 8 to 12 volts pressure with a strength of 1 ampère per square inch of surface is used. After an immersion in the tank for from two to three minutes the work is dipped into hot water, and from there into a tub containing a dip solution. Here the finish of the process takes place, and the beautiful shades of color are produced. A piece of work, such as a lock plate for a store, may be given a green verde smut in the plating tank and then be changed to a light blue background in the dip tub. Gold finishes, rose antique and green, may be produced at will in a few seconds of time, without any gold in the solution.

All of the solutions used in the process are fully protected by patents and are furnished ready for use. They are said to be made up of more than half a dozen elements, the proportions of which are so evenly balanced that a slight variation in the amounts used of each ingredient will throw the entire solution out of gear.

CHAPTER XV.

In the electro-plating industry recourse is frequently had to lacquering in order to make the deposits more resistant against atmospheric influences, or to protect artificially prepared colors, patinas, etc. Thin, colorless shellac solution, which does not affect the color of the deposit or of the patinizing, is, as a rule, employed, while in some cases colored lacquers are used to heighten the tone of the deposit, as, for instance, gold lacquer for brass.

The lacquer is applied with a flat fine fitch brush, the object having previously been heated hand-warm. The brush should be frequently freed from an excess of lacquer, and the lacquer be applied as uniformly as possible without undue pressure of the brush. An excess of lacquer, which may have been applied, is removed by means of a dry brush.

The lacquer for immediate use is kept in a small glass or porcelain pot, across the top of which a string may be stretched. This string is intended for removing by wiping the excess of lacquer taken up by the brush. Crusts of dried-in lacquer should be carefully removed, and the contents of the small pot should under no conditions be poured back into the can, as otherwise the entire supply might be spoiled.

After lacquering, the object is dried in an oven at a temperature of between 140° and 158° F., small irregularities being thereby adjusted, and the layer of lacquer becoming transparent, clear and lustrous.

Electro-plated articles which are to be lacquered must be thoroughly rinsed and dried to remove adhering plating solution from the pores, otherwise ugly stains will form under the coat of lacquer.

(538)

If it becomes necessary to thin a spirit lacquer, only absolute alcohol, *i. e.*, alcohol free from water, should be used for the purpose, since alcohol containing water renders the coat of lacquer muddy and dull.

The development in the art of lacquer-making has advanced with and in a measure kept pace with that made in the electro-deposition of metals. With the use of new metals, the introduction of new and altered formulas and processes for finishing metals, the employment of new and different chemicals, in fact with every change or alteration in the methods of finishing and using metals, changes have been made in the nature of the lacquers employed in their protection.

Lacquers to be acceptable to the metal-worker must be perfectly adapted to each special use, and not only suit the varied metals, finishes and conditions of the work, but also meet and overcome difficulties arising from, for instance, the influence of climatic changes and the use to which the lacquered metal is subjected. Many cases of trouble in the finishing of metal may now be traced to the use of an improper lacquer for the particular metal or finish. Thus ingredients and chemicals which from their nature are antagonistic to a bronze metal and detrimental to it should not be included in a lacquer for bronze, although the same ingredients may be beneficial to a silver, gold or aluminium surface.

The most noted improvements have been effected in lacquers for brass bedsteads, gas and electric fixtures, black lacquers, and the lacquers made with the special object of saving time in their application and money in their use.

A review of all the lacquers made for the above-mentioned purposes is not within the province of this work, and we must therefore confine ourselves to the enumeration of the newest and most important ones for general use, with which we have become familiar.

Pyroxyline lacquers.—These lacquers, known under various names, such as Lastina, Pyramide and Obelisk, etc., were introduced to the trade in America as early as 1876, and were gradually adopted until early in the 80's, when their use became general, and since then they have become known throughout all parts of America and Europe. Pyroxyline lacquer represents a clear, almost colorless fluid, and smells something like fruit-ether, reminding one of bananas. It is chiefly used as a dip lacquer, though there is also a brush lacquer, which is applied with a brush, like spirit lacquer.

The lacquer possesses the following good properties: The transparent, colorless coat obtained with it can be bent with the metallic sheet to which it has been applied without cracking. It is so hard that it can scarcely be scratched with the finger-nail, shows no trace of stickiness, and it is perfectly homogeneous even on the edges. This favorable behavior is very likely due to the slow evaporation of the solvent, and the fact that the lacquer quickly forms a thickish, tenacious layer, which though moved with difficulty is not entirely immobile. Another advantage of the lacquer—especially as regards the metallic objects—is that the coating in consequence of its physical constitution preserves the character of the bases. In accordance with the nature of pyroxyline, the coating is not sensibly affected by ordinary differences in temperature, and does not become dull and non-transparent, as is the case with resins, in consequence of the loss of molecular coherence. It can be washed with water, and protects metals coated with it from the action of the atmosphere. It may also be colored, but of course only with coloring substances—mostly aniline colors—which are soluble in the solvent used.

For lacquering articles by dipping, they should be as clean as for plating, and so arranged that the lacquer will run off properly. Allow them to drip over the drip tank until the lacquer stops flowing. Dry in a temperature of 100° to 120° F., if possible using a thermometer. Dip lacquers will dry in the air, but baking improves the finish.

The receptacle for holding the lacquer and thinner for dipping purposes, should be either of glass, stoneware, chemically enameled iron, or a tin-lined wooden box—the preference being in the order named. Lacquer or thinner should never be placed in copper or galvanized iron tanks.

For thinning the lacquer when it has become too thick by the evaporation of the solvent, use the thinner which is recommended for each particular grade of lacquer.

The appearance of rainbow colors upon objects lacquered with pyroxyline lacquer is due either to insufficient cleanliness, especially to the presence of grease upon the objects, or to the lacquer having been too much diluted. Objects to be lacquered should be freed from grease by the use of platers' compound, rinsed in hot water, dried in thinner and then lacquered. The use of benzine, aside from the danger it entails, is not always effective in removing grease from the pores of the metal. After cleaning, the polished surface of the work should not be touched with the hands. If the rainbow colors are due to the lacquer having been too much thinned, let the vessel containing it stand uncovered for some time in a place free from dust, so that it becomes somewhat more concentrated by the evaporation of the solvent, or correct the tendency to rainbow colors by adding more undiluted lacquer to the mixture. In adding thinner to lacquer it is always advisable to give it plenty of time to act upon the pigment in the lacquer. This can be facilitated by stirring with a wooden paddle.

Very nice shades of color can be produced by coating the objects, previously well cleansed from grease, with lacquer by dipping, allowing the coat to become dry, then suspending the objects for a few seconds in golden-yellow, red, green, etc., dyes, known as dipping colors, next washing in water and finally drying. By mixing the coloring dyes in various proportions nearly every desired tone of color can be obtained.

Special invisible lacquer for ornamental cast and chased interior grille, rail and enclosure work. This lacquer is made in three grades for use, 1, with the brush; 2, with the spraying machine; and 3, as a dip lacquer. Its presence cannot be detected on any of these sensitive finishes, and the fine mat finishes are left without the slightest luster after it has been applied thereto. It can be mixed with the pigment fillings so much used in cast ornamental mountings, figured moulding for the verds, Florentine, rose and antique effects. Sandblasted and brushed plain parts will not take on a sheen from this lacquer and will, therefore, not make a contrast in the lights of the filled and smooth portions of the work. The fine reliefs in these finishes, it has been found, will not be disturbed because of the lacquer softening the pigments when it is applied by spraying. In use this lacquer can be thinned so as to flow away from the various parts that make up a grille or rail without leaving any lines or waves, or causing glossy places or variations of lines. This lacquer is made by The Egyptian Lacquer Manufacturing Co., of New York, and with it the rich subdued effects of dead, mat, sanded and semi-dead finishes can be protected without in the least affecting their appearance.

Satin finish lacquer is made by the same concern just mentioned; it comes in two grades, one for brush and the other for dip work. Its purpose is to maintain the light, but somewhat solid, effect in which body color rather than tints predominate; its deadness gives to these body effects a plastic appearance. It can be used to protect a velvet-like tint resembling the ground gold, frosting or satin finish seen in ormolu and colonial gold, as well as dead and dull surfaces, or unpolished, lusterless and mat gold and mat silver. It can also be used to create a dead luster, or a deadened lustrous surface, for example, on mat designs upon a lustrous ground, where the lacquer lights up the satin finish. Jewelry, silver and novelty manufacturers can use it for general finishing of their work, as it will not alter the sensitive metal colorings, nor will it fill up to a gloss delicately brushed, satined, or chased surfaces or smut tints.

Dip lacquer for pickled castings to be copper-plated and oxidized. Articles made from iron and steel castings that are pickled or water rolled, or from hot-rolled steel, where the scale is pickled off, or any other similar work which is prepared by the same inexpensive method, when copper-plated and oxidized, must be lacquered with a lacquer which will give life to the naturally dead surface of the metal and to the smut left from the oxidation when not scratch-brushed. This is a very rapid and inexpensive process since it does away with the costly operations of polishing, scratch-brushing and cleaning; the finish depends entirely upon the life and luster of the lacquer, hence it is best to use one of the lacquers now designated.

With *helios dip lacquer, special,* which is made by The Egyptian Lacquer Manufacturing Co., of New York, a fine luster is given to the dead backgrounds and a bright and lustrous finish to the smooth parts of the work and in many respects this lacquer renders the work equal to that which has been polished. Many other lacquers which have been tried dry down to the natural deadness of the metal finish and consequently the effect of the plating and oxidizing is lost, or, if not entirely lost, is not brought out in its right color.

Old brass or brush-brass finishes. From 90 to 95 per cent of all brass for gas and electric fixtures, bedsteads and similar work is finished in brush-brass. Lacquers are specially made for the high gloss effects, as well as for the dull, or antique finish. As this finish is more susceptible to tarnish and stain than any other known finish, it is important that precise particulars be given as to the handling of this work preliminary to lacquering. For instance, where this work is finished with pumice, sand, flint, etc., and water, as most of it is, it should, as fast as completed be placed in a tank containing borax solution made by dissolving 1 lb. powdered borax in hot water and adding enough water to make 5 gallons. Use cold. Let the work accumulate in the solution until ready to lacquer. Then rinse the work in hot water and dip it in thinner. It will dry without stain by hanging up for a moment or two, when it should be immediately lacquered.

Where " old brass composition " or emery and oil is used, the work should be cleaned from grease in "plater's compound " or some other non-tarnishing cleaner, and can be placed in the borax-solution as fast as finished on the brush. It is then rinsed in hot (not·too hot) water, dried in thinner, and immediately lacquered.

Wiping the surface with a soft cloth or chamois skin does not remove the moisture from the metal. This is particularly apparent when there is much humidity in the air, and verdigris or oxide rapidly forms in the scratches made by the abrasive materials and causes much subsequent trouble. The heat of the oven converts this moisture, combined with the oxide, into steam which penetrates the lacquer and causes staining of the film. Sawdust should never be used for drying metals given an old-brass finish.

Brush brass finish lacquers. This very sensitive and easily discolored finish is readily marred by the use of an inefficient transparent dip lacquer. No existing finish requires more exacting and careful treatment than the brush brass finish, the finely brushed lines attracting and retaining substances which tarnish it readily. As a rule these substances are not visible, and cannot be easily removed by ordinary cleaning methods. After a time every speck of dirt shows under the lacquer coating, and is the cause of the various discolorations often seen in brush brass finish; they vary from the tints shading into the browns to tints running into the greens, and are in almost every instance caused by the oxidizing influences of contaminating matters left upon or attracted by the metal before it has been lacquered. When work is handled in large quantities these imperfections are especially noticeable, for such work cannot always be inspected one piece at a time. The old method of drying the buffed and smooth-surfaced finishes with sawdust and then rubbing them with a soft muslin material is inadequate as well as uncertain for the brush brass; in fact this process primarily causes the imperfections which it is intended to prevent. At any rate, the result is necessarily doubtful when brush brass is dried in this way and allowed to stand for even a very short time before it is protected with lacquer.

Egyptian brush brass dip lacquer and brush brass thinner, made by the Egyptian Lacquer Manufacturing Co., of New York, meets the necessary and varied conditions called for by this finish. After the brush brass has been washed in plater's compound and well rinsed in cold and hot waters, the work is first dipped into the brush brass thinner, which absorbs all moisture left on the metal and removes whatever impurities may have been attracted to it, and prepares the work for its dip into the brush brass dip lacquer. By the dip into the thinner a chemically pure metal surface is provided for the reception of the lacquer coating, and this guarantees the brush brass finish itself against discoloration, since the lacquer has been applied to a practically chemically clean and pure surface.

Brush brass work which cannot be conveniently dip-lacquered should be spray-lacquered in preference to lacquering with a brush, because the fine irregularities of the brushed surface of the metal retard the free flow of a brush lacquer. In other words, a brush lacquer cannot be applied quite as effectively as on a smooth finish, for, owing to the irregularities of the metal surface spoken of, an obstruction is placed in the way of the flowing of the lacquer when it is applied with a brush, because with the use of the latter the separations in the lacquer, due to the uneven distribution of it from the bristles of the brush, sometimes leave minute parts of the surface unlacquered and the irregularities of the brushed metal surface prevent the lacquer from spreading over these minutely exposed lines. Thus when applying the lacquer with a brush it happens now and then that the exposed and unlacquered portions tarnish and destroy the appearance of the entire work. On large articles the lacquer should be sprayed, and the article itself turned by mechanical means during the application of the lacquer, so as to give momentum to its flow, thereby insuring its even distribution.

Brass bedstead lacquering. Complaints of the same kind, namely, streaks in lacquered work, have been the cause for

35

replacing brush lacquering of brass bedsteads by the spray. Since the vogue for satin and drawn emery finishes have taken the place of the old English gilt bedstead finish the spraying process has become even more necessary. The unusual depth of the cut in the metal surface made by these finishes has created a new problem for lacquer makers. The lacquer used on this work should be unusually heavy, in fact heavy enough and dense enough to fill these abnormally penetrated surfaces, for the lacquer film must in all instances be built up so as to protect the highest exposed points of this finish. The lacquer for this finish must be applied with a spray since it is necessary that a thick and plastic coating should be applied, one indeed which when dry shall be hard and tough enough to resist marring from the usual rough and severe treatment to which a bedstead is subjected.

Dead black lacquers produce imitation dead and mat finishes. These are variously known as imitation Bower Barff, wrought iron, ebony or rubber finishes. If the same preliminary steps are taken in preparing metal goods for the black lacquers as for japan and enamel, just as durable and lasting results will be obtained in a small fraction of the time and at a minimum cost in labor. The best class of japanning on iron castings, hot or cold rolled steel requires two coats of either thin japan or some other similar preparation, each coat requiring several hours' baking, and usually a delay of several days before the surface of the last coat is in condition to be rubbed down.

To get the same results with the black lacquers on *sand pitted cast iron*, two coats of metallic filler, applied with a brush, baked a short time at about 180° F. to harden, and then rubbed down with fine emery cloth or No. 2 garnet paper, followed by one or two air-drying coats of lacquer will be sufficient. On smooth-surfaced metals one or two thin coats of lacquer can be applied in place of the metallic filler as a base for the final coat. In many cases one coat of the lacquer will be found sufficient to give the desired finish, and the entire process may be completed in a few hours, where it will require from one to five days to secure the same finish with japan, and besides all the equipment necessary for the latter will be entirely eliminated.

If desired, the metal can be given a light copper plate and then be oxidized as a base for the finishing coat of black lacquer.

For high luster finishes such as are obtained with enamels, glossy, black lacquers are used, and to increase the brilliancy and high luster the same as with enameled goods which are given a finishing coat of baking varnish, a high grade of transparent lacquer is used over the glossy black lacquer the same as the varnish on the enamel.

On goods made from nonferrous metals such as high-grade optical goods, opera and field glasses, and all classes of instrument work, where sliding tubes and other parts are to be finished with a glossy or dead black lacquer, where both beauty and great durability are the chief essentials, the surface of the brass should be first prepared by chemically blacking the metal with copper-ammonia or any other good black dip. Then a filling coat of any black lacquer, preferably a dead black, should be used. The surface is then in perfect condition for the finishing coat of black lacquer. A black background and very adhesive surface are obtained by this method, and the finish will withstand the hard usage these goods are made for.

Dead black lacquer as a substitute for Bower-Barff. The genuine Bower-Barff is a matted black finish for iron and steel. It is produced by heat and steam liberating the oxygen from the iron and forming magnetic oxide.

The oven and other equipment required for this finish is not practicable in the average factory, as the demand for goods in this finish, outside builders' hardware, is not commensurate with the cost of providing and maintaining a plant for this purpose.

A number of imitation finishes are made, by using solutions of sulphur and linseed oil, sulphur, graphite and turpentine, and other similar solutions. A coating of these mixtures is

applied and the metal heated to a red heat to burn it in or
else the goods are baked in a muffle. But they are all slow
and uncertain processes, and some kind of special equipment
must be provided to do this work.

The method for obtaining this finish most in use, and for
which any plating-room is equipped, is by using an antique
black or Bower-Barff lacquer. Such lacquer has a number of
advantages over the above-described processes, which can
only be used on iron or steel ; the lacquer will give the finish
on any metal.

To get the Bower-Barff on iron or steel the metal should
first be lightly copper-plated and oxidized ; and if brass or
bronze is used it is only necessary to oxidize the metal with
any black dip, or electro-oxidize. Then the antique black
lacquer is applied for the finish. The lacquer can also be
lightly sand-blasted if an increased mat is desired.

In the above-described processes good results are obtained
by the use of the following lacquers, made by the Egyptian
Lacquer Manufacturing Co. of New York : Dead Blacks, Egyptian Antique Blacks, Ebony, and Rubber Finish Lacquers.

Spraying of lacquers.—The application of lacquers by the
pneumatic air spray having for the last few years been gradually adopted, has proved advantageous in finishing various
goods, the success in application depending upon many minor
details of manipulation ; these come readily to the lacquerer
while using the spray.

The spraying machine consists of a pump, called a compressor, generating the air, transferring the air to a storage
tank. If this pump is automatic, copper flexible tubing is
used, if stationary, a gas pipe. The storage tank which holds
the air, has a gauge indicating the number of pounds carried.
A safety valve is also on the tank to control the air. These
tanks vary a great deal ; it depends entirely upon the number
of cups drawing off the air and the air must be regulated accordingly. It will run from 18 lbs., and in some cases as
high as 60. There is a rubber hose of flexible copper tube

connected to the storage tank long enough to cover the entire work-bench to which the cups are fastened. The cup or container is an atomizer throwing a spray very much the same as a perfume atomizer, although it is made in sizes from half a pint to a quart. Cups or containers are made both of glass and metal. Some prefer the glass for the reason that it is possible to see the lacquer in the container at any time. Glass cups have, however, the drawback of liability of breakage which may result from careless or rough usage about the shop, and besides some sprays are so constructed that under certain conditions it is an easy matter for the air pressure to be accidentally switched directly into the container, and with a pressure of 40 lbs. both container and lacquer are destroyed. For these reasons it would seem that metal containers are to be preferred. The spray may be gauged by a small catch on the side of the nozzle. This style of sprayer is considered very practical, although there are many more complicated ones in the market.

The equipment to be used in producing the air, storing it and in forcing it to the spray in a pure condition should be of sufficient capacity and be provided with the proper appliances to guard against fluctuations which in the flow of the lacquer stream itself interferes with the continuous flow of the lacquer. This flow of necessity must be uniform in strength and outflow, else the results cannot fail to be irregular. It naturally follows that the compressor which regulates this must be such as to be capable of sustaining this pressure uninterruptedly. The quality of the lacquering changes with the irregularity of the pressure. The air should be taken from a part of the building which is far removed from the steam exhausts or other localities where the air or atmosphere is impure or moist ; the drier the air, the less condensed water will enter into the pipe line. The air should be stored in a tank close to where the spray is in use, for this helps in the precipitating of impurities just before it goes into the lacquer, and the extra volume close at hand steadies the pressure. A reducing valve in the line between the tank and spray, which can be drained occasionally, is another precaution which may be provided against the admission of water. The addition of a filter will be found to be of great advantage, as it will catch the most minute particles of oil, moisture and dirt just before the air reaches the flexible hose to which the spray is attached.

Assuming that both the pressure and clean air referred to can be relied upon, then the next thing necessary is to use the lacquer in as heavy a condition as possible. By this is meant that it should be neither too heavy nor too light for the air to raise it to the nozzle, atomize it and apply it by flowing it out from the spray upon the work in an even and heavy film.

The lacquer should never be thinned so as to make it easier in the spray, for in that case the lacquer will create runs upon the surface of the work; if unusual thinning is necessary to get an even flow from the spray then either the pressure or the adjustment of the spray, or the spray itself, is at fault. While the lacquer is being applied from the spray the work which is being lacquered should be kept moving in a revolving motion in order to insure an even distribution of the lacquer, and avoid an uneven distribution of it; in other words, to prevent matting.

The high pressure used drives the lacquer onto the object, after which, however, the liquid must take care of itself, and it must then flow together into a smooth surface, or else the whole process is worthless.

The spraying-on of lacquers to be successfully used depends not only upon the nature of the articles sprayed, but upon the lacquer itself. Special lacquers have been made for these purposes, and with them success may readily be obtained. A special lacquer has been made for lamps, chandeliers and gas fixtures; another for silver and white metals; another for builders' hardware. With these when applied uniformly, the lacquer spreads evenly and covers the surface entirely without break, and presents an unusually uniform appearance without disfiguring blotches or patches, indicating an unequal thickness of lacquer.

Most of the lacquers which we have seen tested were made by The Egyptian Lacquer Manufacturing Company of New York. In many instances it will be found that ordinary operators with less skill than the trained lacquerer can do very satisfactory work with these machines.

Spraying black lacquers. By applying the black lacquers with a spray various finishes heretofore made with either baking enamels or japans can now be finished with black lacquer. Whenever great durability and toughness are essential and where the fine finish made with the black lacquer is but a secondary consideration, a priming lacquer should be first applied to the work; after this the black lacquer should be sprayed over it. Such a finish makes up in toughness and tenacity the slight runs in its appearance. A second coating of black lacquer without this priming coat will not be proof against the hard usage to which some of these finishes are frequently exposed.

A coat of priming lacquer is of great advantage in many instances where the metal surface is not of an adhesively magnetic nature, or on a metal that cannot be entirely prevented from taking an oxide if exposed to the air even only during the short time of lacquering. The coat of priming lacquer is also a desirable preventative where large quantities of work are being lacquered and where cleanliness of the work cannot always be absolutely relied upon. Peeling and chipping, either or both, are often caused by the inexperience of the lacquerer in mixing the lacquer; if the body is thinned to the extent that it weakens the binding qualities of the material something of the kind is bound to happen.

Since the advent of antique effects, such as mission and Flemish, and other dark and subdued finishes on furniture, etc., the manufacturers of art metal goods have given close attention to having their goods in conformity with the furniture and trimmings in buildings.

They have found the dead black lacquers the best for this purpose, and the question of application has been solved by the spray, as it was impossible to get the fine results they require by either brushing or dipping the black lacquers, for unless the operator was skilled in the application of lacquers by these methods there would be such imperfections as streaks, laps, runs or drip. With the spray all these difficulties are obviated and the finish cannot be otherwise than perfect, and the lacquer thereby used to the best advantage, with the fine result intended for it by the makers.

To use the spray successfully for this purpose, the base used in the lacquer must be adapted to go through the spray nozzle without clogging and going onto a surface lumpy. With the spray any of the blacks proposed by The Egyptian Lacquer Manufacturing Co. can be applied on all classes of metals, whether of a design with deep indentations or interstices or on the flattest surface, with artistic perfection.

The same can be said about the finishing of other goods, such as slate electric switchboards, gas stoves, heaters, steel or other box enclosures, cast parts, steel or brass stampings, or, in fact, all other articles made from any of the metals or any of the alloys.

The black lacquers will retain their original finish and remain black under heat. And there is no other black made that will adorn this class of goods the same from the points of beauty, durability and salableness. Stoves and heaters have large and porous surfaces as the sheet metal is left in its natural condition just as it comes from the rolling mill, and for this reason it has been found difficult to apply the black lacquer with a brush, but the spray puts it on perfectly, and transforms it from an object of roughness to one of uniformity. To avoid marring the nickel trimmings during the operation of spraying, a mask or other covering is used to protect those parts, and it is then a very rapid process.

This also applies to all other articles constructed from the same material and on the same order of the stoves and heaters.

ı. On other metal goods, where designs or ornaments are to be put on, with black or other colored lacquers, stencils are used, and the lacquers sprayed onto it. With the spray application there will be no runs or other disfigurement of the design.

The spray, with the blacks or other colored lacquers, can be used.

Water-dip lacquers and their use.—These lacquers are not, as often believed, lacquers in which water is used. Their name is derived from the fact that after the metal has been dipped or plated, it may, while wet and without drying it, be dipped into the lacquer without in any way affecting the metal finish.

The advantage of water-dip lacquers is readily appreciated by the manufacturers who are rapidly adopting them for many classes of small work. They are especially well adapted for bright-dipped finishes, such as are usually finished in bulk by basket dipping. Tarnish affects this finish almost instantly if it is allowed to dry and then lacquered in the usual way; therefore the lacquering must be carried out as soon as the dipping process has been finished.

The most flagrant example of tarnishing is in the case of plated work, particularly goods copper-plated, and to prevent tarnishing, such goods must be lacquered at once. Even the customary drying-out will usually not suffice to prevent the tarnishing, and the result is either the increase in labor in handling the goods, or the production of a large amount of imperfect goods.

The method of drying work in sawdust before lacquering, with the consequent carrying of sawdust into the lacquer, and the frequent discoloration of the finish by wet sawdust, can be entirely eliminated, and this one important improvement alone in handling such work has brought about an extensive use for the water-dip lacquers.

Since the advent of mechanical platers water-dip lacquers have another and new field, as the goods plated in this way can be put into a mesh basket as soon as taken from the plater and lacquered, which not only gives the finish immediate protection but is in keeping with the quickness and low cost of this method.

Points to follow when using water-dip lacquers for small work, such as cupboard catches, window fasteners, cheap building and trunk hardware, coat hooks, tacks, furniture nails, and other small specialties and novelties. The work may be placed in a wire mesh basket, rinsed in cold and hot water, dipped into the lacquer, which can be used so thin that there will be no accumulation or drip left, and the work can be dried in bulk without the pieces sticking together. The goods thus lacquered can be put right into a box and sent to the shipping room where they will dry out hard and with a high luster.

The water-dip lacquers, made by The Egyptian Lacquer Manufacturing Company of New York, contain ingredients which permit lacquering in bulk of small brass-plated work, copper acid-dipped or oxidized finishes being rinsed in either hot or cold water, and without drying dipped directly into the lacquer. Electro-deposited copper oxidizes rapidly, and especially so in a humid atmosphere. An instant application of these lacquers, while the work still retains its luster is recommended. It is very desirable for bulk, basket or en-masse lacquering.

A large collection of small articles can be perfectly lacquered by simple immersion.

Syphon out the water every morning from the lacquering tank, either with a rubber hose or by means of a faucet at the bottom of the tank.

A wire screen, nickel-plated and with a coarse mesh, should be placed in the lacquer jar or tank, three or four inches from the bottom. This will prevent the work from being dipped through the lacquer into the precipitated water, which lies at the bottom of the jar. All dirt and foreign matter carried into the lacquer with the work will sift through the screen and can be drawn off with the water.

In but few other branches of industry has the workman so constantly to deal with powerful poisons as well as other substances and vapors, which are exceedingly corrosive in their action upon the skin and the mucous membranes, as in electro-plating. However, with ordinary care and sobriety, all influences injurious to health may be readily overcome.

The necessity of frequently renewing the air in the workshop by thorough ventilation has already been referred to in chapter IV, "Electro-plating Establishments in General." Workmen exclusively engaged in pickling objects are advised to neutralize the action of the acid upon the enamel of the teeth and the mucous membrane of the mouth and throat by frequently rinsing the mouth with dilute solution of bicarbonate of soda. Those engaged in freeing the objects from grease lose, for want of cleanliness, the skin on the portions of the fingers which come constantly in contact with the lime and caustic lyes. This may be overcome by frequently washing the hands in clean water; and previous to each intermission in the work the workman should, after washing the hands, dip them in dilute sulphuric acid, dry them, and thoroughly rub them with cosmoline, or a mixture of equal parts of glycerine and water. The use of rubber gloves by workmen engaged in freeing the objects from grease cannot be recommended, they being expensive and subject to rapid destruction. It is better to wrap a linen rag seven or eight times around a sore finger, many workmen using this precaution to protect the skin from the corrosive action of the lye.

It should be a rule for every employee in the establishment not to drink from vessels used in electro-plating manipula-

(555)

tions ; for instance, porcelain dishes, beer glasses, etc. One workman may this moment use such a vessel to drink from, and without his knowledge another may employ it the next morning for dipping out potassium cyanide solution, and the first using it again as a drinking vessel may incur sickness, or even fatal poisoning.

The handling of potassium cyanide and its solutions requires constant care and judgment. Working with sore hands in such solutions should be avoided as much as possible ; but if it has to be done, and the workman feels a sharp pain in the sore, wash the latter quickly with clean water, and apply a few drops of blue vitriol solution.

Many individuals are very sensitive to nickel solutions, eruptions which are painful and heal slowly breaking out upon the arms and hands, while others may for years come in contact with nickel baths without being subject to such eruptions. In such cases prophylaxis is also the safeguard, *i. e.*, to prevent by immediate thorough washing the formation of the eruption if the skin has been brought in contact with the nickel solution, as, for instance, in taking out with the hand an object which has dropped into a nickel bath.

Poisoning by prussic acid, potassium cyanide and cyanide combinations.—In cases of internal poisoning first aid must be quickly rendered and a physician immediately called. There is but little hope of saving the life of a person poisoned by prussic acid, as well as when potassium cyanide or soluble cyanide combinations in large quantities have been taken into the stomach. In such cases solution of acetate of iron should be quickly administered and the patient made to inhale some chlorine prepared by putting a teaspoonful of chloride of lime in water acidulated with a small quantity of sulphuric acid. Water as cold as possible should at intervals be also poured over the head of the patient.

Poisoning by copper salts.—The stomach should be quickly emptied by means of an emetic or, in want of this, the patient should thrust his finger to the back of his throat and induce vomiting by tickling the *uvula*.   After vomiting, drink milk, white of egg, gum-water, or some mucilaginous decoction.

Poisoning by lead salts requires the same treatment as poisoning by copper-salts.   A lemonade of sulphuric acid, or an alkaline solution containing carbonic acid, such as Vichy water or bicarbonate of soda, is also very serviceable.

Poisoning by arsenic.—The stomach must be quickly emptied by an energetic emetic, when freshly precipitated ferric hydrate and calcined magnesia may be given as an antidote.   Calcined magnesia being generally on hand, mix it with 15 or 20 times the quantity of water, and give of this mixture 5 or 6 tablespoonfuls, every 10 to 15 minutes.

Poisoning by alkalies.—Use weak acids, such as vinegar, lemon juice, etc., and in their absence sulphuric, hydrochloric or nitric acid diluted to the strength of lemonade.   After the pain in the stomach has diminished, it will be well to administer a few spoonfuls of olive oil.

Poisoning by mercury salts.—Mercury salts, and particularly the chloride (corrosive sublimate), form with the white of egg (albumen) a compound very insoluble and inert.   The remedy, albumen, is therefore indicated.   Sulphur and sulphuretted water are also serviceable for the purpose.

Poisoning by sulphuretted hydrogen.—The patient should be made to inhale the vapor of chlorine from chlorine water, Javelle water, or bleaching-powder.   Energetic friction, especially at the extremities of the limbs, should be employed. Large quantities of warm and emollient drinks should be given, and abundance of fresh air.

Poisoning by chlorine, sulphurous acid, nitrous and hyponitric gases.—Admit immediately an *abundance of fresh air*, and administer *light* inspirations of ammonia.   Give plenty of hot drinks and excite friction in order to conserve the warmth and transpiration of the skin.   Employ hot foot-baths to remove the blood from the lungs.   Afterwards maintain in the mouth of the patient some substance which, melting slowly, will keep the throat moist, such as jujube and marshmallow paste, molasses candy, and licorice paste.   Milk is excellent.

CHAPTER XVII.

GALVANOPLASTY (REPRODUCTION).

By galvanoplasty proper is understood the production, with the assistance of the electric current, of copies of articles of various kinds, true to nature, and of sufficient thickness to form a resisting body, which may be detached from the object serving as a mould.

By means of galvanoplasty we are enabled to produce a simple, smooth plate of copper of such homogeneity as never shown by rolled copper, and such copper plates are used for engraving. From a medal, copper-engraving, type or other metallic object, a galvanoplastic copy may be made, which is to be considered as the negative of the original, in so far as it shows the raised portions of the original depressed, and the depressed portions raised. If now from this negative a fresh impression be made by galvanoplasty, the result will be a true copy of the original, possessing the same sharpness and fineness of the contours, lines and hatching.

A true reproduction of plastic works of art can in the same manner be made, but a current-conducting surface is required for effecting the deposit. As seen above, for the reproduction of a metallic original, two galvanoplastic deposits are required, one for the purpose of obtaining a negative, and the other in order to produce from the negative the positive—a copy true to the original.

Jacobi, the inventor of galvanoplasty, already endeavored to avoid the process of two galvanoplastic deposits by making an impression of the original in a plastic mass (melted rosin, wax, or plaster of Paris), rendering this non-metallic negative conductive, and depositing upon it copper, thus obtaining a true copy of the original.

(558)

It is not within the scope of this work to describe the various phases through which the art of galvanoplasty has passed since its invention. In the historical part reference has been made to several facts, such as making non-metallic impressions (moulds or matrices) conductive by graphite, a discovery for which we are indebted to Murray, and which was also made independently by Jacobi ; further, the production of moulds in gutta-percha ; so that in this chapter we have solely to deal with the present status of galvanoplasty.

I. Galvanoplasty in Copper.

Copper is the most suitable metal for galvanoplastic processes, that which is precipitated by electrolysis showing the following valuable properties : It can be deposited chemically pure, and in this state is less subject to change than ordinary commercial copper, or the copper alloys in general use, its tensile strength being 20 per cent. greater than that of smelted copper. Its hardness is also greater, while its specific gravity (8.85) lies between that of cast and rolled copper.

The physical properties of copper deposited by electrolysis are dependent upon the condition of the bath, as well as on the intensity and tension of the current. The bath used for depositing copper is in all cases, a solution of copper sulphate (blue vitriol).

Smee proved by experiments that, with as intense a current-strength as possible without the evolution of hydrogen, the copper is obtained as a tenacious, fine-grained deposit. But when the current-strength is so intense that hydrogen is liberated, copper in a sandy, pulverulent form is obtained, and in coarsely crystalline form when the current-strength is very slight.

At a more recent period, Hübl and Förster have instituted a series of systematic experiments for the determination of the conditions under which deposits with different physical properties are obtained. Förster, in addition, deserves credit for his investigations of the anodal solution-processes.

Hübl worked with 5 per cent. neutral and 5 per cent. acid solutions, as well as with 20 per cent. neutral and 20 per cent. acid solutions. The neutral solutions were prepared by boiling blue vitriol solution with carbonate of copper in excess, and the acid solutions by adding 2 per cent. of sulphuric acid of 66° Bé. The result was that in the neutral 5 per cent. solution less brittle deposits were obtained with a slight current-density than in a more concentrated solution, though the appearance of the deposits was the same. The experiments with acidulated baths confirmed the fact that free sulphuric acid promotes the formation of very fine-grained deposits even with very slight current-densities, and it would seem that the brittleness of copper deposited from the acid baths is influenced less by the concentration than by the current-density used.

With the use of high current-densities, *spongy* deposits of a dark color, but frequently also *sandy* deposits of a red color, are obtained from the neutral as well as from acid blue vitriol solutions. These phenomena are directly traceable to the effect of the hydrogen reduced on the cathodes.

However, such spongy deposits are also obtained with the use of slight current-densities, when the concentration of the electrolyte has become less by the exhaustion of the bath on the cathodes, and Mylius and Fromm have shown that copper reduced under such conditions had absorbed hydrogen, while Lenz, in addition to hydrogen, found in a brittle copper deposit, carbonic oxide and carbonic acid. Soret also found carbonic acid in addition to hydrogen, and attributes to it the unfavorable effect, while he considers a content of hydrogen as unessential for the mechanical properties of the electrolytically deposited copper.

It is impossible to understand where the carbonic acid is to come from, provided there has been no contamination of the electrolyte by organic substances.

A. GALVANOPLASTIC REPRODUCTION FOR GRAPHIC PURPOSES. (ELECTROTYPY.)

The processes used in galvanoplasty may be arranged in two classes, viz., the *deposition* of copper *with, or without, the use of external sources of current*, the first comprising galvanoplastic deposits produced by means of the *single-cell apparatus*, and the other those by the battery, thermo-electric pile, dynamo or accumulator.

1. *Galvanoplastic Deposition in the Cell Apparatus.*

The cell apparatus consists of a vessel containing blue vitriol solution kept saturated by a few crystals of blue vitriol placed in a muslin bag, or a small perforated box of wood, stoneware, etc. In this vessel are placed round or square porous clay cells (diaphragms) which contain dilute sulphuric acid and a zinc plate, the zinc plates being connected with each other and with the objects to be moulded—which may be either metallic or made conductive by graphite—by copper wire or copper rods.

The objects to be moulded play the same role as the copper electrode in the Daniell cell, and the cell apparatus is actually a Daniell cell, closed in itself, in which the internal, instead of the external, current is utilized. As soon as the circuit is closed by the contact of the objects to be copied with the zinc of the porous cell, the electrolytic process begins. The zinc is oxidized by the oxygen and with the sulphuric acid forms zinc sulphate (white vitriol) while the copper is reduced from the blue vitriol solution and deposited in a homogeneous layer upon the objects to be moulded.

Forms of cells. The form and size of the simple cell-apparatus vary very much according to the purpose for which the latter is to be used. While formerly a horizontal arrangement of the objects to be copied and of the zinc plates was generally preferred, because with this arrangement the fluids show a more uniform concentration, preference was later on properly given to the vertical arrangement. Particles becoming detached

36

from the zinc plates get only too easily upon the object to be
reproduced and cause holes in the deposit, while with the
vertical arrangement the progress of deposition can at any
time be controlled by lifting out the objects without taking the
apparatus apart, as in the case with the horizontal arrangement. Hence, only such apparatus in which the zinc plates
and the objects to be moulded are arranged vertically opposite one to the other will here be discussed.

A simple apparatus frequently used by amateurs for moulding metals, reliefs, etc., is shown in Fig. 135.

In a cylindrical vessel of glass or stoneware filled with saturated blue vitriol solution is placed a porous clay cell, and in
the latter a zinc cylinder projecting about 0.039 to 0.79 inch

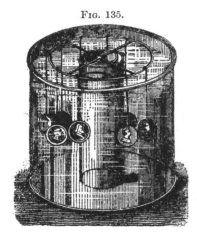

FIG. 135.

above the porous clay cell. To
the zinc is soldered a copper
ring, as plainly shown in the
illustration. The clay cell is
filled with dilute sulphuric acid
(1 acid to 30 water), to which
some amalgamating salt may
be suitably added. The articles
to be moulded are suspended
to the copper ring, care being
had to have the surfaces which
are to be covered near and
opposite to the cell. To supplement the content of copper,
small linen or sail-cloth bags
filled with blue vitriol are attached to the upper edge of the
vessel.

Large apparatus.—To cover large surfaces, large, square
tanks of stoneware, or wood, lined with lead, gutta-percha, or
another substance unacted upon by the bath are used. For
baths up to three feet long, stoneware tanks are to be preferred.

Fig. 136 shows the *French form of cell apparatus.* In the
middle of the vat, and in the direction of its length, is dis-

posed a row of cylindrical cells, close to each other, each provided with its zinc cylinder. A thin metallic ribbon is connected with all the binding screws of the cylinder, and is in
contact at its extremities with two metallic bands on the
ledges of the depositing vat. The metallic rods supporting
the moulds are in contact with the metallic bands of the
ledges, and therefore, in connection with the zincs.

FIG. 136.

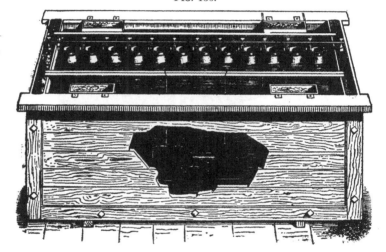

The *German form of cell apparatus* is shown in Fig. 137.
It is provided with long, narrow, rectangular cells of a correspondingly greater height than the column of fluid.

Across the vat are placed three conducting rods connected
with each other by binding screws and copper wire. To the
center rod, which lies over the cells, are suspended the zinc
plates by means of a hook, while the outer two rods serve for
the reception of the moulds.

The zinc surfaces in the simple apparatus should be of a
size about equal to that of the surfaces to be reproduced, if
dilute sulphuric acid (1 acid to 30 water) is to be used.

Copper bath for the cell apparatus.—This consists of a solution of 41 to 44 lbs. of pure blue vitriol free from iron, for a 100-quart bath, with an addition of about $3\frac{1}{2}$ to $4\frac{1}{2}$ lbs. of sulphuric acid of 60° Bé., free from arsenic.

It is not customary to add to the copper bath serving for graphic purposes a larger quantity of sulphuric acid than that mentioned above, because acid diffuses constantly from the fluid in the clay cells into the bath, thus gradually increasing the content of acid in the latter. If the generation

FIG. 137.

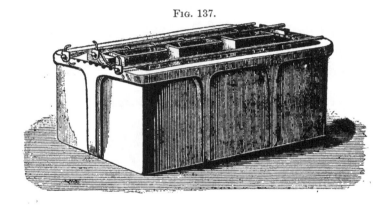

of current is induced by acidulating the water in the clay cells, a further addition of acid for the cells would actually not be required for the progressive development of the current, since the acid-residue formed by the decomposition of the blue vitriol migrates from the zinc of the cells, and brings · fresh zinc-ions into solution. A diffusion of acid from the cells into the bath would in this manner be avoided, and only the zinc-sulphate solution formed would diffuse into the copper bath. It appears, however, that without an occasional addition of a small quantity of sulphuric acid to the cell-solution the process of deposition runs its course very slowly, which is not desirable for the manufacture of clichés.

It may therefore happen that after working the copper bath

for a long time, it contains too much acid, a portion of which has to be removed. For this purpose the bath was formerly mixed with whiting, and the gypsum formed filtered off. This method, however, cannot be recommended, gypsum being not entirely insoluble in water, and it is better to replace it by cupric carbonate or cuprous oxide (cupron). If cupric carbonate be used, it is advisable thoroughly to stir the bath, or what is better, to boil it, so as to remove as completely as possible the carbonic acid.

By the diffusion of zinc sulphate solution from the clay cells, the bath becomes gradually rich in zinc salt, and it will be noticed that a certain limit—with a content of about 10 per cent. zinc sulphate—the copper deposits turn out brittle. The bath has then to be entirely renewed.

The content of copper in the bath decreases in accordance with the copper deposited, and the concentration of the bath would consequently become so low that useful deposits could no longer be obtained, if care were not taken to replace the copper. This is done by suspending perforated baskets of stoneware or lead, filled with blue vitriol crystals, in the bath.

Since directions are frequently found in which the blue vitriol solutions to be used are given according to their weights by volume or degrees of Bé., a table showing the content of blue vitriol is given below.

Degrees Bé.	Weight by volume.	This solution contains crystallized blue vitriol.
5°	1.035	5 per cent.
10°	1.072	11 " "
12°	1.088	13 " "
15°	1.113	17 " "
16°	1.121	18
17°	1.130	19
18°	1.138	20
19°	1.147	21
20°	1.157	23
21°	1.166	24
22°	1.176	25

Electro-motive 'force.—The effective electro-motive force in the cell apparatus amounts to about 0.75 volt. It may be regulated by either bringing the matrices more closely to the diaphragms, or removing them a greater distance from them. In the first case, the resistance of the bath is decreased, the current-density being consequently increased, while in the other, the resistance of the bath is increased and the current density decreased.

For regulating the electro-motive force a rheostat may also be placed in the circuit, between the matrices and the zincs, instead of connecting them directly by a copper wire. Although this method is not in vogue, it is certainly recommendable.

For working on a large scale, the cell apparatus is but seldom used, at least not for the production of electros. It is, however, occasionally employed for the reproduction of objects of art with very high reliefs, so as to cover them as uniformly as possible and quite slowly with copper. The cell is also still liked for the production of matrices.

2. *Galvanoplastic Deposition by the Battery and Dynamo.*

Since it has been shown in the preceding section that a cell apparatus is to be considered as a Daniell cell closed in itself, it will not be difficult to comprehend that in economical respects no advantage is offered by the production of galvanoplastic depositions by a separate battery, because in both cases the chemical work is the same, and the zinc dissolved by the use of the Daniell or Bunsen cell effects no greater quantity of copper deposit in the bath than the same quantity of zinc dissolved in the cells of the single apparatus. In other respects the use of a battery, however, offers great advantages.

The employment of an external source of current requires the same arrangement as shown in Figs. 44 and 45, pp. 140, copper anodes being placed in the bath and connected with the anode pole of the battery. Copper being dissolved in the anodes, the sulphuric acid residue which is liberated is saturated with blue vitriol, the content of copper being thus, if not entirely, at least approximately, kept constant. Furthermore, no foreign metallic salts reach the bath, as is the case in the simple apparatus, by zinc sulphate solution penetrating from the clay cells and causing the formation of rough and brittle deposits of copper. With the use of anodes of chemically pure copper the bath will thus always remain pure.

The current may also be regulated within certain limits by bringing the anodes more closely to the objects, or removing them a greater distance from them. The principal advantage, however, consists in that by placing a rheostat in the circuit the current strength can be controlled as required by the different kinds of moulds.

a. *Depositions with the Battery.*

Cells.—The Daniell cell described on p. 71, yields an electro-motive force of about 1 volt, and is much liked for this purpose. Since the copper bath for galvanoplastic purposes requires for its decomposition an electro-motive force of only 0.5 to 1 volt, it will be best for slightly depressed moulds to couple the elements for quantity (Fig. 19, p. 89) alongside of each other; and only in cases where the particular kind of moulds requires a current of greater electro-motive force to couple two cells for electro-motive force one after the other, an excess of current being rendered harmless by means of the rheostat, or by suspending larger surfaces.

Bunsen or Meidinger cells may, however, be used to great advantage, since the zincs of the Daniell cells become tarnished with copper, and have to be frequently cleansed if the process is not to be retarded or entirely interrupted. The Bunsen cells need only be coupled for quantity, their electro-motive force being considerably greater. To be sure, the running expenses are much greater than with Daniell cells, at least when nitric acid is used for filling. The lasting constancy of the Meidinger cells would actually make them the most suitable of all for continuous working, but by reason of their slight current strength a large number of them would have to be used.

All that has been said under "Installations with Cells," p. 132, in regard to conducting the current, rheostats, conducting rods, anodes, etc., applies also to plants for the galvanoplastic deposition of copper with batteries.

b. *Depositions with the Dynamo.*

The improvements in dynamos have also benefited industrial galvanoplasy, and problems can now be solved in a much shorter time and with much greater ease than in a cell apparatus, without having to put up with the obnoxious vapors which make themselves very disagreeably felt in working on a large scale with a simple apparatus. That the use of a dynamo offers decided advantages is best proved by the fact that no galvanoplastic plant of any importance works at present without one, and there can scarcely be any doubt that establishments which still work exclusively with the simple apparatus will be forced to make use of a dynamo, if they wish to keep up with competition as regards cheapness and rapidity of work.

Dynamos.—It is best to use a dynamo capable of yielding a large quantity of current with an impressed electro-motive force of 2, or, at the utmost, 3 volts, in case it is not to serve for rapid galvanoplasty; for the latter a machine of 5 to 10 volts impressed electro-motive force is required. For the old, slow process, by which deposits for graphic purposes are produced in 5 to 6 hours, an impressed electro-motive force of 2 volts suffices for baths coupled in parallel. If, however, there are to be charged from the dynamo one or more accumulator cells, which are to furnish current to the bath while the steam engine is not running during the intermission of work, or to finish deposits after working hours, the impressed electro-motive force, with cells coupled in parallel, must be 3 volts, and with cells coupled in series in proportion to their number.

It may also happen that in a galvanoplastic plant currents of greatly varying electro-motive force may be required for depositions. For depositing copper according to the old process there should, for instance, be available a large current-strength with only 1 to 1.5 volts, while for a rapid galvano-plastic bath a current of 6 volts is at the same time to be used. If a dynamo of 6 volts impressed electro-motive force were to be used, the excess of electro-motive force would have to be destroyed by rheostats in front of the baths requiring a slight electro-motive force, in case it is not convenient to couple these baths in series (see later on). The destruction of electro-motive force is, however, not economical, and, in such a case, the use of two dynamos with different electro-motive forces is advisable. It is best to combine both dynamos with a motor-generator, if the plant is connected with a power circuit of a

FIG. 138.

central station, the construction being such that the dynamo which is perhaps only temporarily in use can be readily disengaged.

Fig. 138 shows such a double aggregate, built by the firm of Dr. G. Langbein & Co. for the German Imperial Printing Office. The larger dynamo has a capacity of 1000 ampères and 2.5 volts, and the smaller one, one of 250 ampères and 6 volts.

Current-conductors of sufficient thickness, corresponding to the quantities of current have to be provided to prevent loss of current by resistance in the conductors. To avoid repetition, we refer to what has been said on this subject under "Arrangement of Electro-plating Establishments," the directions there given applying also to the galvanoplastic process.

Coupling the baths.—When coupling the baths in parallel, each bath will have to be provided with a rheostat and ammeter, while a voltmeter with a voltmeter switch may be employed in common for several baths. If the baths are of exactly the same composition and the same electrode-distances are maintained in them, regulation of the current by the shunt rheostat of the dynamo will suffice.

Coupling the baths in series may under certain conditions be of advantage. In such a case, a dynamo of adequately higher electro-motive force will of course have to be employed.

With the baths coupled in series the cathode (object) surfaces in all the baths should be of the same size, or at least approximately so. The baths are coupled in series by connecting the anodes of the first bath with the + pole of the dynamo, the cathodes of the first bath with the anodes of the second, the cathodes of the second with the anodes of the third, and so on, and reconducting the current from the cathodes of the last bath to the — pole of the dynamo (Fig. 64).

With this simple coupling in series, the impressed electro-motive force is uniformly distributed in all the baths, so that with four baths coupled in series, and an impressed electro-motive force of 4 volts, an electro-motive force of 1 volt is present in each bath if the conductivity resistance be left out of consideration. Hence, it may be readily calculated how many baths have to be coupled in series to utilize a given impressed electro-motive force, when the electro-motive force required for one bath is known.

If, for instance, there is a dynamo with 6 volts impressed electro-motive force, and the electro-motive force required for one bath is 1.5 volts, then 4 baths have to be coupled in series, since they require $1.5 \times 4 = 6$ volts. If, however, only one volt is required for one bath, then 6 baths will have to be coupled in series, or, in case fewer baths are to be used, the impressed electro-motive force of the dynamo has to be suitably regulated by the shunt rheostat.

Besides, the simple coupling in series, mixed coupling, also called coupling in groups, a combination of coupling in parallel and in series, may be employed. This is effected by combining a number of baths to a group in parallel, and coupling several such groups in series.

Fig. 139.

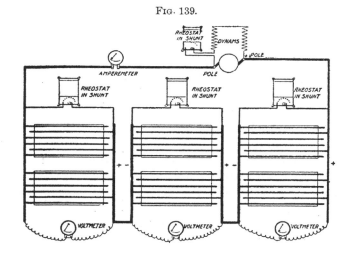

In large galvanoplastic plants the advantages derived from this mixed coupling are as follows:

With the simple couple in series, the electrode-surfaces in all the baths must be of the same size. When finished objects are taken from a bath, the current conditions are changed, until in place of the object taken out fresh surfaces of the same size are suspended in the bath, and as this cannot always be immediately done, irregularities in working will result. If, however, the baths be combined in groups in the manner shown in Fig. 139, only the cathode-surfaces of each of the groups coupled in parallel, need to be of the same size, or approximately so, and with the observation of this condition it is entirely indifferent whether a bath of one group is not at all charged with cathodes.

For the adjustment of any difference in the electro-motive force in the baths of the separate groups, it is advisable to place in each bath a rheostat in shunt.

While with baths coupled in parallel, the electro-motive force of the dynamo corresponds to the requisite electro-motive force of one bath, but the current-strength is calculated from the sum of all the cathodes present in the different baths, with baths coupled in series only the total cathode surface of one bath is decisive as regards the current-strength, the electro-motive force of the machine resulting from the sum of the electro-motive forces of the separate baths. With baths coupled in groups the requisite impressed electro-motive force is calculated from the number of groups of baths coupled in series, but the current-strength from the total cathode-surface of only one group.

The following examples may serve as illustrations : Suppose 3 baths, each with 100 square decimeters cathode-surface, are coupled in parallel, and the electro-motive force for one bath is 1.5 volts. Hence in the three baths there are $100 \times 3 = 300$ square decimeters cathode surface, and if, for instance, one square decimeter requires 2 ampères, then the 3 baths require $300 \times 2 = 600$ ampères. Thus the capacity of the dynamo must be 600 ampères, with an impressed electro-motive force of 1.5 volts, but for practical reasons a machine of 2 volts should be selected.

Suppose 4 baths, each charged with 100 square decimeters cathode-surface, are coupled in series, and the bath electromotive force is 1.25 volts, and the current-density 2 ampères. Hence there will be required, $100 \times 2 = 200$ ampères, and $1.25 \times 4 = 5$ volts.

Suppose 9 baths are coupled, mixed in three groups of 3 baths each, the latter being coupled in parallel, and the three groups coupled in series. Now if each group be charged with 300 square decimeters cathode-surface and the bath electro-

motive force be also 1.25 volts and the current-density 2 ampères, then there will be required, $300 \times 2 = 600$ ampères,
and $1.25 \times 3 = 3.75$ volts, or practically, 4 volts impressed
electro-motive force.

c. *Combined Operation with Dynamo and Accumulators.*

When, as is frequently the case in galvanoplastic plants
working with the slow process of deposition, electrotypes have
to be finished in a hurry, recourse has to be had to night
work. If the dynamo is not driven by a motor-generator fed
from a power circuit of a central station, it will be necessary
to use for night work either a cell apparatus, or to feed the
bath from accumulators.

An interruption in the galvanoplastic deposition of copper
is a great drawback, because an additional deposit made after
the current has been interrupted adheres badly upon the one
previously made, as blisters are readily formed, or the deposit
peels off. The chief object in the use of an accumulator is
that it allows of the work being carried on during the noon
hour when the steam engine is generally stopped, and of finishing matrices which are suspended late in the afternoon,
after working hours.

In order to avoid repetition, the reader is referred to what
has been said on p. 184 *et seq.*, in regard to the use of an accumulator in addition to the dynamo.

For galvanoplasty in copper by the slow process, one accumulator cell of sufficient capacity to supply current for 2 or 3
hours is, as a rule, all that is required. This cell is charged
from the dynamo at the same time, while the latter directly
operates the bath, a machine with an impressed electro-motive
force of 3 volts being required for the purpose.

If several cells have to be used, it has to be decided according to the capacity of the dynamo, whether they are to be
charged in parallel or in series. If great demands are for a
longer time made on the accumulators, it is advisable to use a
separate dynamo for charging purposes.

Copper baths for galvanoplastic depositions with a separate source of current.—The directions for the composition of the bath vary very much, some authors recommending a copper solution of 18° Bé., which is brought up to 22° Bé. by the addition of pure concentrated sulphuric acid. Others again increase the specific gravity of the bath up to 25° Bé. by the addition of sulphuric acid, while some prescribe an addition of 3 to 7 per cent. of sulphuric acid.

It is difficult to give a general formula suitable for all cases, because the addition of sulphuric acid will vary according to the current-strength available, the nature of the moulds, and the distance of the anodes from the objects. The object of adding sulphuric acid is, on the one hand, to render the bath more conductive and, when used in proper proportions, to make the deposit more elastic and smoother, and prevent the brittleness and coarse-grained structure which, under certain conditions, appear.

However, it is also the function of the sulphuric acid to prevent the primary decomposition of the blue vitriol, and to effect in a secondary manner the reduction of the copper. As has been explained on p. 50, acids, bases and salts dissociate in aqueous solution, and only substances which dissociate in aqueous solution are conductors of the electric current, they being the better conductors, the greater their power of dissociation is. The dilute sulphuric acid being much more dissociated, takes charge in a much higher degree of conducting the current than the less strongly dissociated blue vitriol solution. Consequently the cation of the sulphuric acid—the hydrogen-ions—migrates to the cathode, and effects the decomposition of the blue vitriol, an equivalent quantity of copper being reduced upon the cathode.

The addition of a large quantity of sulphuric acid, as recommended by some authors, cannot be approved, it having been found of advantage only in a few cases.

For depositing with a battery, somewhat more sulphuric acid may for economical reasons be added to the bath than when working with the current of a dynamo. The following composition has in most cases been found very suitable for the reproduction of shallow, as well as of deep, moulds

Blue vitriol solution of $19\frac{1}{2}°$ Bé. 100 quarts, sulphuric acid of 66° Bé. free from arsenic $4\frac{1}{2}$ to $6\frac{1}{2}$ lbs.

The bath is prepared as follows: Dissolve $48\frac{1}{2}$ lbs. of blue vitriol in pure warm water, and, to avoid spurting, add gradually, stirring constantly, the sulphuric acid. At the normal temperature of 59° F. the bath may be worked with a current-density of up to 2 ampères, and if the bath be agitated, the current-density may be up to 3 ampères.

Properties of the deposited copper.—As regards elasticity, strength and hardness of galvanoplastic copper deposits, Hübl determined that copper of great tenacity, but possessing less hardness and strength, is deposited from a 20 per cent. solution with the use of a current-density of 2 to 3 ampères.

For copper printing plates a 20 per cent. solution compounded with 3 per cent. sulphuric acid, and current-density of 1.3 ampères was found most suitable by Giesecke. Dissolve for this purpose 50.6 lbs. of blue vitriol for a 100-quart bath and add 6.6 lbs. of sulphuric acid.

Förster and Seidel have shown that the mechanical properties of the copper are materially influenced by the temperature of the electrolyte. From Förster's investigations, with a cathodal current-density of 1 ampère in an electrolyte composed of 150 grammes blue vitriol and 50 grammes sulphuric acid per liter, it appears that the copper obtained with the electrolyte at 140° F., showed the greatest tenacity, it decreasing again at a higher temperature while the strength slightly increased.

The nature, *i. e.*, the composition, of the electrolyte also exerts an influence upon the structure of the deposit. Förster compounded the electrolyte previously used with a quantity of sodium sulphate equivalent to that of the blue vitriol. The result showed that the strength as well as the tenacity was unfavorably influenced at a higher temperature.

Current conditions.—In order to obtain a dense, coherent and elastic deposit in the acid copper bath, it is first of all necessary to bring the current-strength into the proper proportion to the deposition-surface, this applying to depositions in the simple apparatus, as well as to that produced with an external source of current.

The stronger the sulphuric acid in the clay cells of the simple apparatus is, the more rapidly is the copper precipitated upon the moulds. If the zinc-surfaces of the clay cells are very large in proportion to the surfaces of the moulds, the deposition of copper also takes place more rapidly. The rapid reduction of copper, however, must above all be avoided if deposits of desirable qualities are to be obtained, because a deposit of copper forced too rapidly turns out incoherent, spongy, frequently full of blisters and, with a very strong current, even pulverulent.

The color of the deposit is to some extent a criterion of its quality, a red-brown color indicating an unsuitable deposit, while a good, useful one may be counted upon when it shows a beautiful rose color.

For filling the clay cells, it has previously been stated that the acid is to be diluted in the proportions of 1 part of concentrated sulphuric acid of 66° Bé. to 30 parts of water, the zinc surface being supposed to be of about the same size as the matrix-surface. If the zinc-surface should be smaller, stronger acid may be used, and if it be larger, the acid may be more dilute. The proper concentration of the acid in the clay cells is readily ascertained from the progressive result of the deposit and its color.

What has been said in reference to the current-strength applies also to the deposition of copper with a separate source of current (battery or dynamo). The current-strength must be so adjusted by means of a rheostat as to allow of comparatively rapid deposition without detriment to the quality of the deposit.

According to the composition of the bath, a fixed minimum

and maximum current-density corresponds to it, which must not be exceeded if serviceable deposits are to be obtained. There is, however, a further difference according to whether the bath is at rest or agitated. Hübl obtained the following results ·

Blue vitriol solution.	Minimum and maximum current-density per 15.5 square inches.	
	With solution at rest. Ampères.	With solution gently agitated. Ampères.
15 per cent. blue vitriol, without sulphuric acid . .	2.6 to 3.9	3.9 to 5.2
15 per cent. blue vitriol, with 6 per cent. sulphuric acid .	1.5 " 2.3	2.3 " 3.0
20 per cent. blue vitriol, without sulphuric acid	3.4 " 5.1	5.1 " 6.8
20 per cent. blue vitriol, with 6 per cent. sulphuric acid	2.0 " 3.0	3.0 " 4.0

Touching the addition of sulphuric acid, it was shown that no difference in the texture of the deposit is perceptible if the addition of acid varies between 2 and 8 per cent.

The most suitable current-density for the production of good deposits with a bath of the composition given on p. 575, when at rest, is for slow deposition 1 to 2 ampères per square decimeter of matrix surface, and when the bath is agitated 2 to 3 ampères. The current-densities for rapid galvanoplastic baths will be given later on.

Since for ordinary, strongly acid copper baths an electromotive of $\frac{1}{2}$, to at the utmost $1\frac{1}{2}$, volts is required, the more powerful Bunsen cells will have to be coupled alongside each other, while of the weaker Daniell or Lallande cells, two, or of the Meidinger cells, three, will have to be coupled one after the other, and enough of such groups have to be combined to make their active zinc-surfaces of nearly the same size as the surfaces of the matrices. However, for strongly acid baths

37

coupling the separate weaker cells alongside each other also suffices.

When all parts of the matrices, as well as the deeper portions, are covered with copper, the current is weakened in case a deposit of a pulverulent or coarse-grained structure appears on the edges of the moulds, and it is feared that the deposit upon the design or type might also turn out pulverulent. The current need not be weakened more than is necessary to prevent the dark deposits on the edges from progressing further towards the interior of the mould-surfaces. If, by reason of too strong a current, pulverulent copper has already deposited upon the design or type, and the fact is noticed in time and the current suitably weakened, the deposit can generally be saved by the layers being cemented together by the copper which is coherently deposited with the weaker current.

In depositing with the dynamo, the current-density and electro-motive force have to be properly regulated by means of a rheostat.

Brittle copper deposits may be caused, not only by an unsuitable composition of the electrolyte, improper current-densities and impure anodes (see later on), but also by contamination of the bath with non-metallic substances, certain organic substances especially having an unfavorable effect upon the properties of the copper deposit.

Förster found the use of hooks coated with rubber solution in benzol for suspending the cathodes a protection against the attacks of the electrolyte and the air, and smooth copper deposits of a beautiful velvety appearance were obtained, but they were so brittle that they could not be detached without breaking from the basis. The deposit contained small quantities of carbonaceous substances, which could have been derived only from a partial solution of the rubber.

Hübl also describes the fact of having obtained brittle copper by the electrolyte having been contaminated by a small quantity of gelatine which had passed into solution in the preparation of an electrotype upon heliographic gelatine reliefs.

Dr. Langbein had frequently occasion to notice that baths in tanks lined with lead, and provided with a coat of asphalt and mastic dissolved in benzol, yielded brittle copper when the coat of lacquer was not perfectly hard, and it was observed that by reason of an accidental contamination of the electrolyte with gelatine, deposits were formed which showed the branched formation of crystals similar to an *arbor Saturni*, and were extremely brittle.

Erich Müller and P. Behntje * have recently investigated the effects of such organic additions (colloids) on blue vitriol solutions, additions of gelatine, egg albumen, gum, and starch being drawn upon for comparing the effects. After an electrolysis for 15 hours, the deposits obtained from baths compounded with gum and starch solutions showed no material difference in appearance from deposits obtained with the same current-strength from pure acidulated blue vitriol solution. Deposits obtained from baths to which gelatine and egg albumen had been added, however, had lustrous streaks running from top to bottom. The weight of these deposits was, moreover, greater than that of the copper obtained from pure solution, and the presence of gelatine in the deposit could be established by analysis. Further experiments proved that the above-mentioned phenomena were dependent on the current-density, and with smaller additions of gelatine, and 3.5 ampères, thoroughly homogeneous, mirror-bright copper coatings could be obtained, thus rendering it possible to produce, by an addition of gelatine, a lustrous coppering from acid copper baths. The properties of this copper are, however such as are not desirable for galvanoplastic purposes.

It is therefore absolutely necessary to exclude such organic substances, even if only dissolved in traces, from contact with the electrolyte.

Duration of deposition.—The time required for the production of a deposit entirely depends, according to what has been

---

* Zeitschrift für Elektrochemie, XII, 317.

said on p. 124, on the current-density used. One ampère deposits in one hour 1.18 grammes of copper, and from this, when the current-density is known, the thickness acquired by the deposit in a certain time can be readily calculated.

A square decimeter of copper, 1 millimeter thick, weighs about 89 grammes, and to produce this weight with 1 ampère current-density, there are required $\frac{8900}{118} = 75$ hours. By taking the thickness of a deposit as 0.18 millimeter, which suffices for all purposes of the graphic industry, then 1 square decimeter will weigh $89 \times 0.18 = 16.02$ grammes, and for their deposition with 1 ampère current-density will be required, in round numbers, $\frac{1602}{118} = 13\frac{1}{2}$ hours.

Below is given the duration of deposition for electrotypes 0.18 millimeter thick with different current-densities, and in addition the time is stated which would be required for the formation of a deposit 1 millimeter thick, so that the calculation of the time required for depositing a copper film of a thickness different from 0.18 millimeter can be readily made by multiplying the number of hours in the third column by the desired thickness of the copper.

With a current-density of	Duration of deposition for 0.18 millimeter thickness of copper	A deposit of 1 millimeter thickness requires
0.5 ampère	27 hours	150½ hours
0.75	18 "	101⅓ "
1.0 "	13½ "	75 "
1.5	9	50
2 0	6¾	37½
2.5	5½	30
3.0	4½	25
4.0	3⅔	18¾
5.0	2¾	15
6.0	2¼	12½
7.0	2 '	10¾
8.0 "	1 $\frac{7}{10}$ '	9⅜
9.0 "	1½ '	8⅓

Nitrate baths.—To shorten the duration of deposition, baths have been recommended which, in place of blue vitriol, are prepared with cupric nitrate, and which by reason of being more concentrated will bear working with greater current-densities. To further increase their conducting power, ammonium chloride is added. Independent of the fact that deposits obtained in these baths are inferior in quality to those produced in blue vitriol baths, such baths require frequent corrections, they becoming readily alkaline in consequence of the formation of ammonia. Besides, in view of the short time required for deposition by the rapid galvanoplastic process, there is no necessity for nitrate baths.

Agitation of the baths.—From Hübl's table it will be seen that a copper bath in motion can bear considerably higher current-densities, and hence will work more rapidly than a bath at rest. In electrolytically refining copper it was found that, if the process of reducing the copper is to proceed in an unexceptional manner, the bath must be kept entirely homogeneous in all its parts. When a copper bath is at rest, and the operation of deposition is in progress, the following process takes place : The layers of fluid on the anodes having by the solution of copper become specifically heavier, have a tendency to sink down, while layers of fluid which have become poorer in copper, and consequently specifically lighter, rise on the cathodes to the surface. These layers contain more sulphuric acid than the lower ones, hence their resistance is slighter and their conducting power greater, the latter being still further increased by the layers heated by the current also rising to the surface. In consequence of this process there will be a variable growth in thickness of the deposit, and various phenomena may appear which, according to the composition of the layers in question, can be theoretically established. If the intermixture of the electrolyte has not progressed to any great extent, and thus there are no great differences, as regards composition, between the upper and lower layers of the fluid, the deposit will be quite uniformly formed upon the lower as well as upon the upper portions of the cathodes, although it will be somewhat thicker on the lower portions, which dip into the more concentrated copper solution, than on the upper portions. This difference in thickness in favor of the lower cathode-surfaces will become more pronounced as the concentration of the lower layers of the fluid increases, while the growth in thickness on the upper cathode-surface is kept back. The concentration of the upper layers of fluid may finally happen to become so slight that the hydrogen-ions do not meet with sufficient blue vitriol for decomposition, and hydrogen will consequently be separated and the formation of a sandy or spongy deposit noticed.

It may, however, also happen that a current of slight electromotive force cannot overcome the greater resistance of the more concentrated lower layers of fluid, and in consequence passes almost exclusively through the upper layers. So long as the hydrogen-ions find sufficient blue vitriol and the current-density is slight, the growth of the deposit on the upper cathode-surfaces may, in this case, progress, while it comes to a stand-still on the lower ones.

Experiments by Sand * have shown that in consequence of local exhaustion of blue vitriol in an acid copper bath more than 60 per cent. of the current is, notwithstanding natural diffusion, consumed for the evolution of hydrogen. The production of serviceable deposits under such conditions is of course impossible.

By constant agitation of the bath, the layers poorer in metal, which have deposited copper on the cathodes, are rapidly removed, and layers of fluid richer in metal are conveyed to the cathodes, the greatest possible homogeneity of the bath being thus effected, and the operation of deposition becoming uniform.

Baths in motion show less inclination to the formation of buds and other rough excrescences, and hence the current-density may be greater than with solutions at rest, the result

* Zeitschrift für physikalishe Chemie, xxxv, 641.

being that deposition is effected with greater rapidity. These experiences gathered in electro-metallurgical operations on a large scale, have been advantageously applied to galvanoplasty.

Stirring contrivances.—Constant agitation of the copper bath may be effected in various ways. A mechanical stirring contrivance may be provided, or agitation may be effected by blowing in air, or finally, by the flux and reflux of the copper solution.

With the use of a stirring apparatus, stirring rods of hard rubber or glass which are secured to a shaft running over the bath, swing like pendulums, between the electrodes. This motion of the shaft is effected by means of leverage driven from a crank pulley. The stirring rods should not move with too great rapidity, otherwise the slime from the anodes, which settles in the bath, might be stirred up.

Agitation of the bath has also been effected by slowly revolving, by means of a suitable mechanism, cast copper anodes of a square cross-section, this mode of motion having the advantage of very thoroughly mixing the electrolyte without being too violent.

If the bath is to be agitated by blowing in air, the latter is forced in by means of a pump through perforated lead pipes, arranged horizontally about two inches from the bottom of the tank.

It is best to use a small air compressor in connection with an air chamber provided with a safety valve. The quantity of air to be conveyed to the perforated lead pipes is regulated by means of a cock or valve. The number and size of the perforations in the lead coil must be such that the air passes out as uniformly as possible the entire length of the pipe, so that all portions of the electrolyte are uniformly agitated. For smaller baths an ordinary well-constructed air-pump suffices for pressing in the air.

Agitation of the bath by flux and reflux of the solution may be effected in various ways, and is especially suitable where many copper baths are in operation.

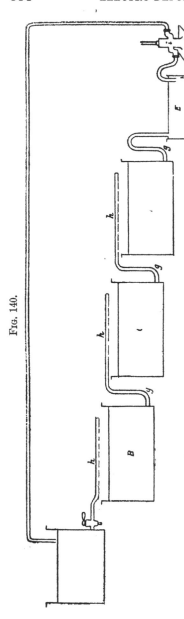

FIG. 140.

The baths are arranged in the form of steps. Near the bottom each bath is provided with a leaden outlet-pipe (Fig. 140), which terminates above the next bath over a distributing gutter, or as a perforated pipe, h. From the last bath the copper solution flows from a reservoir, E, from which it is forced by means of a hard-rubber pump, i, into the reservoir, A, placed at a higher level. From A it again passes through the baths, B, C and D. A leaden steam-coil may, if necessary, be placed in A, to increase the temperature, if it should have become too low. Over A a wooden frame covered with felt may be placed; the copper solution flowing upon the frame and passing through the felt, is thereby filtered.

While agitation of the bath presents great advantages, there is one drawback connected with it, which, however, should not prevent its adoption. With baths at rest, dust and insoluble particles becoming detached from the anodes sink to the bottom and have no injurious effect upon the deposit. On the other hand, in agitated baths, they remain suspended in the electrolyte, and it may happen that they grow into the deposit, giving rise to the formation of roughnesses (buds). Everywhere that such roughness is formed, it increases more rapidly in proportion to the other smooth portions of the cathode, and these excresences frequently attain considerable thickness, which is not at all desirable.

It is, therefore, advisable to make provision for keeping such baths perfectly clean. Baths agitated by flux and reflux can be readily filtered, as described above, previous to their passing into the collecting reservoir. Solutions agitated by a stirring contrivance, or by blowing in air, should be occasionally allowed to rest and settle. The perfectly clear solution is then siphoned off, and the bottom layers are freed from insoluble particles by filtering. With the use of impure anodes, which, however, cannot be by any means recommended, it is best to sew them in some kind of fabric, for instance, muslin, the fibers of which have been impregnated with ethereal paraffine solution to make them more resistant towards the action of the sulphuric acid. In order to keep the electrolyte as clean as possible, it is best to treat chemically pure anodes in the same manner.

In case no means for agitating the bath should be available, good results may, according to Maximowitsch, be obtained by the following arrangement: The electrodes are placed horizontally and in such a manner that in the bath the anodes are over the cathodes. The new solution is thus formed in the upper portions of the bath on the anodes and being specifically heavier than the old exhausted solution sinks to the bottom where it displaces the exhausted solution poor in copper, the latter being by reason of its slighter specific gravity forced upwards. Hence, without the use of any mechanical contrivance the freshly-formed solution is constantly mixed with the old solution.

To prevent small particles of metal from the anodes falling upon the cathodes and there giving rise to the evolution of gas, a frame filled with unbleached, undyed silk is placed between the two electrodes.

For the production of a beautiful, dense and firm deposit, according to this process, Schönbeck recommends the following bath: Crystallized blue vitriol 125 lbs., concentrated sulphuric acid 12½ lbs., water 500 lbs.

Current-density per square decimeter electrode surface 6 to 10 ampères; *electro-motive force for every* ampère 0.8 volt; *electrode distance* 8 centimeters.

Anodes.—Annealed sheets of the purest electrolytic copper should be suspended in the bath. Impure anodes introduce other metallic constituents into the bath, and the result might be a brittle deposit. The use of old copper boiler sheets, so frequently advocated, is decidedly to be rejected.

The more impurities the anodes contain, the darker the residue formed upon them will be, and this residue in time deposits as slime upon the bottoms of the tanks. Anodes of electrolytically deposited, and therefore perfectly pure, copper also yield a residue, which, however, is of a pale brown appearance, and consists of cuprous oxide and metallic copper. It is recommended daily to free the anodes from adhering residues by brushing, so as to decrease the collection of slime in the bath. The anodes of baths in motion are best sewed, as above described, in a close fabric to retain insoluble particles.

In connection with his previously mentioned experiments, Förster ascertained that with the use of ordinary copper, at 0.3 ampère current-density, about 7.4 grammes (4.15 drachms) of red-brown anode slime with 60 to 70 per cent. of copper, partially in the form of cuprous oxide, were at the ordinary temperature obtained from 6.6 lbs. of anode copper. On the other hand, at a temperature of 104° F., only 24 grammes (1.25 drachms) of a pale gray slime consisting chiefly of silver, lead, lead sulphate and antimony combinations with only a slight content of copper were under otherwise equal conditions obtained. By raising the temperature of the electrolyte to 140° F. the quantity of anode slime increased considerably, and at

1 ampère current-density amounted to about twenty times as much. The slime, in addition to a smaller quantity of the above-described pale gray slime, contained larger quantities of well-formed lustrous copper crystals which could scarcely be derived from the rolled copper anodes. Wohlwills made analogous observations in the electrolysis of gold chloride solution containing hydrochloric acid, and based upon these observations, Förster assumes that at the higher temperature the anode copper sends forth augmented univalent cuprous-ions into the copper solution, the cuprous sulphate solution formed being decomposed to cupric sulphate (blue vitriol) while copper is separated, according to the following equation :

$$Cu_2SO_, \quad = \quad CuSO_4 \quad + \quad Cu.$$
$$\text{Cuprous sulphate.} \quad \text{Cupric sulphate.} \quad \text{Copper.}$$

The anode surfaces should be at least equal to that of the moulds, and for shallow moulds the distance between them and the anodes may be from 2 to 3 inches, but for deeper moulds it must be increased.

Tanks.—Acid-proof stoneware tanks serve for the reception of the acid copper baths, or for larger baths, wooden tanks lined with pure sheet-lead about 0.11 to 0.19 inch thick, the seams of which are soldered with pure lead. It should be borne in mind that a coat of lacquer, as previously mentioned, may have an injurious effect.

Rapid galvanoplasty. Thus far galvanoplastic baths with an average content of 22 per cent. of blue vitriol and 2 to 3 per cent. of sulphuric acid have only been referred to. Such baths were exclusively used up to the end of 1899. The current-density employed in practice amounted to scarcely more than 25 ampères, and the customary thickness of 0.15 to 0.18 millimeters for electrotypes was at the best attained in $4\frac{1}{2}$ to 5 hours.

The much-felt want of producing galvanoplastic deposits of sufficient thickness in a materially shorter time gave rise to search for ways and means to attain this object.

Taking into consideration the fact that a larger quantity of copper can in a shorter time be deposited with the use of higher current-densities, the conditions under which the use of higher current-densities becomes possible without leading to the reduction of a useless, brittle or pulverulent deposit had to be ascertained. By the investigations of Hübl it had been shown that the production of good deposits is by no means dependent on a high content of sulphuric acid in the electrolyte, but that acidulating the copper bath only so far as to prevent the formation of basic salts suffices.

It was further known that by the bath containing a large quantity of sulphuric acid, the solubility of blue vitriol is decreased, and since good deposits can with high current-densities be obtained only from highly-concentrated blue vitriol solutions, the reduction of the content of sulphuric acid became an absolute necessity

There is, however, still another reason why copper baths working with high current-densities can only be compounded with small quantities of sulphuric acid. It has previously been mentioned that the sulphuric acid is dissociated into hydrogen-ions and SO_4-ions, and that hydrogen-ions effect the reduction of copper in a secondary manner. By a small addition of sulphuric acid, this secondary reduction is to be largely avoided, and the copper is to be brought to separate chiefly in a primary manner, because by reason of the accelerated process of reduction at high current densities, there is danger of hydrogen-ions being brought to separate as hydrogen gas on the cathodes, which might give rise to the formation of a sandy or spongy deposit.

In a 20 per cent. blue vitriol solution compounded with 1 per cent. of sulphuric acid, the copper solution and the sulphuric acid participate equally, according to Hübl, in conducting the current; while with a content of 5 per cent. of acid, the conduction of the current is almost exclusively taken charge of by the acid-ions. Thus, the smaller the content of free sulphuric acid, the greater the quantity of primarily deposited copper will be, and the less the danger of hydrogen-occlusion, or the formation of a hydrate.

However, a smaller content of sulphuric acid in the electrolyte, together with a greater content of blue vitriol, is by itself not sufficient for removing the possibility of the formation of spongy deposits caused by the layers of fluid on the cathode having become poorer in metal. Provision has to be made for the vigorous agitation of the electrolyte, so that the layers poor in metal on the cathode are constantly replaced by layers richer in metal, a discharge of hydrogen-ions on the cathodes being thus best prevented; and coherent copper-deposits of great hardness and sufficient tenacity for graphic purposes are even with very high current-densities obtained.

This process, based upon the principles above mentioned, was perfected in 1900, and the term *rapid galvanoplasty* has been applied to it.

It is obvious that the term rapid galvanoplastic bath cannot be claimed solely for one composition, but that all acid copper baths which yield deposits in a materially shorter time than was formerly possible may thus be designated. According to the objects the rapidly-working baths are to serve, it would even be rational that their compositions should vary, as will be directly seen.

While shallow impressions, for instance, autotypes, woodcuts, etc., only require a very small addition of sulphuric acid, for deep impressions of set-up type a larger content of sulphuric acid is necessary, especially when the type has been set with low spaces. For the reproduction of moulds of objects of art in very high relief, rapid galvanoplasty is only within certain limits applicable.

Below will be given two compositions of rapid galvanoplastic baths, which are considered the highest and lowest limits, though it is not to be understood that good results cannot be obtained with baths containing more or less blue vitriol and sulphuric acid. These two baths, however, have proved reliable in practical rapid galvanoplasty, and the necessity for other compositions will scarcely arise.

For shallow impressions of autotypes, wood-cuts, etc.—In a 100 quart bath : 74.8 lbs. of blue vitriol, 0.44 lb. of sulphuric acid of 66° Bé.

Dissolve the blue vitriol with the assistance of heat.

This bath being oversaturated with blue vitriol, crystals would be formed, which must by all means be avoided, and for this purpose the bath has to be constantly kept at a temperature of about 78.8° to 82.4° F. At this temperature the bath shows about 25° Bé., and at 64.4° F., 27° Bé.

Heating the bath is best effected by means of a lead coil on the bottom of the lead-lined tank, through which steam is introduced until the bath shows the desired temperature. Since, by reason of the high current-densities, the temperature of the bath is still further increased, which might be detrimental with the use of wax moulds, the lead coil should be furnished with an additional branch for the introduction of cold water in case the temperature becomes excessive.

It is not likely that a larger bath of the above-mentioned composition will cool off enough over night for the crystallization of blue vitriol, especially if it is covered and the workroom is not exceedingly cold. There is danger of the crystallization of blue vitriol if the work-room is not kept at an even temperature, or the bath is not worked for one or more days in succession. In the latter case it is advisable, the evening before work is stopped, to heat the bath more than usual and dilute it with water. The quantity of the latter which has to be added to make up what may in a certain time be lost by evaporation will soon be learned by experience. If, for the sake of precaution, the bath is covered, it will be found ready for work when operations are resumed.

In order to obtain deposits of good quality with high current-densities, vigorous agitation of the bath is required. This is most uniformly effected by blowing in air by means of an air compressor. The bath may also be agitated, though less uniformly so in all portions, by means of a copper paddle fitted to the front of the tank and driven by means of a band from a transmission. It is placed about six inches above the bottom of the tank and, the paddles being set at an angle of 45°, a vigorous motion of the lower layers towards the surface is effected.

If the above-mentioned conditions be observed, the *current-density* for this bath may amount up to 6 ampères per square decimeter, and with a distance of about 6 centimeters of the cathodes from the anodes, the electro-motive force will approximately be 6 volts. When working on an average with 6 ampères per square decimeter, a deposit of 0.15 millimeter thick will in this bath be obtained in $1\frac{1}{2}$ to $1\frac{3}{4}$ hours.

With the use of gutta-percha matrices, the bath may be somewhat more heated than when working with wax moulds, and still higher current-densities than those given above may be employed, the deposit being then finished in a still shorter time. It is, however, advisable not to carry the work of the current to an excess, otherwise the copper might readily show properties not at all desirable.

It may, under certain circumstances be advantageous, nay even necessary, to face the black-leaded matrices with copper at a somewhat slighter current-density, while the bath is at rest, *i. e.*, not agitated by a stirring contrivance, or by blowing in air, and to resume agitation and increase the current-strength only after the matrices are coated with copper. Thus, according to the size of the galvanoplastic plant, it may be desirable to have a smaller coppering bath not furnished with a stirring contrivance, from which the matrices, after having been faced with copper, are transferred to the agitated bath.

It may here be remarked that Knight's process of coppering the matrices with neutral blue vitriol solution and iron filings, which is much liked, is not applicable in rapid galvanoplasty. In suspending such matrices coated with copper in the rapid bath, the slight copper-film is, so to say, burnt, and a proper deposit can no longer be effected.

In a bath of the composition given above, it is sometimes

difficult to obtain with the above-mentioned high current-densities unexceptionable electrotypes from matrices produced from deep and steep set-up type. The shallow portions, to be sure, copper well, but the copper does not spread into the deeper portions, and holes are left. By the addition of certain substances, for instance, alcohol, this drawback can, to be sure, be somewhat improved, but not entirely removed, and for this reason such matrices are further worked in a bath, the composition of which is given below. It is, however, preferable to preparatively copper such type-compositions, especially when low spaces have been used, and after about $\frac{1}{2}$ hour to transfer the matrices to the rapid galvanoplastic bath. By working in this manner, the electrotypes will be free from holes, and finishing even the largest customary forms will not require more than 2 hours.

For deep impressions.—In a 100-quart bath : 57.2 lbs. of blue vitriol, 1.76 lb. of sulphuric acid.

It is recommended not to deposit at a lower temperature of the bath than 68° F., though with this concentration the danger of crystallization is less. For heating and cooling the electrolyte, a lead coil, as previously described, is advantageously used, and provision for thorough agitation has to be made. This bath is generally allowed to work with 4.5 to 5 ampères current-density, the electro-motive force, with a distance of 6 centimeters of the anodes from the cathodes, amounting then to about $4\frac{1}{2}$ volts. The copper deposit attains in $2\frac{1}{4}$ hours a thickness of 0.15 millimeter, and in $2\frac{3}{4}$ hours one of 0.18 millimeter. Higher current-densities are also permissible, and the operator will soon find out how far he can go in this respect.

Deeper forms become well covered, especially if, according to Rudholzner's proposition, about 1 lb. of alcohol is added. But, nevertheless, it is recommended to preparatively copper in the ordinary acid copper bath impressions of very steep set-up type with low spaces, as with the use of high current-densities the streaks which are temporarily formed upon the printing faces of the electrotypes are thus most surely avoided.

Heating the baths may be omitted in plants lacking the necessary contrivances. The blue vitriol solution must then be of such a composition as to preclude all danger of blue vitriol crystallizing out even at the lowest temperature of the work-room. Somewhat lower current-densities corresponding to the slighter concentration have of course to be used.

Regarding the quality of the copper deposit effected with high current-densities, it may be said that its tenacity is good, better in the second bath than in the one first mentioned, but in all cases sufficient for the electrotypes. The copper is however, decidedly somewhat harder than that deposited from the ordinary baths as proved by its slight wear in printing.

The treatment of the rapid galvanoplastic baths will be readily understood from what has been said above. On the one hand, the baths must not be allowed to cool to a temperature at which the blue vitriol would no longer be held in solution, but would crystallize ; and, on the other, the reaction has from time to time to be tested with red congo paper which must acquire a plainly-perceptible blue color. If such is not the case, no, or too little, free sulphuric acid is present in the bath, and brittle deposits will be formed which cannot be detached whole from the matrices. When this is noticed add 0.44 lb. of sulphuric acid per 100 quarts of bath, or 1.76 lb. to the bath for deep impressions.

The excess of acid is very rapidly consumed with the use of copper plates which have been electrically deposited and, without recasting and rolling, suspended as anodes in the bath. The use of rolled anodes is therefore absolutely necessary, and, as previously described, they should be sewed in a close fabric to avoid contamination of the bath by the anode-slime formed, and by small copper crystals.

Special attention should be paid to furnish the matrices with conductors of sufficiently large cross-section corresponding to the great current-strengths. This will later on be referred to.

Examination of the Acid Copper Baths.

The copper withdrawn from the bath by deposition is only partially restored, but not entirely replaced, by the anodes, and hence the content of copper will in time decrease, and the content of free acid increase. The deficiency of copper can, however, be readily replaced by suspending bags filled with blue vitriol in the bath, while too large an excess of acid is removed by the addition of copper carbonate or cuprous oxide (cupron).

However, in order not to grope in the dark in making such corrections of the bath, it is necessary to determine from time to time the composition of the copper solution as regards the content of copper and acid, for which purpose the methods described below may be used.

Determination of Free Acid.—The free acid is determined by titrating the copper solution with standard soda solution, congo-paper being used as an indicator. Bring by means of a pipette, 10 cubic centimeters of the copper bath into a beaker, dilute with the same quantity of distilled water, and add drop by drop from a burette standard soda solution, stirring constantly, until congo-paper is no longer colored blue when moistened with a drop of the solution in the beaker. Since 1 cubic centimeter of standard soda solution is equal to 0.049 gramme of free sulphuric acid, the cubic centimeters of standard soda solution used multiplied by 4.9 give the number of grammes of free sulphuric acid per liter of bath.

Volumetric determination of the content of copper according to Haën's method.—This method is based upon the conversion of blue vitriol and potassium iodide into copper iodide and free iodine. By determining the quantity of separated free iodine by titrating with solution of sodium hyposulphite of known content, the content of blue vitriol is found by simple calculation. The process is as follows: Bring 10 cubic centimeters of the copper bath into a measuring flask holding $\frac{1}{10}$ liter, neutralize the free acid by the addition of dilute soda lye until a precipitate of bluish cupric hydrate, which does not disappear even with vigorous shaking, commences to separate. Now add, drop by drop, dilute sulphuric acid until the precipitate just dissolves; then fill the measuring flask up to the mark with distilled water, and mix by vigorous shaking. Of this solution bring 10 cubic centimeters by means of a pipette into a flask of 100 cubic centimeters' capacity and provided with a glass stopper; add 10 cubic centimeters of a 10 per cent. potassium iodide solution ; dilute with some water, and allow the closed vessel to stand about 10 minutes. Now add from a burette, with constant stirring, a decinormal solution of sodium hyposulphite until starch-paper is no longer colored blue by a drop of the solution in the flask. Since 1 cubic centimeter of decinormal solution corresponds to 0.0249 gramme of blue vitriol ($= 0.0003$ gramme of copper), the content of blue vitriol in one liter of the solution is found by multiplying the number of cubic centimeters of decinormal solution used by 24.9. For the correctness of the result it is necessary that the copper bath should be free from iron.

The electrolytic determination of the copper being more simple, it is to be preferred to the volumetric method. Bring by means of the pipette 10 cubic centimeters of the copper bath into the previously weighed platinum dish, add 2 cubic centimeters of strong nitric acid, fill the dish up to within 1 centimeter of the rim with distilled water, and electrolyze with a current-strength ND 100 $= 1$ ampère.

Deposition of copper is finished when a narrow strip of platinum sheet placed over the rim of the dish and dipping into the fluid shows in 10 minutes no trace of a copper deposit, which is generally the case in $3\frac{1}{2}$ hours. The deposit is then washed without interrupting the current, rinsed with alcohol and ether, and dried for a short time at 212° F. in the air-bath. The increase in weight of the platinum dish multiplied by 100 gives the content of metallic copper in grammes per 1 liter of bath. To find the content of blue vitriol, multiply the found content of copper per liter by 3.92, or multiply the content of copper determined in 10 cubic centimeters of bath by 3.92.

If now the content of free acid and of the blue vitriol in the bath has been ascertained, a comparison with the contents originally present in preparing the bath will show how many grammes per liter the content of acid has increased, and how many grammes the content of copper has decreased. Then by a simple calculation it is found how much dry pure copper carbonate has to be added per liter of solution to restore the original composition. For each gramme more of sulphuric acid than originally present, 1.26 grammes of copper carbonate have to be added, and each gramme of copper carbonate increases the content of blue vitriol 2.02 grammes per liter of bath. By reference to these data the operator is enabled to calculate whether the quantity of copper carbonate added for the neutralization of the excess of free acid suffices to restore the original content of blue vitriol, or whether, and how much, blue vitriol per liter has to be added.

With the use of baths in which the solutions circulate, the additions are best made in the reservoir placed at a higher level, into which the solution constituting the bath is raised by means of a pump. The composition of such baths, connected one with the other, is the same, and a single determination of the content of copper and free sulphuric acid will suffice. However, with baths, the contents of which do not circulate and are not mixed, a special determination has to be made for each bath, and the calculated additions have to be made to each separate bath.

Operations in Galvanoplasty for Graphic Purposes.

The manipulations for the production of galvanoplastic deposits for printing books and illustrations will first be described.

1. *Preparation of the moulds (matrices) in plastic material.* If a negative of the original for the production of copies is not to be made by direct deposition upon a metallic object, it has to be prepared by moulding the original either in a plastic mass which, on hardening, will retain the forms and

lines of the design to the finest hatchings, or in a material, which plastic itself, retains the impression unaltered. Suitable materials for this purpose are : Gutta-percha, wax (stearine, etc.), and lead.

The preparation of moulds in gutta-percha and wax will first be described, and the production of metallic matrices will be referred to in the next section.

a. *Moulding in gutta-percha.*—For the reproduction of the fine lines of a wood-cut or copper-plate, pure gutta-percha, freed by various cleansing processes from the woody fibers, earthy substances, etc., found in the crude product, is very suitable. Besides the requisite degree of purity, the gutta-percha should possess three other properties, viz., it must become *highly plastic* by heating, without, however, becoming *sticky*, and finally it should *rapidly harden*.

The most simple way of softening gutta-percha is to immerse it in water of 170° to 190° F. When thoroughly softened no hard lumps should be felt on kneading with the hands, which should be kept thoroughly moistened with water during the operation. A fragment of the gutta-percha corresponding to the size of the object to be moulded is then rolled into a plate about $\frac{1}{8}$ to $\frac{3}{4}$ inch thick. To facilitate the detachment of the mould after cooling, the surface of the gutta-percha which is to receive the impression should be well brushed with black-lead (plumbago or graphite), an excess of it being removed by blowing.

The original (wood-cut, autotype, set-up type, etc.) must be firmly locked in the usual manner, and the surface is then cleansed from dirt and stale ink by brushing with benzine. When dry it is brushed over with plumbago, an excess of it being removed by means of a bellows.

The black-leaded surface of the warm gutta-percha plate is then placed upon the black-leaded face of the original, and after gently pressing the former with the hand upon the latter, the whole is placed in the press.

b. *Moulding in wax.*—Beeswax is a very useful material for

preparing moulds, but, like stearine, it is according to the temperature now softer and now harder, which must be taken into consideration. In the cold state pure beeswax is quite brittle, and apt to become full of fissures in pressing. To decrease the brittleness certain additions are made to the wax; various formulas for such compositions recommended by different authors are here given:

a. White wax 120 parts, stearin 50, tallow 30, Syrian asphalt 40, elutriated graphite 5. (G. L. von Kress).

b. Yellow beeswax 700 parts, paraffin 100, venetian turpentine 55, graphite 175; or, cake wax 50 parts, yellow wax 50, ceresin 15, venetian turpentine 5. (Karl Kempe).

c. Wax 20 parts, thick turpentine 20, rosin 10, graphite 50. (Hackewitz). By reason of its large content of graphite, this composition which is excellent in every respect, can be recommended for taking moulds from objects which can be black-leaded only with difficulty.

d. Yellow wax 900 parts, venetian turpentine 135, graphite 22. (Urquhart).

e. Pure beeswax 850 parts, crude turpentine 100, elutriated graphite 50. (Furlong). The mixture is to be freed from all moisture by boiling in a steam pot for 2 hours. In the hot season of the year it is recommended to add 50 parts of burgundy pitch to impart greater hardness to the wax.

ƒ. Pfanhauser recommends the following composition especially for taking moulds from undercut objects. The mass is very elastic and objects with quite wide projecting portions can, with care, be moulded with it.

Yellow beeswax 400 parts, ozocerite 300, paraffine 100, venetian turpentine 60, elutriated graphite 100. For use in the summer months the composition of the mass is as follows: Yellow beeswax 250 parts, ozocerite 450, paraffin 50, Venetian turpentine 35, elutriated graphite 180.

The proportions given in the formulas cannot always be strictly adhered to and one has to be guided by prevailing conditions. If the wax turns out rather brittle, somewhat more tallow or turpentine has to be added and, on the other hand, in the hot season of the year when the wax is too soft, a smaller quantity of turpentine or tallow will have to be used.

To avoid overheating it is advisable not to melt the wax mixture over an open fire, and a jacketed kettle heated by steam or gas is generally used. With the use of steam, the latter passes through a valve into the jacket while the condensed water is discharged through another valve. When gas is used the space between the jacket and kettle is filled with water, the latter being from time to time replenished as evaporation progresses.

Two wax-melting kettles will be required, because the wax which has been in contact with the bath, has to be entirely freed from water in the one kettle before it can be again used for moulding. The dehydrated wax is then transferred to the other kettle.

To prepare the wax for receiving the impression, pour the melted composition in the mould-box, which is a tray of sufficient size with shallow sides about ¼ inch in depth all round, and with a continuation of the bottom plate on one of the shorter sides for about 3 inches beyond the box, to allow of its being supported by hooks from the conducting rods of the bath. The moulding-box is placed upon a level surface and filled to the brim. Air bubbles and other impurities forming on the surface are at once removed by a touch with a hot iron rod.

The surface of the wax, while still luke-warm, is then dusted over with the finest plumbago. The black-leaded original is then placed, face downwards, upon the wax surface and submitted to intense pressure. When black-leading has been carefully done, the original can be readily and perfectly detached from the mould. Some operators apply a light coat of oil to the original in place of black-leading it, but care must be taken not to leave any considerable portion of oil upon the original.

In this country, before the impression is taken, the wax plate or wax mould is frequently treated as follow : Black-lead and water are mixed to the consistency of cream. The mixture is carefully and uniformly applied to the wax plate and rubbed dry with the hand.

The method above described, according to which the melted wax is poured in the moulding-box is constantly more and more abandoned, the work being generally done as follows :

Lead plates, the size of the original to be moulded, are cast, laid upon the wax-moulding table, and enclosed by a rim of the depth of the required thickness of the wax plate. The box thus formed is then filled to the brim with melted wax, air-bubbles and other impurities being removed, any excess of wax cut off, and the mould black-leaded by means of a soft brush. In some galvanoplastic plants the moulded wax plates, previous to making the impressions, are planed perfectly level by a shaving machine. While gutta-percha matrices will bear quite vigorous treatment with the brush, care must in this respect be exercised with wax matrices to prevent in- jury.

The wax plates prepared according to the process just described are black-leaded and laid upon the originals to be moulded, the whole being then placed under the press.

2. *Presses.*—For making the impressions of the form in the moulding composition, a moulding press is used which is cap- able of giving a gradual and powerful pressure. Fig. 141 represents a form of moulding press in common use, and known as the " toggle " press. It consists of a massive frame having a planed, movable bed, over which a head is moved on pivots and counter-balanced by a heavy weight, as shown, so that it can be readily thrown up, having the bed exposed, the black-leaded type form being placed on the bed. The well black-leaded case is attached by clamps to the movable head, or the form (also black-leaded) is laid face down on the case, and the head is then turned down and held in place by the swinging bar (shown turned back in the cut). All being

ready, the toggle-pressure is put on by means of the hand-wheel and screw, the result being to raise the bed of the press with an enormous pressure, causing the face of the type form to impress itself into the exposed moulding surface.

FIG. 141.

Fig. 142 represents a form of "hydraulic press" less commonly used than that just described. It is provided with projecting rails and sliding plate, on which the form and case are arranged before being placed in the press. The pump, which is worked by hand, is supported by a frame-work on

the cistern below the cylinder, and is furnished with a graduated adjustable safety-valve to give any desired pressure.

Metal matrices.—Attempts have for many years been made
to mould originals in lead, since lead matrices possess many
advantages over gutta-percha and wax matrices as they do
not require to be rendered conductive by black-leading, and
no changes in dimensions take place in consequence of the
transition from the heated into the cold state.   However,

FIG. 142.

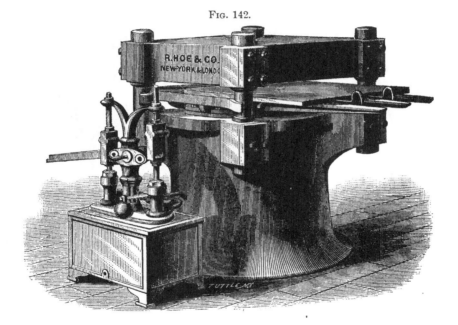

objects readily liable to injury, such as wood cuts, compositions, etc., could not withstand the pressure required for impression in lead plates, and were demolished ; steel plates at
the utmost were capable of standing the high pressure.
Serviceable results were not obtained, even with the use of
very thin lead foil backed, in pressing, with moist paste-board
or gutta-percha, because the portions of the lead foil subject to
the most severe demands would tear.

To Dr. E. Albert of Munich is due the credit of having discovered the cause to which these failures were due, and of having devised a method for the rational preparation of metal matrices.

Dr. Albert says in reference to this matter * : ." Every galvanoplastic operator knows that in making impressions of forms of mixed composition and illustration, that the composition down to the quads is impressed before the shades, for instance, of a wood cut or an autotype, are finished. In making impressions, the moist paste-board referred to above acted exactly in the same manner as wax or gutta-percha softened by heating ; *i. e.* by the moist paste-board the lead foil had to be pressed first into the deeper, and finally into the more shallow depressions. Notwithstanding the enormous ductility of lead, the lead foil could not satisfy these demands on extension and, in consequence of this over-demand, tore in many places. Hence this process was not available for general practice, it being at the utmost suitable only for forms with very slight differences in level, and even not for this purpose with the large forms now in general use.

It must be borne in mind that, for instance, upon a square millimeter of an autotype there are 36 depressions into which the lead foil has to be pressed and to 144 side-walls per square millimeter of which it has to attach itself. Especially with under-etched printing forms considerable force is required to detach the matrix from the moulding material, and it is therefore impossible with larger forms to manipulate the lead foil which, for the sake of decreasing the pressure, has to be very thin so as to maintain at the same time a level surface.

This method of impression by which the parts corresponding to the dark portions of the original can only be impressed when the moulding material has been forced into the last corner of the deepest depressions of a printing form, is not premeditated nor one by choice, but is conditioned on the physical

* Zur Theorie und Praxis der Metall-matrize, 1905.

properties of the material itself. The pressure required to force
the moulding material into the smallest depressions cannot
be applied so long as the moulding material has a chance to
escape into an empty space.

In consequence of this property the matrices have to undergo extensive manipulations, since the large angular elevations
which correspond to the depressions of the printing form would
prevent the further development of the electro, especially also
the formation of the copper-deposit upon the matrix. Hence
the prominent portions have to be removed in the known
manner.

This necessary after-manipulation would of course be impracticable with matrices of thin lead foils, and for this reason
also the method is not available for line-etching, wood-cut and
composition.

In the preceding it has been specified as characteristic of the
bodies hitherto used for the preparation of matrices that the
impression of the deepest depressions takes place before that of
the more shallow ones; with soft metals, particularly with lead,
just the reverse is the case. The interior coherence of the
body-molecules is so much greater in comparison with wax
and gutta-percha mass, or moistened paste-board, that at the
commencement of the pressure the lateral escape is avoided,
whereby the moulding material yields first in the direction of
the pressure and fills the smallest depressions. Only with increasing pressure, which is necessary for forcing the lead into
the deeper depressions of the printing form, the lead also
begins to yield laterally in the region of the portions pressed
first.

Independent of the fact that the small points already impressed, which correspond to the smallest impressions of the
printing form, are again impressed, this pushing of the lead
has the further drawback that the lead firmly settles in these
smallest depressions, thus rendering the original useless.

Besides, there is no type composition, no wood-cut, etc., the
printing elements of which, especially when standing isolated,

could withstand the enormous pressure which has to be used for forcing a lead plate at least 5 millimeters thick into the large depressions. However, such a thickness of the lead plate would be necessary just as with wax and gutta-percha impressions, since the difference in height between printing and justifying surface is about 1 cicero = 4.5 millimeters.

Hence, with the means hitherto available, the production of matrices, either with thin or thick metal plates, was impracticable, and until lately recourse had to be had to the old and qualitatively inferior wax and gutta-percha matrices, till Dr. Albert, in 1903, succeeded in finding a method for the rational production of metal matrices.

This method is based upon a number of inventions patented in all civilized countries, and the characteristic features of the process will here be briefly given.

The basis for the solution of the problem rested upon the adoption of such a thickness of the metal plate, that the manipulations required for the production of the matrix and its after-manipulations without deformation could be effected by the hand of any workman; as well as upon a new method of impressing which would render it possible for the thickness of the plate to be materially less than the relief difference of the printing form.

While in the production of medals and coins by means of galvanoplasty, the problem consists in a perfectly detached reproduction of all the differences in level of the original, with an electro for graphic purposes, the impression of the matrix in the large depressions is only a matter of technical necessity so that in the subsequent use of the electro for printing the white portion will not smear. This knowledge led to the expedient of pressing or bending by means of a support of a soft body, the about 2 millimeters thick lead plates only so far into the above-mentioned depressions as required for technical reasons.

Hence this method of impression is based upon a combination of pressing and bending. The lead is bent to a greater extent the larger and wider the sunk surface is, the electro automatically receiving thereby all the white portions of such depth that they do not smear in printing.

The process may be explained by Figs. 143 and 144.

Fig. 143 represents the arrangement of the platen, lead plate, and soft intermediate layer previous to the moment of impression. The material used for the intermediate layer must possess certain properties and must be softer than the moulding material. It should be compressible without materially yielding laterally under pressure and, by reason of elasticity or internal friction, also oppose a certain resistance to compression in order to bend with this resisting power of the lead-plate where the latter lies hollow. On the other hand, it

FIG. 143.

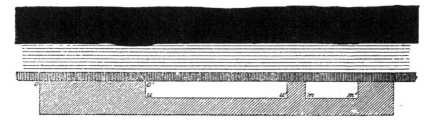

must not be too soft in the sense of its affinity to a liquid aggregate state, as, for instance, heated wax, but it should be more porously soft either in conformity with its nature or its arrangement. In principle the latter is generally based upon the production of many empty intermediate spaces in the material (wood shavings and snow are softer than wood and ice), or upon placing many thin layers of the material one above the other. Such bodies can be compressed without yielding too much laterally. If the character of the body approaches more the liquid state, more elastic properties have to be added, which by their tendency to equalize the change suffered in form counteract the lateral yielding, or other checks have to be arranged. Besides, a certain degree of elasticity is useful for bending the lead plate on the free-lying places.

Such an intermediate layer may appropriately consist of a number of layers of paper. Such a layer, by reason of the character of the paper fiber itself, as well as of the intermediate layer of air, is soft and elastic as regards the direction vertical to the impression-plane, while on the other hand the texture of the paper-stuff affords the necessary checks in the direction parallel to the impression-plane to prevent, after the commencement of pressure, the lateral yielding of the intermediate layer. The latter important property was in former experiments neutralized by moistening the paper.

In Fig. 144 the platen has sunk so that the intermediate

FIG. 144.

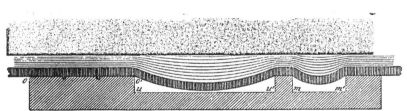

layer opposite to the places $o\ o'$, from which the first counter-pressure emanates, is compressed to one-half of its original volume. At the moment when the intermediate layer has by compression acquired the degree of hardness of the moulding material, it is forced by the next increase in pressure into the small depressions of the plane $o\ o'$. On the places opposite to $u\ u'$, the lead, which lies here perfectly free, and hence exerts no counter-pressure, is at the same time pressed into the hollow space $u\ u'$ by the resisting force of the intermediate layer.

The same is also the case opposite to the places $m\ m'$, but the bending takes place in a less degree, just as a board resting upon supports 6 feet apart is more bent by a weight than one whose supports are only 3 feet apart.

This also answers technical requirements, since the white portions smear the more readily in the press, the greater their dimensions are.

Thus there had always been made the gross error of treating according to the same principles which had proved good for wax and gutta-percha, a body, such as lead, of an entirely different physical character. The process of pressing had in the main to be excluded, and a bending process substituted for it. This was rendered possible by a suitable thickness of the moulding material, and by backing it with a soft and yielding body, which, as regards its extensibility parallel to the impression-plane, was checked by its texture or otherwise.

By this bending process the pressure required for impression was under certain circumstances reduced to one-tenth of its former magnitude, so that metal matrices could also be produced from wood-cuts and composition.

This reduction in pressure is least manifest with printing forms with many very fine and crowded printing elements, for instance, autotypes, for which, according to the character of the picture, a pressure of 500 to 1000 kilogrammes per square centimeter is required ; this is more than hitherto used for wax and gutta-percha.

The problem of the production of metal matrices was thus solved only for forms of moderate size, since, although the pressure had been largely reduced by the selection of a correct thickness of the lead plate and by backing the latter with a soft, elastic body, it was nevertheless much greater than that required for wax and gutta-percha. The ordinary hydraulic presses, with some few hundred atmospheres, were therefore not available for impressing larger forms.

By the use of successive partial pressure with the simultaneous introduction of side-pressure, Dr. Albert has succeeded in increasing, at a small expense, about twenty times the capacity of every press now in use.

The gradual progression of a limited pressure over the entire printing form also prevents the extremely troublesome phenomena appearing in other methods of impressing, namely, that it is impossible for the process of impression being affected by occluded air, the latter having at any time a chance to escape.

The impressions being automatically effected, there is no loss of time worth speaking of with this method. Thus, for instance, only 55 seconds were required for impressing a form of the " Woche," and not quite two minutes for one of the " Berliner Illustrierte Zeitung." For impressing illustration-forms of the same size without letters, only half the above-mentioned time was necessary.

Thus there is no difficulty whatever in executing impressions of any size.

Fischer endeavors to attain the same object as Dr. Albert by the use of lead plates with corrugated backs, small pyramids

about 2 to 3 millimeters high being thus formed. These corrugations act like Albert's elastic intermediate layer in so far that the lead plates are not pressed, but bent, into the deep portions of the printing form, a reduction in the otherwise high pressure required being thus effected. Now, suppose in Fig. 144, instead of an elastic intermediate layer, a lead plate with corrugated back is placed upon the form, the small pyramids which are opposite to the portion $o\ o'$ of the printing form are first compressed, while the part of the lead plate corresponding to the portion $u\ u'$ is bent through by the pressure exerted by the platen upon the points of the corrugations, the latter being thereby not very much flattened. If now the pressure be increased the lead plate is first flattened at $o\ o'$, and then the actual impression, $i.\ e.$, pressing the lead into the design of the original or into the composition begins.

Kunze does not use corrugated lead plates, but provides the platen with corrugations, and combines therewith a process of successive partial pressure invented by him. (German patent applied for.) As the patent has not yet been granted, details of the process cannot be given.

39

3. *Further manipulation of the moulds.*—The moulds when detached from the original show in addition to the actual impression certain inequalities which have to be removed.

With gutta-percha moulds such inequalities in the shape of elevations, are carefully pared away with a sharp knife, while with wax moulds they are melted down. For this purpose serves a brass tube about 4 inches long, drawn out to a fine point and connected by means of a rubber tube with a gas jet. By opening the gas-cock more or less, the gas burns with a larger or smaller pointed flame, and the brass tube is guided by the hand, so that the elevations are melted down and the deeper portions of the electrotype will present a smooth appearance. A more modern instrument for this purpose is so arranged that the flame can be regulated by the finger pressing upon a rubber bulb. However, not only the inequalities are melted down, but the upper edges, of the steep contours of the impression are melted together, and melted wax is built up all around in order to enlarge the depressions in the electrotype and avoid cutting. The wax is readily built up by holding in one hand a thin stick of wax at a distance of about 0.19 inch from the edge of the impression and at about the same distance above the mould, and melting off the wax, drop by drop, by means of a pointed flame guided by the other hand. One drop is placed close alongside the other, and when the entire edge of wax is thus completed it is made perfectly smooth by again melting with the pointed flame.

The next process is

4. *Making the moulds conductive*, without which a galvanoplastic deposit would be impossible. Black-lead is almost exclusively used for this purpose, and must be of the purest quality and in a most minute state of division. The best material for this purpose is prepared from the purest selected Ceylon graphite, which is ground by rolling with heavy iron balls until it is reduced to a dead black, impalpable powder.

Black-leading the moulds is performed either by hand or more commonly by machines.

Fig. 145 shows one of these machines with its cover removed to exhibit its construction. It has a traveling carriage holding one or more forms, which passes backward and forward, under a laterally vibrating brush. Beneath the machine

FIG. 145.

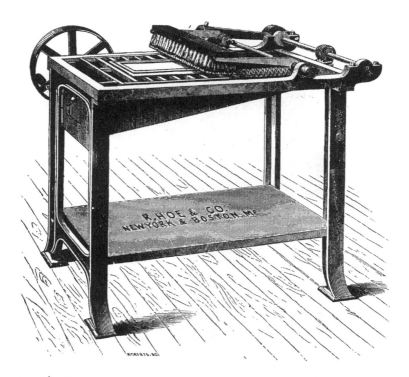

is placed an apron which catches the powder, which is again used.

Another construction of a black-leading machine is shown in Fig. 146, the details of which will be understood without lengthy description. The moulds are placed upon the slowly revolving, horizontal wheel, upon which the brush moves rapidly up and down with a vertical, and at the same time

lateral, vibrating motion.   The black-leading space being
closed air-tight, scattering of black-lead dust is entirely prevented, the excess of black-lead collecting in a vessel placed
in the pedestal.

On account of the dirt and dust caused by the dry process
of black-leading, some electrotypers prefer the wet process
as it is claimed to work more quickly and neatly, producing
moulds that are thinly, evenly and perfectly covered.   The

Fig. 146.

moulds are placed upon a shelf in a suitable receptacle, and a
rotary pump forces an emulsion of graphite and water over
their surface through a traveling fine rose-nozzle.

Black-leading machines have recently been introduced, their
action being based upon the principle of the blast.   The graphite powder is by means of a current of strongly-compressed
air carried with considerable force towards the surface of the
mould to be black-leaded.   The process of making the moulds

conductive according to this system, is claimed to be thorough and complete and quickly accomplished. However, many operators prefer black-leading by hand, especially moulds of autotypes, the lines remaining sharper.

5. *Electrical contact.*—The black-leaded moulds have now to be provided with contrivances for conducting the current upon the black-leaded surface.

With gutta-percha moulds, the edges are trimmed off to within 0.19 to 0.31 inch of the impression. In two places on the edges of the mould holes are made by means of an awl. Through these holes stout copper wires doubled together are drawn, so that after twisting them together they lie firmly on the edge of the mould. These wires serve for suspending the mould to the conducting rod, and previous to twisting them together, two fine copper wires, the so-called *feelers*, are placed between them and the edge of the mould. The object of these thin wires being to effect the conduction of the current to the lower portions of the mould, they must be firmly secured in twisting together the suspension-wires.

However, before allowing these feelers to rest upon the black-leaded surface, the place of contact of the wire with the mould is again thoroughly brushed with black-lead, in order to be sure that the current will not meet with resistance on these points. With very large moulds it is advisable to use more than two feelers and to arrange them especially in deeper depressions. The thickness of the feelers should be about that of horse-hair.

No black-lead should get on the edges or back of the mould, otherwise copper would also be deposited on them.

In place of the wires for suspending the mould, the method for wax moulds described below may also be applied, a small, hot copper plate being melted in on the edge of the mould and the latter secured to the conducting rod by means of a hook.

Gutta-percha moulds, being specifically lighter than the copper bath, would float in it, and have, therefore, to be loaded by securing heated pieces of lead to the backs.

For black-leaded wax moulds the process is as follows : A bright copper plate about 1.18 inches square and 0.039 inch thick is melted in on the upper edge of the mould, and the edges are leveled by means of a pointed flame, so as to produce a smooth joint between the copper plate and wax surface. This place is again thoroughly black-leaded with the hand, and the edges, having been first beveled, are then melted together with the flame. The wax over the hole in the lead plate through which the hook of the mould-holder is pushed is finally removed with a knife. The shape of the mould-holder is shown in the accompanying illustration, Fig. 147. The hook to which the mould is suspended is insulated from the rest of the holder by hard rubber plates, and the screw-

FIG. 147.

threads by hard rubber boxes, so that the lead plate which comes in contact with the hook receives no current, and no copper can deposit upon it. The small, square block cast on the holder lies perfectly level upon the copper plate in the mould, a good and abundant conduction of current being thus effected, such as is absolutely required, for instance, for rapid galvanoplasty.

To prevent the copper deposit from spreading much beyond the impression towards the edge, it has been proposed to cover these portions of the mould with strips of glass, hard rubber, or celluloid. For this purpose heated glass strips, 0.15 inch wide and 0.19 inch high, are pressed about 0.079 inch deep into the wax mould so as to form a closed frame around the impression. Strips of hard rubber or celluloid of the above-mentioned width and height, are fastened together with copper pins. By these means the object in view is perfectly attained.

With very deep forms of type, it is sometimes of advantage to first coat the black-leaded surface with copper, in order to obtain a uniform deposit in the bath. The process is as follows: Pour alcohol over the black-leaded form, let it run off, and then place the form horizontally over a water trough. Now pour over the form blue vitriol solution of 15° to 16° Bé., dust upon it from a pepper-box some impalpable fine iron filings and brush the mixture over the whole surface, which thus becomes coated with a thin, bright, adherent film of copper. Should any portion of the surface after such treatment remain uncoppered, the operation is repeated. The excess of copper is washed off and the form, after being provided with the necessary conducting wires, is ready for the bath.

Gilt or silvered black-lead is also sometimes used for very deep forms. It is, however, cheaper to mix the black-lead with $\frac{1}{8}$ its weight of finest white bronze powder from finely divided tin. When forms thus black-leaded are brought into the copper bath, the particles of tin become coated with copper, also causing a deposit upon the black-lead particles in contact with them.

6. *Suspending the mould in the bath.* Previous to suspending the mould in the copper bath, it has to be perfectly freed from every particle of black lead which might give rise to defects in the deposit.

Strong alcohol is then poured over the mould, the object of this being to remove any traces of greasy impurities, which are readily dissolved and removed by the alcohol. Moulds thus treated at once become uniformly wet in the bath, which, if this precaution be omitted, is not the case, and causes an irregular formation of the deposit (by air-bubbles).

The moulds are suspended in the bath in the manner above described, special attention being paid to having them hang parallel to the anodes so that all portions of them may receive a uniform deposit.

Before being suspended in the bath, the backs of *lead matrices* should be provided with a protecting layer of celluloid or other suitable material to prevent them from becoming coppered.

7. *Detaching the deposit or shell from the mould. a. From*

gutta-percha moulds. When the mould has acquired a deposit of sufficient thickness, it is taken from the bath, rinsed in water, and all edges which might impede the detachment of the deposit from the mould are removed with a knife. The deposit is then gradually lifted by inserting under one corner a flat horn plate, or a thin dull brass blade, and applying a very moderate pressure. Particles of gutta-percha which may still adhere to the deposit, are carefully burnt off over a flame.

b. *From wax moulds.* Wax moulds are placed level upon

Fig. 148.

a table, and hot water is several times poured over them. By pushing the finger-nail under one corner of the deposit, it can readily and without bending be detached from the softened wax. If not successful at first, continue pouring hot water over the mould until the deposit can be detached without difficulty.

In larger establishments, a cast-iron moulding and melting table, such as is shown in Fig. 148, is used for wax moulds. The planed table plate is hollow, and by means of tongues cast to the plate the steam which is introduced is forced to uniformly heat the entire plate. The electrotypes are placed upon the plate, the wax side down. The wax melts and runs through stop-cocks on the side into a jacketed copper kettle, which can be heated by steam for melting the wax. The iron ledges screwed upon the table plate are made tight with asbestos paper, so that the wax cannot run off except through the stop-cocks.

If the table is to be used for moulding the wax plates, cold water, instead of steam, is allowed to circulate through the hollow table plate, whereby rapid congealing of the wax is effected.

Two such kettles are required, since the wax which has been in contact with the bath has to be for several hours heated in one of the kettles to render it free from water before it can be again used for moulding. The wax freed from water is brought into the kettle and used for moulding wax plates.

c. From metal-matrices. If the matrix has been free from fat, the deposit adheres very firmly, and cannot be lifted off in the ordinary manner as with gutta-percha matrices; nor can the deposit be separated from the lead by melting the latter, as with the temperature required for this purpose, the copper shell might be damaged.

Albert found that by allowing the metal matrix together with the copper deposit to float upon readily fusible metallic alloys with many free calories, the deposit, in consequence of the unequal expansion of the metals, can completely and without injury be separated. By detaching the deposit in this manner, Albert succeeded in using the lead matrix freed from the deposit four times for the preparation of electros, the last electro thus made being not inferior in quality to the first one.*

8. *Backing the deposit or shell.* The face of the electro is first freed from all residues by careful burning off over a flame

* Dr. E. Albert, "Zur Theorie und Praxis der Metall-Matrize," p. 10.

and washing with benzine, and scoured bright with whiting and hydrochloric acid. The edges are then trimmed with shears to the width of a finger from the picture. The tinning of the back of the shell is the next operation, and has for its object the strengthening of the union between the shell and the backing metal. For this purpose the back of the shell is cleansed by brushing with " soldering fluid," made by allowing hydrochloric acid to take up as much zinc as it will dissolve, and diluting with about one-third of water, to which some ammonium chloride is sometimes added. Then the shell, face down, is heated by laying it upon an iron soldering plate, floated upon a bath of melted stereotype metal, and, when hot enough, melted solder (half lead and half tin) is poured over the back, which gives it a clean, bright metallic covering. Or the shell is placed downward in the backing-pan, brushed over the back with the soldering fluid, alloyed tinfoil spread over it, and the pan floated on the hot backing metal until the foil melts and completely covers the shell. When the foil is melted the backing-pan is swung on to a leveling stand, and the melted backing metal is carefully poured on the back of the shell from an iron ladle, commencing at one of the corners and gradually running over the surface until it is covered with a backing of sufficient thickness. Another method is as follows : After tinning the shell it is allowed to take the temperature of the backing metal on the floating iron plate. The plate is then removed from the melted metal, supported in a level position on a table having projecting iron pins, on which it is rested, and the melted stereotype metal is carefully ladled to the proper thickness on the back of the tinned shell. This process is called " backing." The thickness of the metal backing is about an eighth of an inch. A good composition for backing metal consists of lead 90 parts, tin 5 and antimony 5. An alloy of lead 100 parts, tin 3 and and antimony 4 is also recommended as very suitable.

9. *Finishing.*—For this purpose the plates go first to the saw table (Fig. 149) for the removal of the rough edges by means

of a circular saw. The plates are then shaved to take off any roughness from the back and make them of even thickness. In large establishments this portion of the work, which is very laborious, is done with a power planing or shaving machine, types of which are shown in Figs. 150 and 151, Fig. 150 being a shaving machine with steam one way, and Fig. 151 one with steam both ways. By means of a straight-edge, the

Fig. 149.

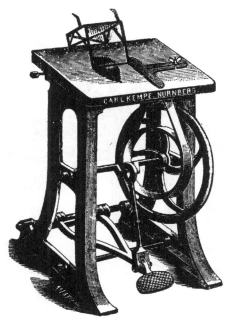

plates are then tested as to their being level, and any unevenness is rectified by gentle blows with a polished hammer, care being taken not to damage the face. The plate then passes to the hand-shaving machine, where the back is shaved down to the proper thickness, smooth and level. The edges of the plate are then planed down square and to a proper size, and finally the plates are mounted on wood type-high.

Book-work is generally not mounted on wood, the plates being left unmounted and finished with beveled edges, by which they are secured on suitable plate-blocks of wood or iron supplied with gripping pieces, which hold them firmly at the proper height, and enable them to be properly locked up.

Fig. 150.

Copper deposits from metallic surfaces.—It remains to say a few words about the process, by which a copy may be directly made from a metallic surface without the interposition of wax or gutta-percha. If the metallic surface to be moulded were free from grease and oxide, the deposit would adhere so firmly as to render its separation without injury almost impossible.

Hence, the metallic original must first undergo special preparation, so as to bring it into a condition favorable to the detachment of the deposit. This is done by thoroughly rubbing the original with an oily rag, or, still better, by lightly silvering it and exposing the silvering for a few minutes to an atmosphere of sulphuretted hydrogen, whereby silver sulphide is formed, which is a good conductor, but prevents the adherence of the deposit to the original. For the purpose of silvering, free the

Fig. 151.

surface of the metallic original (of brass, copper, or bronze) from grease, and pickle it by washing with dilute potassium cyanide solution (1 part potassium cyanide to 20 water). Then brush it over with a solution of $4\frac{1}{2}$ drachms of silver nitrate and 1 oz. 6 drachms of potassium cyanide (98 per cent.) in one quart of water; or, still better, immerse the original for a few seconds in this bath, until the surface is uniformly coated with a film of silver. The production of the layer of silver sulphide is effected according to the process described later on. The negative thus obtained is also silvered, made black with sulphuretted hydrogen, and a deposit of copper is then made, which represents an exact copy of the original. Instead of sulphurizing the silvering with sulphuretted hydrogen, it may also be iodized by washing with dilute solution of iodine in alcohol. The washed plate, prior to bringing it into the copper bath, is for some time exposed to the light.

To prevent the reduction of copper on the back of the metallic original to be copied, it is coated with asphalt lacquer, which must be thoroughly dry before bringing into the bath. When the deposit of copper is of sufficient thickness, the plate is taken from the bath, rinsed in water, and dried. The edges are then trimmed off by filing or cutting to facilitate the separation of the shell from the original.

Of course only metals which are not attacked by the acid copper solution can be directly brought into the bath. Steel plates must therefore first be thickly coppered in the alkaline copper bath, and even this precaution does not always protect them from corrosion. It is therefore better to produce in a silver bath (formula I., p. 368) a copy in silver of sufficient thickness to allow of the separation of both plates. The silver plate is iodized, and from it a copy in copper is made by the galvanoplastic process. The copper plate thus obtained is an exact copy of the original, and after previous silvering, the desired number of copies may be made from it.

Other operations which may have to be done in galvano-plastic plants, for instance, coppering of zinc etchings, and of stereotypes, and nickeling and cobalting the latter, as well as electrotypes, have already been described in the part devoted to electro-plating, so that few words will here suffice.

Stereotypes are, as a rule, coppered in the acid copper bath, stereotype metal being not attacked by it. The bath, how-ever, should not have a large content of free sulphuric acid. In order to have the copper adhere well the plates, previous to being brought into the bath must, of course, be thoroughly freed from grease by brushing with warm soda solution and whiting.

Zinc plates are thoroughly freed from grease, and then coppered or brassed. Nickeling is effected according to the process given under " Deposition of Nickel."

Preparation of type-matrices.—The process varies according to whether the originals consist of zinc or of a material (lead-antimony-bismuth alloy) indifferent towards the acid copper bath.

It is best to brass *zinc originals*, and to give the brass deposit higher lustre by polishing with Vienna lime powder upon a small flannel bob. They are then freed from grease by brushing with quicklime, silvered by the method previously given, and iodized. The surfaces which are to remain free from deposit are stopped off with wax, and the originals placed in the acid copper bath, care being taken to bring them in contact with the current-carrying conducting rod before immersion in the bath.

Originals of hard lead or similar alloys, after having been suitably prepared, may be directly suspended in the copper bath, since a heavy copper deposit can be quite readily detached from them, though slightly oiling them will do no harm.

The current-density for depositing must be slight to prevent formation of buds. The deposit is generally made 0.079 to 0.098 inch thick, when it is detached from the original, and after filing the edges backed with zinc or brass. The matrix is finally justified.

Regarding *nickel matrices*, see " Galvanoplasty in Nickel."

Electro-etching.—It is in place here to discuss the process of electro-etching, it being chiefly applied in the graphic industries, and a few methods of etching, which are not executed by electrical means, will first be referred to.

Methods of dissolving the various metals by acids were probably known many centuries ago, it being beyond doubt shown by the notable productions of the goldsmiths, as well as of the armorers, about the year 1400, that they possessed a knowledge of etching. It may also be supposed that the niello work of the goldsmiths was the forerunner of copper engraving, an art still highly appreciated at the present day, and the earliest impression of which dates from the year 1446.

There are four different methods of copper engraving, but that in which etching plays an important rôle, would seem to be the most interesting.

To protect separate portions of metallic surfaces from the action of the acid, a so-called covering or etching ground is used, which consists of a mixture of 2½ parts asphalt, 2 parts wax, 1 part rosin and 2 parts black pitch, applied hot.

The copper engraver uses for his work another composition of resins, and it is here given because this covering ground has proved capable of resisting 25 per cent. nitric acid. Yellow wax 4 parts, Syrian asphalt 4, black pitch 1, and white Burgundy pitch 1. Melt the ingredients, and when the mixture boils, gradually add, whilst stirring constantly, 4 parts more of pulverized Syrian asphalt. Continue boiling until a sample poured upon a stone and allowed to cool breaks in bending. Then pour the mixture into cold water and shape it into small balls, which for use are dissolved in oil of turpentine.

Upon a heated plate, ground perfectly level, the copper engraver then applies the above-mentioned covering ground so thin that the metallic surface appears golden-yellow. The covering ground is next blackened by means of a wax torch, and the outlines of the picture to be made are then sketched.

Now commences the work which shows the artistic talent of the engraver. With a fine etching-needle he scratches the contours of the picture into the covering ground, without, however, injuring the metal, and finishes his work by narrower and wider lines until the desired effect is believed to be produced.

However, to make this work fit to be printed, the lines of the picture must lie depressed in the metal plate. For this purpose the plate is surrounded with a wax rim and subjected to etching with nitric acid or, more recently, with ferric chloride. After the at first weak acid has acted for a short time, the finest lines have acquired the required depth. The fluid is then poured off and the fine lines are stopped off, when etching is recommenced. Thus progressing, a picture with lines becoming constantly deeper, as well as broader, is formed, the result finally showing the artistic talent of the engraver. The plate is cleansed and handed to the printer, or it may be steeled or manifolded by galvanoplasty.

While speaking of this process of copper-engraving, our attention is involuntarily directed to a very interesting achievement, which deserves mention in connection with the work of the etcher and of the operator in galvanoplasty. The process is *Photo-engraving*, by means of which copper plates, as well as small and also very extensive pictures, of such high artistic value can be produced that they form at present an important branch of the art business.

Former investigators have shown :

1. That of all the varieties of glue, gelatine possesses the greatest swelling capacity.

2. That when mixed with potassium dichromate and exposed to the action of light, gelatine becomes insoluble, *i. e.*, it loses entirely its power of swelling.

Upon this is based the following process: Take a sheet of well-sized paper and make a rim around it, about 0.39 inch high, by turning up the sides. The paper thus prepared, which now forms a sort of dish, is placed upon a perfectly level surface and a solution, consisting mostly of gelatine colored black, is poured over it. Such paper is found in commerce under the name of *black pigment paper*. It is immersed in solution of ammonium dichromate, dried in a dark room and stored for use.

A perfect diapositive of the original is placed in a copying frame and, after covering it with the prepared pigment paper, the frame is closed.

By the rays of light which strike the prepared paper through
40

the diapositive, the layer of chromium and gelatine is hardened, the process taking place in the same gradations of tone as conditioned by the diapositive. After sufficient exposure to the light, the pigment paper is placed in a water bath and a quite perceptible picture in relief will in a short time appear. The portions which had not been exposed to the light, swell up very much and lose the greater part of the coloring matter mixed with the gelatine. The result is, therefore, the reverse of the diapositive used.

By means of an ingenious contrivance, a layer of impalpable asphalt powder has in the meanwhile been applied to a finely ground copper-plate, and melted upon it. The above-mentioned chrome-gelatine picture is now placed upon the plate and is made to adhere by rubbing. The paper can now be readily detached, while the picture adheres to the copperplate. The gelatine-layer forms the protection from the effect of the etching with ferric chloride.

It will be readily understood that for this, and all the preceding manipulations, great skill and years of experience are required in order to produce such results as we have occasion to admire in the art stores.

If galvanoplasty is to be employed for the production of such copper-plates, a glass or metal plate is used and coated with the chrome-gelatine above described. It is then exposed to the light under a photographic glass negative, allowed to swell up, and for a short time laid in a weak chrome alum solution. The layer is then so hard as to allow of making a wax mould and an electrotype. The process is called *photogalvanography*.

The swelling power of gelatine, as well as its insolubility, has led to the production of *collographic printing*. The manipulations for the preparation of the printing plates required for this purpose differ but little from those for photo-galvanography.

Pour over a glass plate, 0.19 to 0.27 inch thick, a layer of chrome-gelatine, which, however, must not be colored, and place the plate in a drying-oven heated to 113° F. The plate is then exposed to the light under a photographic negative and the layer of gelatine allowed to swell up.

Another property shown by chrome-gelatine is that the portions which have become insoluble by exposure to light are very susceptible to fat colors. If now such a glass plate be wiped over with a moist sponge and then blackened all over by means of a suitable color with the use of a roller, a picture showing all the details of the negative used appears upon the glass plate. By placing upon this picture a sheet of printing-paper, and drawing both through the collographic printing-press, the color adheres to the paper.

An etching process which includes all the improvements made in metal etching, and which, by reason of the great progress made in photography, has won a great field of activity, is

Zincography.—All plates produced by this process are intended for book printing, and must show all the lines and points of the picture in relief, while the parts which in printing result in the white portions of the picture should be as deep as possible. It is obvious that this requirement makes the highest demands on the etching process, and that long experience and perseverance are required to achieve excellency in this respect.

All former experiments will here be omitted, and only the process which has proved of practical value will be described.

Freshly-made impressions are reprinted upon fine zinc plates ground perfectly level, drawings executed with suitable ink upon prepared paper being used in the same manner.

When the reprints have been successfully made and any defects removed by retouching, very finely powdered rosin is poured upon the metal plate and rubbed with a brush into the points and lines of the drawing. Since no rosin powder adheres to the portions of the plate not printed on, the plate may at once be laid upon the hot-plate and highly heated. The rosin powder combines intimately with the printing ink, a layer which well resists weak nitric acid being thus formed.

After etching for a short time with dilute nitric acid, fine silvery edges produced by the washing away of the dissolved metal appear on all the lines and points of the reprint. If etching would now be continued, the lines and points would also be laterally attacked by the acid. Over-etching would thus take place, and all the fine portions of the picture disappear. Hence a fresh protecting cover has to be applied, which protects from corrosion, not only the surfaces of the lines and points, but also the above-mentioned silvery edges. For this purpose, the etcher uses a lithographic roller and a suitable etching color. The drawing is then dusted over, and the plate heated as previous to the first etching. Proceeding in the same manner, the manipulations are repeated until the plate has the necessary depth for printing. Finally, all unnecessary metal is cut away with a fret-saw, and the etching having been mounted on wood, is ready to be given to the printer.

If, however, the original handed in for reproduction, is to be enlarged or reduced, a photographic negative of it is first made and copied directly upon the zinc plate. For this purpose, a coating of asphalt solution, or of a mixture of egg or glue with ammonium dichromate, is applied to the zinc plate, and the negative having been placed upon the latter, it is exposed to the light. The result is the same as has been described under photo-engraving, a picture being obtained which is exactly treated as the reprinted drawings, *i. e.*, powdered, heated, and etched.

Another process of transferring is effected by reprinting:

A sheet of paper coated with chrome-gelatine is dried in a dark room, placed in a copying frame under a negative and exposed to the light until a beautiful, chestnut-brown picture is perceptible. The chromium salt is dissolved in the water bath, the picture inked with reprinting ink and, after drying, transferred to the metal plate.

Up till now we have only spoken of points and lines, because the originals have to be composed of such to be suitable for the reproduction process. Photography, however, makes it also possible to transform water-color paintings, photographs, India-ink sketches, etc , into book-printing plates.

For this purpose the photographer uses a glass plate provided with a network of very fine lines, places it between the sensitive glass plate and the original, and thus produces a negative, which, though composed of millions of small points, nevertheless gives all the shadings of the original. This process is called *autotypy*, and is at present used to such an extent and has been brought to such a state of perfection, as to make it difficult to say when the limit of what can be done by the etching process in connection with photography may be reached.

The achievements in photography widen almost daily the field of activity of the etcher, and it may be anticipated that printing plates will in this manner be produced which, when printed in three colors, will yield impressions such as could formerly only be attained by the lithographer with the use of many stones. It is by no means impossible that the electric current may before long be utilized in the execution of the above-mentioned etching processes, and for this reason a few hints will here be given which may be of use to the galvano-plastic operator.

In etching steel, copper or zinc plates, in the ordinary way, a covering ground, as previously mentioned, is applied to the plate to be etched. The drawing is then transferred to the covering ground and traced with the graver, taking care that the tool lays bare the metal in all the lines. A rim of wax is then made around the plate, and dilute nitric acid or another solution poured over it. The basis-metal is attacked by the acid and the drawing is thus etched.

The injurious acid vapors evolved thereby and the lateral corrosion of the lines, as well as other drawbacks, have brought about the execution of etching with the assistance of the elec tric current, the above-mentioned drawbacks being thereby obviated, and more rapid and reliable working rendered possible. The plate is treated in exactly the same manner as for ordinary etching, but instead of furnishing it with a wax rim and pouring acid over it, it is suspended in a suitable solution as anode—hence connected with the positive pole—a metal plate of the same size connected with the negative pole being suspended parallel to it. The metal is dissolved by the acid-residue appearing on the positive pole.

For *copper-plates* which are to be etched, the ordinary acid copper bath is used ; for *zinc-plates,* solution of zinc sulphate ; for *steel-plates,* solution of copperas or of ammonium chloride ; for *brass,* solution of ferric chloride. In place of the baths of metallic salts, pure water slightly acidulated with sulphuric, hydrochloric, or nitric acid may be used.

As covering or etching ground, the previously mentioned mixture of rosin and wax, or the acid-proof varnish is used.

Since the current-strength is under perfect control, the etching may be carried to any depth desired. Some portions may be less etched than others by taking the plate from the bath, and, after washing and drying, coating the portions which are not to be further etched with lacquer, and returning the plate to the bath.

Printing plates in relief may in this manner be prepared by slightly etching the bared design of a copper plate in the galvanoplastic copper bath, and then bringing the plate as object in contact with the negative pole, while a plate of chemically pure copper serves as anode. The deposited copper unites firmly with the rough copper of the etched plates, and after removing the etching ground with benzine or oil of turpentine, the design appears in relief.

Heliography.—The heliographic process, invented by Pretsch, and improved by Scamoni, consists in taking by photography a good negative of the engraving or other object to be reproduced, developing with green vitriol, reinforcing with pyrogallic acid and silver solution, and then fixing with sodium hyposulphite solution in the same manner as customary for photographic negatives. A further reinforcement with chloride of mercury solution then takes place until the layer appears light gray. Now wash thoroughly and intensely blacken the light portions by pouring upon them dilute potassium cyanide solution. As in the photographic process, the solution must be applied in abundance and without stopping, as otherwise streaks and stains are formed. After washing, the plate is dried, further reinforced, and finally coated with a colorless negative varnish. From this negative a positive collodion picture is taken, which is in the same manner developed, reinforced and fixed, the reinforcement with pyrogallic acid being continued until the picture is quite perceptibly raised. After careful washing, pour upon the plate quite concentrated chloride of mercury solution, which has to be frequently renewed. until the picture, at first deep black, acquires a nearly white color, and the lines are perceptibly strengthened. Now wash with distilled water, next with dilute potassium iodide solution, and finally with ammoniacal water, whereby the picture acquires first a greenish, then a brown, and finally a violet-brown, color. After draining, the plate may be progressively treated with solutions of platinum chloride, gold chloride, green vitriol and pyrogallic acid, the latter exerting a solidifying effect upon the pulverulent metallic deposits. The metallic relief is now ready; the layer is slowly dried over alcohol, and the plate, when nearly cold, quickly coated with a thin rosin varnish, which, after momentary drying, remains sufficiently sticky to retain a thin layer of black lead, which is applied with a tuft of cotton. The edge of the plate is finally surrounded with wax, and, after being wired, the plate is brought into the galvanoplastic copper bath to be reproduced.

Electro-engraving.—Below an outline of Rieder's patented process is given, it being supposed that the subject under discussion is the production of a die by means of which reliefs are to be stamped in metal plates.

The relief is first produced in a material readily worked, for instance, wood, wax, etc., and a copy of it made in plaster of

Paris. The plaster of Paris plate, which is about $\frac{1}{2}$ to $\frac{3}{4}$ inch or more thick, is placed in a metal cylinder in such a manner that a plaster of Paris surface of 0.11 to 0.15 inch depth projects above the edge of the cylinder. This cylinder containing the plaster of Paris model is secured in a vessel containing solution of ammonium chloride and a metal spiral connected with the negative pole of the source of current. By a suitable mechanical contrivance the vessel, together with the cylinder containing the model, is pressed against the steel plate connected with the positive pole.

The process is now as follows: The porous plaster of Paris absorbs to saturation ammonium chloride solution. The steel plate first comes in contact with the highest points of relief, and the current becoming active, dissolves the steel on the point of contact. The ferrous chloride solution which is formed penetrates downwards into the capillaries of the plaster of Paris, so that fresh quantities of the electrolyte constantly act upon the steel plate. Etching thus progresses, and gradually every portion of the plaster of Paris model comes in contact with the steel plate, when etching is finished.

However, the practical execution of the work is not so simple as the theoretical process above described. The carbon in the steel and other admixtures, such as silicon, etc., prevent uniform etching and must, therefore, from time to time, be mechanically removed from the etching surface. For this purpose the vessel containing the electrolyte, together with the model, has to be lowered, the steel plate taken from the apparatus and cleansed. It will, therefore, be readily understood that accurate etching corresponding to the metal can only take place when the principal parts, namely, the steel plate and model, after cleansing, mathematically occupy exactly the same place and position as before, so that the model presses accurately against the same parts of the steel plate as in the beginning of the etching operation.

Conjointly with Dr. Geo. Langbein & Co., Rieder has constructed an apparatus which works with such precision as to fulfill all the above-mentioned conditions, and may be briefly described as follows : * The plaster mould is secured by two conical wedges to a cast-iron frame upon a vertically moving table, the latter being set in motion by an eccentric. Above this table is the clamping plate for the steel anode to be etched. This clamping plate is also adjustable, and by a suitable contrivance can be set exactly parallel to the model. Cleansing of the steel plate is effected by means of a carriage carrying a revolving brush and worked by an eccentric ; the brush receives water through a perforated pipe, and in addition a sponge roller is carried over the model for the purpose of acidulating the latter.

The machine works as follows : By means of the movable table the model is without shock and as elastically as possible placed against the plate to be etched. After the plate and model have been in contact for 15 seconds, the model is lifted off and the cleaning process by brushing, etc., is effected. As soon as the cleaning carriage is withdrawn, the model is again brought against the steel-plate and the operation repeated.

Each electro-engraving machine is supplied with a model casting arrangement, the frames of the machine being utilized for the purpose.

The dynamo used has an impressed electro-motive force of 12 to 15 volts, and the current-strength for a plate of 200 x 300 millimeters is about 50 ampères, when the whole surface of the plaster model has been brought in contact with the steel-plate. An electro-engraving plant of this kind was exhibited at the Paris Exposition in 1900.

The depth of the etching depends on the time of contact, but it may be laid down as a rule that, according to the fineness of the model 4 or 5 hours are required for a depth of 1 millimeter. The cleaning process above described may eventually be effected with the assistance of an air compressor.

* Pfanhauser, Die Herstellung von Metallgegenständen auf elektrolytischen Wege und die Elektrogravüre, 1903.

Allowing 12 seconds as the duration of etching, about 600 to 800 etching periods must take place in order to etch to the depth of 1 millimeter. Experiments made on a large scale have proved this method to be very suitable for many purposes, even if it does not make hand-engravers superfluous. For the latter, however, it is an excellent auxiliary for the purpose of obtaining engravings absolutely true to nature from models in wax, etc., and it allows of the preparation in a very short time of engraved dies, plates, etc. The last retouching and polishing have to be done by the hand of the engraver, because in accordance with the nature of the porous plaster-of-Paris model, the etched surface shows but little luster. Further details would not come within the compass of this work, and interested parties are referred to the firm " Elektro-gravüre," Leipsic, Saxony, Germany, who has secured the patents and constructs the electro-engraving machines.

B. Galvanoplastic Reproduction of Plastic Objects.

The reproduction of *busts, vases, etc.*, requires an entirely different process of preparing the moulds than that described as applied in the graphic arts, the material for moulding depending on the nature of the original. Besides gutta-percha and wax, readily fusible metals, oil gutta-percha, plaster of Paris, and glue will have to be considered. If the original bears heating to about 230° F., a copy in one of the readily fusible alloys given later on may be made. If it will stand heat and pressure, it is best to mould in gutta-percha, but if no pressure and only slight heat can be used, recourse may be had to oil gutta-percha. If neither heat nor pressure can be applied, the moulds will have to be executed in plaster of Paris or in glue. The manner of moulding and the material to be chosen furthermore depend on whether surfaces in high relief or round plastic bodies are to be copied, whether projecting portions are undercut, and whether the mould can be directly detached, or, if this is not the case, whether the original has to be dissected and moulded in separate parts.

Regarding the practice of moulding, the reader is referred to special works on that subject. Only the main points for the most frequently occurring reproductions will here be given.

Surfaces in relief and not undercut are readily moulded in an elastic mass, such as gutta-percha or wax ; however, undercut reliefs, and especially round plastic objects, mostly require a plaster-of-Paris mould and are generally dissected. The dissection, of course, is not carried further than absolutely necessary, because the separate parts must be united by a soldering seam, which requires careful work, and the seam itself must be worked over and made invisible. Hence the section should as much as possible be made through smooth surfaces, edges, etc., where the subsequent union by a soldering seam will prove least troublesome ; cutting through ornaments or through portions, the accurate reproduction of which is of the utmost importance, should be avoided. Heads and busts are always executed in a core mould and in portions, unless the entire figure is to be deposited in one piece in a closed mould. The section is made either through the center line of the head through the nose, which, however, makes the subsequent union very troublesome, if the copy is to be an exact reproduction of the original, or the mould is divided from ear to ear, which has the disadvantage that the deepest part of the mould corresponding to the nose receives the thinnest deposit. It has, therefore, been proposed to make two cuts so that three portions are formed ; one cut from one ear at the commencement of the growth of hair to the other ear ; and the second cut from one ear in a downward direction below the lower jaw in the joint of the head and neck, through this joint below the chin, and then upwards to the other ear, and in front of it to where the hair begins. In bearded male heads the cut follows the contour of the beard and not the joint on the neck behind the beard.

Moulding with oil gutta-percha.—Oil gutta-percha has the advantage of allowing moulding without any pressure of the largest shield-shaped or semi-circular objects with all the under-cuts, which otherwise can only be accomplished with glue. The mould can be readily detached from the original as well as from the deposit, which is of great advantage. But on the other hand, oil gutta-percha deteriorates by frequent use, and sticks to the mould when worked too hot, the result being that it is difficult to detach from the original, and, besides, air bubbles are formed. However, the heat must neither be too slight, otherwise the sharpness of the impression would suffer.

. Oil gutta-percha is prepared by heating on the water-bath 100 parts of gutta-percha, 10 parts of olive oil and 2 parts of stearine.

The original, preferably of copper, should be slightly oiled. It is laid upon an iron plate and the latter heated by a flame until the original can be just for a moment retained in the hand. The oil gutta-percha, previously heated on a sand-bath and thoroughly stirred, is then brought in a slow stream upon the original. After allowing the oil gutta-percha to congeal superficially, the original, together with the heating plate, is brought into cold water, where complete congealing soon takes place.

For moulding in the press or by hand with oil gutta-percha, the heated mass is poured into cold water and then kneaded to the consistency of stiff dough.

Moulding with gutta-percha.—To mould round articles in gutta-percha, the softened gutta-percha is kneaded with wet hands upon the oiled original, or, in order to avoid some portions receiving a stronger pressure than others, and to insure a layer of gutta-perch of uniform thickness upon all parts, moulding may also be executed in a ring or frame of iron or zinc under a press. For the rest, all that has been previously said in regard to moulding in gutta-percha is also applicable.

Metallic moulds.—The following *metallic alloys* have been proposed for the preparation of moulds:

I. Lead 2 parts, tin 3, bismuth 5 ; fusible at 212° F.

II. Lead 5, tin 3, bismuth 8 ; fusible at 183° F.

III. Lead 2, tin 2, bismuth 5, mercury 1 ; fusible at 158° F.

IV. Lead 5, tin 3, bismuth 5, mercury 2; fusible at 127.5° F.

The advantage of metallic moulds consists in the metal being a good conductor of electricity, in consequence of which heavy deposits of greater uniformity can be produced than with non-metallic moulds which have been made conductive by black lead. Nevertheless, they are but seldom employed, on account of the crystalline structure of the alloys and the difficulty of avoiding the presence of air bubbles. Böttger claims that a mixture of lead 8 parts, tin 3, and bismuth 8, which is fusible at 227° F., shows a less coarse-grained structure.

Fusible alloys containing mercury should not be used for taking casts of metallic objects—iron excepted—as these will amalgamate with the mercury and be injured. Moreover, copper deposits produced upon such alloys are very brittle, this being due to the combination of the mercury with the deposited copper.

For moulding with metallic alloys, place the oiled object at the bottom of a shallow vessel and pour the liquid metal upon it; or pour the liquid metal into a box, remove the layer of oxide with a piece of thick paper, and when the metal is just beginning to congeal firmly press the object upon it.

Plaster-of-Paris moulds are used for making casts of portions from originals which are so strongly undercut that a mould consisting of one piece could not be well detached from them. For taking casts from metallic coins and medals, or from small plaster reliefs, it is a very convenient material. The mode of procedure is as follows: After the original model, say a medal, has been thoroughly soaped or black-leaded, wrap round the rim a piece of sufficiently stout paper or thin lead foil, and bind it in such a manner by means of sealing-wax that the face of the medal is at the bottom of the receptacle thus formed. Then place the whole to a certain depth in a layer of fine sand, which prevents the escape of the semi-fluid plaster of Paris between the rim of the medal and the paper.

Now mix plaster of Paris with water to a thin paste, take up a small quantity of this paste with a pencil or brush and spread it in a thin film carefully and smoothly over the face of the medal, then pour on the remainder of the paste up to a proper height, and allow it to set. After a few minutes the plaster heats and solidifies. Then remove the surrounding paper, scrape off with a knife what has run between the paper and the rim of the medal, and carefully separate the plaster cast from the model. If, instead of applying the first layer with a brush, the whole of the plaster were run at once into the receptacle, there would be great risk of imprisoning air bubbles between the model and the mould, which would consequently be worthless. The mould is finally made impervious and conductive according to one of the methods to be described later on.

The moulding in plaster of Paris in portions, when casts from large plastic objects with undercut surfaces and reliefs are to be taken, is troublesome work, because each separate mould must not only be so that it can be readily separated without injury to the original, but must also fit closely to its neighbors. Hence thought and judgment are required to see of which parts separate moulds are to made, or, in other words, in how many parts the mould is to be made. After determining on the plan of the work, the mode of procedure is as follows: Oil a portion of the object, if it consists of metal, or soap it, if of plaster-of-Paris, marble, wood, etc., and apply by means of a brush a thinly-fluid paste of plaster-of-Paris, taking care that no air bubbles are formed by the strokes of the brush. When this thin coat is hard, continue the application of plaster-of-Paris with a horn spatula until the coat has acquired a thickness of $\frac{3}{4}$ to 1 inch, and allow it to harden. Then separate the mould, and after cutting or sawing the edges square and smooth, replace it upon the portion of the original model corresponding to it. Now oil or soap the neighboring portions of the model, and at the same time the smooth edges of the first mould which come in contact with the mould now to be made, and then proceed to make the second mould in precisely the same manner as the first. When the second mould is hard, trim the edges and replace it upon the model ; the same process being continued until the entire original model is reproduced in moulds fitting well together. To prevent the finished moulds from falling off, and to retain them in a firm position upon the original model, they are tied with lead wire or secured with catches of brass wire or sheet. When the moulds of the larger portion of the model, for instance, one-half of a statue, are finished, the so-called case or shell is made, i. e., the backs of all the moulds are coated with a layer of plaster-of-Paris which holds them together. This case is best made not too thin in order to attain a better resisting. power.

The entire model having been cast in the manner above described, and the moulds provided with the case, the whole is completely dried in an oven.

Rendering plaster-of-Paris moulds impervious.—The next operation is to make the plaster-of-Paris impervious to fluids, as otherwise by the moulds absorbing the acid copper bath, copper would be deposited in the pores of the plaster and the moulds be spoiled, while the copy would turn out rough instead of having the smooth exterior of the model. To render plaster-of-Paris and other porous substances impervious, they are saturated with wax or stearine, or covered with a coat of varnish, the latter process being generally employed for large moulds. Apply a coat of thick linseed-oil varnish to the face of the mould, and, after drying, repeat the process until the mould is considered to be sufficiently impervious.

Rendering the mould impervious with wax or stearine is a better and more complete method. For this purpose place the heated mould in a vat filled with melted wax or stearine, so that the face does not come in contact with the wax but absorbs it by capillarity from the bath. However, as this cannot be done in every case, the mould, if necessary, may be entirely submerged in the melted wax until no more air-bubbles escape.   It is then taken from the bath and laid, face up, in a drying oven, whereby the wax in melting oozes down in consequence of its gravity, the face of the mould being thus freed from an excess of wax.

To prevent the removal of too much wax from the face, the mould is cooled off with cold water the moment the excess of wax is noticed to have penetrated from the face into the interior.   After drying the mould, the face is coated with *gutta-percha lacquer*, in order to make the high reliefs, which may have been too much freed from wax, impervious.   Gutta-percha lacquer is prepared as follows ·

Bring into a wide-mouthed glass bottle provided with a glass-stopper, gutta-percha cut up in small pieces, and fill the bottle with a mixture of equal parts by volume of ether and benzol.   The bottle is allowed to stand for several weeks in a moderately warm room, the contents being frequently shaken. In this time as much gutta-percha as the solvent can absorb will be dissolved.

For rendering impervious porous, non-metallic moulds upon which copper is later on to be deposited, Greif has patented the following process : The impregnating agent consists of about 70 parts coal-tar pitch, 20 parts retene (methylpropyl phenanthrene), and 10 parts naphthalene.   The mixture of the ingredients having been melted by steam, the body to be impregnated is immersed in the liquid mass, and allowed to remain in it a short time to become throughout impregnated. An excess of the impregnating agent is readily removed by allowing it to drain off.

Metallizing or rendering the moulds conductive.—Metallization by the dry way.   The moulds thus varnished or impregnated with wax are next rendered conductive with black lead, the operation being the same as that for moulds for the graphic arts.

In some cases *metallization by metallic powders* is, however, to be preferred to black-leading or metallizing by the wet way. Metallic or bronze powders are metals in an exceedingly fine state of division, of which, for galvanoplastic purposes, pure copper and brass powders only are of interest. Since such metallic powders adhere badly to waxed surfaces, the mould must be provided with a quick-drying coat of lacquer, upon which, before it is completely dry, the powder is scattered or sifted. When the lacquer is hard a smooth surface is produced by going over the mould with a soft brush dipped in the metallic powder, an excess being removed by a thin jet of water.

For many undercut or very deep portions which cannot be thoroughly manipulated with the brush, metallization with black-lead proves insufficient, and recourse will have to be had to

Metallization by the wet way.—This method consists in the deposition of certain metallic salts upon the moulds and their reduction to metal or conversion to conductive sulphur combinations. The process in general use is as follows: Apply with a brush upon the mould a not too concentrated solution of silver nitrate in a mixture of equal parts of distilled water and 90 per cent. alcohol. When the coat is dry expose it in a closed box to an atmosphere of sulphuretted hydrogen. The latter converts the silver nitrate into silver sulphide, which is a good conductor of the current. For the production of the sulphuretted hydrogen, place in the box, which contains the mould to be metallized, a porcelain plate or dish filled with dilute sulphuric acid (1 acid to 8 water), and add five or six pieces of iron pyrites the size of a hazelnut. The development of the gas begins immediately, and the box should be closed with a well-fitting cover to prevent inhaling the poisonous gas; if possible the work should be done in the open air or under a well-drawing chimney. The formation of the layer of silver sulphide requires but a few minutes, and if not many moulds have to be successively treated, the acid is poured off from the iron pyrites and clean water poured upon the latter so as not to cause useless development of gas.

It has also been recommended to decompose the silver salt by vapors of phosphorus and to convert it into silver phosphide,

41

a solution of phosphorus in carbon disulphide being used for the purpose. The layer of silver salt is moistened with the solution or exposed to its vapors. This method possesses, however, no advantage over the preceding one, because, on the one hand, the phosphorus solution takes fire spontaneously, and, on the other, the odor of the carbon disulphide is still more offensive than that of sulphuretted hydrogen.

A somewhat *modified method* is given by Parkes as follows: Three solutions, A, B, C, are required. Solution A is prepared by dissolving 0.5 part of caoutchouc cut up in fine pieces in 10 parts of carbon disulphide and adding 4 parts of melted wax; stir thoroughly, then add a solution of 5 parts of phosphorus in 60 of carbon disulphide, together with 5 of oil of turpentine and 4 of pulverized asphalt; then thoroughly shake this mixture, A. Solution B consists of 2 parts by weight of silver nitrate in 600 of water; and solution C of 10 parts of gold chloride in 600 of water. The mould to be metallized is first provided with wires and then brushed over with or immersed in solution A, and after draining off, dried. The dry mould is then poured over with the silver solution (B) and suspended free for a few minutes until the surface shows a dark luster. It is then rinsed in water and treated in the same manner with the chloride of gold solution (C), whereby it acquires a yellowish tone, when, after drying, it is sufficiently prepared for the reception of the deposit. Care must be taken in preparing solution A, carbon disulphide which contains phosphorus readily taking fire.

However, in some cases, either one of the above-mentioned methods may leave the operator in the lurch. On the one hand, a small accumulation of silver salt solution in the deeper places cannot be well prevented, a slightly crystalline layer of salt being consequently formed and, on the other, it may happen that the layer of silver sulphide becomes without discernible reason detached from the mould in the copper bath, thus necessitating a repetition of the process.

In many cases the following method has been successfully

used : Dilute iodized collodion solution, such as is used for photographic purposes, with an equal volume of ether-alcohol, and pour this solution quickly and without intermission over the mould, the latter being inclined so that all portions of it come in contact with the collodion solution, when the mould is turned face down to allow an excess to run off. By manipulating with sufficient rapidity a film of collodion solution remains upon the mould. This film, at the moment of congealing, is exposed for 2 to 3 minutes to the action of a weak solution of silver nitrate in water, the operation being best effected in a darkened room. The collodion containing potassium iodide forms with the silver bath, silver iodide, the previously clear collodion layer becoming yellowish. In this state the mould is taken from the silver bath, washed with a weak jet of water to remove an excess of silver solution, and then for a few minutes exposed to the sun. By this means a reduction of the silver salt takes place, which is rendered still more intense by laying the mould in a solution of copperas in water, alcohol and glacial acetic acid, in the proportion of 1.76 ozs. copperas, 1 oz. glacial acetic acid, 0.7 oz. alcohol per quart of water. The mould is then rinsed in water and immediately brought into the copper bath, the conduction of the current to the layer of silver having been first effected by means of a few feelers.

In applying this method it must be borne in mind that the collodion layer will not bear rough handling, and injury of it, by touching it with the hands or a strong jet of water, or by careless application of the conducting wires (feelers), must be avoided. When operating with the care required, the results are very satisfactory and sure.

In place of iodized collodion, a mixture of equal parts of white of egg and saturated common salt solution may be used, the process for the rest being the same as above described.

Lenoir's process—Galvanoplastic method for originals in high relief.—Lenoir's method for reproducing statues in a manner approaches in principle to that of the foundry. He begins by making with gutta-percha a mould in several pieces, which are united together so as to form a perfect hollow mould of the original. This having been done, cover all the parts carefully with black-lead. Make a skeleton with platinum wire, following the general outline of the model, but smaller than the mould, since it must be suspended in it without any point of contact. If the skeleton thus prepared is enclosed in the metallized gutta-percha mould, and the whole immersed in the galvanoplastic bath, it will be sufficient to connect the inner surface of the mould with the negative pole of the battery, and the skeleton of platinum wires (which should have no points of contact with the metallized surfaces of the mould) with the positive pole, in order to decompose the solution of sulphate of copper which fills the mould. When the metallic deposit has reached the proper thickness the gutta-percha mould is removed by any convenient process, and a faithful copy of the original will be produced. Lead wires may be substituted for the expensive platinum wires. This method requires a knowledge of the moulder's art, so that good results can only be obtained by an experienced hand.

Gelatine moulds.—Under certain conditions the elasticity of gelatine allows of the possibility of its removal from undercut or highly-wrought portions of the model, when it reassumes the shape and position it had before removal therefrom. But gelatine requires that the deposit shall be made rapidly, otherwise it will swell and be partially dissolved by too long an immersion in the copper bath

To make a good gelatine mould, proceed as follows: Allow white gelatine (cabinet-maker's glue) to swell for about 24 hours in cold water, then drain off the water, and heat the swollen mass in a water-bath until completely dissolved. Compound the glue solution with pure glycerine in the proportion of 5 to 10 cubic centimeters (0.24 to 0.3 cubic inch) of glycerine to 30 grammes (1.05 ozs.) of gelatine, which prevents the gelatine from shrinking in cooling. When somewhat cooled off, apply the gelatine to the oiled original, which must be surrounded with a rim of plaster of Paris or wax, to prevent the gelatine from running off; when cold, lift the gelatine mould from the model. Before metallizing and suspending in the copper bath, the mould has to be prepared to resist the action of the latter, as otherwise it would at once swell and be partially dissolved before being covered with the deposit. This is effected by placing the mould in a highly concentrated solution of tannin, which possesses the property of making gelatine insoluble.

Brandley gives the following directions for preparing gelatine solution with an addition of tannin, which renders the moulds impervious to water : Dissolve 20 parts of the best gelatine in 100 of hot water, add ½ part of tannic acid and the same quantity of rock candy, then mix the whole thoroughly, and pour it upon the model.

The same object is attained by the use of potassium dichromate solution in place of tannin solution. In this case, the potassium dichromate solution must be allowed to act in a dark room, the mould being then for some time exposed to the action of the sun. The chrome gelatine layer formed upon the surface does not swell up, and is insoluble, at least for the time required to cover the mould with copper. Rendering glue moulds conductive by means of black-lead is, as a rule, impracticable, and metallization has to be accomplished by the wet way in order to effect a rapid formation of the deposit. Moulds rendered conductive by black-lead should be rapidly covered with copper while the bath is being agitated, because a bath in which a considerable quantity of gelatine has been dissolved, yields brittle copper, while by a very small quantity of it, the density of the deposit is increased.

Special Applications of Galvanoplasty.

Nature printing, so named by Mr. v. Auer, Director of the Imperial Printing Office at Vienna, has for its object the galvanoplastic reproduction of leaves and other similar bodies. The leaf is placed between two plates, one of polished steel, the other of soft lead, and is then passed between rollers, which exert a considerable pressure. The leaf thus imparts an exact impression of itself and of all its veins and markings to the lead, and this impression may be electrotyped, and the copper plate produced used for printing in the ordinary way. Instead of taking the impression in lead, it is advisable to use gutta-percha or wax for delicate objects, which should previously be black-leaded or oiled. In the same manner galvanoplastic copies of laces, etc., may be obtained.

Elmore produces *copper tubes* by galvanoplastic deposition by allowing the metallic core-bar to revolve slowly between the anodes, while a polishing steel is by means of a mechanical contrivance carried with strong pressure over the deposit, whereby the latter is made dense and any roughness removed.

It would seem that the process for the production of copper tubes, profiled hollow copper bodies, etc., patented by Ignaz Klein, is better than Elmore's method. The black-leaded or metallic core-bars are allowed to roll to and fro upon smooth or profiled plates, the so-called milling plates, or the core-bars are concentrically arranged around a cylindrical anode and allowed with pressure to roll on an exterior round milling surface. According to this method, the space in the baths can be better utilized than in the Elmore process, and the deposit shows excellent properties as regards uniform density and power of resistance.

According to Dieffenbach and Limpricht's method (German patent 125404) tubes and hollow bodies of copper of great toughness and strength are obtained by allowing the metal cores, or other cores rendered conductive in a suitable manner, to revolve in an acid bath to which fine sand or, better, infusorial earth has been added. The infusorial earth exerts a scouring effect and removes any hydrogen bubbles which may have been separated. Experiments made with this process yielded copper tubes which on being tested showed excellent values as regards strength.

Corvin's niello.—Corvin has invented a process of producing inlaid work by galvanoplasty. The process is as follows: A matrice of metal whose surface is finely polished is first made. This matrice may be used for the production of numerous duplicates of the same kind of object. The incrustations (mother-of-pearl, glass, ivory, amber, etc.) are then shaped by means of a saw, files and other tools to the form corresponding to that which they are to occupy in the design. The side of the incrustation which is laid upon the matrice is, as a rule, smooth. The shaped incrustations, smooth side down, are pasted on to the parts of the model they are to occupy in the design. The latter being thus produced, the backs of the non-metallic laminæ are metallized, and the portions of the metallic plate left free are slightly oiled. By now placing the matrice thus prepared in the galvanoplastic bath, the copper is deposited, not only upon the metallic matrice, but also upon the back of the inlaid pieces, the latter being firmly inclosed by the deposited metal. When the deposited metal has acquired the desired thickness, it is detached from the matrice, and incrustations with the right side polished are thus obtained. The laminæ are more accurately and evenly laid in than would be possible by the most skilled hand-work.

Plates for the production of imitations of leather. The demand for alligator and similar leathers is at the present time greater than the supply, and, therefore, imitations are made by pressing ox-leather, the plate being prepared by galvanoplasty, as follows: A large piece of the natural skin or leather is made impervious to the bath by repeated coatings with lacquer, and, when completely dry, secured with asphalt lacquer to a copper or brass plate. The leather is then black-leaded, and, after being made conductive by copper wire or small lead plates, brought into the copper bath. When the copper deposit has acquired the desired thickness, the plate is further strengthened by backing with stereotype metal.

Incrusting galvanoplasty.—This term may be applied to the process by which a thick coat of copper, or of another metal, is deposited upon an article. This deposit, however, is not detached from the original, as in reproduction-galvanoplasty, but remains upon it, the object being, as a rule, to embellish the article or to give non-metallic articles the appearance of metallic ones.

Non-metallic objects to be coated have also to be rendered impervious to the electrolyte, great care being required in this respect, since in case the acid copper bath penetrates into the article to be coated, the deposit would later on effloresce and peel off.

The objects may be rendered impervious by one of the methods mentioned above, it being best, however, first to heat them, as for instance, terra-cotta busts, and then place them in melted wax, or mixtures of wax and paraffine. By heating the greater portion of the air is expelled from the pores, the wax thus penetrating better and closing the pores. An excess of wax is removed by draining off in a warm room (air-bath), and when cold the objects are coated with gutta-percha lacquer and metallized with black-lead.

However, it will frequently be impracticable to reach every portion with the black-lead brush, and in this case it is recommended to effect metallization by the wet way, as follows :

Coat the articles, previously rendered impervious, with a thickly-fluid solution of shellac in alcohol, and allow them to become thoroughly dry. Then immerse them for one minute in saturated silver-nitrate solution in 4 parts of water and 6 parts of alcohol, and allow them to drain off. Bring the articles, while still moist, into a vessel which can be closed air-tight, and introduce sulphuretted hydrogen. A thin layer of silver sulphide will in a few minutes be formed, when the articles are taken from the vessel and allowed to dry, the same manipulation being once or twice repeated. By operating in this manner, non-success is next to impossible, because when the object is immersed in the alcoholic silver-nitrate solution, the coat of lacquer is superficially softened and absorbs silver nitrate, which after its conversion into silver sulphide adheres very firmly after drying, so that the layer of it becomes very seldom detached in the bath. Should this nevertheless happen, it is generally caused by the objects not having been thoroughly dried between the separate operations.

Large objects which cannot be immersed will have to be carefully brushed over with the solutions, or the latter be poured over them. According to the nature of the objects to be coated, the process may have to be somewhat modified, or one of the methods already described will have to be employed. This will soon be learned by experience.

Copper bath and current conditions.—Deposits for incrusting galvanoplasty should be effected with very slight current-densities in order to avoid roughness and a coarsely crystalline structure of the deposit. For many delicate objects to be coated with copper, the use of the cell-apparatus is therefore advisable.

Neubeck has investigated the work in the cell-apparatus and its application to encrusting galvanoplasty, and has found a copper bath which contains in 100 quarts, 44 lbs. of crystallized blue vitriol and 13.2 lbs. of sulphuric acid of 66° Bé., to be especially valuable for the purpose.

In place of zinc, he used iron plates or iron tubes, and as solution, one of sodium sulphate with the addition of a few drops of sulphuric acid. The current generated by this arrangement produces a very finely crystalline deposit free from roughness and efflorescences.

Additional manipulation of the deposits.—The deposits of copper produced by one or the other method require, as a rule, additional mechanical manipulation to give the outlines greater sharpness, and to the whole a more pleasing appearance. This is done by chiseling to make, for instance, with busts, the eyes, nose, ears, etc., more prominent. Scratch-brushing, brushing, and polishing are applied if luster is to be imparted to the deposits as is required for sufficiently effective nickeling, gilding, etc.

Laces and tissues are, according to Philip, impregnated with melted wax, and after removing an excess with blotting-paper, they are made conductive by black-leading with a brush. It is, however, preferable to metallize such delicate objects by the wet way, employing one of the methods previously described.

Grasses, leaves, flowers, etc., are first dried and their former shape and elasticity restored by placing them for a considerable time in glycerine. They are then several times immersed in gutta-percha lacquer, and metallized with silver nitrate solution and sulphuretted hydrogen, or according to one of the other processes described.

Wooden handles of surgical instruments are provided with a galvanoplastic deposit of copper to adapt them to antiseptic rules. The wood has to be protected from the bath-fluid penetrating into it, by remaining for a considerable time in melted wax or in a solution of wax or paraffine in ether, as otherwise the copper-deposit formed will be broken by the wood swelling. Metallization is effected by dusting the wood thus prepared with black lead or bronze powder, or by the wet way. The deposit is, as a rule, ground, polished, and then nickeled.

Busts and other objects of terra-cotta, stoneware, clay, etc., are immersed in melted wax and, after removing an excess of wax, coated with gutta-percha lacquer. If there is no obstacle to black-leading, this method may be used for metallization, otherwise recourse will have to be had to one of the processes previously described.

When the copper deposit is finished the objects should be thoroughly soaked in water, and then for a few hours placed in a 5 per cent. yellow prussiate of potash solution for the neutralization of any bath-residue which may still remain in some of the pores.

To bring out sharp outlines, especially when the deposit is quite thick, the coppered busts and other objects are chiseled, scratch-brushed, and polished. They receive, as a rule, a patina, according to one of the methods given in Chapter XIV, or are brassed, silvered, or gilded.

The mercury vessels of thermometers for vacuum and distilling apparatus are as a protection given a galvanoplastic deposit of copper. This is effected in the most simple manner by coating the glass with copal lacquer and black-leading the layer of lacquer, or applying bronze powder. The glass may also be matted by the sand blast, or with fluoric acid, and directly black-leaded, the black-lead adhering well to the matted glass surface.

Mirrors are coppered to protect the thin layer of silver from injury. For the success of coppering in the acid copper bath without danger of the film of silver becoming detached, it is necessary to use a weaker bath of at the utmost 8° to 10° Bé. with 1 per cent. of free sulphuric acid, and to deposit with a very slight current-density.

Glass and porcelain ware, for instance, tumblers, bowls, coffee and tea sets, when furnished with galvanoplastic decorations in copper or silver, produce beautiful effects. However, metallization by one of the processes thus far described is not practicable, the high reliefs having to adhere firmly to the base, so as not to become detached by cleansing or wear.

Metallization of the portions to be coated is effected by painting the arabesques, flowers, monograms, etc., with solution of Dutch gold, factitious silver, or platinum, and after drying, burning-in in a muffle at a dark red heat. A lustrous layer of metal firmly fused together with the glaze is thus obtained and is without further preparation fit for coppering or silvering. Still greater solidity is attained by triturating conducting silver enamel with lavender oil upon a palette to a mass of the consistency of paint, applying the latter with a brush and burning-in in the muffle. In addition to pure silver, the enamel contains fluxing agents which effect a firm union of the silver with the porcelain or glass. After burning-in, the decorations are gone over with a fine copper brush and, the conduction of the current to the metallized portions having been effected by means of fine copper wires, the objects are brought into the galvanoplastic bath.

Mr. A. A. Le Fort * gives the following process for "silver deposit" on glass and china, which, if followed according to directions, will be found satisfactory in every way. The color will be found a clear white which is necessary on glass objects, when the work is fixed properly and being fused into the glass or china, will make a firm body that will hold the deposited silver, so that there will be but a very small loss in the plating and finishing operations due to the silver (failing) or peeling off, while the loss is very large when the proper metallic paint or painting solutions are not employed. The formula for the *metallic paint*, which must be weighed accurately, is as follows :

"Metallized silver" 4 ozs., boracic acid 4 dwt., potassium nitrate 4 dwt., powdered flint 4 dwt., powdered glass 4 dwt., soda ash 4 dwt., red lead 4 dwt., calcined borax 8 dwt.

While the named ingredients are used in all paints, the weight of each may vary in different formulas, but the above will be found to be one of the best in use. In order to get the silver ready for use in the paint, it must be treated as follows, which is called metallizing : Either take fine silver chloride, or else cut down your own silver in the usual way, namely, one part water and one part nitric acid, in a hot water bath, precipitate in the usual way with common salt, wash the chloride four or five times, then put the chloride in a glass or porcelain bowl, and cover with a solution of two parts sulphuric acid to seven parts of water. Then cut common sheet zinc into sirips of about one-half inch in width and four or five inches in length, and put them in the silver chloride and the diluted sulphuric acid, stir occasionally and keep on adding zinc until a gray or brownish-colored precipitate of metallic silver is formed. · Wash thoroughly until free from acid, testing with blue litmus paper to make sure. After the silver is washed, it is necessary to dry it perfectly, by drying it in a porcelain or other suitable vessel over a sand-bath. Make

* Metal Industry, February, 1913.

sure that it is perfectly dry, so that there will be no error in weighing when making up the paint. If the silver is damp the weight will not be correct to conform with the other ingredients in the formula.

After weighing all the ingredients named in the formula, mix together, then thin down with oil of turpentine, grind in a paint mill, which is made for that purpose, or grind by hand, with a glass "muller" on a heavy glass or stone, which can be procured from most art or paint dealers. The grinding by hand is a long and tedious operation, as it takes a long time to do the work properly, for the paint has to be ground very fine—the finer, the better the finish will be and the easier to handle, when putting on the designs. At any rate, the ingredients must be fine enough to be mixed thoroughly, as if too coarse the ingredients in the paint would divide; that is, the silver, glass and flint being very coarse would not combine correctly with the other powdered ingredients. A good way to grind the paint, which saves the cost of a paint mill and is just as satisfactory, is to make a small barrel with wooden strips, that will hold a common quart fruit jar, and is made in the form of a tumbling barrel. This is run by simply placing a belt over the shafting and around the barrel. Place the paint in the jar with about fifteen or sixteen glass marbles of different sizes, from one-half to one inch in diameter. Do not make the paint too thin; cover the jar tightly, place it inside of the barrel and pack around with old cloth or rags, to stop the jarring or breaking of the jar. Let the barrel run for a day or two. The marbles will grind the paint to a fineness, without any cost for labor, that could not be obtained by hand in several hours. After the paint is ground fine enough, let it settle to bottom of jar, and pour off the surplus turpentine. This paint (take only what is required for four or five hours' work, so that there will be no waste) is then mixed with "fat oil of turpentine" (about one part of oil to five or six parts of paint) on a glass surface or artist's palette with a palette knife; then thin to proper consistency with oil of turpentine by dip-

ping the decorating brush in same and working it out on the
palette. The paint should be mixed or stirred up regularly
while in the act of decorating, to keep it uniform ; as the oil
comes to the top, and as the color of the paint is brownish, it
would be hard to tell whether the lines were painted with
simply the oil or the paint body.

Firing.—The second operation is the firing, which proceeds
as follows : After the work is decorated let it stand five or six
hours before firing, so that it is partly dry ; then place in the
oven, light one burner of the gas oven (which is generally used
for this class of work); light a second burner in about five
minutes, and the rest of the burners at about the same intervals, so that the heat will start gradually, and not make the
paint blister or crack, which would be the case if a strong heat
was applied too quickly. After all the burners are started, it
will take from one to one and one-half hours to fire work,
according to size of oven used, but the operator will have to
try a few pieces, then use his own judgment as to when work
is properly fired. After four or five trials he should be able
to get correct results. The oven should be in a dark place, or
in a room that can be darkened when the oven is in use, so
that the operator can get the right light (when looking at the
work being fired) through the openings in door of oven to see
if proper heat is at hand. Heat the oven until the bottom,
top and sides are at a cherry-red heat, when the paint is then
properly fused into the glass. Turn off the gas quickly by
shutting off the cock in the feed-pipe, then let the work cool
off gradually. Do not open the door of the oven for some
time—two or three hours—as the cold current of air would
crack the hot glass. When cool enough to handle, take out
carefully ; keep the fingers or hands off the painted parts, as
they would leave marks on the paint, which would be liable
to blister when plating.

Now wire the work by making connections, directly on the
painted parts. Some designs which are not connected will,
of course, have to have a separate wire on each part, but most

designs are made so that one connection is all that is required. After the work is wired, it is then ready for the plating bath without any other process of cleaning, brushing or striking up. As the silver must be deposited very slowly, it takes between twelve to twenty-four hours for a proper deposit, according to the grade of work, the articles which have to be engraved to bring out the designs taking the longer time, as all that is required for plain or scroll designs is just enough silver to stand the finishing operations. The plating solution should be composed of not less than 6 ounces silver, and not more than 1 or 1½ ounces of free cyanide to the gallon. If the solution is too light in metal or too strong in cyanide, the work will be very hard to finish, and the silver being too brittle would raise up from the edges of the painted surface in the solution or in the finishing. As it requires such a long time for plating, it is really necessary to run plating baths with storage batteries during the night to make any headway, that is, run the dynamo while the power is going, then switch on to the batteries until the work is fully plated. Thus the work does not have to be removed from the plating tank at night and returned to solution the next day, as that would require the cleaning, rewiring and striking up of the work in order not to have any failed or peeled pieces. Every particle of space in the plating tank should be utilized in order to get in as many pieces of work as possible in every batch, and twenty-five to thirty ampères are all that are required for a batch of seventy-five or eighty pieces of mixed work; as the plating surface is very small even with a full batch of work. After the work is plated it is then sand-buffed, cut down with tripoli or other cutting-down rouge, washed in benzine, engraved, then colored in the usual way; that is, finished, the same as sterling or silver-plated hollow ware. The quality of glass used must also be considered, for some glass will turn yellow before the proper heat is reached to fuse the paint, or else will get out of shape during the firing. Imported glass always gives best results, while thin articles of American glass are fairly satisfactory. The heavy, thick glass gives very poor results. China will stand and requires more heat in firing than glass, so it is advisable to fire china and glass separately to obtain the best results.

Umbrella and cane handles of celluloid are decorated with a metallic deposit by means of galvanoplasty. The simplest mode of metallizing them is to paint the decorations with a mixture of bronze powder and acetone. On the point of contact, the acetone dissolves a small quantity of celluloid, the latter thus becoming quite firmly united with the bronze powder. After drying, an excess of non-adhering powder is removed with a brush, and the objects are then brought into the copper bath.

Baby-shoes for a keep-sake, are coppered by galvanoplasty, and the deposit is patinized or silvered. Metallization is best effected by applying several coats of copal lacquer, and black-leading the layer of lacquer on the outside, while the inside, which is more difficult of access, is made conductive by means of bronze powder or by the wet way.

Carbon pins and carbon blocks for electro-technical purposes are frequently coated with copper in order to effect a more sure metallic contact in their mountings. The carbons are impregnated with wax so as to prevent the blue vitriol solution from penetrating. They are then brushed with quick-lime and without further preparation brought into the bath.

Rolls of steel and cast-iron, pump-pistons, etc., are first provided with as thick a deposit as possible in the cyanide copper bath and then brought into the galvanoplastic acid copper bath. With a bath at rest the current-density should not exceed 30 ampères per square meter, but with an agitated bath up to 120 ampères may be used.

Steel gun barrels for marine purposes are treated in the same manner, after all the portions which are not to receive a deposit of copper, have been thoroughly covered with a mixture of wax, mastic and red lead.

Candelabra, stairs and structural parts of buildings of rough castings require a somewhat modified treatment. While the rolls and steel gun-barrels previously referred to are always turned smooth, rough iron castings are used for the above-mentioned purposes, and the production in the potassium cyanide bath of a copper deposit of such thickness as not to corrode the basis-metal in the acid copper bath is frequently connected with difficulties. In a plant recently furnished by Dr. Geo. Langbein & Co. for coppering rust-proof, and then brassing structural parts of a postoffice building in Mexico, the following plan was adopted : The rough castings were first carefully cleaned by means of a sand blast and then heavily nickeled ; upon this nickel coating the 1 millimeter thick copper deposit could without risk be produced in the acid copper bath. The parts were then thoroughly rinsed, dried, brightened with a brush and emery and, after carefully freeing from grease, electrolytically provided with a heavy deposit of bronze, and patinized.

It is believed sufficient examples of the uses of galvanoplasty have here been given. It allows of the most varied applications, and by studying the special processes described above, the reader will be in a position to find out the most suitable method for every other contingency.

II. Galvanoplasty in Iron (Steel).

Under " Deposition of Iron," the galvanoplastic production of heavy detachable deposits of iron has already been referred to.

Serviceable iron electrotypes were first produced about 1870, by Klein of St. Petersburg, and used for printing Russian bank notes. Their preparation was, and is still, very troublesome, success depending on the fulfillment of many conditions, so that, notwithstanding continued experiments and the expense of much labor, the former expectation of entirely supplanting electrotypes in copper by clichés in steel has thus far not been realized.

The bath used by Klein, and still employed for this purpose,
42

consists of a 10 per cent. solution of a mixture of equal parts of ferrous sulphate (green vitriol) and magnesium sulphate (Epsom salt). The solution has a specific gravity of 1.05. To obtain successfully a serviceable electrotype from an original, for instance, from a copper plate, which should previously be silvered and coated with a thin layer of silver sulphide by means of sulphuretted hydrogen, the following conditions have to be fulfilled, according to Klein's statement: The bath must be kept absolutely neutral, which is effected by suspending in it linen bags filled with magnesium carbonate, and the current-strength must be so regulated that absolutely no evolution of hydrogen is perceptible on the anodes.   Further, the plates are every half hour to be taken from the bath and rinsed with a powerful jet of water to remove any adhering gas-bubbles. Care must be taken during this process that the plates do not become dry, since fresh layers do not adhere well upon plates which have become dry.

For the removal of adhering gas-bubbles, it has also been proposed frequently to pass a feather over the plates.

It may here be mentioned that Lenz found a not inconsiderable content of hydrogen in iron deposits, and also carbonic acid, carbonic oxide and nitrogen in varying quantities. However, examinations made by Dr. Geo. Langbein established positively only a content of hydrogen, and it would seem that this hydrogen, which is absorbed and tenaciously retained by the deposit, is the cause of all the difficulties encountered in the production of heavy iron deposits.

If, however, the occlusion of hydrogen is regarded as the cause of the mischief, ways and means to counteract it as much as possible may be found in the fact that iron deposited with greater current-density is more brittle, shows a greater tendency to peel off in the bath, and contains a larger quantity of hydrogen than a deposit produced with slighter current-density.

In this respect experience gained in the electrolytic refining of copper shows us the way in so far that for the production of heavy deposits of iron, the bath must be kept in constant, vigorous agitation, to remove, on the one hand, layers of fluid poorer in metal from the cathode, and, on the other, to force, by the agitation, the gas-bubbles adhering to the cathode to escape. Further, deposition must be effected with so slight a current-density that no evolution of hydrogen is perceptible on the cathode, and a current-density of 0.25 ampère may be designated as the maximum per $15\frac{1}{2}$ square inches, with which heavy deposits of iron can be produced.

To counteract the spoiling of the deposits, further precautionary measures are, however, necessary, especially heating the electrolyte, and from time to time interrupting the current. In heated baths the escape of the gas is facilitated, especially when the electrolyte is agitated, and hence adhering gas-bubbles cannot remain long in one place. A constantly-repeated interruption of the current is of advantage and effective, because metallic parts covered with a minimum quantity of hydrogen cannot be coated with a fresh deposit until the hydrogen is removed by the agitation of the heated electrolyte. Hence the interruption of the process of deposition would give opportunity and time for the removal of the gas molecules before further deposition takes place, and without a knowledge of the more intimate processes, Klein succeeded in affecting the interruption of the deposit, by taking the plates at short intervals from the bath and removing the adhering gas by a powerful jet of water.

With the present state of galvanoplasty it is not necessary to follow Klein's primitive method, and it would be more practical to provide the positive conducting rod of the bath with a contrivance which mechanically effects the interruption of the current. Suppose upon such a metallic conducting rod is mounted a copper or brass wheel, which is secured to a pulley and revolves around the conducting rod, and half of the periphery of which is insulated, and that upon the rod drags a metallic brush which effects the transmission of the positive current. Now, it will be seen that while the contact-wheel is revolving, current is introduced only one-half the time and not during the other half, and that by the rapidity of revolution of the contact-wheel, the number of interruptions of the current can be varied at will.

Since Neubeck has in a relatively very short time produced in hot baths, deposits 1 to 2 millimeters thick, of coherent form and good quality, the possibility is presented of making steel electrotypes in an indirect way by obtaining first from the impression a copper electrotype, from this a negative in copper, and after silvering the latter, producing a heavy deposit of steel upon it.

It may also be expected that by complying with the above-mentioned conditions and the discovery of new methods, it will be possible directly to produce steel electrotypes upon moulds of gutta-percha or wax as is now successfully done with nickel.

It is well known that electrolytically deposited iron possesses great hardness, and that such deposits well deserve the name of steel deposits, their hardness being greater than that of iron, and approaching that of steel. This feature cannot be explained otherwise than by the hydrogen absorbed by the deposit. Hence it will be seen that, on the one hand, this absorption of hydrogen has an injurious effect upon the separation of iron, while, on the other, it imparts to the deposits the most valuable property of great hardness. It would seem that the quantities of iron first deposited upon the mould are, and can be, richer in hydrogen in order to impart to the printing surface the utmost possible hardness. However, in further strengthening and augmenting the deposit, our efforts must be directed, by the reduction of the current, to deposit strengthening layers as free from hydrogen as possible.

The question now arises, whether it is of greater advantage to steel a copper electrotype in order to increase its power of resistance, or whether it is better to produce an iron electrotype and to strengthen its back in the acid copper bath. If the above expressed view that the layers of iron first deposited are

richer in hydrogen, and therefore harder, is correct, the preference must be given to iron electrotypes, because with steeled
copper electrotypes the softer layers are exposed to wear, while
the harder layers lie upon the copper plate. The reverse
is the case with an iron electrotype, the first deposit, rich in
hydrogen, forming the printing face.

However, on the other hand, steeled copper electrotypes
have the advantage that, when worn, the old deposit of iron
can be readily removed by dilute sulphuric acid, and the electrotypes resteeled, while worn iron electrotypes have to be
renewed.

III. GALVANOPLASTY IN NICKEL.

Although by the electro-deposition of nickel, electrotypes are
rendered fit for printing with metallic colors, which attack copper, and their power of resisting wear is increased, the latter
advantage can to the fullest extent be obtained only by a thick
deposit. However, this always alters the design somewhat,
especially the fine hatchings, this being the reason why in
nickel-plating electrotypes a deposit of medium thickness is as
a rule not exceeded. If a hard nickel surface is desired, without injury to the fine lines of the design, the layer of nickel
has to be produced by galvanoplasty, and the deposit of nickel
strengthened in the copper bath.

But upon black-leaded gutta-percha or wax moulds a nickel
deposit can only be obtained in fresh baths. The deposit,
however, is faultless only in rare cases, it generally showing
holes in the depressions. Hence the object has to be attained
in a round about way, the mode of procedure being as follows:
An impression of the original is taken in gutta-percha or wax
and from this impression a positive cliché in copper is made.
The latter is then silvered, the silvering iodized as previously
described, and a negative in copper is then prepared from this
positive. The negative is again silvered, iodized, and then
brought into a nickel bath, where it receives a deposit of the
thickness of stout writing paper. It is then rinsed in water,

and the deposit immediately strengthened in the acid copper bath. For the rest, it is treated like ordinary copper deposits.

If for the production of the nickel electrotype, a nickel bath of the composition given on p. 266 and heated to between 185° and 194° F. is used, deposition may be made with high current-densities—5 ampères and eventually more—so that a thickness of 0.2 millimeter is in about 2½ hours attained. This deposit is slightly coppered in the acid copper bath and backed.

In this manner nickel electrotypes of 15.74 × 11.81 inches have been produced, but as will be seen for the purposes of printing houses, the process is too troublesome and time-consuming, by reason of the necessary production of the copper matrices. Hence the direct method of deposition upon black-leaded gutta-percha or wax moulds is decidedly to be preferred.

This direct method requires a cold nickel bath, which yields heavy deposits without the nickel rolling off, and for deposits of a thickness suitable for printing, a few contrivances to prevent spontaneous detachment of the nickel from the matrix. The electrolyte, according to patent No. 134736 given on page 267 is applicable to this purpose. With this electrolyte, deposits of 6 millimeters' thickness were without trouble produced at the ordinary temperature upon gutta-percha. In testing other nickel baths described or patented, not a single one was found which allowed of obtaining useful nickel deposits directly upon the matrices, the nickel always rolling off, and when the latter drawback was prevented by suitable means, it was impossible to obtain a deposit of more than 0.05 millimeter thickness, cracks being formed in the center.

In working with this direct process of deposition, it is absolutely necessary always to keep the nickel bath slightly acidulated, because in a neutral or alkaline electrolyte the deposit becomes readily rough and forms with a dark color, which is an indication of the formation of sponge. It is also of advantage to keep the electrolyte constantly agitated. By reason of the oxidation of the ethyl-sulpho combinations which takes place, agitation, however, must not be effected by blowing in air, but by mechanical means, or eventually by blowing in carbonic acid.

An electro-motive force of 2.2 volts and a current-density of 0.2 to 0.3 ampère proved most suitable for the production of the above-mentioned nickel electrotypes of 6 millimeters in thickness. The current output—about 70 per cent.—was not particularly favorable, this being, however, of little importance as compared with the advantages offered by the use of this electrolyte.

It has previously been mentioned that in order to obtain

Fig. 152.   Fig. 154.   Fig. 155.

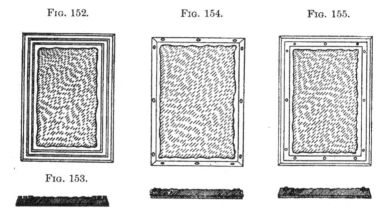

Fig. 153.

deposits of greater thickness upon gutta-percha or wax, a few contrivances are required to prevent the deposit from rolling off. With the use of gutta-percha matrices there is less danger of rolling off than with that of wax moulds, the tendency to rolling off being much earlier shown by the latter. However, in order to prevent failures, it is advisable not to omit these devices even when working with gutta-percha matrices.

According to the patent specification, a groove undercut towards the design and at a distance of about 3 millimeters from it, is made all round, and another such groove at a further distance of about 3 millimeters, as shown in Fig. 152 in front view, and in Fig. 153 in section.

The object of this contrivance is that upon the carefully black-leaded grooves, nickel continuous with the nickel upon the design is also deposited, and by reason of the undercutting towards the design the nickel is thus prevented from rolling off or becoming detached if, in consequence of the occlusion of hydrogen, the deposit shows a tendency to bend up.

According to the above-mentioned patent, the same effect may also be attained by firmly securing a metallic edge all round the design. While the nickel deposit cannot become intimately attached to the black-leaded, but otherwise non-metallic surfaces of the gutta-percha or wax matrices, it adheres very firmly to the metallic edge, rolling off being thus prevented. Dr. Langbein used thin brass strips, 0.1 millimeter thick and 5 millimeters wide which, as shown in Figs. 154 and 155, were secured by small pins either to the impressed surface or to the sides. Wires wound round the four sides of the matrix and lying everywhere closely upon its black-leaded surface, may be used in place of metal strips. It is advisable to place the metal strips in a heated state upon the matrix and press them gently into the matrix-material so that their surfaces lie perfectly level with the surface of the impression. The matrix and the metal edge having been carefully black-leaded, the outside of the latter is brushed over with a rag moistened with potassium cyanide solution, care being taken not to damage the black-leading of the metal towards the design. Then rinse the matrix with alcohol and suspend it at once in the nickel bath, the latter being kept at rest until the matrix is covered, and then agitated.

Nickel matrices.—In casting type from copper matrices, the latter oxidize quite rapidly, in consequence of which the edges and lines especially lose sharpness, while the surfaces become scarred. As early as 1883, Weston mentions in his English patent 4784, the possibility of obtaining heavy deposits of solid nickel, and that this invention is valuable for the production of *electrotypes*, which without doubt includes electrolytically prepared matrices for casting type, the influence of the temperature of the liquid metal upon such nickel matrices being so slight that they do not warp, etc.

Notwithstanding the fact that thus the employment of an electrolytically-produced casting matrix of nickel was known, the "Aktien-Gesellschaft für Schriftgiesserei" obtained a patent, the characteristic feature of which is that zinc can be directly cast around the face "without further galvanoplastic reinforcement."

Hence the above-mentioned patent cannot include such nickel matrices in which by the deposition of nickel the face is produced of a thickness which by itself is insufficient to allow of the deposit being detached from the original without fear of bending or breaking, the deposit requiring absolutely to be reinforced to the customary thickness by a galvanoplastic deposit of copper. It is obvious that a thickness of 0.1 to 0.25 millimeter of nickel suffices to withstand the effect of temperature and, when reinforced by copper, also the pressure in casting in the machine. Reinforcement of the casting of the back with copper has, however, the further advantage that in casting zinc around the face, the zinc alloys to a certain degree with the copper casting, thus uniting firmly with it, which is not the case when zinc is cast around the pure nickel face not enveloped by copper, nickel not entering into a solid combination with zinc.

Matrices electrolytically produced from cobalt also cannot be claimed under the above-mentioned patent. In hardness, cobalt is equal to nickel and resists the hot type-metal as well, possessing therefore all the properties required for casting matrices.

The most suitable material for such matrices would be an alloy of nickel and cobalt such as has been described on p. 312 as hard nickel alloy.

In order to effect an intimate union of the copper casing with the nickel, the nickel deposit when taken from the nickel bath has to be brushed over with nitric acid, rinsed, and without delay brought into the copper bath.

The omission of these manipulations, which require dexterity, may have been the cause why no more favorable results were obtained by former experiments to reinforce thinner nickel deposits by copper to a thickness of 2 millimeters, the nickel deposits becoming detached from the copper when the matrix was in use. If, however, in accordance with the suggestions given above, the nickel deposit is made 0.1 to 0.25 millimeter thick, and the back, which is to be reinforced, cleansed with nitric acid and rinsed, and then as rapidly as possible brought into the copper bath to be reinforced to 2.5 or 3 millimeters in thickness, the copper will adhere firmly and a durable matrix will be obtained.

By means of galvanoplasty matrices of massive nickel or cobalt for use in the casting machine may even be produced. However, by reason of their hardness, such massive nickel matrices are justified with difficulty, and besides they are too expensive.

While no experiments for the production of nickel matrices have been made with Weston's baths, nickel deposits several millimeters in thickness can without doubt be made with them by slightly changing their composition and heating them to between 176° and 194° F. In the experiments made baths of the composition given on p. 266 were used. They contained in 100 quarts, 77 lbs. of nickel sulphate and 39.6 lbs. of magnesium sulphate, and were always kept slightly acid with acetic acid, the temperature during deposition being as constantly as possible maintained at 194° F.

The originals have to be prepared in a manner different from that for matrices in copper. In place of wax for insulating the surfaces which are to receive no deposit, a material which does not soften at 194° F. has to be used. For this purpose, it was found most suitable to cast plaster of Paris around the original, or a paste of asbestos meal and waterglass. By treatment for 10 hours in the hot nickel bath during which time the current must in no wise be interrupted, and the original, especially in the beginning, be vigorously shaken, a nickel deposit about 0.25 millimeter in thickness is obtained. This deposit, as previously described, is reinforced in the acid copper bath to about 1.75 to 2.25 millimeters in thickness, and zinc having in the usual manner been cast around it, is justified for the casting machine.

The production of nickel matrices may also be effected with the use of the cold nickel bath described on p. 267, but much more time is required. In this case the originals may of course be insulated with wax.

IV. GALVANOPLASTY IN SILVER AND GOLD.

The preparation of reproductions in silver and gold presents many difficulties. While copper is reducible in a compact state from its sulphate solution, silver and gold have to be reduced from their double salt solutions—potassium-silver cyanide and potassium-gold cyanide. However, these alkaline solutions attack moulds of fatty substances, such as wax and stearine, consequently also, plaster-of-Paris moulds impregnated with these substances, as well as gutta percha and gelatine. Hence, only metallic moulds can be advantageously used, unless the end is to be attained in a round-about way ; that is, by first coating the mould with a thin film of copper, reinforcing this in the silver or gold bath, and finally dissolving the film of copper with dilute nitric acid.

The double salt solutions mentioned above require a well-conducting surface such as cannot be readily prepared by black-leading, a further reason why metallic moulds are to be preferred.

The simplest way for the galvanoplastic reproduction in gold or silver of surfaces not in too high relief or too much undercut, is to cover the object with lead, silver or gold foil, and pressing softened gutta-percha upon it. The foil yields to the pressure without tearing, and adheres to the gutta-percha so firmly that it can be readily separated together with it. This method is of course only applicable if the originals to be moulded can bear the pressure of the press.

With originals which cannot stand pressure, or have portions in very high relief, or much undercut, oil gutta-percha may be used. The original secured to a brass plate, having been heated to between 122° and 140° F. and slightly oiled, the oil gutta-percha in small cubes is applied so that one cube is first placed upon the original, and, when soft, pressed firmly down with the moistened finger, other cubes being then in the same manner applied until the entire surface of the original is covered, when the whole is allowed to cool, which may be accelerated by placing it in very cold water. This impression can be detached in good shape from the original by the use of gentle force, the oil gutta-percha being in a hardened state sufficiently pliable to allow of its being readily taken out from the undercut portions. The face of the mould is next freed from oil by means of alcohol, or by brushing with liquid ammonia, and then dried. Now powder the mould with fine silver powder, thoroughly rubbing the latter with a brush into the depressions, so that it adheres firmly to the gutta-percha, and after blowing off an excess, bring the mould into the silver bath.

The most suitable composition of the *galvanoplastic silver bath* is as follows:

Fine silver (in the form of silver cyanide) 1¾ ozs., 99 per cent. potassium cyanide 4¼ ozs., water 1 quart.

Maximum current-density, 0.3 ampère.

A slighter current-density than that given above can only be beneficial, and the electro-motive force should be as low as possible, the best deposits having been obtained with 0.5 volt and an electrode-distance of 10 centimeters.

For *galvanoplasty in gold*, the same process as described above is used. Good results are obtained with a bath composed as follows:

Fine gold (in the form of neutral chloride of gold or fulminating gold) 1 oz., 99 per cent. potassium cyanide 3½ ozs., water 1 quart.

Current-density, 0.1 ampère.

Electro-motive force at 10 cm. electrode-distance, 0.4 volt.

CHAPTER XVIII.

CHEMICALS USED IN ELECTRO-PLATING AND GALVANOPLASTY.

BELOW the characteristic properties of the chemical products employed in the workshop will be briefly discussed, and the reactions indicated which allow of their recognition. It frequently happens that the labels become detached from the bottles and boxes, thus rendering the determination of their contents necessary.

I. *Acids.*

1. *Sulphuric acid (oil of vitriol).*—Two varieties of this acid are found in commerce, viz., fuming sulphuric acid (disulphuric acid) and ordinary sulphuric acid. The first is a thick oily fluid, generally colored yellowish by organic substances. and emits dense, white vapors in the air. Its specific gravity is 1.87 to 1.89. The only purpose for which fuming sulphuric acid is used in the electro-plating art, is as a mixture with nitric acid for stripping silvered objects.

Ordinary sulphuric acid has a specific gravity of 1.84. Diluted with water it serves for filling the Bunsen elements and as a pickle for iron ; in a concentrated state it is used in the preparation of pickles and as an addition to the galvanoplastic copper bath. The crude commercial acid generally contains arsenic, hence care must be had to procure a pure article. In diluting the acid with water, it should in all cases be added to the water in a very gentle stream and with constant stirring, as otherwise a sudden generation of steam of explosive violence might result, and the dangerous corrosive liquid be scattered in all directions. Concentrated sulphuric acid vigorously attacks all organic substances, and hence has

(669)

to be kept in bottles with glass stoppers, and bringing it in contact with the skin should be carefully avoided.

Recognition.—One part of the acid mixed with 25 parts of distilled water gives, when compounded with a few drops of barium chloride solution, a white precipitate of barium sulphate.

2. *Nitric acid (aqua fortis, spirit of nitre).*—It is found in trade of various degrees of strength. For our purposes, acid of 40° and 30° Bé. is generally used. The acid is usually a more or less deep yellow, and frequently contains chlorine. The vapors emitted by nitric acid are poisonous and of a characteristic odor, by which the concentrated acid is readily distinguished from other acids. It is used for filling the Bunsen elements (carbon in nitric acid), and for pickling in combination with sulphuric acid and chlorine. On coming in contact with the skin it produces yellow stains.

Recognition.—By heating the not too dilute acid with copper, brown-red vapors are evolved. For the determination of dilute nitric acid, add a few drops of it to green vitriol solution, when a black-brown coloration will be produced on the point of contact.

3. *Hydrochloric acid (muriatic acid).*—The pure acid is a colorless fluid which emits abundant fumes in contact with the air, and has a pungent odor by which it is readily distinguished from other acids. The specific gravity of the strongest hydrochloric acid is 1.2. The crude acid of commerce has a yellow color, due to iron, and contains arsenic. Dilute hydrochloric acid is used for pickling iron and zinc.

Recognition.—On adding to the acid, very much diluted with distilled water, a few drops of solution of nitrate of silver in distilled water, a heavy white precipitate is formed, which becomes black by exposure to the light.

4. *Hydrocyanic acid (prussic acid).*—This extremely poisonous acid exists in nature only in a state of combination in certain vegetables and fruits, and especially in the kernels of the latter, as, for instance, in the peach, the berries of the

cherry laurel, bitter almonds, the stones of the apricot, of plums, cherries, etc. It may be obtained anhydrous, but in this state it is useless, and very difficult to preserve from decomposition. Diluted hydrocyanic acid is colorless, with a bitter taste and the characteristic smell of bitter almonds. It is employed in the preparation of gold immersion baths, and for the decomposition of the potassa in old silver baths. The inhalation of the vapors of this acid may have a fatal effect, as also its coming in contact with wounds.

Recognition.—By its characteristic smell of bitter almonds. Or mix it with potash lye until blue litmus paper is no longer reddened, then add solution of green vitriol which has been partially oxidized by standing in the air, and acidulate with hydrochloric acid. A precipitate of Berlin blue is formed.

5. *Citric acid.*—Clear colorless crystals of 1.542 specific gravity, which dissolve with great ease in both hot and cold water. It is frequently employed for acidulating nickel baths, and, combined with sodium citrate, in the preparation of platinum baths.

Recognition.—Lime-water compounded with aqueous solution of citric acid remains clear in the cold, but on boiling deposits a precipitate of calcium citrate. The precipitate is soluble in ammonium chloride, but on boiling is again precipitated, and is then insoluble in sal ammoniac.

6. *Boric acid (boracic acid).*—This acid is found in commerce in the shape of scales with nacreous luster and greasy to the touch ; when obtained from solutions by evaporation, it forms colorless prisms. Its specific gravity is 1.435; it dissolves with difficulty in cold water (1 part of acid requiring at 64.4° F. 28 of water), but is more rapidly soluble in boiling water (1 part of acid requiring 3 of water at 212° F.). According to Weston's proposition, boric acid is employed as an addition to nickel baths, etc.

Recognition.—By mixing solution of boric acid in water with some hydrochloric acid and dipping turmeric paper in the solution, the latter acquires a brown color, the color becoming more intense on drying. Alkalies impart to turmeric paper a similar coloration, which, however, disappears on immersing the paper in dilute hydrochloric acid.

7. *Arsenious acid (white arsenic, arsenic, ratsbane).*—It generally occurs in the shape of a white powder, and sometimes in vitreous-like lumps, resembling porcelain. For our purposes the white powder is almost exclusively used. It is slightly soluble in cold water, and more readily so in hot water and hydrochloric acid. Notwithstanding its greater specific gravity (3.7), only a portion of the powder sinks to the bottom on mixing it with water, another portion being retained on the surface by air bubbles adhering to it. It is employed as an addition to brass baths, further, in the preparation of arsenic baths, for blacking copper alloys, and in certain silver whitening baths.

Recognition.—When a small quantity of arsenious acid is thrown upon glowing coals an odor resembling that of garlic is perceptible. By mixing solution of arsenious acid, prepared by boiling with water, with a few drops of ammoniacal solution of nitrate of silver, a yellow precipitate of arsenate of silver is obtained. The ammoniacal solution of nitrate of silver is prepared by adding ammonia to solution of nitrate of silver until the precipitate at first formed disappears.

8. *Chromic acid.*—It forms crimson-red needles, and also occurs in commerce in the shape of a red powder. It is readily soluble in water, forming a red fluid, which serves for filling batteries.

Recognition.—Chromic acid can scarcely be mistaken for any other chemical product employed by the electro-plater. A very much diluted solution of it gives, after neutralization with caustic alkali and adding a few drops of nitrate of silver solution, a crimson-red precipitate of chromate of silver.

9. *Hydrofluoric acid.*—A colorless, corrosive, very mobile liquid of a sharp, pungent odor. The anhydrous acid fumes strongly in the air and attracts moisture with avidity. Hydrofluoric acid is used for etching glass and for pickling alumin-

ium dead white. Great care must be observed in working with the acid, since not only the aqueous solution, but also the vapors, have an extremely corrosive effect upon the skin and respiratory organs.

Recognition.—By covering a small platinum dish containing hydrofluoric acid with a glass plate free from grease, the latter in half an hour appears etched.

II. *Alkalies and Alkaline Earths.*

10. *Potassium hydrate (caustic potash).*—It is found in commerce in various degrees of purity, either in sticks or cakes. It is very deliquescent, and dissolves readily in water and alcohol ; by absorbing carbonic acid from the air it rapidly becomes converted into the carbonate, and thus loses its caustic properties. It should, therefore, be kept in well-closed vessels. Substances moistened with solution of caustic potash give rise to a peculiar soapy sensation of the skin when touched. It should never be allowed to enter the mouth, as even dilute solutions almost instantaneously remove the lining of tender skin. Should such an accident happen, the mouth should at once be rinsed several times with water and then with very dilute acetic acid. Pure caustic potash serves as an addition to zinc baths, gold baths, etc. For the purpose of freeing objects from grease the more impure commercial article is used.

11. *Sodium hydrate (caustic soda).*—It also occurs in commerce in various degrees of purity, either in sticks or lumps. It is of a highly caustic character, resembling potassium hydrate (see above) in properties and effects. It is employed for freeing objects from grease, for the preparation of alkaline tin and zinc baths, etc.

12. *Ammonium hydrate (ammonia or spirits of hartshorn).*— It is simply water saturated with ammonia gas. By exposure ammonia gas is gradually evolved, so that it must be kept in closely-stoppered bottles, in order to preserve the strength of the solution unimpaired. Four qualities are generally found in commerce, viz., ammonia of 0.910 specific gravity (contain-

43

ing 24.2 per cent. of ammonia gas); of 0.920 specific gravity (with 21.2 per cent. of ammonia gas); of 0.940 specific gravity (with 15.2 per cent. of ammonia gas); and 0.960 specific gravity (with 9.75 per cent. of ammonia gas). It is employed for neutralizing nickel and cobalt baths when too acid, in the preparation of fulminating gold, and as an addition to some copper and brass baths.

Recognition.—By the odor.

13. *Calcium hydrate* (*burnt or quick lime*).—It forms hard, white to gray pieces, which on moistening with water crumble to a light white powder, evolving thereby much heat. Vienna lime is burnt lime containing magnesia. Lime serves for freeing objects from grease, and for this purpose is made into a thinly-fluid paste with chalk and water, with which the objects to be freed from grease are brushed. Vienna lime is much used as a polishing agent.

III. *Sulphur Combinations.*

14. *Sulphuretted hydrogen* (*sulphydric acid, hydrosulphuric acid*).—A very poisonous, colorless gas with a fetid smell resembling that of rotten eggs. Ignited in the air, it burns with a blue flame, sulphurous acid and water being formed. At the ordinary temperature water absorbs about three times its own volume of the gas, and then acquires the same properties as the gas itself. Sulphuretted hydrogen serves for the metallizing of moulds as described in the preceding chapter, where the manner of generating it is also given. It is sometimes employed for the production of "oxidized" silver. Care should be taken not to bring metallic salts, gilt or silvered articles, or pure gold and silver in contact with sulphuretted hydrogen, they being rapidly sulphurized by it.

Recognition.—By its penetrating smell; further, by a strip of paper moistened with sugar of lead solution becoming black when brought into a solution of sulphuretted hydrogen or an atmosphere containing it.

15. *Potassium sulphide* (*liver of sulphur*).—It forms a hard

green-yellow to pale-brown mass, with conchoidal fracture. It readily absorbs moisture, whereby it deliquesces and smells of sulphuretted hydrogen. It is employed for coloring copper and silver black.

Recognition.—On pouring an acid over liver of sulphur, sulphuretted hydrogen is evolved with effervescence, sulphur being at the same time separated.

16. *Ammonium sulphide (sulphydrate or hydrosulphate of ammonia)*.—When freshly prepared it forms a clear and colorless fluid, with an odor of ammonia and sulphuretted hydrogen; by standing it becomes yellow, and, later on, precipitates sulphur. It is used for the same purpose as liver of sulphur.

17. *Carbon disulphide or bisulphide.*—Pure carbon disulphide is a colorless and transparent liquid which is very dense, and exhibits the property of double refraction. Its smell is characteristic and most disgusting, and may be compared to that of rotten turnips. It burns with a blue flame of sulphurous acid, carbonic acid being at the same time produced. It is used as a solvent for phosphorus and rubber in metallizing moulds according to Parkes' method. This solution should be very carefully handled.

18. *Antimony sulphide.*—a. *Black sulphide of antimony (stibium sulfuratum nigrum)* is found in commerce in heavy, gray and lusterless pieces or as a fine black-gray powder, with slight luster. It serves for the preparation of antimony baths, and for coloring copper alloys black.

b. *Red sulphide of antimony (stibium sulfuratum aurantiacum)* forms a delicate orange-red powder without taste or odor; it is insoluble in water, but soluble in ammonium sulphide, spirits of hartshorn and alkaline lyes. In connection with ammonium sulphide or ammonia it serves for coloring brass brown.

19. *Arsenic trisulphide or arsenious sulphide (orpiment)*.—It is found in commerce in the natural, as well as artificial, state, the former occurring mostly in kidney-shaped masses of a lemon color, and the latter in more orange-red masses, or as a

dull yellow powder. Specific gravity 3.46. It is soluble in the alkalies and spirits of sal ammoniac.

20. *Ferric sulphide.*—Hard, black masses generally in flat plates, which are only used for the generation of sulphuretted hydrogen.

IV. *Chlorine Combinations.*

21. *Sodium chloride (common salt, rock salt).*—The pure salt should form white, cubical crystals, of which 100 parts of cold water dissolve 36, hot water dissolving slightly more. The specific gravity of sodium chloride is 2.2. In electroplating sodium chloride is employed as a conducting salt for some gold baths, as a constituent of argentiferous pastes, and for precipitating the silver as chloride from argentiferous solutions.

Recognition.—An aqueous solution of sodium chloride on being mixed with a few drops of lunar caustic solution, yields a white caseous precipitate, which becomes black by exposure to light, and does not disappear by the addition of nitric acid, but is dissolved by ammonia in excess.

22. *Ammonium chloride (sal ammoniac).*—A white substance found in commerce in the shape of tough fibrous crystals. It has a sharp saline taste, and is soluble in $2\frac{1}{4}$ parts of cold, and in a much smaller quantity of hot water. By heat it is sublimed without decomposition. It serves for soldering and tinning, and as a conducting salt for many baths.

Recognition.—By sublimation on heating. By adding to a saturated solution of the salt a few drops of solution of platinum chloride, a yellow precipitate of platoso-ammonium chloride is formed.

23. *Antimony trichloride (butter of antimony).*—A crystalline mass which readily deliquesces in the air. Its solution in hydrochloric acid yields the *liquor stibii chlorati*, also called liquid butter of antimony. It has a yellowish color, and on mixing with water yields an abundant white precipitate, soluble in potash lye. The solution serves for coloring brass steel-gray, and for browning gun-barrels.

24. *Arsenious chloride.*—A thick, oily fluid, which evaporates in the air with the emission of white vapors.

25. *Copper chloride.*—Blue-green crystals readily soluble in water. The concentrated solution is green, and the dilute solution blue. On evaporating to dryness, brown-yellow copper chloride is formed. It is employed in copper and brass baths as well as for patinizing.

26. *Tin chloride.*—a. *Stannous chloride or tin salt.* A white crystalline salt readily soluble in water, but its solution on exposure to the air becomes turbid; by adding, however, hydrochloric acid, it again becomes clear. On fusing the crystallized salt loses its water of crystallization, and forms a solid non-transparent mass of a pale yellow color—the fused tin salt. The crystallized, as well as the fused, salt serves for the preparation of brass, bronze and tin baths.

Recognition.—By pouring hydrochloric acid over a small quantity of tin salt and adding potassium chromate solution, the solution acquires a green color. By mixing dilute tin salt solution with some chlorine water and adding a few drops of gold chloride solution, purple of Cassius is precipitated; very dilute solutions acquire a purple color.

b. *Stannic chloride* occurs in commerce in colorless crystals. In the anhydrous state it forms a yellowish, strongly fuming caustic liquid known as the "fuming liquor of Libadius."

27. *Zinc chloride (hydrochlorate or muriate of zinc; butter of zinc).*—A white crystalline or fused mass which is very soluble and deliquescent. The salt prepared by evaporation generally contains some zinc oxychloride, and hence does not yield an entirely clear solution. It serves for preparing brass and zinc baths, and its solution in nickeling by immersion, soldering, etc.

Recognition.—Solution of caustic potash separates a voluminous precipitate of zinc oxyhydrate, which redissolves in an excess of the caustic potash solution. By conducting sulphuretted hydrogen into a solution of a zinc salt acidulated with acetic acid, a precipitate of white zinc sulphide is formed.

28. *Chloride of zinc and ammonia.*—This salt is a combination of zinc chloride and ammonium chloride, and forms very deliquescent crystals. Its solution in water serves for soldering, and zincking by contact.

29. *Nickel chloride.*—It is found in commerce in the shape of deep green crystals and of a pale green powder. The latter contains considerably less water and less free acid than the crystallized article, and is to be preferred for electro-plating purposes. The crystallized salt dissolves readily in water, and the powder somewhat more slowly. Should the solution of the latter deposit a yellow precipitate, consisting of basic nickel chloride, it has to be brought into solution by the addition of a small quantity of hydrochloric acid. Nickel chloride is employed for nickel baths.

Recognition—By mixing the green solution of the salt with some spirits of sal ammoniac, a precipitate is formed, which dissolves in an excess of spirits of sal ammoniac, the solution showing a deep blue color.

30. *Cobaltous chloride.*—It forms small rose-colored crystals, which, on heating, yield their water of crystallization, and are converted into a blue mass. The crystals are readily soluble in water, while the anhydrous blue powder dissolves slowly. Cobalt chloride is employed for the preparation of cobalt baths.

Recognition.—Caustic potash precipitates from a solution of cobalt chloride a blue basic salt which is gradually converted into a rose-colored hydrate, and, with the access of air, into green-brown cobaltous hydrate. The aqueous solution yields with solution of yellow prussiate of potash, a pale gray-green precipitate.

31. *Silver chloride.*—A heavy white powder which by exposure to light becomes gradually blue-gray, then violet, and finally black. When precipitated from silver solutions, a caseous precipitate is separated. At 500° F. it fuses, without being decomposed, to a yellowish fluid which, on cooling, congeals to a transparent, tenacious, horn-like mass. Silver chloride is practically insoluble in water, but dissolves with

ease in liquid ammonia and in potassium cyanide solution. It is employed in the preparation of baths for silver-plating, for silvering by boiling, and in the pastes for silvering by friction.

Recognition.—By its solubility in ammonia, pulverulent metallic silver being separated from the solution by dipping in it bright ribands of copper.

32. *Gold chloride (terchloride of gold, muriate of gold, auric chloride).*—This salt occurs in commerce as crystallized gold chloride of an orange-yellow color, and as a brown crystalline mass which is designated as neutral gold chloride, or as gold chloride free from acid, while the crystallized articles always contains acid, and, hence, should not be used for gold baths. Gold chloride absorbs atmospheric moisture and becomes resolved into a liquid of a fine gold color. On being moderately heated, yellowish-white aurous chloride is formed, and on being subjected to stronger heat, it is decomposed to metallic gold and chlorine gas. By mixing its aqueous solution with ammonia, a yellow-brown powder consisting of *fulminating gold* is formed. In a dry state this powder is highly explosive, and, hence, when precipitating it from gold chloride solution for the preparation of gold baths, it must be used while still moist.

Recognition.—By the formation of the precipitate of fulminating gold on mixing the gold chloride solution with ammonia. Further, by the precipitation of brown metallic gold powder on mixing the gold chloride solution with green vitriol solution.

33. *Platinic chloride.*—The substance usually known by this name is *hydroplatinic chloride*. It forms red-brown, very soluble—and in fact deliquescent—crystals. With ammonium chloride it forms platoso-ammonium chloride. Both combinations are used in the preparation of platinum baths. The solution of platinic chloride also serves for coloring silver, tin, brass and other metals.

Recognition.—By the formation of a precipitate of yellow

platoso-ammonium chloride on mixing concentrated platinic chloride solution with a few drops of saturated sal ammoniac solution.

V. *Cyanides.*

34. *Potassium cyanide (white prussiate of potash).* — For electro-plating purposes pure potassium cyanide with 98 to 99 per cent., as well as that containing 80, 70 and 60 per cent., is used, whilst for pickling the preparation with 45 per cent. is employed. For the preparation of alkaline copper and brass baths, as well as silver baths, the pure 98 to 99 per cent. product is generally employed. However, for preparing gold baths the 60 per cent. article is mostly preferred, because the potash present in all potassium cyanide varieties with a lower content renders fresh baths more conductive. However, gold baths may also be prepared with 98 per cent. potassium cyanide without fear of injury to the efficiency of the baths, while, under ordinary circumstances, a preparation with less than 98 per cent. may safely be used for the rest of the baths. However, when potassium cyanide has to be added to the baths, as is from time to time necessary, only the pure preparation free from potash should be used, because the potash contained in the inferior qualities gradually thickens the bath too much.

No product is more important to the electro-plater than potassium cyanide. The pure 98 to 99 per cent. product is a white, transparent, crystalline mass, the crystalline structure being plainly perceptible upon the fracture. In a dry state it is odorless, but when it has absorbed some moisture it has a strong smell of prussic acid. It is readily soluble in water, and should be dissolved in *cold* water only, since when poured into hot water it is partially decomposed, which is recognized by the appearance of an odor of ammonia. Potassium cyanide solution in cold water may, however, be boiled for a short time without suffering essential decomposition. Potassium cyanide must be kept in well-closed vessels, since when exposed to the air it becomes deliquescent, and is decomposed by the carbonic acid of the air, whereby potassium carbonate is formed while prussic acid escapes. It is a deadly poison and must be used with the utmost caution.

While pure fused potassium cyanide of 98 to 99 per cent. could formerly be everywhere obtained in commerce, the present commercial product consists, as a rule, of a mixture of potassium cyanide and sodium cyanide. The reason for this is that the dried yellow prussiate of potash was formerly fused by itself, whereby one-third of its content of cyanogen was lost, while, for the purpose of fixing this quantity of cyanogen, it is now fused with metallic sodium. The resultant product contains 78 per cent. potassium cyanide and 21 per cent. sodium cyanide.

While for many electro-plating purposes, this mixture may take the place of pure potassium cyanide, its use for some processes, for instance, in the preparation of more concentrated gold baths, is connected with certain drawbacks. While the double salt—potassium-gold cyanide—dissolves very readily, sodium-gold cyanide is less soluble and separates in the form of a pale-yellow powder. Sodium-copper cyanide shows a similar behavior, it being less soluble than the potassium double salt and as the electro-motive forces for decomposing the potassium and sodium double salts vary, the use of a mixture of potassium cyanide and sodium cyanide is, to say the least, not rational. For certain purposes the electro-plater should demand from his dealer pure potassium cyanide free from sodium cyanide.

Potassium cyanide with 80, 70, 60 or 45 per cent. forms a gray-white to white mass with a porcelain-like fracture. A pale gray coloration is not a proof of impurities, being due to somewhat too high a temperature in fusing. These varieties are found in commerce in irregular lumps or in sticks, the use of the latter offering no advantage. Their behavior towards the air and in dissolving is the same as that of the pure product.

Recognition.—By the bitter almond smell of the solution. By mixing potassium cyanide solution with ferric chloride and then with hydrochloric acid until the latter strongly predominates, a precipitate of Berlin blue is formed.

The pure salt free from potash does not effervesce on adding dilute acid, which is, however, the case with the inferior qualities.

To facilitate the use of potassium cyanide with a different content than that given in a formula for preparing a bath, the following table is here given :

Potassium cyanide with

98 per cent.	80 per cent.	70 per cent.	60 per cent.	45 per cent.
By weight.	By weight.	By weight.	By weight.	By weight.
1 part $=$	1.2: 0 parts $=$	1.400 parts $=$	1.660 parts $-$	2.180 parts.
0.820 part $=$	1 part $=$	1.143 parts $=$	1.333 parts $-$	1.780 parts.
0.714 part $=$	0.875 part $=$	1 part $=$	1.170 parts $=$	1.550 parts.
0.615 part $=$	0.740 part $=$	0.857 part $=$	1 part $=$	1.450 parts.
0.460 part $=$	0.562 part $=$	0.643 part $=$	0.750 part $=$	1 part.

35. *Copper cyanide.*—There is a cuprous and a cupric cyanide ; that used for electro-plating purposes being a mixture of both. It is a green-brown powder, which should not be entirely dried, since in the moist state it dissolves with greater ease in potassium cyanide than the dried product.

It is chiefly used in the form of a double salt potassium-copper cyanide, *i. e.*, a combination of copper cyanide with potassium cyanide, in the preparation of copper, brass, tombac, and red gold baths.

Recognition.—By evaporating a piece of copper cyanide the size of a pea, or its solutions, in hydrochloric acid, to dryness on a water-bath, wherein care must be taken not to inhale the vapors, and dissolving the residue in water, a green-blue solution is obtained which acquires a deep blue color by the addition of ammonia in excess.

36. *Zinc cyanide (hydrocyanate of zinc, prussiate of zinc).*—A white powder insoluble in water, but soluble in potassium cyanide, ammonia and the alkaline sulphites. The fresher it is, the more readily it dissolves, the dried product dissolving with difficulty. Its solution in potassium cyanide forms potassium-zinc cyanide, which is used for brass baths.

Recognition.—By evaporating zinc cyanide, or its solution, with an excess of hydrochloric acid on the water-bath, zinc chloride remains behind, which is recognized by the same reaction given under zinc chloride.

37. *Silver cyanide (prussiate or hydrocyanate of silver).*—A white powder which slowly becomes black when exposed to light. It is insoluble in water and cold acids, which, however, will dissolve it with the aid of heat. At 750° F. it melts to a dark red fluid, which, on cooling, forms a yellow mass with a granular structure. It is readily dissolved by potassium cyanide, but is only slightly soluble in ammonia, differing in this respect from silver chloride. It forms a double salt with potassium cyanide—potassium-silver cyanide—and as such is employed in the preparation of silver baths.

38. *Potassium ferro-cyanide (yellow prussiate of potash).*—It occurs in the shape of yellow semi-translucent crystals with mother-of-pearl luster, which break without noise. Exposed to heat they effloresce, losing their water of crystallization, and crumbling to a yellowish-white powder. For the solution of 1 part of the salt, 4 of water of medium temperature are required, the solution exhibiting a pale yellow color. It precipitates nearly all the metallic salts from their solutions, some of the precipitates being soluble in an excess of the precipitating agent. This salt is not poisonous. It serves for the preparation of silver and gold baths; its employment, however, offering over potassium cyanide no advantages, unless the non-poisonous properties be considered as such.

Recognition.—When the yellow solution is mixed with ferric chloride, a precipitate of Berlin blue is formed; by blue vitriol solution a brown-red precipitate is obtained.

VI. *Carbonates.*

39. *Potassium carbonate (potash).*—It is found in commerce in gray-white, bluish, yellowish pieces, the colorations being due to admixtures of small quantities of various metallic oxides. When pure it is in the form of a white powder, or in pieces the size of a pea. The salt, being very deliquescent, has to be kept in well-closed receptacles. It is readily soluble, and if pure, the solution in distilled water should be clear. It serves as an addition to some baths, and in an impure state for freeing objects from grease.

Recognition.—The solution effervesces on the addition of hydrochloric acid. When neutralized with hydrochloric acid it gives with platinum chloride a heavy yellow precipitate of platinic potassium chloride, provided it be not too dilute.

40. *Acid potassium carbonate or monopotassic carbonate, commonly called bicarbonate of potash.*—Colorless, transparent, crystals, which at a medium temperature dissolve to a clear solution in 4 parts of water. It is not deliquescent; however, on boiling its solution it loses carbonic acid, and contains then only potassium carbonate. It is employed for the preparation of certain baths for gilding by simple immersion.

41. *Sodium carbonate (washing soda).*—It occurs in commerce as crystallized or calcined soda of various degrees of purity. The crystallized product forms colorless crystals or masses of crystals, which, on exposure to air, rapidly effloresce and crumble to a white powder. By heating, the crystals also lose their water, a white powder, the so-called calcined soda, remaining behind. Soda dissolves readily in water, and serves as an addition to copper and brass baths, for the preparation of metallic carbonates, and for freeing objects from grease, the ordinary impure soda being used for the latter purpose.

The directions for additions of sodium carbonate to baths generally refer to the crystallized salt. If calcined soda is to be used instead, 0.4 part of it will have to be taken for 1 part of the crystallized product.

42. *Sodium bicarbonate (baking powder).*—A dull white

powder soluble in 10 parts of water of 68° F. On boiling, the solution loses one-half of its carbonic acid, and then contains sodium carbonate only.

43. *Calcium carbonate (marble, chalk).*—When pure it forms a snow-white crystalline powder, a yellowish color indicating a content of iron. It is insoluble in water, but soluble, with effervescence, in hydrochloric, nitric and acetic acids. In nature, calcium carbonate occurs as marble, limestone, chalk.

In the form of *whiting* (ground chalk carefully freed from all stony matter) it is used for the removal of an excess of acid in acid copper baths, and mixed with burnt lime, as an agent for freeing objects from grease.

44. *Copper carbonate.*—Occurs in nature as malachite and allied minerals. The artificial carbonate is an azure-blue substance, insoluble in water, but soluble, with effervescence, in acids. Copper carbonate precipitated from copper solution by alkaline carbonates has a greenish color. Copper carbonate is employed for copper and brass baths and for the removal of an excess of acid in acid copper baths.

Recognition.—Dissolves in acids with effervescence ; on dipping a ribband of bright sheet-iron in the solution, copper separates upon the iron. On compounding the solution with ammonia in excess, a deep blue coloration is obtained.

45. *Zinc carbonate.*—A white powder, insoluble in water. The product obtained by precipitating a zinc salt with alkaline carbonate is a combination of zinc carbonate with zinc oxyhydrate. It serves for brass baths in connection with potassium cyanide.

Recognition.—In a solution in hydrochloric acid, which is formed with effervescence, according to the reactions given under zinc chloride (27).

46. *Nickel carbonate.*—A pale apple-green powder, insoluble in water, but soluble, with effervescence, in acids. It is employed for neutralizing nickel baths which have become acid.

Recognition.—In hydrochloric acid, it dissolves, with effervescence, to a green fluid. By the addition of a small quantity of ammonia, nickel oxyhydrate is precipitated, which, by adding ammonia in excess, is redissolved, the solution showing a blue color.

47. *Cobaltous carbonate.*—A reddish powder, insoluble in water, but soluble in acids, the solution forming a red fluid.

VII. *Sulphates and sulphites.*

48. *Sodium sulphate (Glauber's salt).*—Clear crystals of a slightly bitter taste, which effloresce by exposure to the air. They are readily soluble in water. On heating, the crystals melt in their water of crystallization, and when subjected to a red heat, calcined Glauber's salt remains behind. It is used as an addition to some baths.

49. *Ammonium sulphate.*—It forms a neutral, colorless salt, which is constant in the air, readily dissolves in water, and evaporates on heating. It serves as a conducting salt for nickel, cobalt and zinc baths.

Recognition.—By its evaporating on heating. A concentrated solution compounded with platinic chloride gives a yellow precipitate of platoso-ammonium chloride, while a solution mixed with a few drops of hydrochloric acid gives with barium chloride a precipitate of barium sulphate.

50. *Potassium-aluminium sulphate (potash-alum).*—Colorless crystals or pieces of crystals with an astringent taste. It is soluble in water, 12 parts of it dissolving in 100 parts of water at the ordinary temperature. On heating, the crystals melt, and are converted into a white, spongy mass, the so-called burnt alum. Potash-alum serves for the preparation of zinc baths and for brightening the color of gold.

Recognition.—On adding sodium phosphate to the solution of this salt a jelly-like precipitate of aluminium phosphate is formed, which is soluble in caustic potash, but insoluble in acetic acid.

51. *Ammonium-alum* is exactly analogous to the above, the potassium sulphate being simply replaced by ammonium sulphate. It is for most purposes interchangeable with potash-

alum.   On exposing ammonium-alum to a red heat, the ammonium sulphate is lost, pure alumina remaining behind. Ammonium-alum is used for preparing a bath for zincking iron and steel by immersion.

Recognition.—The same as potash-alum.   On heating the comminuted ammonium-alum with potash-lye, an odor of ammonia becomes perceptible.

52. *Ferrous sulphate (sulphate of iron, protosulphate of iron, copperas, green vitriol).*—Pure ferrous sulphate forms bluish-green, transparent crystals of a sweetish, astringent taste, which readily dissolve in water, and effloresce and oxidize in the air. The crude article forms green fragments frequently coated with a yellow powder.   It generally contains, besides ferrous sulphate, the sulphate of copper and of zinc, as well as ferric sulphate.   Ferrous sulphate is employed in the preparation of iron baths, and for the reduction of gold from its solutions.

Recognition.—By, compounding the green solution with a few drops of concentrated nitric acid, a black-blue ring is formed on the point of contact.   On mixing the lukewarm solution with gold chloride, gold is separated as a brown powder, which by rubbing acquires the luster of gold.

53. *Iron-ammonium sulphate.*—Green ·crystals which are constant in the air and do not oxidize as readily as green vitriol.   100 parts of water dissolve 16 parts of this salt.   It is used for the same purposes as green vitriol.

54. *Copper sulphate (cupric sulphate, blue vitriol, or blue copperas).*—It forms large, blue crystals, of which 190 parts of cold water dissolve about forty parts, and the same volume of hot water about 200 parts.   Blue vitriol which does not possess a pure blue color but shows a greenish luster, is contaminated with green vitriol, and should not be used for electro-plating purposes.   Blue vitriol serves for the preparation of alkaline copper and brass baths, acid copper baths, etc.

Recognition.—By its appearance, as it can scarcely be mistaken for anything else.   A content of iron is recognized by boiling blue vitriol solution with a small quantity of nitric acid, and adding ammonia in excess; brown flakes indicate iron.

55. *Zinc sulphate (white vitriol).*—It forms small colorless prisms of a harsh metallic taste, which readily oxidize on exposure to the air. By heating the crystals melt, and by heating to a red heat they are decomposed into sulphurous acid and oxygen, which escape, while zinc oxide remains behind as residue. 100 parts of water dissolve about 50 parts of zinc sulphate in the cold, and nearly 100 parts at the boiling-point. Zinc sulphate is employed for the preparation of brass and zinc baths, as well as for mat pickling.

Recognition.—By mixing zinc sulphate solution with acetic acid and conducting sulphuretted hydrogen into the mixture, a white precipitate of zinc sulphide is formed. A slight content of iron is recognized by the zinc sulphate solution, made alkaline by ammonia, giving with ammonia sulphide a somewhat colored precipitate instead of a pure white one. However, a slight content of iron does no harm.

56. *Nickel sulphate.*—Beautiful dark green crystals, readily soluble in water, the solution exhibiting a green color. On heating the crystals to above 536° F., yellow anhydrous nickel sulphate remains behind. Like the double salt described below, it serves for the preparation of nickel baths and for coloring zinc.

Recognition.—By compounding the solution with ammonia the green color passes into blue. Potassium carbonate precipitates pale green basic nickel carbonate, which dissolves on adding ammonia in excess, the solution showing a blue color. A content of copper is recognized by the separation of black-brown copper sulphide on introducing sulphuretted hydrogen into a heated solution previously strongly acidulated with hydrochloric acid.

57. *Nickel-ammonium sulphate.*—It forms green crystals of a somewhat paler color than nickel sulphate. This salt dissolves with more difficulty than the preceding, 100 parts of water dissolving only 5.5 parts of it. It is used for the same pur-

poses as the nickel sulphate, and is also recognized in the same manner. The following reaction serves for distinguishing it from nickel sulphate : By heating nickel sulphate in concentrated solution with the same volume of strong potash or soda lye, no odor of ammonia is perceptible, while nickel-ammonium sulphate evolves ammoniacal gas which forms dense clouds on a glass rod moistened with hydrochloric acid.

58. *Cobaltous sulphate.*—Crimson crystals of a sharp metallic taste. They are constant in the air and readily dissolve in water, the solution showing a red color. By heating the crystals lose their water of crystallization without, however, melting, and become thereby transparent and rose-colored. The salt is used for cobalt baths for the electro-deposition of cobalt and for cobalting by contact.

Recognition.—In the presence of ammoniacal salts, caustic potash precipitates a blue basic salt, which on heating changes to a rose-colored hydrate and, by standing for some time in the air, to a green-brown hydrate. By mixing a concentrated solution of the salt strongly acidulated with hydrochloric acid, with solution of potassium nitrate, a reddish-yellow precipitate is formed.

59. *Cobalt-ammonium sulphate.*—This salt forms crystals of the same color as cobalt sulphate, which, however, dissolve more readily in water.

60. *Sodium sulphite and bisulphite.*—a. *Sodium sulphite.* Clear, colorless, and odorless crystals, which are rapidly transformed into an amorphous powder by efflorescence. The salt readily dissolves in water, the solution showing a slight alkaline reaction due to a small content of sodium carbonate. It is employed in the preparation of gold, brass, and copper baths, for silvering by immersion, etc.

Recognition.—The solution when mixed with dilute sulphuric acid has an odor of burning sulphur.

b. *Sodium bisulphite.*—Small crystals, or more frequently in the shape of a pale yellow powder with a strong odor of sulphurous acid and readily soluble in water. The solution

44

shows a strong acid reaction and loses sulphurous acid in the air. * It is employed in the preparation of alkaline copper and brass baths.

Both the sulphite and bisulphite must be kept in well-closed receptacles, as by the absorption of atmospheric oxygen they are converted to sulphate.

61. *Cuprous sulphite.*—A brownish-red crystalline powder formed by treating cuprous hydrate with sulphurous acid solution. It is insoluble in water, but readily soluble in potassium cyanide, with only slight evolution of cyanogen. It serves for the preparation of alkaline copper baths in place of basic acetate of copper (verdigris), blue vitriol, or cuprous oxide.

VIII. *Nitrates.*

62. *Potassium nitrate (saltpetre, nitre).*—It forms large, prismatic crystals, generally hollow, but also occurs in commerce in the form of a coarse powder, soluble in 4 parts of water at a medium temperature. The solution has a bitter, saline taste and shows a neutral reaction. Potassium nitrate melts at a red heat, and on cooling congeals to an opaque, crystalline mass. It is employed in the preparation of desilvering pickle and for producing a mat luster upon gold and gilding. For these purposes it may, however, be replaced by the cheaper *sodium nitrate*, sometimes called *cubic nitre* or *Chile saltpetre*.

Recognition.—A small piece of coal when thrown upon melting saltpetre burns fiercely. When a not too dilute solution of saltpetre is compounded with solution of potassium bitartrate saturated at the ordinary temperature, a crystalline precipitate of tartar is formed.

63. *Sodium nitrate (cubic nitre* or *Chile saltpetre).*—Colorless crystals, deliquescent and very soluble in water; the solution shows a neutral reaction. It is used for the same purposes as potassium nitrate.

64. *Mercurous nitrate.*—It forms small, colorless crystals, which are quite transparent and slightly effloresce in the air. On heating, they melt and are transformed, with the evolution

of yellow-red vapors, into yellow-red mercuric oxide, which, on further heating, entirely evaporates. With a small quantity of water, mercurous nitrate yields a clear solution. By the further addition of water it shows a milky turbidity, which, however, disappears on adding nitric acid. It is employed for quicking the zincs of the cells, and the objects previous to silvering, and for brightening (with subsequent heating) gilding. For the same purpose is also used:

65. *Mercuric nitrate* (*nitrate of mercury*).—This salt is obtained with difficulty in a crystallized form. It is generally sold in the form of an oily, colorless liquid which, in contact with water, separates a basic salt. This precipitate disappears upon the addition of a few drops of nitric acid, and the liquid becomes clear.

Recognition.—A bright ribband of copper dipped in solution of mercurous or mercuric nitrate becomes coated with a white amalgam, which disappears upon heating.

66. *Silver nitrate* (*lunar caustic*).—This salt is found in commerce in three forms: Either as crystallized nitrate of silver in thin, rhombic, and transparent plates; or in amorphous, opaque, and white plates of fused nitrate; or in small cylinders of a white, or gray, or black color, according to the nature of the mould employed, in which form it constitutes the lunar caustic for surgical uses. For our purposes only the pure, crystallized product, free from acid, should be employed. The crystals dissolve readily in water. In making solutions of this and other silver salts, only distilled water should be used; all other waters, owing to the presence of chlorine, produce a cloudiness or even a distinct precipitate of silver chloride. When subjected to heat the crystals melt to a colorless, oily fluid, which, on cooling, congeals to a crystalline mass. Silver nitrate is employed in the preparation of chloride and cyanide of silver for silver baths. The solution in potassium cyanide may also be used for silver baths. The alcoholic solution is employed for metallizing non-conductive moulds for galvanoplastic deposits.

Recognition.—Hydrochloric acid and common salt solution precipitate from silver nitrate solution silver chloride, which becomes black on exposure to the light, and is soluble in ammonia.

IX. *Phosphates and Pyrophosphates.*

67. *Sodium Phosphate.*—Large, clear crystals, which readily effloresce, and whose solution in water shows an alkaline reaction. It is employed in the preparation of gold baths and for the production of metallic phosphates for soldering.

Recognition.—The dilute solution compounded with silver nitrate yields a yellow precipitate of silver phosphate.

68. *Sodium pyrophosphate.*—It forms white crystals, which are not subject to efflorescence, and are soluble in 6 parts of water at a medium temperature ; the solution shows an alkaline reaction. Sodium pyrophosphate also occurs in commerce in the form of an anhydrous white powder, though it may here be said that the directions for preparing baths refer to the crystallized salt. It is employed in the preparation of gold, nickel, bronze, and tin baths.

Recognition.—The dilute solution compounded with silver nitrate yields a *white* instead of a *yellow* precipitate.

69. *Ammonium phosphate.*—A colorless crystalline powder quite readily soluble in water ; the solution should be as neutral as possible. A salt smelling of ammonia, as well as one showing an acid reaction, should be rejected. It is employed in the preparation of platinum baths.

X. *Salts of Organic Acids.*

70. *Potassium bitartrate (cream of tartar).*—The pure salt forms small transparent crystals, which have an acid taste, and are slightly soluble in water. The commercial crude tartar or *argol*, which is a by-product in the wine-industry, forms gray or dirty-red crystalline crusts. In a finely powdered state, purified tartar is called *cream of tartar*. It is employed for the preparation of the whitening silver baths,

for those of tin, and for the silvering paste for silvering by friction, and in scratch-brushing different deposits.

71. *Potassium-sodium tartrate (Rochelle or Seignette salt).*— Clear colorless crystals, constant in the air. of a cooling, bitter, saline taste, and soluble in 2.5 parts of water of a medium temperature. The solution shows a neutral reaction. This salt is employed in the preparation of copper baths free from cyanide, as well as of nickel and cobalt baths, which are to be decomposed in the single cell apparatus.

Recognition.—By the addition of acetic acid the solution yields an abundant precipitate of tartar.

72. *Antimony-potassium tartrate (tartar emetic).*—A white crystalline substance, of which 100 parts of cold water dissolve 5 parts, while a like volume of hot water dissolves 50 parts. The solution shows a slight acid reaction. The only use of this salt is for the preparation of antimony baths.

Recognition.—The solution of the salt compounded with sulphuric, nitric, or oxalic acid yields a white precipitate, insoluble in an excess of the cold acid. Sulphuretted hydrogen imparts to the dilute solution a red color. Hydrochloric acid effects a precipitate, which is redissolved by the acid in excess.

73. *Copper acetate (verdigris).*—It is found in the market in the form of dark green crystals showing an acid reaction, or as a neutral bright green powder.

The crystallized copper acetate forms opaque dark green prisms, which readily effloresce, becoming thereby coated with a pale green powder. They dissolve with difficulty in water, but readily in ammonia, forming a solution of a blue color. They dissolve readily also in potassium cyanide and alkaline sulphites.

The neutral copper acetate forms a blue-green crystalline powder, imperfectly soluble in water, but readily soluble in ammonia, forming a solution of a blue color.

Copper acetate is used for preparing copper and brass baths, for the production of artificial patinas, for coloring, gilding, etc.

Recognition.—On pouring sulphuric acid over copper acetate, a strong odor of acetic acid is noticed ; with ammonia it yields a blue solution.

74. *Lead acetate (sugar of lead).*—Colorless lustrous prisms or needles of a nauseous sweet taste, and poisonous. The crystals effloresce in the air, melt at 104° F., and are readily soluble in water, yielding a slightly turbid solution. Lead acetate is employed for preparing lead baths (Nobili's rings) and for coloring copper and brass.

Recognition.—By compounding lead acetate solution with potassium chromate solution, a heavy yellow precipitate of lead chromate is formed.

75. *Sodium citrate.*—Colorless crystals, presenting a moist appearance, which are readily soluble in water ; the solution should show a neutral reaction. This salt is employed in the preparation of the platinum bath according to Böttger's formula, and as conducting salt for nickel and zinc baths.

APPENDIX.

CONTENTS OF VESSELS.

To find the number of gallons a tank or other vessel will hold, divide the number of cubic inches it contains by 231.

If rectangular, multiply together the length, breadth and depth.

If cylindrical, multiply the square of the diameter by 0.7854, and the product by the depth.

If conical, add together squares of diameters of top and bottom, and the product of the two diameters. Multiply their sum by 0.7854, and the resulting product by the depth. Divide the product by 3.

If hemispherical, to three times the square of the radius at top add the square of the depth. Multiply this sum by the depth and the product by 0.5236.

Avoirdupois Weight.

	= Ounces.	= Drams.	= Grains.	= Grams.
1 Pound	16	256	7,000	453.25
1 Ounce	1	16	437.5	28.33
1 Dram	0.062	1	27.34	1.77

Troy Weight.

	= Ounces.	= Dwt.	= Grains.	= Grams.
1 Pound	12	240	5,760	372.96
1 Ounce	1	20	480	31.08
1 Pennyweight	0.05	1	24	1.55

Imperial Fluid Measure.

	‖ Quarts.	‖ Pints.	‖ Fluid Ounces.	‖ Fluid Drams.	‖ Minims.	‖ Weight in Grains.	‖ Cubic Inches.	‖ Liters.	‖ Cubic Centimeters.
1 Gallon	4	8	160	1280	76,800	70,000	277.276	4.541	4,541
1 Quart	1	2	40	320	19,200	17,500	69 310	1.135	1,135.2
1 Pint............	0.5	1	20	160	9,600	8,750	34.659	0.567	576.6
1 Fluid Ounce ...	0.025	0.05	1	8	· 480	437.5	1.733	0.0284	283.8
1 Fluid Dram ...	0.0031	0.0062	0.125	1	60	54.7	0.217	0.0035	35.5
1 Minim	0.00005	0.0001	0.0021	0.0167	1	0.91	0.0036	0.00006	0.59

Table of Useful Numerical Data.

1 millimeter equals	.03937 inches.
1 centimeter "	.39370 "
1 decimeter "	3.93700 "
1 meter "	39.37000 "
1 cubic centimeter of water equals }	1 gram.
1 liter "	1000 "
1 liter " {	35.275 } ounces by measure.
1 gallon (or 160 fluid ounces) equals }	4.536 liters.
1 gallon "	277.276 cubic ins.
1 pint (or 20 fluid ounces equals }	34.659 "
1 fluid ounce "	1.733 "
1 liter "	61.024 "
1 avoirdupois pound equals }	7000. grains.

1 troy pound equals	5760. grains.
1 avoirdupois ounce equals }	437.5 "
1 troy ounce equals	480.
1 avoirdupois drm. equals }	27.34
1 troy pennyweight equals }	24.
1 gram equals	15.43
1 kilogram equals	15432. "
1 liter of water equals	15432. "
1 cubic inch of water equals }	252.5
1 cubic centimeter of water equals }	1. "
1 kilogram equals	35.274 avoirdupois ozs

To convert Fahrenheit thermometer degrees (F.) *to Centigrade degrees* (C.), **first** subtract 32, then multiply by 5, and divide by 9.

$$C = \frac{5(F. - 32)}{9}$$

To convert Centigrade degrees to Fahrenheit degrees, multiply by 9, divide by 5, then add 32.

$$F - \frac{9C}{5} + 32$$

Value to be Converted.	LENGTH		AREA		VOLUME						WEIGHT	
	Of Inches equals Millimeters.	Of Millimeters equals Inches.	Of Square Inches equals Square Decimeters.	Of Square Decimeters equals Square Inches.	Of Cubic Inches equals Cubic Centimeters.	Of Cubic Centimeters equals Cubic Inches.	Of Gallons equals Liters.	Of Liters equals Gallons.	Of Cubic Inches equals Pints.	Of Pints equals Cubic Inches.	Of Grains equals Grammes.	Of Grammes equals Grains.
0.05	1.2	0.002	.003	0.77	0.82	0.003	0.227	.011	.0015	1.73	.003	0.772
.1	2.5	.004	.006	1.55	1.64	.06	.454	.022	.003	3.47	.06	1.543
.2	5.1	.008	.013	3.10	3.28	.012	.909	.044	.006	6.93	.013	3.086
.3	7.6	.012	.019	4.65	4.92	.018	1.363	.066	.009	10.40	.019	4.630
.4	10.2	.016	.026	6.20	6.55	.024	1.817	.088	.011	13.86	.026	6.173
.5	12.7	.020	.032	7.75	8.19	.031	2.272	.110	.014	17.33	.032	7.716
.6	15.2	.024	.039	9.30	9.83	.037	2.726	.132	.017	20.79	.039	9.259
.7	17.8	.028	.045	10.85	11.47	.043	3.180	.154	.020	24.26	.045	10.803
.8	20.3	.031	.051	12.40	13.11	.049	3.635	.176	.023	27.73	.052	12.346
.9	22.9	.035	.058	13.95	14.75	.055	4.089	.198	.026	31.19	.058	13.889
1.0	25.4	.039	.064	15.50	16.39	.061	4.543	.220	.029	34.66	.065	15.432
2	50.8	.079	.129	31.00	32.77	.122	9.087	.440	.058	69.32	.130	30.865
3	76.2	.118	.193	46.50	49.16	.183	13.630	.660	.087	103.98	.194	46.297
4	101.6	.157	.258	62.00	65.54	.244	18.174	.880	.115	138.64	.259	61.729
5	127.0	.197	.323	77.50	81.93	.305	22.717	1.100	.144	173.29	.324	77.162
6	152.4	.236	.387	93.00	98.31	.366	27.261	1.321	.173	207.95	.389	92.594
7	177.8	.275	.452	108.50	114.70	.427	31.804	1.541	.202	242.61	.454	108.026
8	203.2	.315	.516	124.00	131.09	.488	36.348	1.761	.231	277.27	.518	123.459
9	228.6	.354	.581	139.50	147.47	.549	40.891	1.981	.260	311.93	.583	138.891
10	254.0	.394	.645	155.01	163.86	.610	45.434	2.201	.289	346.59	.648	154.323
20	508.0	.787	1.290	310.01	327.72	1.221	90.869	4.402	.577	693.18	1.296	308.647
30	762.0	1.181	1.935	465.02	491.58	1.831	136.304	6.603	.866	1039.77	1.944	462.970
40	1016.0	1.575	2.581	620.02	655.44	2.441	181.738	8.804	1.154	1386.36	2.592	617.204
50	1270.0	1.969	3.226	775.03	819.30	3.051	227.173	11.005	1.443	1732.95	3.240	771.617
60	1523.9	2.362	3.871	930.04	983.15	3.662	272.607	13.206	1.731	2079.5	3.888	925.941
70	1777.9	2.756	4.516	1085.04	1147.01	4.272	318.042	15.407	2.020	2426.1	4.536	1080.264
80	2031.9	3.150	5.161	1240.05	1310.87	4.882	363.477	17.608	2.308	2772.7	5.184	1234.588
90	2285.9	3.543	5.806	1395.05	1474.73	5.493	408.912	19.809	2.596	3119.3	5.832	1388.911
100	2539.9	3.937	6.451	1550.06	1638.59	6.103	454.346	22.010	2.885	3465.9	6.480	1543.235
200	5079.8	7.874	12.903	3100.12	3277.18	12.206	908.692	44.019	5.770	6931.8	12.960	3086.470
300	7619.8	11.811	19.354	4650.18	4915.78	18.308	1363.037	66.029	8.656	10397.5	19.440	4629.704
400	10159.7	15.748	25.806	6200.24	6554.37	24.411	1817.383	88.039	11.541	13863.6	25.920	6172.930
500	12699.7	19.685	32.257	7750.3	8192.96	30.514	2271.729	110.048	14.426	17329.5	32.399	7716.174
1000	25399.4	39.371	64.514	15500.6	16385.92	61.028	4543.458	220.097	28.852	34659.	64.799	15543.348

From the above table any ordinary conversions up to 2000 units may be readily made. For example: It is required to find the number of cubic centimeters equal to 1728 cubic inches.

$$1728 = 1000 + 700 + 20 + 8 \text{ cubic inches.}$$

But a reference to the sixth column shows that

1000 cubic inches	= 16,385.92	cubic centimeters.		
700 "	= 11,470.10	"	"	
20 "	= 327.72	"	"	
8 ..	= 131.09	"	"	
Add together, and 1728 "	= 28,314.83	"	"	

Table of Solubilities of Chemical Compounds Commonly Used in Electro-Technics

NAMES.	Soluble in 100 parts by weight of water at	
	50° F. Parts by weight.	212° F. Parts by weight.
Acid potassium carbonate (bicarbonate of potash........	23	45 at 158° F.
Aluminium chloride	400	very soluble.
Aluminium sulphate (calculated to anhydrous salt)..................................|	35	1130
Ammonium alum	9	422
Ammonium chloride......................	33	73
Ammonium sulphate	73.6	97.5
Antimony-potassium tartrate (tartar emetic)..	5.2	28 at 167° F.
Arsenious acid............................	4	9.5
Boric acid	2.7	29
Cadmium chloride, crystallized............	140	149
Cadmium sulphate........................	95	80
Chromic acid...........................	very soluble	very soluble.
Citric acid	133	very soluble.
Cobalt-ammonium sulphate (calculated to anhydrous salt)........................	11.6	43.3 at 167° F.
Cobalt sulphate (calculated to anhydrous salt)	30.5	63.7 at 158° F.
Copper acetate (verdigris) neutral..........	7.4	20
Copper chloride	soluble	very soluble.
Copper sulphate (blue vitriol) crystallized ...	37	203
Ferrous sulphate (green vitriol, copperas)....	61	333
Gold chloride...........................	soluble	soluble.
Gold cyanide	very soluble	very soluble.
Lead acetate (sugar of lead)...............	45.35	very soluble.
Lead nitrate............................	48	139
Magnesium sulphate (Epsom salt)..........	31.5	71.5
Mercuric chloride (corrosive sublimate)	6.57	54
Mercuric nitrate	decomposable	decomposable.
Mercuric sulphate.......................	decomposable	decomposable.

Table of Solubilities of Chemical Compounds Commonly Used in Electro-Technics.—Continued.

NAMES.	Soluble in 100 parts by weight of water at	
	50° F. Parts by weight.	212° F. Parts by weight.
Mercurous nitrate	slightly soluble	slightly soluble.
Mercurous sulphate	very slightly soluble	decomposable.
Nickel-ammonium sulphate (calculated to anhydrous salt)	3.2	28.6
Nickel chloride, crystallized	50 to 66	very soluble.
Nickel nitrate, crystallized	slightly soluble	slightly soluble.
Nickel sulphate (calculated to anhydrous salt)	37.4	62 at 158° F.
Platinic chloride	soluble	very soluble.
Platoso-ammonium chloride	c.65	1.25
Potassium-aluminium sulphate (potash alum), crystallized	9.8	357.5
Potassium bitartrate (cream of tartar)	0.4	6.9
Potassium carbonate (potash)	109	156
Potassium-copper cyanide	94	154
Potassium cyanide	soluble	decomposable.
Potassium dichromate	8.0	98
Potassium ferrocyanide (yellow prussiate of potash)	28	50
Potassium-gold cyanide	soluble	soluble.
Potassium nitrate (saltpetre)	21.1	247
Potassium permanganate	6.45	very soluble.
Potassium-silver cyanide	12.5	100
Potassium-sodium tartrate (Rochelle or Seignette salt)	58	very soluble.
Potassium sulphide (liver of sulphur)	very soluble	very soluble.
Potassium-zinc cyanide	42	78.5
Silver nitrate (lunar caustic)	122 at 32° F. 227 at 67° F.	714 at 185° F. 1111 at 230° F.
Sodium bisulphite	very soluble	very soluble.
Sodium carbonate, anhydrous (calcined soda)	12	45
Sodium carbonate (crystallized soda)	40	540 at 219.2° F.
Sodium chloride (common salt)	36	40.7 at 215.6° F.
Sodium dichromate	108.5	163
Sodium hydrate (caustic soda)	96.1	213
Sodium hyposulphite, sodium thiosulphate (anhydrous salt)	65	102 at 140° F.
Sodium phosphate	20	150
Sodium pyrophosphate	6.8	93
Sodium sulphate (Glauber's salt)	9	42.5
Sodium sulphite (neutral), crystallized	25	100
Stannic chloride	soluble	soluble.
Stannous chloride (tin salt)	271	decomposable.
Tartaric acid	125.7	343.3
Zinc chloride	300	very soluble.
Zinc sulphate (white vitriol), crystallized	138.2	653.6

Content of Metal in the Most Commonly Used Metallic Salts.

METALLIC COMBINATION.	FORMULA.	Content of metal in per cent.
Cobalt ammonium sulphate, crystallized...	$(NH_4)_2Co(SO_4)_2+6H_2O$	14.62
Cobalt chloride..	$CoCl_2+6H_2O$	24.68
Cobalt sulphate, crystallized............	$CoSO_4+7H_2O$	20.92
Copper acetate, crystallized (verdigris)...	$Cu(C_2H_3O_2)_2+H_2O$	31.87
Copper carbonate.....................	$2CuCO_3(CuOH)_2$	55.20
Copper chloride, crystallized............	$CuCl_2+2H_2O$	37.07
Copper cyanide......................	$Cu_3(CN)_4+5H_2O$	56.50
Copper oxide, black	CuO	79.83
Copper sulphate (blue vitriol), crystallized.	$CuSO_4+5H_2O$	25.40
Cuprous oxide.......................	Cu_2O	88.79
Ferrous sulphate (green vitriol), crystallized	$FeSO_4+7H_2O$	20.14
Gold chloride (brown), technical........	$AuCl_3+x$ aq	50 to 52
Gold chloride (orange), technical	$AuCl_3+x$ aq	48 to 49
Iron-ammonium sulphate, crystallized.....	$(NH_4)Fe(SO_4)_2+6H_2O$	14.62
Lead acetate (sugar of lead), crystallized.	$Pb(C_2H_3O_2)_2+3H_2O$	54.57
Lead nitrate, crystallized..............	$Pb(NO_3)_2$	62.51
Mercuric chloride	$HgCl_2$	73.87
Mercurous nitrate	$Hg_2(NO_3)_2$	79.36
Nickel-ammonium sulphate, crystallized...	$(NH_4)_2Ni(SO_4)_2+6H_2O$	14.94
Nickel carbonate, basic (separated at 212° F.)	$NiCO_34NiO, 5H_2O$	57.87
Nickel chloride, crystallized............	$NiCl_2+6H_2O$	24.63
Nickel chloride, anhydrous.............	$NiCl_2$	45.30
Nickel hydrate.....	$\left\{\begin{array}{l} Ni(OH)_2+H_2O \text{ (separated} \\ \text{at 212° F.)} \end{array}\right\}$	63.34
Nickel nitrate, crystallized............	$Ni(NO_3)_2+6H_2O$	18.97
Nickel oxide.........................	Ni_2O_3	71.00
Nickel sulphate, crystallized	$NiSO_4+7H_2O$	22.01
Platinic chloride.....................	$PtCl_4+5H_2O$	45.66
Platoso-ammonium chloride	$(NH_4)_2PtCl_6$	43.91
Potassium-copper cyanide, crystallized, technical.........................	$K_4Cu_2(CN)_6$	28.83
Potassium mercuric cyanide............	$K_2Hg(CN)_4$	53.56
Potassium-silver cyanide, crystallized....	$KAg(CN)_2$	54.20
Potassium-zinc cyanide, crystallized.....	$K_2Zn(CN)_4$	26.35
Silver chloride......................	$AgCl$	68.20
Silver cyanide......................	$AgCN$	80.57
Silver nitrate, crystallized.............	$AgNO_3$	64.98
Stannous chloride (tin-salt)	$SnCl_2+2H_2O$	52.45
Zinc-ammonium chloride...............	$NH_4ZnCl_3+2H_2O$	28.98
Zinc carbonate......................	$ZnCO_3Zn(OH)_2$	29.05
Zinc chloride	$ZnCl_2$	47.84
Zinc cyanide	$Zn(CN)_2$	56.59
Zinc sulphate (white vitriol), crystallized..	$ZnSO_4+7H_2O$	22.73

Table Showing the Electrical Resistance of Pure Copper Wire of Various Diameters.

No. of wire, Birmingham wire gauge.	Resistance of 1 foot in ohms.	Number of feet required to give resistance of 1 ohm.	No. of wire, Birmingham wire gauge.	Resistance of 1 foot in ohms.	Number of feet required to give resistance of 1 ohm.
0000	0.0000516	19358	17	0.00316	316.1
000	0.0000589	16964	18	0.00443	225.5
00	0.0000737	13562	19	0.00603	165.7
0	0.0000922	10857	20	0.00869	115.1
1	0.000118	8452.6	21	0.01040	96.2
2	0.000132	7575.1	22	0.01358	73.6
3	0.000159	6300.1	23	0.01703	58.7
4	0.000188	5319.9	24	0.02200	45.5
5	0.000220	4545.9	25	0.02661	37.6
6	0.000258	3870.3	26	0.03286	30.4
7	0.000329	3043.4	27	0.04159	24.0
8	0.000391	2557.1	28	0.05432	18.4
9	0.000486	2057.7	29	0.06300	15.9
10	0.000593	1686.5	30	0.07393	13.5
11	0.000739	1352.5	31	0.10646	9.4
12	0.000896	1116.0	32	0.13144	7.6
13	0.001180	847.7	33	0.16634	6.0
14	0.001546	647.0	34	0.21727	4.6
15	0.002053	487.0	35	0.42583	2.4
16	0.002520	396.8	36	0.66537	1.5

Resistance and Conductivity of Pure Copper at Different Temperatures.

Centigrade temperature.	Resistance.	Conductivity.	Centigrade temperature.	Resistance.	Conductivity.
0°	1.00000	1.00000	16°	1.06168	.94190
1	1.00381	.99624	17	1.06563	.93841
2	1.00756	.99250	18	1.06959	.93494
3	1.01135	.98878	19	1.07356	.93148
4	1.01515	.98508	20	1.07742	.92814
5	1.01896	.98139	21	1.08164	.92452
6	1.02280	.97771	22	1.08553	.92121
7	1.02663	.97406	23	1.08954	.91782
8	1.03048	.97042	24	1.09365	.91445
9	1.03435	.96679	25	1.09763	.91110
10	1.03822	.96319	26	1.10161	.90776
11	1.04199	.95970	27	1.10567	.90443
12	1.04599	.95603	28	1.11972	.90113
13	1.04990	.95247	29	1.11882	.89784
14	1.05406	.94893	30	1.11782	.89457
15	1.05774	.94541			

Table of Hydrometer Degrees according to Baumé, at 63.5° F.,
and their Weights by volume.

Degrees Bé.	Weight by volume.	Degrees Bé.	Weight by volume.	Degrees Bé.	Weight by volume.	Degrees Bé.	Weight by volume.
0	1.0600	19	1.1487	38	1.3494	57	1.6349
1	1.0068	20	1.1578	39	1.3619	58	1.6533
2	1.0138	21	1.1670	40	1.3746	59	1.6721
3	1.0208	22	1.1763	41	1.3876	60	1.6914
4	1.0280	23	1.1858	42	1.4009	61	1.7111
5	1.0353	24	1.1955	43	1.4143	62	1.7313
6	1.0426	25	1.2053	44	1.4281	63	1.7520
7	1.0501	26	1.2153	45	1.4421	64	1.7731
8	1.0576	27	1.2254	46	1.4564	65	1.7948
9	1.0653	28	1.2357	47	1.4710	66	1.8171
10	1.0731	29	1.2462	48	1.4860	67	1.8398
11	1.0810	30	1.2569	49	1.5012	68	1.8632
12	1.0890	31	1.2677	50	1.5167	69	1.8871
13	1.0972	32	1.2788	51	1.5325	70	1.9117
14	1.1054	33	1.2901	52	1.5487	71	1.9370
15	1.1138	34	1.3015	53	1.5652	72	1.9629
16	1.1224	35	1.3131	54	1.5820		
17	1.1310	36	1.3250	55	1.5993		
18	1.1398	37	1.3370	56	1.6169		

Table of Bare Copper Wire for Low Voltage.

B. & S. gauge.	Diameter.	Safe carrying capacity. Ampères.	Weight per 90 feet.	Resistance per 1,000 feet in Ohms.	B. & S. gauge.	Diameter.	Safe carrying capacity. Ampères.	Weight per 100 feet.	Resistance per 1,000 feet in Ohms.
0000	.460″	300	63 lbs.	.0487	4	.290″	100	13 lbs.	.2624
000	.400″	245	50 "	.0682	6	.160″	80	8 "	.4264
00	.360″	215	40 "	.0853	8	.130″	60	5 "	.5686
0	.320″	190	31 "	.1066	10	.100″	40	3 "	1.0662
1	.290″	160	25 "	.1218	12	.080″	30	2 "	1.706
2	.260″	135	20 "	.1550	14	.060″	22	1.2 "	2.843
3	.230″	115	16 "	.2007	16	.050″	15	.78 "	4.264

INDEX.

(703)

45

MULTIPOLAR TYPE DYNAMOS

W E manufacture and are prepared to furnish this type dynamo in sizes ranging from 50–10,000 amperes, either shunt or compound wound, or with fields wound for separate excitation. We can absolutely recommend this type machine as the most modern and complete plating dynamo on the market.

We can supply low voltage dynamos direct connected to a motor of suitable size, the whole outfit mounted on a substantial iron sub-base

Motors can be furnished in any voltage to suit conditions.

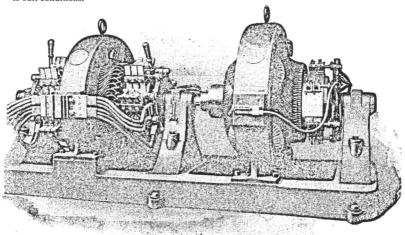

THE HANSON & VAN WINKLE COMPANY

U. S. A.

Main Office, No. 269 Oliver St. NEW YORK, CHICAGO, ILL.
NEWARK, N. J. No. 79 Walker St. No. 108 North Clinton St.

Canada Office, Canadian Hanson & Van Winkle Co., Limited
Morrow Ave., Toronto, Ont.

NICKEL ANODES

WITH the advances made in nickel plating in the past years the tendency has been to use larger containers, which naturally require anodes of increased dimensions. To meet this necessity various devices have been tried in the way of crowding a larger number of plates into the tank or using irregular shapes, sometimes with cumbersome attachments. After exhaustive experiments we have at last solved the question, and now offer to the trade our patented Elliptic anode.

For many years it has been customary to use flat nickel Anodes, only because there was nothing else obtainable, and these flat plates are still in general use in many large establishments, which have not taken time to investigate the decided advantage and economy in our late developments.

The patented Elliptic Anodes possess many points of real merit; they are the result of careful experiments covering several years, and they overcome the disadvantages in all other shapes.

Elliptic Anodes are 2½ inches wide by 1½ inches thick, and are cast in any ordinary length.

Experience shows that all Anodes work more from the edges than from the centre, showing conclusively that circulation around the Anode is necessary to get the greatest amount of corrosion or disintegration.

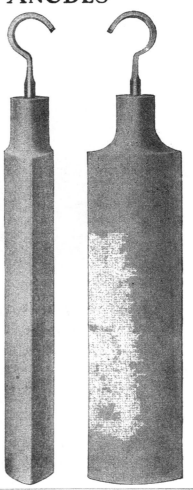

THE HANSON & VAN WINKLE CO.

U. S. A.

Newark, N. J.	New York	Chicago, Ill.
269 Oliver Street	79 Walker Street	110 North Clinton St.

Canada Office: Canadian Hanson & Van Winkle Co.
Morrow Ave., Toronto, Ont.

Aluminum Dipping Baskets.

We illustrate herewith our patented perforated sheet aluminum dipping baskets, also aluminum wire baskets which we are prepared to furnish in any style or shape.

Aluminum dipping baskets are practically acid proof *but must not be used in potash, muriatic or hydrofluoric acid.* They are particularly adap- ted for use in washing and dipping. After a thorough test we do not hesitate to recommend them. They are very light, very durable and will outlast the ordinary dipping baskets.

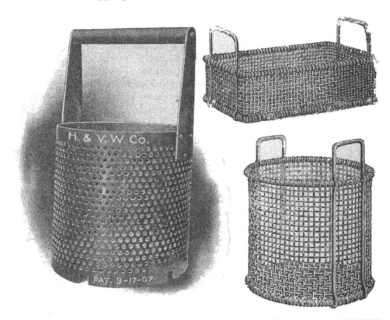

THE HANSON & VAN WINKLE CO.

U. S. A.

Newark, N. J.	New York	Chicago, Ill.
269 Oliver Street	79 Walker Street	110 North Clinton St.

Canada Office: Canadian Hanson & Van Winkle Co.
Morrow Ave., Toronto, Ont.

CATALOGUE

OF

Practical and Scientific Books

PUBLISHED BY

Henry Carey Baird & Co.

INDUSTRIAL PUBLISHERS, BOOKSELLERS AND IMPORTERS

810 Walnut Street, Philadelphia.

☞ Any of the Books comprised in this Catalogue will be sent by mail, free of postage, to any address in the world, at the publication prices.

☞ A Descriptive Catalogue, 94 pages, 8vo, will be sent free and free of postage, to any one in any part of the world, who will furnish his address.

☞ Where not otherwise stated, all of the Books in this Catalogue are bound in muslin.

AMATEUR MECHANICS' WORKSHOP:
A treatise containing plain and concise direction for the manipulation of Wood and Metals, including Casting, Forging, Brazing, Soldering and Carpentry. By the author of the "Lathe and Its Uses." Ninth edition. Illustrated. 8vo...$1.50

ARLOT.—A Complete Guide for Coach Painters:
Translated from the French of M. ARLOT, Coach Painter, for eleven years Foreman of Painting to M. Eherler, Coach Maker, Paris By A. A. FESQUET, Chemist and Engineer. To which is added an Appendix, containing Information respecting the Materials and the Practice of Coach and Car Painting and Varnishing in the United States and Great Britain 12mo..................................$1.25

ARMENGAUD, AMOROUX, AND JOHNSON.—**The Practical Draughtsman's Book of Industrial Design, and Machinist's and Engineer's Drawing Companion:**
Forming a Complete Course of Mechanical Engineering and Architectural Drawing. From the French of M. Armengaud the elder, Prof. of Design in the Conservatoire of Arts and Industry, Paris, and M. Armengaud the younger, and Amoroux, Civil Engineers. Rewritten and arranged with additional matter and plates, selections from and examples of the most useful and generally employed mechanism of the day. By WILLIAM JOHNSON, Assoc. Inst. C. E. Illustrated by fifty folio steel plates, and fifty wood-cuts. A new edition, 4to., cloth.................................$5.00

ARROWSMITH.—**The Paper-Hanger's Companion**
Comprising Tools, Pastes, Preparatory Work; Selection and Hanging of Wall-Papers; Distemper Painting and Cornice-Tinting; Stencil Work; Replacing Sash-Cord and Broken Window Panes; and Useful Wrinkles and Receipts. By JAMES ARROWSMITH. A New, Thoroughly Revised, and Much Enlarged Edition. Illustrated by 25 engravings, 162 pages. (1905).................................$1.00

ASHTON.—**The Theory and Practice of the Art of Designing Fancy Cotton and Woolen Cloths from Sample:**
Giving full instructions for reducing drafts, as well as the methods of spooling and making out harness for cross drafts and finding any required reed; with calculations and tables of yarn. By FREDERIC T. ASHTON, Designer, West Pittsfield, Mass. With fifty-two illustrations. One vol. folio.....$4.00

ASKINSON.—**Perfumes and Cosmetics:**
A Comprehensive Treatise on Perfumery, containing Complete Directions for Making Handkerchief Perfumes, Smelling-Salts, Sachets, Fumigating Pastils; Preparations for the Care of the Skin, the Mouth, the Hair; Cosmetics, Hair Dyes, and other Toilet Articles. By G. W. ASKINSON. Translated from the German. Revised by W. L. DUDLEY. 32 illustrations. 8vo.....................................$5.00

BAIRD.—**The American Cotton Spinner, and Manager's and Carder's Guide:**
A Practical Treatise on Cotton Spinning; giving the Dimensions and Speed of Machinery, Draught and Twist Calculations, etc; with notices of recent Improvements; together with Rules and Examples for making changes in the size and numbers of Roving and Yarn. Compiled from the papers of the late ROBERT H. BAIRD. 256 pp., 12mo..........$1.50

BEANS.—A Treatise on Railway Curves and Location of Railroads:

By E. W. BEANS, C. E. Illustrated.   12mo.   Morocco $1.00

BELL.—Carpentry Made Easy:

Or, The Science and Art of Framing on a New and Improved System. With Specific Instructions for Building Balloon Frames, Barn Frames, Mill Frames, Warehouses, Church Spires, etc. Comprising also a System of Bridge Building, with Bills, Estimates of Cost, and valuable Tables. Illustrated by forty-four plates, comprising nearly 200 figures. By WILLIAM E. BELL, Architect and Practical Builder. 8vo...$5.00

BERSCH.—Cellulose, Cellulose Products, and Rubber Substitutes:

Comprising the Preparation of Cellulose, Parchment-Cellulose, Methods of Obtaining Sugar, Alcohol, and Oxalic Acid from Wood-Cellulose; Production of Nitro-Cellulose and Cellulose Esters; Manufacture of Artificial Silk, Viscose, Celluloid, Rubber Substitutes, Oil-Rubber, and Faktis. By Dr. JOSEPH BERSCH. Translated by WILLIAM T. BRANNT. 41 Illustrations. (1904)....................................$3.00

BILLINGS.—Tobacco:

Its History, Variety, Culture, Manufacture, Commerce, and Various Modes of Use. By E. R. BILLINGS. Illustrated by nearly 200 engravings. 8vo......................$3.00

BIRD.—The American Practical Dyers' Companion:

Comprising a Description of the Principal Dye-Stuffs and Chemicals used in Dyeing, their Nature and Uses; Mordants and How Made; with the best American, English, French and German processes for Bleaching and Dyeing Silk, Wool, Cotton, Linen. Flannel, Felt, Dress Goods, Mixed and Hosiery Yarns, Feathers. Grass, Felt, Fur, Wool, and Straw Hats, Jute Yarn, Vegetable Ivory. Mats, Skins, Furs, Leather, etc., etc., by Wood, Aniline, and other Processes, together with Remarks on Finishing Agents, and Instructions in the Finishing of Fabrics; Substitutes for Indigo, Water-Proofing of Materials, Tests and Purification of Water. Manufacture of Aniline and other New Dye Wares, Harmonizing Colors, etc., etc.,; embracing in all over 800 Receipts for Colors and Shades, accompanied by 170 Dyed Samples of Raw Materials and Fabrics. By F. J. BIRD, Practical Dyer, Author of "The Dyers' Hand-Book." 8vo...........$4.00

BLINN.—A 'Practical Workshop Companion for Tin, Sheet-Iron, and Copper-plate Workers:

Containing Rules for describing various kinds of Patterns

used by Tin, Sheet-Iron and Copper-plate Workers; Practical Geometry; Mensuration of Surface and Solids; Tables of the Weights of Metals, Lead-pipe, etc.; Tables of Areas and Circumferences of Circles; Japan, Varnishes, Lacquers, Cements, Compositions, etc., etc. By LEROY J. BLINN, Master Mechanic. With One Hundred and Seventy Illustrations. 12mo..$2.50

BOOTH.—Marble Worker's Manual:

Containing Practical Information respecting Marbles in general, their Cutting, Working and Polishing; Veneering of Marble; Mosaics; Composition and Use of Artificial Marble, Stuccos, Cements, Receipts, Secrets, etc., etc. Translated from the French by M. L. BOOTH. With an Appendix concerning American Marbles. 12mo., cloth $1.50

BRANNT.—A Practical Treatise on Animal and Vegetable Fats and Oils:

Comprising both Fixed and Volatile Oils, their Physical and Chemical Properties and Uses, the Manner of Extracting and Refining them, and Practical Rules for Testing them; as well as the Manufacture of Artificial Butter and Lubricants, etc., with lists of American Patents relating to the Extraction, Rendering, Refining, Decomposing and Bleaching of Fats and Oils. By WILLIAM T. BRANNT, Editor of the "Techno-Chemical Receipt Book." Second Edition, Revised and in great part Rewritten. Illustrated by 302 Engravings. In Two Volumes. 1304 pp. 8vo..................$10.00

BRANNT.—A Practical Treatise on Distillation and Rectification of Alcohol:

Comprising Raw Materials; Production of Malt, Preparation of Mashes and of Yeast; Fermentation; Distillation and Rectification and Purification of Alcohol; Preparation of Alcoholic Liquors, Liqueurs, Cordials, Bitters, Fruit Essences, Vinegar, etc.; Examination of Materials for the Preparation of Malt as well as of the Malt itself; Examination of Mashes before and after Fermentation; Alcoholometry, with Numerous Comprehensive Tables; and an Appendix on the Manufacture of Compressed Yeast and the Examination of Alcohol and Alcoholic Liquors for Fusel Oil and other Impurities. By WILLIAM T. BRANNT, Editor of "The Techno-Chemical Receipt Book." Second Edition. Entirely Rewritten. Illustrated by 105 engravings. 460 pages. 8vo. (Dec., 1903).......................................$10.00

BRANNT.—India Rubber, Gutta-Percha and Balata:

Occurrence, Geographical Distribution, and Cultivation, Obtaining and Preparing the Raw Materials, Modes of Working and Utilizing them, including Washing, Maceration, Mixing, Vulcanizing, Rubber and Gutta-Percha Compounds, Utilization of Waste, etc. By WILLIAM T. BRANNT. Illustrated. 12mo. A new edition in preparation.

BRANNT.—A Practical Treatise on the Manufacture of Vinegar and Acetates, Cider, and Fruit-Wines:

Preservation of Fruits and Vegetables by Canning and Evaporation; Preparation of Fruit-Butters, Jellies, Marmalades, Catchups, Pickles, Mustards, etc. Edited from various sources. By WILLIAM T. BRANNT. Illustrated by 101 Engravings. 575 pp. 8vo; 3d edition..........Net, $6.00

BRANNT.—The Metallic Alloys: A Practical Guide:

For the Manufacture of all kinds of Alloys, Amalgams, and Solders, used by Metal Workers: together with their Chemical and Physical Properties and their Application in the Arts and the Industries; with an Appendix on the Coloring of Alloys and the Recovery of Waste Metals. By WILLIAM T. BRANNT. 45 Engravings. Third, Revised, and Enlarged Edition. 570 pages. 8vo.....................Net, $5.00

BRANNT.—The Metal Worker's Handy-Book of Receipts and Processes:

Being a Collection of Chemical Formulas and Practical Manipulations for the working of all Metals; including the decoration and Beautifying of Articles Manufactured therefrom, as well as their Preservation. Edited from various sources. By WILLIAM T. BRANNT. Illustrated. 12mo.$2.50

BRANNT.—Petroleum:

Its History, Origin, Occurrence, Production, Physical and Chemical Constitution, Technology, Examination and Uses; Together with the Occurrence and Uses of Natural Gas. Edited chiefly from the German of Prof. Hans Hoefer and Dr. Alexander Veith by Wm. T. BRANNT. Illustrated by 3 Plates and 284 Engravings. 743 pp. 8vo...........$12.50

BRANNT.—The Practical Dry Cleaner, Scourer and Garment Dyer:

Comprising Dry, Chemical, or French Cleaning; Purification of Benzine; Removal of Stains, or Spotting; Wet Cleaning; Finishing Cleaned Fabrics; Cleaning and Dyeing Furs, Skin Rugs and Mats; Cleaning and Dyeing Feathers; Cleaning and Renovating Felt, Straw and Panama Hats; Bleaching and Dyeing Straw and Straw Hats; Cleaning and Dyeing Gloves; Garment Dyeing; Stripping; Analysis of Textile Fabrics. Edited by WILLIAM T. BRANNT, Editor of "The Techno-Chemical Receipt Book." Fourth Edition, Revised and Enlarged. Illustrated by Forty-One Engravings. 12 mo. 371 pp.....................................$2.50

CONTENTS: I. Dry Chemical or French Cleaning. II. Removal of Stains or Spotting. III. Wet Washing. IV. Finishing Cleaned Fabrics. V. Cleaning and Dyeing Furs, Skin Rugs and Mats. VI. Cleaning and Dyeing Feathers. VII. Cleaning and Renovating Felt, Straw and Panama Hats; Bleaching and Dyeing Straw and Straw Hats. VIII. Cleaning and Dyeing Gloves. IX. Garment Dyeing. X. Stripping Colors from Garments and Fabrics. XI. Analysis of Textile Fabrics. Index.

BRANNT.—The Soap Maker's Hand-Book of Materials, Processes and Receipts for every description of Soap; including Fats, Fat Oils and Fatty Acids; Examination of Fats and Oils; Alkalies; Testing Soda and Potash; Machines and Utensils; Hard Soaps; Soft Soaps; Textile Soaps; Washing Powders and Allied Products; Toilet Soaps, Medicated Soaps, and Soap Specialties; Essential Oils and other Perfuming Materials; Testing Soaps. Edited chiefly from the German of DR. C. DEITE, A. ENGELHARDT, F. WILTNER, and numerous other Experts. With Additions by WILLIAM T. BRANNT, Editor of "The Techno-Chemical Receipt Book." Illustrated by Fifty-four Engravings. Second edition, Revised and in great part Re-Written. 535 pp. 8vo....$6.00

BRANNT.—Varnishes, Lacquers, Printing Inks and Sealing Waxes:

Their Raw Materials and their Manufacture, to which is added the Art of Varnishing and Lacquering, including the Preparation of Putties and of Stains for Wood, Ivory, Bone, Horn, and Leather. By WILLIAM T. BRANNT. Illustrated by 39 Engravings, 338 pages. 12mo.......... ..$3.00

BRANNT-WAHL.—The Techno-Chemical Receipt Book:

Containing several thousand Receipts covering the latest, most important, and most useful discoveries in Chemical Technology, and their Practical Application in the Arts and the Industries. Edited chiefly from the German of Drs. Winckler, Elsner, Heintze, Mierzinski, Jacobsen, Koller and Heinzerling, with additions by WM. T. BRANNT and WM. H. WAHL, Ph. D. Illustrated by 78 engravings. 12mo. 495 pages...$2.00

BROWN.—Five Hundred and Seven Mechanical Movements:

Embracing all those which are most important in Dynamics, Hydraulics, Hydrostatics, Pneumatics, Steam Engines, Mill and other Gearing, Presses, Horology, and Miscellaneous Machinery; and including many movements never before published, and several of which have only recently come into use. By HENRY T. BROWN......................$1.00

BULLOCK.—The Rudiments of Architecture and Building:

For the use of Architects, Builders, Draughtsmen, Machinists, Engineers and Mechanics. Edited by JOHN BULLOCK, author of "The American Cottage Builder." Illustrated by 250 Engravings. 8vo.........................$2.50

BYRNE.—Hand-Book for the Artisan, Mechanic, and Engineer:

Comprising the Grinding and Sharpening of Cutting Tools, Abrasive Processes, Lapidary Work, Gem and Glass Engraving, Varnishing and Lacquering, Apparatus, Materials and Processes for Grinding and Polishing, etc. By OLIVER BYRNE. Illustrated by 185 wood engravings. 8vo....$4.00

BYRNE.—Pocket-Book for Railroad and Civil Engineers:

Containing New, Exact and Concise Methods for Laying out Railroad Curves, Switches, Frog Angles and Crossings; the Staking out of work; Levelling; the Calculation of Cuttings; Embankments; Earthwork, etc. By OLIVER BYRNE. 18mo., full bound, pocketbook form...:......................$1.50

BYRNE.—The Practical Metal-Worker's Assistant:

Comprising Metallurgic Chemistry; the Arts of Working all Metals and Alloys; Forging of Iron and Steel; Hardening and Tempering; Melting and Mixing; Casting and Founding; Works in Sheet Metals; the Process Dependent on the Ductility of the Metals; Soldering; etc. By JOHN PERCY. The Manufacture of Malleable Iron Castings, and Improvements in Bessemer Steel. By A. A. FESQUET, Chemist and Engineer. With over Six Hundred ·Engravings, Illustrating every Branch of the Subject. 8vo..................$3.50

CABINET MAKER'S ALBUM OF FURNITURE:

Comprising a Collection of Designs for various Styles of Furniture. Illustrated by Forty-eight Large and Beautifully Engraved Plates. Oblong, 8vo....................

CALLINGHAM.—Sign Writing and Glass Embossing:

A complete Practical Illustrated Manual of the Art. By JAMES CALLINGHAM. To which are added Numerous Alphabets and the Art of Letter Painting Made Easy. By JAMES C. BADENOCH. 258 pages. 12mo........ $1.50

CAREY.—A Memoir of Henry C. Carey:

By DR. WM. ELDER. With a portrait. 8vo., cloth...... 75

CAREY.—The Works of Henry C. Carey:

Manual of Social Science. Condensed from Carey's "Principles of Social Science." By KATE McKEAN 1 vol. 12mo...$2.00

Miscellaneous Works. With a Portrait. 2 vols. 8vo. $10.00
Past, Present and Future. 8vo..................$2.50
Principles of Social Science. 3 volumes, 8vo......$10.00
The Slave-Trade, Domestic and Foreign; Why it Exists, and How it may be Extinguished (1853). 8vo.......$2.00
The Unity of Law: As Exhibited in the Relations of Physical, Social, Mental and Moral Science (1872). 8vo....$2.50

COOLEY.—A Complete Practical Treatise on Perfumery:
Being a Hand-book of Perfumes, Cosmetics and other Toilet Articles, with a Comprehensive Collection of Formulæ. By ARNOLD COOLEY. 12mo.........................$1.00

COURTNEY.—The Boiler Maker's Assistant in Drawing, Templating, and Calculating Boiler Work and Tank Work, etc.
Revised by D. K. CLARK. 102 ills. Fifth edition.......80

COURTNEY.—The Boiler Maker's Ready Reckoner:
With Examples of Practical Geometry and Templating. Revised by D. K. CLARK, C. E. 37 illustrations. Fifth edition..$1.60

CRISTIANI.—A Technical Treatise on Soap and Candles:
With a Glance at the Industry of Fats and Oils. By R. S. Cristiani, Chemist. Author of "Perfumery and Kindred Arts." Illustrated by 176 Engravings. 581 pages, 8vo
$15.00

CROSS.—The Cotton Yarn Spinner:
Showing how the Preparation should be arranged for Different Counts of Yarns by a System more uniform than has hitherto been practiced; by having a Standard Schedule from which we make all our Changes. By RICHARD CROSS. 122 pp. 12mo...75

DAVIDSON.—A Practical Manual of House Painting, Graining, Marbling, and Sign-Writing:
Containing full information on the processes of House Painting in Oil and Distemper, the Formation of Letters and Practice of Sign-Writing, the Principles of Decorative Art, a Course of Elementary Drawing for House Painters, Writers, etc., and a Collection of Useful Receipts. With nine colored illustrations of Woods and Marbles, and numerous wood engravings. By ELLIS A. DAVIDSON. 12mo...........$2.00

DAVIES.—A Treatise on Earthy and Other Minerals and Mining:
By D. C. DAVIES. F. G. S., Mining Engineer, etc. Illustrated by 76 Engravings. 12mo.........................$5.00

DAVIES.—A Treatise on Metalliferous Minerals and Mining:

By D. C. DAVIES, F. G. S., Mining Engineer, Examiner of Mines, Quarries and Collieries. Illustrated by 148 engravings of Geological Formations, Mining Operations and Machinery, drawn from the practice of all parts of the world. Fifth Edition, thoroughly Revised and much Enlarged by his son, E. Henry Davies. 12mo. 524 pages.. ..$5.00

DAVIS.—A Practical Treatise on the Manufacture of Brick, Tiles and Terra-Gotta:

Including Stiff Clay, Dry Clay, Hand Made, Pressed or Front, and Roadway Paving Brick, Enamelled Brick, with Glazes and Colors, Fire Brick and Blocks, Silica Brick, Carbon Brick, Glass Pots, Retorts, Architectural Terra-Gotta, Sewer Pipe, Drain Tile, Glazed and Unglazed Roofing Tile, Art Tile, Mosaics, and Imitation of Intrarsia or Inlaid Surfaces. Comprising every product of Clay employed in Architecture, Engineering, and the Blast Furnace. With a Detailed Description of the Different Clays employed, the Most Modern Machinery, Tools, and Kilns used, and the Processes for Handling Disintegrating, Tempering, and Moulding the Clay into Shape, Drying, Setting, and Burning. By CHARLES THOMAS DAVIS. Third Edition. Revised and in great part rewritten. Illustrated by 261 engravings. 662 pages.. (Scarce.)

DAVIS.—The Manufacture of Paper:

Being a Description of the various Processes for the Fabrication, Coloring and Finishing of every kind of Paper, Including the Different Raw Materials and the Methods for Determining their Values, the Tools, Machines and Practical Details connected with an intelligent and a profitable prosecution of the art, with special reference to the best American Practice. To which are added a History of Paper, complete Lists of Paper-Making Materials, List of American Machines, Tools and Processes used in treating the Raw Materials, and in Making, Coloring and Finishing Paper. By CHARLES T. DAVIS. Illustrated by 156 Engravings. 608 pages. 8vo.$6.00

DAWIDOWSKY-BRANNT.—A Practical Treatise on the Raw Materials and Fabrication of Glue, Gelatine, Gelatine Veneers and Foils, Isinglass, Cements, Pastes, Mucilages, etc.:

Based upon Actual Experience. By F. DAWIDOWSKY, Technical Chemist. Translated from the German, with extensive additions, including a description of the most Recent American Processes, by WILLIAM T. BRANNT. 2d revised edition, 350 pages. (1905) Price........................$3.00

DEITE.—A Practical Treatise on the Manufacture of Perfumery:

Comprising directions for making all kinds of Perfumes, Sachet Powders, Fumigating Materials, Dentrifices, Cosmetics, etc., with a full account of the Volatile Oils, Balsams, Resins, and other Natural and Artificial Perfume-substances, including the Manufacture of Fruit Ethers, and tests of their purity. By DR. C. DEITE, assisted by L. BORCHERT, F. EICHBAUM, E. KUGLER, H. TOEFFNER, and other experts. From the German, by WM. T. BRANNT. 28 Engravings. 358 pages. 8vo................................$3.00

DE KONINCK-DIETZ.—A Practical Manual of Chemical Analysis and Assaying:

As applied to the Manufacture of Iron from its Ores, and to Cast Iron, Wrought Iron, and Steel, as found in Commerce. By L. L. DEKONINCK, Dr. Sc., and E. DIETZ, Engineer. Edited with Notes, by ROBERT MALLET, F. R. S., F. S. G., M. I. C. E., etc. American Edition, Edited with Notes and an Appendix on Iron Ores, by A. A. FESQUET, Chemist and Engineer. 12mo................   ....$1.00

DIETERICHS.—A Treatise on Friction, Lubrication, Oils and Fats:

The Manufacture of Lubricating Oils, Paint Oils, and of Grease, and the Testing of Oils. By E. F. DIETERICHS, Member of the Franklin Institute; Member National Association of Stationary Engineers; Inventor of Dietrichs' Valve-Oleum Lubricating Oils. Second Edition Revised. 12mo. A practical book by a practical man..   ....$1.25

DUNCAN.—Practical Surveyor's Guide:

Containing the necessary information to make any person of common capacity, a finished land surveyor, without the aid of a teacher. By ANDREW DUNCAN. Revised. 72 Engravings. 214 pp. 12mo............................$1.50 ·

DUPLAIS.—A Treatise on the Manufacture and Distillation of Alcoholic Liquors:

Comprising Accurate and Complete Details in Regard to Alcohol from Wine, Molasses, Beets, Grain, Rice, Potatoes, Sorghum, Asphodel, Fruits, etc.; with the Distillation and Rectification of Brandy, Whiskey, Rum, Gin, Swiss Absinthe, etc., the Preparation of Aromatic Waters, Volatile Oils or Essences, Sugars, Syrups, Aromatic Tinctures, Liqueurs, Cordial Wines, Effervescing Wines, etc., the Ageing of Brandy and the Improvement of Spirits, with Copious Directions and Tables for Testing and Reducing Spirituous Liquors, etc.,

etc. Translated and Edited from the French of MM. Du-
PLAIS. By M. McKENNIE, M. D. Illustrated. 743 pp.
8vo..$15.00

EDWARDS.—A Catechism of the Marine Steam-Engine:
For the use of Engineers, Firemen, and Mechanics. A Practical Work for Practical Men. By EMORY EDWARDS, Mechanical Engineer. Illustrated by sixty-three Engravings,
including examples of the most modern Engines. Third
edition, thoroughly revised, with much additional matter.
12mo. 414 pages.................................$1.50

**EDWARDS.—American Marine Engineer, Theoretical
and Practical:**
With Examples of the latest and most approved American
Practice. By EMORY EDWARDS. 85 Illustrations. 12mo. $1.50

EDWARDS.—Modern American Locomotive Engines:
Their Design, Construction and Management. By EMORY
EDWARDS. Illustrated. 12mo.....................$1.50

**EDWARDS.—Modern American Marine Engines, Boilers,
and Screw Propellers:**
Their Design and Construction. 146 pp. 4to.......$2.00

EDWARDS.—900 Examination Questions and Answers:
For Engineers and Firemen (Land and Marine) who desire
to obtain a United States Government or State License.
Pocket-book form, gilt edge............... ...$1.50

EDWARDS.—The American Steam Engineer:
Theoretical and Practical, with examples of the latest and
most approved American practice in the design and construction of Steam Engines and Boilers. For the use of
Engineers, machinists, boiler-makers, and engineering students. By EMORY EDWARDS. Fully illustrated. 419 pages.
12mo...$1.50

EDWARDS.—The Practical Steam Engineer's Guide:
In the Design, Construction, and Management of American
Stationary, Portable, and Steam Fire-Engines, Steam Pumps,
Boilers, Injectors, Governors, Indicators, Pistons and Rings,
Safety Valves and Steam Gauges. For the use of Engineers,
Firemen, and Steam Users. By EMORY EDWARDS. Illus-
trated by 119 engravings. 420 pages. 12mo.......$2.00

**ELDER.—Conversations on the Principal Subjects of
Political Economy:**
By DR. WILLIAM ELDER. 8vo.....................$1.50

ELDER.—Questions of the Day:

Economic and Social.   By Dr. William Elder.   8vo..$3.00

ERNI AND BROWN.—Mineralogy Simplified:

Easy Methods of Identifying Minerals, including Ores, by Means of the Blow-pipe, by Flame Reactions, by Humid Chemical Analysis, and by Physical Tests.   By Henri Erni, A. M., M. D.   Fourth Edition, revised, re-arranged and with the addition of entirely new matter, including Tables for the Determination of Minerals by Chemicals and Pyrognostic Characters, and by Physical Characters.   By Amos P. Brown, A. M., Ph. D.   464 pp.   Illustrated by 123 Engravings, pocket-book form, full flexible morocco, gilt edges.
$2.50

FAIRBAIRN.—The Principles of Mechanism and Machinery of Transmission:

Comprising the Principles of Mechanism, Wheels, and Pulleys, Strength and Proportion of Shafts, Coupling of Shafts, and Engaging and Disengaging Gear.   By Sir William Fairbairn, Bart., C E.   Beautifully illustrated by over 150 wood-cuts.   In one volume.   12mo.....   ...$2.00

FLEMING.—Narrow Gauge Railways in America:

A Sketch of their Rise, Progress, and Success.   Valuable Statistics as to Grades, Curves, Weight of Rail, Locomotives, Cars, etc.   By Howard Fleming.   Illustrated.   8vo..$1.00

FLEMMING.—Practical Tanning:

A Handbook of Modern Processes, Receipts, and Suggestions for the Treatment of Hides, Skins, and Pelts of Every Description.   By Lewis A. Flemming, American Tanner.   Third Edition Revised and in great part rewritten, over 600 pp.   8vo   1916.................................$6.00

FORSYTH.—Book of Designs for Headstones, Mural, and other Monuments:

Containing 78 Designs.   By James Forsyth, With an Introduction by Charles Boutell, M. A.   4to.   Cloth..$3.00

GARDNER.—Everybody's Paint Book:

A Complete Guide to the Art of Outdoor and Indoor Painting   38 Illustrations.   12mo.   183 pp...   ....$1.00

GARDNER.—The Painter's Encyclopedia:

Containing Definitions of all Important Words in the Art of Plain and Artistic Painting, with Details of Practice in Coach, Carriage, Railway Car, House, Sign, and Ornamental Painting, including Graining, Marbling, Staining, Varnishing, Polishing, Lettering, Stenciling, Gilding, Bronzing, etc.   By Franklin B. Gardner   158 illustrations.   12mo.   427 pp
$2.00

GEE.—The Goldsmith's Handbook:

Containing full instructions for the Alloying and Working of Gold, including the Art of Alloying, Melting, Reducing, Coloring, Collecting, and Refining; the Processes of Manipulation, Recovery of Waste; Chemical and Physical Properties of Gold; with a New System of Mixing its Alloys; Solders, Enamels; and other Useful Rules and Recipes. By GEORGE E. GEE. 12mo......................................$1.25

GEE.—The Jeweler's Assistant in the Art of Working in Gold:

A Practical Treatise for Masters and Workmen. 12mo $3.00

GEE.—The Silversmith's Handbook:

Containing full instructions for the Alloying and Working of Silver, including the different modes of Refining and Melting the Metal; its Solders; the Preparation of Imitation Alloys; Methods of Manipulation; Prevention of Waste; Instructions for Improving and Finishing the Surface of the Work; together with other Useful Information and Memoranda. By GEORGE E. GEE. Illustrated. 12mo......................$1.25

GOTHIC ALBUM FOR CABINET-MAKERS:

Designs for Gothic Furniture   Twenty-three plates. Oblong..$1.00

GRANT.—A Handbook on the Teeth of Gears:

Their Curves, Properties, and Practical Construction   By GEORGE B. GRANT   Illustrated. Third Edition, enlarged. 8vo...$1.00

GREGORY.—Mathematics for Practical Men:

Adapted to the Pursuits of Surveyors, Architects, Mechanics, and Civil Engineers. By OLINTHUS GREGORY. 8vo., plates..$3.00

GRISWOLD.—Railroad Engineer's Pocket Companion for the Field:

Comprising Rules for Calculating Deflection Distances and Angles, Tangential Distances and Angles and all Necessary Tables for Engineers; also the Art of Levelling from Preliminary Survey to the Construction of Railroads, intended Expressly for the Young Engineer, together with Numerous Valuable Rules and Examples. By W. GRISWOLD   12mo Pocketbook form............   ....$1.50

GRUNER.—Studies of Blast Furnace Phenomena:

By M. L. GRUNER, President of the General Council of Mines of France, and lately Professor of Metallurgy at the Ecole des Mines. Translated, with the author's sanction, with an Appendix, by L. D. B. GORDON, F R. S. E., F. G. S. 8vo. $2.50

Hand-Book of Useful Tables for the Lumberman, Farmer and Mechanic:

Containing Accurate Tables of Logs Reduced to Inch Board Measure, Plank, Scantling and Timber Measure; Wages and Rent, by Week or Month; Capacity of Granaries, Bins and Cisterns; Land Measure, Interest Tables with Directions for finding the Interest on any sum at 4, 5, 6, 7 and 8 per cent., and many other Useful Tables. 32mo., boards. 186 pages...25

HASERICK.—The Secrets of the Art of Dyeing Wool, Cotton and Linen:

Including Bleaching and Coloring Wool and Cotton Hosiery and Random Yarns. A Treatise based on Economy and Practice   By E. C. HASERICK.   Illustrated by 323 Dyed Patterns of the Yarns or Fabrics   8vo..............$4.50

HATS AND FELTING:

A Practical Treatise on their Manufacture.   By a Practical Hatter.   Illustrated by Drawings of Machinery, etc.   8vo.
$1.00

HAUPT.—A Manual of Engineering Specifications and Contracts:

By LEWIS M. HAUPT, C. E.   Illustrated with numerous maps.   328 pp.   8vo.............................$2.00

HAUPT.—The Topographer, His Instruments and Methods:

By LEWIS M. HAUPT, A. M., C. E.   Illustrated with numerous plates, maps and engravings.   247 pp.   8vo......$2.00

HAUPT.—Street Railway Motors:

With Descriptions and Cost of Plants and Operation of the various systems now in use.   12mo...................$1.50

HULME.—Worked Examination Questions in Plane Geometrical Drawing:

For the Use of Candidates for the Royal Military Academy, Woolwich; the Royal Military College, Sandhurst; the Indian Civil Engineering College, Cooper's Hill; Indian Public Works and Telegraph Department; Royal Marine Light Infantry; the Oxford and Cambridge Local Examinations, etc. By F. EDWARD HULME, F. L. S., F. S. A., Art-Master Marlborough College.   Illustrated by 300 examples.   Small quarto...$1.00

KELLEY.—Speeches, Addresses, and Letters on Industria. and Financial Questions:

By HON. WILLIAM D. KELLEY, M. C.   544 pages.   8vo $.200

KEMLO.—Watch Repairer's Hand-Book:
Being a Complete Guide to the Young Beginner, in Taking Apart, Putting Together, and Thoroughly Cleaning the English Lever and other Foreign Watches, and all American Watches. By F. KEMLO, Practical Watchmaker. With Illustrations. 12mo....................................$1.25

KICK.—Flour Manufacturer:
A Treatise on Milling Science and Practice By FREDERICK KICK, Imperial Regierungsrath, Professor of. Mechanical Technology in the Imperial German Polytechnic Institute, Prague. Translated from the second enlarged and revised edition with supplement by H. H. P. POWLES, Assoc. Memb. Institution of Civil Engineers. Illustrated with 28 Plates, and 167 Wood-cuts. 367 pages. 8vo...... ..$7.50

KINGZETT.—The History, Products, and Processes of the Alkali Trade:
Including the most Recent Improvements. By CHARLES THOMAS KINGZETT, Consulting Chemist. With 23 illustrations. 8vo.......................................$2.00

KIRK.—A Practical Treatise on Foundry Irons:
Comprising Pig Iron, and Fracture Grading of Pig and Scrap Irons; Scrap Irons; Mixing Irons; Elements and Metalloids; Grading Iron by Analysis; Chemical Standards for Iron; Castings; Testing Cast Iron; Semi-Steel; Malleable Iron; Etc., Etc. By EDWARD KIRK, Practical Moulder and Melter, Consulting Expert in Melting. Illustrated. 294 pages. 8vo. 1911....................................$3.00

KIRK.—The Cupola Furnace:
A Practical Treatise on the Construction and Management of Foundry Cupolas. By EDWARD KIRK, Practical Moulder and Melter, Consulting Expert in Melting. Illustrated by 106 Engravings. Third Edition, revised and enlarged. 482 pages. 8vo. 1910..............................$3.50

KOENIG.—Chemistry Simplified:
A Course of Lectures on the Non-Metals, Based upon the Natural Evolution of Chemistry. Designed Primarily for Engineers. By GEORGE AUGUSTUS KOENIG, Ph. D., A. M , E. M., Professor of Chemistry, Michigan College of Mines, Houghton. Illustrated by 103 Original Drawings. 449 pp. 12mo. (1906)..................................$2.25

LANGBEIN.—A Complete Treatise on the Electro-Deposition of Metals:
Comprising Electro-Plating and Galvanoplastic Operations, The Deposition of Metals by the Contract and Immersion

Processes, the Coloring of Metals, the Methods of Grinding and Polishing, as well as the Description of the Voltaic Cells, Dynamo-Electric Machines, Thermopiles, and of the Materials and Processes Used · in Every Department of the Art. Translated from the latest German Edition of DR. GEORGE LANGBEIN, Proprietor of a Manufactory for Chemical Products, Machines, Apparatus and Utensils for Electro-Platers, and of an Electro-Plating Establishment in Leipzig. With Additions by WILLIAM T. BRANNT, Editor of "The Techno-Chemical Receipt Book." Seventh Edition, Revised and Enlarged. Illustrated by 163 Engravings. 8vo. 725 pages. 1913..$5.00

LARKIN.—The Practical Brass and Iron Founder's Guide:

A Concise Treatise on Brass Founding, Moulding, the Metals and their Alloys, etc.; to which are added Recent Improvements in the Manufacture of Iron, Steel by the Bessemer Process, etc., etc. By JAMES LARKIN, late Conductor of the Brass Foundry Department in Reany, Neafie & Co.'s Penn Works, Philadelphia. New edition, revised, with extensive additions. 414 pages. 12mo......................$2.50

LEHNER.—The Manufacture of Ink:

Comprising the Raw Materials, and the Preparation of Writing, Copying and Hektograph Inks, Safety Inks, Ink Extracts and Powders, etc. Translated from the German of SIGMUND LEHNER, with additions by WILLIAM T. BRANNT. Illustrated. 12mo...............................$2.00

LEROUX.—A Practical Treatise on the Manufacture of Worsteds and Carded Yarns:

Comprising Practical Mechanics, with Rules and Calculations applied to Spinning; Sorting, Cleaning, and Scouring Wools; the English and French Methods of Combing, Drawing, and Spinning Worsteds, and Manufacturing Carded Yarns. Translated from the French of CHARLES LEROUX, Mechanical Engineer and Superintendent of a Spinning-Mill, by HORATIO PAINE, M. D.; and A. A. FESQUET, Chemist and Engineer. Illustrated by twelve large Plates. 8vo....$3.00

LESLIE.—Complete Cookery:

Directions for Cookery in its Various Branches. By MISS LESLIE. Sixtieth thousand. Thoroughly revised, with the additions of New Receipts. 12mo..................$1.00

LE VAN.—The Steam Engine and the Indicator:

Their Origin and Progressive Development; including the Most Recent Examples of Steam and Gas Motors, together with the Indicator, its Principles, its Utility, and its Application. By WILLIAM BARNET LE VAN. Illustrated by 205 Engravings, chiefly of Indicator-Cards. 469 pp. 8vo. $2.00

LIEBER.—Assayer's Guide:

Or, Practical Directions to Assayers, Miners, and Smelters, for the Tests and Assays, by Heat and by Wet Processes, for the Ores of all the principal Metals, of Gold and Silver Coins and alloys, and of Coal, etc. By OSCAR M. LIEBER. Revised. 283 pp. 12mo..........................$1.50

Lockwood's Dictionary of Terms:

Used in the Practice of Mechanical Engineering, embracing those Current in the Drawing Office, Pattern Shop, Foundry, Fitting, Turning, Smith's and Boiler Shops, etc., etc., comprising upwards of Six Thousand Definitions. Edited by a Foreman Pattern Maker, author of "Pattern Making." 417 pp. 12mo............................... $3.75

LUKIN.—The Lathe and Its Uses:

Or Instruction in the Art of Turning Wood and Metal. Including a Description of the Most Modern Appliances for the Ornamentation of Plane and Curved Surfaces, an Entirely Novel Form of Lathe for Eccentric and Rose-Engine Turning. A Lathe and Planing Machine Combined; and Other Valuable Matter Relating to the Art. Illustrated by 462 engravings. Seventh Edition. 315 pages 8vo.....$4.25

MAUCHLINE.—The Mine Foreman's Hand-Book:

Of Practical and Theoretical Information on the Opening, Ventilating, and Working of Collieries. Questions and Answers on Practical and Theoretical Coal Mining. Designed to Assist Students and Others in Passing Examinations for Mine Foremanships. By ROBERT MAUCHLINE. 3d Edition. Thoroughly Revised and Enlarged by F. ERNEST BRACKETT. 134 Engravings. 8vo. 378 pages. (1905.).........$3.75

MOLESWORTH.—Pocket-Book of Useful Formulæ and Memoranda for Civil and Mechanical Engineers:

By GUILFORD L. MOLESWORTH, Member of the Institution of Civil Engineers, Chief Resident Engineer of the Ceylon Railway. Full-bound in Pocketbook form. ..$1.00

MOORE.—The Universal Assistant and the Complete Mechanic:

Containing over one million Industrial Facts, Calculations, Receipts, Processes, Trades Secrets, Rules, Business Forms, Legal Items, etc., in every occupation, from the Household to the Manufactory. By R. MOORE. Illustrated by 500 Engravings. 12mo...............................$2.50

NAPIER.—A System of Chemistry Applied to Dyeing:

By JAMES NAPIER, F. C. S.   A New and Thoroughly Revised Edition.   Completely brought up to the present state of the Science, including the Chemistry of Coal Tar Colors, by A. A. FESQUET, Chemist and Engineer.   With an Appendix on Dyeing and Calico Printing, as shown at the Universal Exposition, Paris, 1867.   Illustrated.   8vo.   422 pages...$2.00

NICHOLLS.—The Theoretical and Practical Boiler-Maker and Engineer's Reference Book:

Containing a variety of Useful Information for Employers of Labor, Foremen and Working Boiler-Makers, Iron, Copper, and Tinsmiths, Draughtsmen, Engineers, the General Steam-using Public, and for the Use of Science Schools and classes By SAMUEL NICHOLLS.   Illustrated by sixteen plates.   12mo.
$2.50

NYSTROM.—On Technological Education and the Construction of Ships and Screw Propellers

For Naval and Marine Engineers.   By JOHN W. NYSTROM, late Acting Chief Engineer, U. S. N.   Second Edition, Revised, with additional matter.   Illustrated by seven Engravings.   12mo......$1.00

O'NEILL.—A Dictionary of Dyeing and Calico Printing:

Containing a brief account of all the Substances and Processes in use in the Art of Dyeing and Printing Textile Fabrics; with Practical Receipts and Scientific Information.   By CHARLES O'NEILL, Analytical Chemist.   To which is added an Essay on Coal Tar Colors and their application to Dyeing and Calico Printing.   By A. A. FESQUET, Chemist and Engineer.   With an appendix on Dyeing and Calico Printing, as shown at the Universal Exposition, Paris, 1867.   8vo. 491 pages......$2.00

ORTON.—Underground Treasures:

How and Where to Find Them.   A Key for the Ready Determination of all the Useful Minerals within the United States.   By JAMES ORTON, A. M., Late Professor of Natural History in Vassar College, N. Y.; author of the "Andes and the Amazon," etc.   A New Edition, with An Appendix on Ore Deposits and Testing Minerals.   (1901.)   Illustrated.
$1.50

OSBORN.—A Practical Manual of Minerals, Mines and Mining:

Comprising the Physical Properties, Geologic Position; Local Occurrence and Associations of the Useful Minerals, their Methods of Chemical Analysis and Assay; together with Various Systems of Excavating and Timbering, Brick and

Masonry Work, during Driving, Lining, Bracing and other
Operations, etc.   By PROF. H. S. OSBORN, LL. D., Author of
"The Prospector's Field-Book and Guide." ·171 Engravings.
Second Edition, Revised. 8vo..................... $4.50

OSBORN.—The Prospector's Field Book and Guide:
In the Search For and the Easy Determination of Ores and
Other Useful Minerals. By PROF. H. S. OSBORN, LL. D.
Illustrated by 66 Engravings. Eighth Edition. Revised
and Enlarged. 401 pages. 12mo (1910.)........... $1.50

OVERMAN.—The Moulder's and Founder's Pocket Guide:
A Treatise on Moulding and Founding in Green-sand, Dry-
sand, Loam, and Cement; the Moulding of Machine Frames,
Mill-gear, Hollow Ware, Ornaments, Trinkets, Bells, and
Statues; Description of Moulds for Iron, Bronze, Brass, and
other Metals; Plaster of Paris, Sulphur, Wax, etc.; the Construction of Melting Furnaces, the Melting and Founding of
Metals; the Composition of Alloys and their Nature, etc.,
etc   By FREDERICK OVERMAN, M. E.   A new Edition, to
which is added a Supplement on Statuary and Ornamental
Moulding, Ordnance, Malleable Iron Castings, etc. By A.
A. FESQUET. Chemist and Engineer. Illustrated by 44
engravings. 12mo.............................. $2.00

PAINTER, GILDER, AND VARNISHER'S COMPANION:
Comprising the Manufacture and Test of Pigments, the Arts
of Painting, Graining, Marbling, Staining, Sign-writing,
Varnishing, Glass-staining, and Gilding on Glass; together
with Coach Painting and Varnishing, and the Principles of
the Harmony and Contrast of Colors. Twenty-seventh
Edition. Revised, Enlarged, and in great part Rewritten.
By WILLIAM T. BRANNT, Editor of "Varnishes, Lacquers,
Printing Inks and Sealing Waxes." Illustrated. 395 pp.
12mo... $1.50

PERCY.—The Manufacturing of Russian Sheet-Iron:
By JOHN PERCY, M. D., F. R. S. Paper.................. 25

POSSELT.—Cotton Manufacturing:
Part I. Dealing with the Fibre, Ginning, Mixing, Picking,
Scutching and Carding. By E. A. POSSELT. 104 Illustra-
tions, 190 pp.................................... $3.00
Part II. Combing, Drawing, Roller Covering and Fly Frame,
$3.00

POSSELT.—The Jacquard Machine Analysed and Ex-
plained:
With an Appendix on the Preparation of Jacquard Cards, and
Practical Hints to Learners of Jacquard Designing. By E.
A. POSSELT. With 230 Illustrations and numerous diagrams.
127 pp. 4to.................................. $3.00

POSSELT.—Recent Improvements in Textile Machinery Relating to Weaving:
Giving the Most Modern Points on the Construction of all Kinds of Looms, Warpers, Beamers, Slashers, Winders, Spoolers, Reeds, Temples, Shuttles, Bobbins, Heddles, Heddle Frames, Pickers, Jacquards, Card Stampers, Etc., Etc. By E. A. POSSELT. 4to.   Part I, 600 ills.; Part II, 600 ills. Each part...... ....$1.50

POSSELT.—Recent Improvements in Textile Machinery, Part III:
Processes Required for Converting Wool, Cotton, Silk, from Fibre to Finished Fabric, Covering both Woven and Knit Goods; Construction of the most Modern Improvements in Preparatory Machinery, Carding, Combing, Drawing, and Spinning Machinery, Winding, Warping, Slashing Machinery, Looms, Machinery for Knit Goods, Dye Stuffs, Chemicals, Soaps, Latest Improved Accessories Relating to Construction and Equipment of Modern Textile Manufacturing Plants By E. A. POSSELT.   Completely Illustrated.   4to.....$5.00

POSSELT.—Technology of Textile Design:
The Most Complete Treatise on the Construction and Application of Weaves for all Textile Fabrics and the Analysis of Cloth.   By E. A. POSSELT.   1,500 Illustrations.   4to..$5.00

POSSELT.—Textile Calculations:
A Guide to Calculations Relating to the Manufacture of all Kinds of Yarns and Fabrics, the Analysis of Cloth, Speed, Power and Belt Calculations.   By E. A. POSSELT.   Illustrated.   4to..$2.00

REGNAULT.—Elements of Chemistry:
By M. V. REGNAULT.   Translated from the French by T. FORREST BETTON, M. D., and edited, with Notes, by JAMES C. BOOTH, Melter and Refiner U. S. Mint, and WILLIAM L. FABER, Metallurgist and Mining Engineer.   Illustrated by nearly 700 wood-engravings   Comprising nearly 1,500 pages. In two volumes, 8vo., cloth....................................$5.00

RICH.—Artistic Horse-Shoeing:
A Practical and Scientific Treatise, giving Improved Methods of Shoeing, with Special Directions for Shaping Shoes to Cure Different Diseases of the Foot, and the Correction of Faulty Action in Trotters.   By GEORGE E. RICH.   362 Illustrations. 217 pages.   12mo..............................$2.00

RICHARDSON.—Practical Blacksmithing:
A Collection of Articles Contributed at Different Times by Skilled Workmen to the columns of "The Blacksmith and Wheelwright," and Covering nearly the Whole Range of Blacksmithing, from the Simplest Job of Work to some of the

most Complex Forgings   Compiled and Edited by M. T.
RICHARDSON.
Vol.   I.   210 Illustrations.   224 pages.   12mo......$1.00
Vol. II.   230 Illustrations.   262 pages.   12mo......$1.00
Vol. III.   390 Illustrations.   307 pages.   12mo......$1.00
Vol. IV.   226 Illustrations.   276 pages.   12mo......$1.00
RICHARDSON.—Practical Carriage Building:
Comprising Numerous Short Practical Articles upon Carriage
and Wagon Woodwork; Plans for Factories; Shop and Bench
Tools; Convenient Appliances for Repair Work; Methods of
Working; Peculiarities of Bent Timber; Construction of
Carriage Parts; Repairing Wheels; Forms of Tenons and Mor-
tises; Together with a Variety of Useful Hints and Sugges-
tions to Woodworkers. Compiled by M. T. RICHARDSON.
Vol. I. 228 Illustrations. 222 pages.. ....$1.00
Vol. II. 283 Illustrations. 280 pages.....   ..$1.00
RICHARDSON.—The Practical Horseshoer:
Being a Collection of Articles on Horseshoeing in all its
Branches which have appeared from time to time in the columns of "The Blacksmith and Wheelwright," etc. Compiled
and edited by M. T. RICHARDSON. 174 Illustrations, $1.00
RIFFAULT, VERGNAUD, and TOUSSAINT.—A Practical
    Treatise on the Manufacture of Colors for Painting:
Comprising the Origin, Definition, and Classification of Colors,
the Treatment of the Raw Materials; the best Formulae and
the Newest Processes for the Preparation of every description
of Pigment, and the Necessary Apparatus and Directions for
its use; Dryers; the Testing, Application, and Qualities of
Paints, etc., etc. By MM. RIFFAULT, VERGNAUD, and
TOUSSAINT, Revised and Edited by M. F. MALPEYRE, Trans-
lated from the French by A. A. FESQUET. Illustrated by
Eighty Engravings. 659 pp. 8vo.................$5.00
ROPER.—Catechism for Steam Engineers and Elec-
    tricians:
Including the Construction and Management of Steam Engines, Steam Boilers and Electric Plants. By STEPHEN
ROPER   Twenty-first edition, rewritten and greatly enlarged
by E. R. KELLER and C. W. PIKE. 365 pages. Illustrations.
18mo., tucks, gilt................................$2.00
ROPER.—Engineer's Handy Book:
Containing Facts, Formulæ, Tables and Questions on Power,
its Generation, Transmission and Measurement; Heat, Fuel,
and Steam; The Steam Boiler and Accessories; Steam Engines
and their Parts; Steam Engine Indicator; Gas and Gasoline
Engines; Materials; their Properties and Strength; Together
with a Discussion of the Fundamental Experiments in Electricity, and an Explanation of Dynamos, Motors, Batteries,
etc.. and Rules for Calculating Sizes of Wires. By STEPHEN

ROPER   15th edition. Revised and Enlarged by E. R.
KELLER, M. E., and C. W. PIKE, B. S. With numerous
Illustrations. Pocket-book form. Leather...........$3.50

ROPER.—Hand-Book of Land and Marine Engines:
Including the Modeling, Construction, Running, and Management of Land and Marine Engines and Boilers. With
Illustrations. By STEPHEN ROPER, Engineer. Sixth Edition.
12mo., tucks, gilt edge.................... .$3.50

ROPER.—Hand-Book of the Locomotive:
Including the Construction of Engines and Boilers, and the
Construction, Management, and Running of Locomotives.
By STEPHEN ROPER. Eleventh Edition. 18mo., tucks, gilt
edge..$2.50

ROPER.—Hand-Book of Modern Steam Fire-Engines;
With Illustrations. By STEPHEN ROPER, Engineer. Fourth
Edition, 12mo., tucks, gilt edge..   ....$3.50

ROPER.—Instructions and Suggestions for Engineers and
    Firemen:
By STEPHEN ROPER, Engineer. 18mo., Morocco.....$2.00

ROPER.—Questions and Answers for Stationary and
    Marine Engineers and Electricians:
With a Chapter of What to Do in Case of Accidents. By
STEPHEN ROPER, Engineer. Sixth Edition, Rewritten and
Greatly Enlarged by EDWIN R. KELLER, M. E., and CLAYTON
W. PIKE, B. A. 306 pp. Morocco, pocketbook form, gilt
edges...$2.00

ROPER.—The Steam Boiler: Its Care and Management:
By STEPHEN ROPER, Engineer. 12mo., tuck, gilt edges. $2.00

ROPER.—Use and Abuse of the Steam Boiler:
By STEPHEN ROPER, Engineer. Ninth Edition, with Illustrations. 18mo., tucks, gilt edge.................$2.00

ROPER.—The Young Engineer's Own Book: .
Containing an Explanation of the Principle and Theories on
which the Steam Engine as a Prime Mover is based. By
STEPHEN ROPER, Engineer. 160 Illustrations, 363 pages.
18mo., tuck.......................................$2.50

ROSE.—The Complete Practical Machinist:
Embracing Lathe Work, Vise Work, Drills and Drilling, Taps
and Dies, Hardening and Tempering, the Making and Use of
Tools, Tool Grinding, Marking out work, Machine Tools, etc.
By JOSHUA ROSE. 395 Engravings. Nineteenth Edition,
greatly Enlarged with New and Valuable Matter. 12mo.,
504 pages...$2.50

ROSE.—Mechanical Drawing Self-Taught:
Comprising Instructions in the Selection and Preparation of
·Drawing Instruments, Elementary Instruction in practical

Mechanical Drawing, together with Examples in Simple Geometry and Elementary Mechanism, including Screw Threads, Gear Wheels, Mechanical Motions, Engines and Boilers. By JOSHUA ROSE, M. E. Illustrated by 330 Engravings. 8vo. 313 pages........................$3.50

ROSE.—The Slide-Valve Practically Explained:
Embracing simple and complete Practical Demonstrations of the operation of each element in a Slide-valve Movement, and illustrating the effects of Variations in their Proportions by examples carefully selected from the most recent and successful practice. By JOSHUA ROSE, M. E. Illustrated by 35 Engravings................................$1.00

ROSE.—Steam Boilers:
A Practical Treatise on Boiler Construction and Examination, for the Use of Practical Boiler Makers, Boiler Users, and Inspectors; and embracing in plain figures all the calculations necessary in Designing or Classifying Steam Boilers. By JOSHUA ROSE, M. E. Illustrated by 73 Engravings. 250 pages. 8vo......................................$2.00

ROSS.—The Blowpipe in Chemistry, Mineralogy and Geology.
Containing all Known Methods of Anhydrous Analysis, many Working Examples, and Instructions for Making Apparatus. By LIEUT COLONEL W A. ROSS, R. A.. F. G. S. With 120 Illustrations. 12mo.............................$2.00

SCHRIBER.—The Complete Carriage and Wagon Painter:
A Concise Compendium of the Art of Painting Carriages, Wagons, and Sleighs, embracing Full Directions in all the Various Branches, including Lettering, Scrolling, Ornamenting, Striping, Varnishing, and Coloring, with numerous Recipes for Mixing Colors. 73 Illustrations. 177 pp. 12mo.
$1.00

SHAW.—Civil Architecture:
Being a Complete Theoretical and Practical System of Building, containing the Fundamental Principles of the Art. By EDWARD SHAW, Architect. To which is added a Treatise on Gothic Architecture, etc. By THOMAS W. SILLOWAY and GEORGE M. HARDING, Architects. The whole illustrated by 102 quarto plates finely engraved on copper. Eleventh Edition 4to......................................$5.00

SHERRATT.—The Elements of Hand-Railing:
Simplified and Explained in Concise Problems that are Easily Understood. The whole illustrated with Thirty-eight Accurate and Original Plates, Founded on Geometrical Principles, and showing how to Make Rail Without Centre Joints, Making Better Rail of the Same Material, with Half the Labor,

and Showing How to Lay Out Stairs of all Kinds. By R. J.
SHERRATT. Folio.................................$2.50

SHUNK.—A Practical Treatise on Railway Curves and
Location for Young Engineers:
By W. F. SHUNK, C. E. 12mo. Full bound pocket-book
form...$2.00

SLOANE.—Home Experiments in Science:
By T. O'CONOR SLOANE, E. M.. A M., Ph. D. Illustrated
by 91 Engravings. 12mo........................$1.00

SLOAN.—Homestead Architecture:
Containing Forty Designs for Villas, Cottages, and Farm-
houses, with Essays on Style, Construction, Landscape Gar-
dening, Furniture, etc., etc. Illustrated by upwards of 200
Engravings. By SAMUEL SLOAN, Architect. 8vo.....$2.00

SMITH.—The Dyer's Instructor:
Comprising Practical Instructions in the Art of Dyeing Silk,
Cotton, Wool, and Worsted, and Woolen Goods; containing
nearly 800 Receipts. To which is added a Treatise on the
Art of Padding; and the Printing of Silk Warps, Skeins, and
Handkerchiefs, and the various Mordants and Colors for the
different styles of such work. By DAVID SMITH, Pattern
Dyer. 12mo....................................$1.00

SMITH.—A Manual of Political Economy:
By E. PESHINE SMITH. A New Edition. to which is added
a full Index. 12mo.............................$1.25

SMITH.—Parks and Pleasure-Grounds:
Or Practical Notes on Country Residences, Villas, Public
Parks, and Gardens. By CHARLES H. J. SMITH, Landscape
Gardener and Garden Architect, etc., etc. 12mo.....$2.00

SNIVELY.—The Elements of Systematic Qualitative
Chemical Analysis:
A Hand-book for Beginners. By JOHN H. SNIVELY, Phr. D.
16mo..$2.00

STOKES.—The Cabinet Maker and Upholsterer's Companion:
Comprising the Art of Drawing, as applicable to Cabinet
Work; Veneering, Inlaying, and Buhl-Work; the Art of Dyeing and Staining Wood, Ivory, Bone, Tortoise-Shell, etc.
Directions for Lacquering, Japanning. and Varnishing; to
make French Polish, Glues, Cements, and Compositions;
with numerous Receipts, useful to workmen generally. By
J. STOKES. Illustrated. A New Edition, with an Appendix
upon French Polishing, Staining, Imitating, Varnishing, etc.,
etc. 12mo.....................................$1.25

STRENGTH AND OTHER PROPERTIES OE METALS:
Reports of Experiments on the Strength and other Properties

of Metals for Cannon   With a Description of the Machines
for Testing Metals, and of the Classification of Cannon in
service.   By Officers of the Ordnance Department, U. S.
Army.   By authority of the Secretary of War.   Illustrated
by 25 large steel plates.   Quarto........   ....$3.00

SULZ.—A Treatise on Beverages:
Or the Complete Practical Bottler.   Full Instructions for
Laboratory Work with Original Practical Recipes for all
kinds of Carbonated Drinks, Mineral Waters, Flavoring
Extracts, Syrups, etc.   By CHARLES HERMAN SULZ, Tech-
nical Chemist and Practical Bottler.   Illustrated by 428
Engravings.   818 pp.   8vo........................$7.50

SYME.—Outlines of an Industrial Science:
By DAVID SYME.   12mo........................$2.00

**TABLES SHOWING THE WEIGHT OF ROUND, SQUARE
AND FLAT BAR IRON, STEEL, ETC.**
By Measurement.   Cloth..........................63

**TEMPLETON.—The Practical Examinator on Steam and
the Steam-Engine:**
With Instructive References relative thereto, arranged for
the Use of Engineers, Students, and others.   By WILLIAM
TEMPLETON, Engineer   12mo.....................$1.00

THALLNER.—Tool-Steel:
A Concise Hand-book on Tool-Steel in General.   Its Treat-
ment in the Operations of Forging, Annealing, Hardening,
Tempering, etc., and the Appliances Therefor.   By OTTO
THALLNER, Manager in Chief of the Tool-Steel Works, Bis-
marckhutte, Germany.   From the German by WILLIAM T.
BRANNT.   Illustrated by 69 Engravings.   194 pages   8vo.
1902...$2.00

**THAUSING.—The Theory and Practice of the Preparation
of Malt and the Fabrication of Beer:**
With especial reference to the Vienna Process of Brewing.
Elaborated from personal experience by JULIUS E. THAUSING,
Professor at the School for Brewers, and at the Agricultural
Institute, Modling, near Vienna.   Translated from the German by WILLIAM T. BRANNT.   Thoroughly and elaborately
edited, with much American matter, and according to the
latest and most Scientific Practice, by A. SCHWARZ and DR.
A. H. BAUER   Illustrated by 140 Engravings.   8vo.   815
pages..$10.0

TOMPKINS.—Cotton and Cotton Oil:
Cotton: Planting, Cultivating, Harvesting and Preparation
for Market.   Cotton Seed Oil Mills: Organization, Construction and Operation.   Cattle Feeding: Production of Beef
and Dairy Products, Cotton Seed Meal and Hulls as Stock

Feed. Fertilizers: Manufacture, Manipulation and Uses.
By D. A. TOMPKINS. 8vo. 494 pp. Illustrated......$7.50

TOMPKINS.—Cotton Mill, Commercial Features:
'A Text-Book for the Use of Textile Schools and Investors.
With Tables showing Cost of Machinery and Equipments
for Mills making Cotton Yarns and Plain Cotton Cloths. By
D. A. TOMPKINS. 8vo. 240 pp. Illustrated.......$5.00

TOMPKINS.—Cotton Mill Processes and Calculations:
An Elementary Text-Book for the Use of Textile Schools and
for Home Study. By D. A TOMPKINS. 312 pp. 8vo.
Illustrated. $5.00

TURNER'S (THE) COMPANION:
Containing Instructions in Concentric, Elliptic, and Eccen-
tric Turning; also various Plates of Chucks, Tools, and In-
struments; and Directions for using the Eccentric Cutter,
Drill, Vertical Cutter, and Circular Rest; with Patterns and
Instructions for working them. 12mo...............$1.00

VAN CLEVE.—The English and American Mechanic:
Comprising a Collection of Over Three Thousand Receipts,
Rules, and Tables, designed for the Use of every Mechanic
and Manufacturer. By B. FRANK VAN CLEVE. Illustrated.
500 pp. 12mo.................................$2.00

VAN DER BURG.—School of Painting for the Imitation
 of Woods and Marbles:
A Complete, Practical Treatise on the Art and Craft of Graining and Marbling with the Tools and Appliances. 36 Plates.
Folio, 12x20 inches.............................$6.00

VILLE.—The School of Chemical Manures:
Or, Elementary Principles in the Use of Fertilizing Agents
From the French of M. GEO. VILLE, by A. A. FESQUET,
Chemist and Engineer. With Illustrations. 12mo....$1.25

VOGDES.—The Architect's and Builder's Pocket-Com-
        panion and Price-Book:
Consisting of a Short but Comprehensive Epitome of Deci-
mals, Duodecimals, Geometry and Mensuration; with Tables
of United States Measures, Sizes, Weights, Strength, etc., of
Iron, Wood, Stone, Brick, Cement and Concretes, Quanti-
ties of Materials in given Sizes and Dimensions of Wood,
Brick and Stone; and full and complete Bills of Prices for
Carpenter's Work and Painting; also, Rules for Computing
and Valuing Brick and Brick Work, Stone Work, Painting,
Plastering, with a Vocabulary of Technical Terms, etc. By
FRANK W. VOGDES, Architect, Indianapolis, Ind. Enlarged,
Revised and Corrected. In one volume 368 pages, full-
bound, pocketbook form, gilt edges.................$2.00
Cloth..$1 50

WAHNSCHAFFE.—A Guide to the Scientific Examination of Soils:
Comprising Select Methods of Mechanical and Chemical Analysis and Physical Investigation. Translated from the German of DR F. WAHNSCHAFFE. With additions by WILLIAM T. BRANNT. Illustrated by 25 Engravings. 12mo. 177 pages......................... ....$1.50

WARE.—The Sugar Beet:
Including a History of the Beet Sugar Industry in Europe, Varieties of the Sugar Beet, Examination, Soils, Tillage Seeds and Sowing, Yield and Cost of Cultivation, Harvesting, Transportation, Conservation, Feeding Qualities of the Beet and of the Pulp, etc. By LEWIS S. WARE, C. E, M. E. Illustrated by ninety Engravings. 8vo.... $2.00

WARN.—The Sheet-Metal Worker's Instructor:
For Zinc, Sheet-Iron, Copper, and Tin-Plate Workers, etc. Containing a selection of Geometrical Problems; also Fractical and Simple Rules for Describing the various Patterns required in the different branches of the above Trades. By REUBEN H. WARN, Practical Tin-Plate Worker. To which is added an Appendix, containing Instructions for Boiler-Making, Mensuration of Surfaces and Solids, Rules for Calculating the Weights of different Figures of Iron and Steel, Tables of the Weights of Iron, Steel, etc. Illustrated by thirty-two Plates and thirty-seven Wood Engravings. 8vo...$2.00

WARNER.—New Theorems, Tables, and Diagrams, for the Computation of Earth-work:
Designed for the use of Engineers in Preliminary and Final Estimates, of Students in Engineering and of Contractors and other non-professional Computers. In two parts, with an Appendix. Part I. A Practical Treatise; Part II. A Theoretical Treatise, and the Appendix Containing Notes to the Rules and Examples of Part I.; Explanations of the Construction of Scales, Tables, and Diagrams, and a Treatise upon Equivalent Square Bases and Equivalent Level Heights. By JOHN WARNER, A. M., Mining and Mechanical Engineer. Illustrated by 14 Plates. 8vo.....................$3.00

WATSON —A Manual of the Hand-Lathe:
Comprising Concise Directions for Working Metals of all kinds, Ivory, Bone and Precious Woods; Dyeing, Coloring, and French Polishing; Inlaying by Veneers, and various methods practised to produce Elaborate work with dispatch, and at Small Expense. By EGBERT P. WATSON, Author of "The Modern Practice of American Machinists and Engineers." Illustrated by 78 Engravings $1.00

WATSON.—The Modern Practice of American Machinists and Engineers:
Including the Construction, Application, and Use of Drills,

Lathe Tools, Cutters for Boring Cylinders, and Hollow-work generally, with the most economical Speed for the same; the Results verified by Actual Practice at the Lathe, the Vise, and on the floor. Together with Workshop Management, Economy of Manufacture, the Steam Engine, Boilers, Gears, Belting, etc., etc. By EGBERT P. WATSON   Illustrated by eighty-six Engravings. 12mo......................$2.00

WEATHERLY.—Treatise on the Art of Boiling Sugar, Crystallizing, Lozenge-making, Comfits, Gum Goods: And other processes for Confectionery, including Methods for Manufacturing every Description of Raw and Refined Sugar Goods. A New and Enlarged Edition, with an Appendix on Cocoa, Chocolate, Chocolate Confections, etc. 196 pages. 12mo.......................................$1.50

WILL.—Tables of Qualitative Chemical Analysis: With an Introductory Chapter on the Course of Analysis By PROFESSOR HEINRICH WILL, of Giessen, Germany. Third American, from the eleventh German Edition. Edited by CHARLES F. HIMES, Ph. D , Professor of Natural Science Dickinson College, Carlisle, Pa. 8vo...............$1.00

WILLIAMS.—On Heat and Steam: Embracing New Views of Vaporization, Condensation and Explosion. By CHARLES WYE WILLIAMS, A. I. C. E. Illustrated. 8vo.......................................$2.00

WILSON.—The Practical Tool-Maker and Designer: A Treatise upon the Designing of Tools and Fixtures for Machine Tools and Metal Working Machinery, Comprising Modern Examples of Machines with Fundamental Designs for Tools for the Actual Production of the work; Together with Special Reference to a Set of Tools for Machining the Various Parts of a Bicycle. Illustrated by 189 Engravings (1898)...$2.50

CONTENTS: Introductory. Chapter I. Modern Tool Room and Equipment. II. Files, Their Use and Abuse. III. Steel and Tempering. IV. Making Jigs. V. Milling Machine Fixtures. VI. Tools and Fixtures for Screw Machines. VII. Broaching. VIII. Punches and Dies for Cutting and Drop Press. IX. Tools for Hollow-ware. X. Embossing: Metal, Coin and Stamped Sheet-Metal Ornaments. XI. Drop Forging. XII. Solid Drawn Shells or Ferrules; Cupping or Cutting and Drawing; Breaking Down Shells. XIII. Annealing, Pickling and Cleaning. XIV. Tools for Draw Bench. XV. Cutting and Assembling Pieces by Means of Ratchet Dial Plates at One Operation. XVI. The Header. XVII. Tools for Fox Lathe. XVIII. Suggestions for a set of Tools for Machining the Various Parts of a Bicycle. XIX. The Plater's Dynamo. XX. Conclusion—With a few Random Ideas. Appendix. Index.

WORSSAM.—On Mechanical Saws: From the Transaction of the Society of Engineers, 1869. By S. W. WORSSAM, JR. Illustrated by Eighteen large lal s. 8vo..P.$1.50

THE

·SOAP MAKER'S HAND BOOK

OF

MATERIALS, PROCESSES AND RECEIPTS FOR EVERY DESCRIPTION OF SOAP

INCLUDING

FATS, FAT OILS, AND FATTY ACIDS; EXAMINATION OF FATS AND OILS;
ALKALIES; TESTING SODA AND POTASH; MACHINES AND UTENSILS;
HARD SOAPS; SOFT SOAPS; TEXTILE SOAPS; WASHING POWDERS
AND ALLIED PRODUCTS; TOILET SOAPS, MEDICATED SOAPS,
AND SOAP SPECIALTIES; ESSENTIAL OILS AND OTHER
PERFUMING MATERIALS; TESTING SOAPS.

· EDITED CHIEFLY FROM THE GERMAN OF

DR. C. DEITE, A. ENGELHARDT, F. WILTNER,

AND NUMEROUS OTHER EXPERTS.

WITH ADDITIONS

BY

WILLIAM T. BRANNT,

EDITOR OF "THE TECHNO CHEMICAL RECEIPT BOOK."

ILLUSTRATED BY FIFTY-FOUR ENGRAVINGS.

SECOND EDITION, REVISED AND IN GREAT PART RE-WRITTEN.

PHILADELPHIA:
HENRY CAREY BAIRD & CO.,
INDUSTRIAL PUBLISHERS, BOOKSELLERS, AND IMPORTERS,
810 WALNUT STREET.
1912

·KIRK'S CUPOLA FURNACE.

An Eminently, Practical, Up-to-Date Book, by an Expert.
Third Thoroughly Revised and Partly Re-written Edition.
In one volume, 8vo., 482 pages, illustrated by one hundred
and six engravings. Price $3.50. Free of Postage to any
Address in the World, or by Express C. O. D., freight paid to
any Address in the United States or Canada.

PUBLISHED AUGUST, 1910.

THE CUPOLA FURNACE

A PRACTICAL TREATISE ON THE

CONSTRUCTION AND MANAGEMENT

OF

FOUNDRY CUPOLAS:

COMPRISING

IMPROVEMENTS IN CUPOLAS AND METHODS OF THEIR CONSTRUCTION AND MANAGE-
MENT; TUYERES; MODERN CUPOLAS; CUPOLA FUELS; FLUXING OF IRON; GETTING
UP CUPOLA STOCK; RUNNING A CONTINUOUS STREAM; SCIENTIFICALLY
DESIGNED CUPOLAS; SPARK-CATCHING DEVICES; BLAST-PIPES AND
BLAST; BLOWERS; FOUNDRY TRAM RAIL, ETC., ETC.

BY

EDWARD KIRK,

PRACTICAL MOULDER AND MELTER, CONSULTING EXPERT IN MELTING.
Author of " The Founding of Metals," and of Numerous Papers on Cupola Practice.

ILLUSTRATED BY ONE HUNDRED AND SIX ENGRAVINGS.

THIRD THOROUGHLY REVISED AND PARTLY RE-WRITTEN EDITION.

PHILADELPHIA ·
HENRY CAREY BAIRD & CO.,
INDUSTRIAL PUBLISHERS, BOOKSELLERS, AND IMPORTERS,
810 WALNUT STREET.
1912

KIRK'S FOUNDRY IRONS.

A Practical, Up-to-Date Book, by the well known Expert.
In one volume, 8vo, 294 pages, illustrated. Price $3.00 net.
Free of Postage to any Address in the World, or by Express
C. O. D., freight paid to any Address in the United States or
Canada.

PUBLISHED JUNE, 1911.

A PRACTICAL TREATISE

ON

FOUNDRY IRONS:

COMPRISING

PIG IRON, AND FRACTURE GRADING OF PIG AND SCRAP IRONS;
SCRAP IRONS; MIXING IRONS; ELEMENTS AND METALLOIDS;
GRADING IRON BY ANALYSIS; CHEMICAL STANDARDS
FOR IRON CASTINGS; TESTING CAST IRON; SEMI-
STEEL; MALLEABLE IRON; ETC., ETC.

BY

EDWARD KIRK,

PRACTICAL MOULDER AND MELTER; CONSULTING EXPERT IN MELTING.
AUTHOR OF "THE CUPOLA FURNACE," AND OF NUMEROUS
PAPERS ON CUPOLA PRACTICE.

ILLUSTRATED

PHILADELPHIA:

HENRY CAREY BAIRD & CO.,

INDUSTRIAL PUBLISHERS, BOOKSELLERS AND IMPORTERS,

810 WALNUT STREET.

1911

BRANNT'S DRY CLEANER.

The only book including Hat Cleaning and Reno-vating in any language, in one volume, 12mo, 371 pages, illustrated. Price $2.50 net. Free of postage to any address in the world, or by express freight paid to any address in the United States or Canada.

PUBLISHED OCTOBER, 1911.

THE PRACTICAL
DRY CLEANER, SCOURER, AND GARMENT DYER:

COMPRISING

DRY, CHEMICAL, OR FRENCH CLEANING; PURIFICATION OF BENZINE;
REMOVAL OF STAINS, OR SPOTTING; WET CLEANING; FINISHING
CLEANED FABRICS; CLEANING AND DYEING FURS, SKIN RUGS
AND MATS; CLEANING AND DYEING FEATHERS; CLEANING
AND RENOVATING FELT, STRAW AND PANAMA HATS;
BLEACHING AND DYEING STRAW AND STRAW HATS;
CLEANING AND DYEING GLOVES; GARMENT
DYEING; STRIPPING; ANALYSIS OF
TEXTILE FABRICS.

EDITED BY

WILLIAM T. BRANNT,
EDITOR OF "THE TECHNO-CHEMICAL RECEIPT BOOK."

FOURTH EDITION, REVISED AND ENLARGED.

ILLUSTRATED BY FORTY-ONE ENGRAVINGS.

PHILADELPHIA:
HENRY CAREY BAIRD & CO.,
INDUSTRIAL PUBLISHERS, BOOKSELLERS AND IMPORTERS,
810 WALNUT STREET.
1911·

Lightning Source UK Ltd.
Milton Keynes UK
UKOW06f2106111115